SONS OF

CAMELOT

ALSO BY LAURENCE LEAMER

The Kennedy Men: 1901–1963

Three Chords and the Truth: Hope and Heartbreak and Changing Fortunes in Nashville

The Kennedy Women: The Saga of an American Family

King of the Night: The Life of Johnny Carson

As Time Goes By: The Life of Ingrid Bergman

Make-Believe: The Story of Nancy and Ronald Reagan

Ascent: The Spiritual and Physical Quest of Willi Unsoeld

Assignment: A Novel

Playing for Keeps: In Washington

The Paper Revolutionaries: The Rise of the Underground Press

An Imprint of HarperCollins*Publishers*

SONS OF

CAMELOT

The Fate of an American Dynasty

LAURENCE LEAMER

HarperCollins books may be purchased for educational, business, or sales promotional use. For information, please write: Special Markets Department, HarperCollins Publishers, Inc., 10 East 53rd Street, New York, NY 10022.

FIRST EDITION

Designed by Claire Naylon Vaccaro

Printed on acid-free paper

Library of Congress Cataloging-in-Publication Data

Leamer, Laurence.
    The sons of Camelot : the fate of an American dynasty / Laurence Leamer.—1st ed.
      p. cm.
    ISBN 0-06-620965-X (alk. paper)
    1. Kennedy family. 2. Kennedy, Joseph P. (Joseph Patrick), 1888-1969—Family. 3. Kennedy, John F. (John Fitzgerald), 1917-1963—Family.
4. Politicians—United States—Biography. 5. Legislators—United States—Biography. 6. Children of presidents—United States—Biography.

E843.L44 2004
072'.929'0904—dc22

04 05 06 07 08 WBC/QWF 10 9 8 7 6 5 4 3 2 1

In memory of
Mirko Obradovic
1921–2001

Contents

The Sons of Camelot

John Fitzgerald Kennedy   (b. 1917   d. 1963)
married 1953   Jacqueline Lee Bouvier   (b. 1929   d. 1994)
   John Fitzgerald Kennedy Jr.   (b. 1960   d. 1999)
   married 1996   Carolyn Bessette   (b. 1966   d. 1999)

Eunice Mary Kennedy   (b. 1921)
married 1953   R. Sargent Shriver Jr.   (b. 1915)
   Robert Sargent Shriver III   (b. 1954)
   Timothy Perry Shriver   (b. 1959)
   married 1986   Linda S. Potter   (b. 1956)
      Sophia Rose Potter Shriver   (b. 1987)
      Timothy Potter Shriver Jr.   (b. 1988)
      Samuel Kennedy Potter Shriver   (b. 1992)
      Kathleen Potter Shriver   (b. 1994)
      Caroline Potter Shriver   (b. 1997)
   Mark Kennedy Shriver   (b. 1964)
   married 1992   Jeannie Ripps   (b. 1965)
      Mary Elizabeth Shriver   (b. 1998)
      Thomas Kennedy Shriver   (b. 1999)
   Anthony Paul Kennedy Shriver   (b. 1965)
   married 1993 Alina Mojica   (b. 1965)
      Jorge Edward Nuñez   (b. 1988)
      Eunice Julia Shriver   (b. 1994)
      Francesca Maria Shriver   (b. 1994)

Patricia Kennedy   (b. 1924)
married 1954   Peter Lawford   (b. 1923   dv. 1962   d. 1984)
   Christopher Kennedy Lawford   (b. 1955)
   married 1984   Jeannie Ollson   (b. 1955)

DAVID CHRISTOPHER KENNEDY LAWFORD   *(b. 1987)*

SAVANNAH ROSE LAWFORD   *(b. 1990)*

MATTHEW PETER VALENTINE LAWFORD   *(b. 1995)*

JEAN ANN KENNEDY   *(b. 1928)*

married 1956   STEPHEN EDWARD SMITH   *(b. 1927)*

STEPHEN EDWARD SMITH JR.   *(b. 1957)*

WILLIAM KENNEDY SMITH   *(b. 1960)*

ROBERT FRANCIS KENNEDY   *(b. 1925   d. 1968)*

married 1950   ETHEL SKAKEL   *(b. 1928)*

JOSEPH PATRICK KENNEDY II   *(b. 1952)*

married 1979 dv. 1991   SHEILA BREWSTER RAUCH   *(b. 1949)*

married (2nd) 1993   BETH KELLY   *(b. 1957)*

MATTHEW RAUCH KENNEDY   *(b. 1980)*

JOSEPH PATRICK III   *(b. 1980)*

ROBERT FRANCIS KENNEDY JR.   *(b. 1954)*

married 1982 dv. 1994   EMILY RUTH BLACK   *(b. 1957)*

married (2nd) 1994   MARY RICHARDSON   *(b. 1959)*

ROBERT FRANCIS KENNEDY III   *(b. 1984)*

KATHLEEN ALEXANDRA KENNEDY   *(b. 1988)*

JOHN CONOR KENNEDY   *(b. 1994)*

KYRA LEMOYNE KENNEDY   *(b. 1995)*

WILLIAM FINBAR KENNEDY   *(b. 1997)*

AIDAN VIEQUES KENNEDY   *(b. 2001)*

DAVID ANTHONY KENNEDY   *(b. 1955   d. 1984)*

MICHAEL LE MOYNE KENNEDY   *(b. 1958)*

married 1981   VICTORIA GIFFORD   *(b. 1957)*

MICHAEL LEMOYNE KENNEDY JR.   *(b. 1983)*

KYLE FRANCIS KENNEDY   *(b. 1984)*

RORY GIFFORD KENNEDY   *(b. 1987)*

MATTHEW MAXWELL TAYLOR KENNEDY   *(b. 1965)*

married 1991   VICTORIA STRAUSS   *(b. 1964)*

   MATTHEW MAXWELL TAYLOR KENNEDY JR.   *(b. 1993)*

   CAROLINE SUMMER ROSE KENNEDY   *(b. 1994)*

CHRISTOPHER GEORGE KENNEDY   *(b. 1963)*

married 1987   SHEILA BERNER   *(b. 1962)*

   KATHERINE KENNEDY   *(b. 1990)*

   CHRISTOPHER KENNEDY JR   *(b. 1992)*

DOUGLAS HARRIMAN KENNEDY   *(b. 1967)*

married 1998   MOLLY STARK   *(b. na)*

   RILEY ELIZABETH   *(b. 1999)*

EDWARD MOORE KENNEDY   *(b. 1932)*

married 1958   dv. 1983   VIRGINIA JOAN BENNETT   *(b. 1936)*

married (2nd) 1992   VICTORIA REGGIE   *(b.1954)*

EDWARD MOORE KENNEDY JR.   *(b. 1961)*

married 1993   KATHERINE GERSHMAN   *(b. 1959)*

   KILEY ELIZABETH KENNEDY   *(b. 1994)*

   EDWARD MOORE KENNEDY III   *(b. 1998)*

PATRICK KENNEDY   *(b. 1967)*

VICTORIA REGGIE'S CHILDREN FROM 1ST MARRIAGE

   CURRAN RACHLIN   (B. 1983)

   CAROLINE RACHLIN   (B. 1985)

SONS OF

CAMELOT

A Soldier's Salute

O n his third birthday, John F. Kennedy Jr. stood holding his mother's hand as the caisson pulled by six gray horses rolled by, bearing the body of his father. It was a cold day, and John was wearing shorts and a cloth coat. His mother, Jacqueline Bouvier Kennedy, whispered to her son, and John saluted his father. This was not a little boy making a stab at a military greeting, but a young actor performing a soldier's salute. Practically everyone in America who viewed the funeral of President John F. Kennedy on television or saw the picture in the newspapers felt a poignant identity with the fatherless child. It was an indelible image, forever frozen in that moment.

After they buried the president on November 25, 1963, the Kennedys returned to the White House to celebrate John's birthday. The party was a masquerade of joyousness within the somber patterns of this day. It was both a retreat into the safe harbor of family and an assertion that they would go on as they always had. Seated at the table with John were many of the same energetic children who the summer before had clambered onto the president's electric cart at the Kennedy summer estate on Cape Cod. Robert Francis Kennedy and his wife, Ethel Skakel Kennedy, were there with their seven children. Alongside them were Patricia Kennedy Lawford and Peter Lawford's daughter, Sydney Maleia.

Several of these children were old enough to know that a terrible event had occurred. Bobby's eight-year-old son David was a boy of immense sensitivity. When he had been picked up by one of his father's aides from parochial school only minutes after his uncle's death, he presumably had no way to know what had transpired in Dallas, but somehow he had figured it out. "Jack's hurt," he said, after dialing numbers on his toy phone. "Why did somebody shoot him?"

Senator Edward Moore Kennedy had been presiding over the Senate when he learned that his brother had been shot in Dallas. His first reaction was to worry about the safety of his wife, Joan Bennett Kennedy. He had driven back to his home in Georgetown, running traffic lights and honking other vehicles out of his way. He then flew up to Hyannis Port, Massachusetts, to tell his father, Joseph Patrick Kennedy, that the president had been assassinated, but he broke into sobs before entering the room and his sister, Eunice Kennedy Shriver, gave Joe the news.

Ted returned immediately to Washington, where this evening he stood at the birthday party next to his brother Bobby. Ted managed to keep up a facade of good cheer in front of the children, but his surviving brother wore a gray mask of mourning. Bobby had been the president's alter ego and protector. He could finish his brother's sentences and complete a task that Jack signaled with no more than a nod or a gesture. He had loved his brother so intensely and served him so well that within the administration it was hard to tell where one man ended and the other began.

Now Jack was dead. That was grief enough to buckle the knees of most men, but that was only the beginning of Bobby's agonies. He was the attorney general of the United States, and John F. Kennedy had died on his watch. Bobby may have feared that his responsibility went even further, that the man or men who murdered the president—be they CIA agents, Cuban exiles, mobsters, or a strange lone man enraged at the attack on Castro's Cuba—had been egged on by a policy that the attorney general himself had instituted.

When Jack died, Bobby's immediate reaction was to try to discover who might have killed his brother, first looking within his own government. Then he protected the president's secrets by locking up his papers and files. Bobby's grief was sharpened further by the fact that Vice President Lyndon

Johnson was now president. Bobby considered Johnson a vulgar usurper who, he believed, would turn away from his brother's principles and ideals.

One of Bobby's first acts after his brother's assassination was to write a letter to his eldest son, reminding eleven-year-old Joseph Patrick Kennedy II of the obligations of his name. "You are the oldest of all the male grandchildren," he wrote. "You have a special and particular responsibility now which I know you will fulfill. Remember all the things that Jack started—be kind to others that are less fortunate than we—and love our country." Young Joe was the oldest of all the Kennedy grandchildren, and if it was not burden enough to be faced with the violent death of his beloved uncle, he now was being given another, even heavier load to lift.

Bobby sent the letter to Joe, but the message was meant for all his sons and nephews. More than anything else, Jack willed to his brothers, son, and nephews a treasure chest of promise, golden nuggets of what might have been and what might yet be. Just as the forty-six-year-old leader would be forever young, his administration would be forever unfulfilled. Historians would endlessly debate the qualities of distinction he had shown in the Oval Office, but he would stand high in the minds of his fellow citizens, remembered by most Americans as one of the greatest of presidents.

As they attempted to fulfill the mandate that Jack had left them, Bobby and Ted had an immense capital of goodwill and feeling unlike anything an American political family had known before. Americans had worn the black crepe of mourning for Abraham Lincoln and Franklin D. Roosevelt, but they did not seek to elevate their heirs or to see their presidencies as part of an ongoing family endeavor in which a brother or a son might rightfully assume that same mantle of high power.

A s his son was buried, Joe lay bedridden in the family home in Hyannis Port. The family patriarch was not so much sad and prayerful as angry and belligerent toward a God who had let his second son die. He raged blasphemously, frightening his wife Rose, who sought in God a solace and understanding she found nowhere else.

Joe had considered himself a man of what he called "natural cynicism." He believed that in each generation a few powerful men were the rightful leaders of their generation. He thought that he and his sons were part of this natural aristocracy. Acting as if there were few limits to his conduct, few moral parameters inside of which he had to live, he tramped over borders that held back lesser men, scaled walls into territories where laws were whatever a man imposed.

Joe's greatest goal in life had been to see that one of his four sons became president of the United States. To advance Jack's career, he had been ready to do what had to be done, whether it was providing money, manipulating behind the scenes, or meeting with mobsters in 1960 when Jack was running for the presidency.

Joe may have been a man of massive hypocrisies, but in his own mind, and the minds of his sons, his life came seamlessly together. Joe knew that he had achieved so much in America because of the liberty and opportunities. He believed that sons of privilege and wealth had an obligation to serve their country and to return something of the bounty that they had inherited. Joe taught that blood ruled and that they must trust each other and venture out into a dangerous world full of betrayals and uncertainty, always returning to the sanctuary of family. His sons took on part of Joe's psychological makeup, the sense of lives without boundaries and ambitions without restrictions.

By any measure, Joe was an extraordinary man, the shrewdest, the most focused, the most willful of human beings. Joseph P. Kennedy created one great thing in his life, and that was his family. With acumen as great as his wealth, and limitless purpose, he built a family of sons who sought to reach the peak of American political life.

Jack had the deepest insight into the family of any of his siblings. He understood how difficult it was to survive his father's admonitions and directives. He saw that there was a dichotomy between a life of personal happiness and a life spent in the pursuit of power. In the geography of Jack's life, he could take only one road. With the death of his brother, Joseph Patrick Kennedy Jr., in World War II, he took the narrow pathway to power, forsaking the quest for happiness and personal fulfillment.

E ven though Joe was bedridden and unable to be the firm voice of the family, his presence still shadowed the Kennedy sons and grandsons, as it would even after his death. With the assassination of Bobby in 1968, Joe would give three of his sons to the nation. Jack and Bobby would be revered by millions as heroes who had been cut down before they finished their great work. Whatever other goals Ted, his sons, and his nephews may have had, they were morally obligated to try to complete what his brothers had begun.

The survivors were overwhelmed by what faced them. Even families that do not reach the heights scaled by the Kennedys falter and fall in the next generation, disappearing from public consciousness and squandering much of their wealth. That is true throughout the world. Americans talk about "going from shirtsleeve to shirtsleeve." To the Irish it is "from clogs to clogs." The Chinese say "from rice paddy to rice paddy."

A family begins in poverty. The first generation struggles upward against great odds. The second generation enjoys the fruits of success, while the third generation often dissipates most of what their predecessors had gained. It is not always three generations, but the pattern is a recurring one.

What is implicit is the idea of regression to the mean, which is not simply a statistical law but a rule of nature. A hungry, ambitious man of energy and willful determination is likely to have heirs who are ambitious and talented, but short of the founding father. The wealth and status that the founding generations will to their heirs carry within them the seeds of their loss. Not only are the heirs probably less talented and forceful, but they have grown up in indulgent circumstances that hone their frailties.

The story of the Kennedy family began in 1849 when Patrick Kennedy arrived in East Boston from County Wexford in southeastern Ireland. On the boat he met Bridget Murphy, like himself an impoverished Irish peasant. They married and had four surviving children, three daughters and a son. When Patrick died in 1858, Bridget worked for a time as a servant and later in a notions shop in East Boston that she eventually purchased. She sent two daughters off to work—one as a dressmaker, the other as a millworker—and gave her son, Patrick Joseph "P.J.," the money to open a tavern. P.J. became a liquor wholesaler, a member of the Massachusetts legislature, a political boss, and one of the most successful men in East Boston. He and his wife, Mary Augusta Hickey, had two surviving daughters and a son, Joseph Patrick.

In a classic second-generation pattern, P.J. was content to have his son stay in East Boston consolidating all that he had gained. Joe's mother, Mary Augusta, had far greater ambitions. She pushed Joe to attend Boston Latin School, the best public school in America, and then Harvard College. Joe had the sense of place and confidence of the second generation and the ambition, energy, and initiative of a man raised up from abject poverty. At twenty-five, he became president of a Boston bank in which his father had a major interest, and then he married Rose Fitzgerald, the mayor's daughter.

Joe went on to make millions as a stockbroker and investor. He was a financier of bootleg liquor. He was the leading Catholic layman in America. When he went to Hollywood to produce films, he was celebrated as a family man who would bring morality back to the industry. He did so while living in Beverly Hills with the actress Gloria Swanson. He believed that history was made by great men, not great peoples. As ambassador to the Court of St. James's, Joe was not so much pro-Nazi as pro-power. He considered Hitler a powerfully commanding figure who had mobilized the German people to overwhelm an effete, declining Great Britain. Joe was the beloved father of nine children and a satyr who preyed on his sons' dates. He seduced a virginal secretary and kept her as his mistress for ten years as she worked in the Kennedy home in the presence of Joe's wife, Rose. Joe's infidelities "just tore at the human fundamentals," reflects Chuck Spalding, one of Jack's closest friends. "It left them with vulnerability in that area—it was like a contagious disease."

J oe oversaw the marriages of his sons and daughters, which were blessed by the leaders of the Church, celebrated in the society pages, and attended by the American elite. The marriages of the surviving families would almost all end in failures. Despite her Catholic faith, Patricia Kennedy Lawford left her actor husband, Peter Lawford, making her son Christopher Kennedy Lawford the first Kennedy son of divorce. After years of difficulties, Ted and Joan's marriage also ended in a divorce that was deeply troubling to their two sons, Edward Moore "Teddy" Kennedy Jr. and Patrick Joseph Kennedy. Jean Ann Kennedy Smith and her husband, Stephen Edward Smith, appeared to have a stable marriage, but Steve kept a series of mistresses. That did not make it easy for their two sons, Stephen Edward "Steve" Smith Jr. and William Kennedy "Willie" Smith. Of all the families, only Eunice Mary Kennedy Shriver and R. Sargent "Sarge" Shriver Jr. provided a traditional family setting for their four sons, Robert Sargent "Bobby" Shriver III, Timothy Perry "Tim" Shriver, Mark Kennedy Shriver, and Anthony Paul Kennedy Shriver, but that did not inoculate them from the psychological dilemmas that faced all the grandchildren.

As the world knew, the grandsons of the other two families lost their fathers. After the assassination of President Kennedy, Jackie attempted to provide a strong, emotionally healthy home for her son. Much of the psychological drama in John's life would be about his quest for a father figure, and then for fatherly ideals, a search that could never be fully successful. Bobby and Ethel's seven sons, Joseph Patrick II, Robert Francis Jr., David Anthony, Michael Le Moyne, Christopher George, Matthew Maxwell Taylor, and Douglas Harriman, had an intense, sometimes emotionally overwrought mother who seemed to find balance only in her bottomless devotion to her husband.

The younger Kennedys did not have their grandfather's brutal determination, and many of their advantages proved illusory. Although several of them did important work in politics, for the environment, and in social service, none of them approached their fathers' achievements at a comparable age. Their grandfather had shown business acumen and daring, but the one forthrightly profit-making business that one of his grandsons managed to start, the magazine *George*, had as its crucial asset the Kennedy name, and it lost millions. One grandson became the first Kennedy to retire from political office. A second became the first Kennedy to refuse to seek a higher office largely because he feared he could not handle the legacy. A third grandson, after raising one of the largest war chests ever accumulated for a congressional race, became the first Kennedy to lose a congressional primary. And a fourth grandson, the one with perhaps the most political promise, died in a plane crash as he was contemplating running for office.

In the years after President Kennedy's assassination in 1963, among these two surviving sons and seventeen grandsons, four would die violently: one by assassination, one by drug overdose, one by playing a dangerous game on the ski slopes, and one in a private plane crash brought on largely by overconfidence. These men would be the drivers in automobile accidents that killed one young woman and crippled another. Five others would die traveling with these men. At least seven of these Kennedy men would be for a time drug addicts or self-proclaimed alcoholics, while several others would have publicly unacknowledged problems with alcohol or drugs. Two of them would be accused of rape, one involving a fourteen-and-a-half-year-old. Although there were Kennedy men who lived lives without public troubles and pursued causes of immense social goodness, the sad litany above includes sons from five of the six Kennedy families. It is a startling tally, and one that speaks to a complex tale of hidden lives and often untoward conduct leading to tragic consequences.

These many tragedies and misfortunes have been seen by some almost as divine retribution for the hubris and limitless ambition of Joseph P. Kennedy, a clan doomed by its heritage and its genes to endless tragedy and mishap. What is lost in such analysis, and is equally the most inspiring and most devastating aspect of this history, is that these Kennedy men made their own lives. Yes, they bore a heavy load of heritage. Yes, they had the rich blood of inheritance, but they were not puppets pulled by the strings of history.

The son of Camelot who rode at night down the rutted road to Chappaquiddick also spent the next decades building one of the most distinguished careers in the Senate. The son of Camelot with a limitless future who flew his plane into a murky sky to his death was not the passive bystander to his fate, living out a scripted life. The son of Camelot who built Special Olympics to a place beyond even his mother's exalted dreams was no mere inheritor, surviving off the wages of his forebears. The son of Camelot who rose out of heroin addiction to become a leader of the environmental movement did so in his own name.

Each Kennedy man made his own choices. If it is devastating to see how several of them squandered their fortune of goodwill and opportunity, then the monumental impact several of them have made on our country and our world is inspiring. They lived their lives, they made their choices, and they determined the fate of an American dynasty.

CHAPTER TWO

Sheep Without a Shepherd

oon after she left the White House, Jackie purchased a 170-year-old house at 3017 N Street in Georgetown. Although she hoped to avoid constant public scrutiny, she and her two children were among the capital's leading tourist attractions. Every day the tour buses wheezed their way past N Street, the tourists with their faces pressed up against the windows while scores of others stood on the corner, kept back by D.C. police. To avoid onlookers, Secret Service agents would take the family out the back door, through a neighbor's yard, and into cars waiting on Dumbarton Avenue.

During the day John scampered through the Georgetown house, his eyes glistening with anticipation, a font of endless mischief, seemingly unaware that his father was gone. This was no different from the way the boy had been in the White House, rushing through the rooms with proprietary zeal. He had nothing of his six-year-old sister Caroline's shy, taciturn manner but posed for photographers with professional aplomb and bounced up and down on the laps of his father's friends.

Caroline was old enough to have crystal clear and impregnable memories of her father. The loss of her father, as great as it was, remained something she could comprehend. For John it was different. As a young man, he would have memories of his father that he talked about only with his closest friends. He remembered how he scurried into the Oval Office and pulled his father's newspaper down from his face, as the president grimaced in seeming fear. This was a real memory, not something taken unconsciously from newsreels and documentaries or from tales told to him. He had an uncanny ability to separate what he had seen and felt himself from what others told him had happened, to distinguish between his own truths about his family and the attempts of others to re-create and interpret his world.

John was attuned to the nonverbal signals around him and had a growing sense of how much his life had changed. The people around him were different. His surroundings were different. His mother, as much as she tried to act the same, was different. There were no longer those daily trips down from the family quarters of the White House with his father, no jaunts into the Oval Office, no more games with a newspaper.

John might have seemed a carefree child, but as he grew up he had a greater awareness of what was going on around him than he displayed to the world, and from an early age he sensed that his father was gone. John's friends learned never to ask about the loss of his father. He searched for his father everywhere, trying to replace him by choosing several of the Secret Service agents as surrogate fathers. He even searched for his father through the friends he chose, some of whom had also lost their fathers or suffered some other great loss unknown to most youths. He chose male friends of tough spirit and unquestioned masculinity. He played rough, manly sports, trying to create his own sense of manhood without a father to guide him.

John's mother attempted to replicate the security and serenity of the White House, and yet the home was full of emotional subterfuge. Jackie pretended that things had not changed when in fact they would never be the same. She looked away from the darkness and chose to remember the years in the White House as Camelot. John also learned about his father from his older cousins, who had their own recollections of those White House years: the joyous summers at Hyannis Port, the romps on the White House grounds, the rides in helicopters and giant planes. As these Kennedy children grew older, it was natural for them to accept this light-filled vision as the truth that they were supposed to carry forth.

In trying to re-create the familiar atmosphere of the White House,

Jackie brought in Mike Howard, a Secret Service agent who had been assigned to the first lady and her children. The amiable Texan regaled Jackie by singing ribald blues and gave piggyback rides to little John. On this new detail in the first months after the death of the president, Howard never saw Jackie sitting in gloom or misery, letting on to the unhappiness she must have felt.

One night Howard was working the overnight shift, sitting in a chair in the living room. It was tedious, and he had to will himself to stay alert. A scream punctuated the silence. Howard leapt up the narrow staircase, his pistol raised, and flung open the former first lady's bedroom door. Mrs. Kennedy was sitting up in bed, her small reading light on.

"Ma'am, are you all right?" he asked.

"Yes."

"Must have had a nightmare."

"Yes. I just saw some bad things."

B obby Kennedy saw bad things in the night as well, so much so that he often could not sleep. The deepest part of his grief he tried to hide, but the visible layer was shocking enough to behold. At times he appeared shriveled up, the life drained out of him, his face an advertisement of his anguish.

Of all the Kennedy brothers, Bobby had always had the deepest faith, but now in scriptural passages where he had once found truth, he saw only platitudes. Bobby found his deepest solace in Jackie, and he often visited her. Howard had seen Bobby making his frequent visits to the Georgetown house. The Secret Service agent knew of the rumors of the relationship between them, but he did not know what to make of them. Jackie and Bobby spent many hours together in the house in Georgetown and in New York. In March 1964 they flew off to Sun Valley, Idaho, for a skiing vacation. For Easter they traveled together with Jackie's sister, Lee, her husband, Prince Stanislaus Radziwill, and family friend Charles "Chuck" Spalding for a vacation at the Antigua estate of Bunny Mellon. "We were far from home and down on this island, and the beauty of the whole thing contrasted with the loss and everybody's terrible sense of dejection," recalls Spalding. "It was so paralyzing that unless you were going to forfeit your own [life], eventually each of us had to simply say, you can't go on like this."

Bobby sought through Jackie to touch his brother once again, to hear his laughter, to feel his life again. Bobby found with Jackie things that he would never find with Ethel. Only Jackie spoke the same expressive language that Bobby spoke and seemed to understand the nuances of his every gesture. Whatever the specifics of their relationship, it was a mark of how much Bobby missed his brother and how desperate he was that he could be so unconcerned about the inevitable rumors and gossip.

Ethel, meanwhile, was the most loyal and loving of wives, a cheerleader and a comrade in arms. She idealized her husband and considered Bobby's critics to be traitors and fools. In attempting to outrun the shadow of desolation and depression that haunted her husband, she was acting more Kennedy than Bobby. When he had been sick as a boy and when his big brother Joe had died, his parents had ordered him out of the house to run his sorrow away. Ethel whirled from one activity to another. Painting a happy face on every day, she could hardly stop or slow down enough to contemplate the emotional depths to which her husband had fallen.

At a dinner party one evening at their home, Bobby sat at one end of the long table, musing endlessly about his departed brother, the only subject that stirred him from his malaise, yet it always led him downward again. At the other end of the table Ethel was her eternally exuberant self, talking incessantly. Bobby looked up and said words he must have thought a thousand times since the assassination. "There was never anyone quite like Jack," he said solemnly. "Nobody."

No one wanted to break the silence that hung over the table. "We don't have to worry," Ethel said finally, with cheery certitude. "We know that Jack is up there in heaven, Bobby, and he's looking down on us and taking care of everything."

Bobby looked back down the table as if his wife was someone he did not know. "Those words were spoken by the wife of the attorney general of the United States," he said, dismissing her platitudinous boilerplate with disdain.

T he six-acre estate in McLean, Virginia, where Bobby's family lived was called Hickory Hill. The mansion was a spirited admixture of the Kennedy compound in Hyannis Port and the Skakel mansion in Greenwich, Connecticut, where Ethel had grown up. Her family was as wealthy and iconoclastic as the Kennedys, yet whereas the Kennedys strove for social acceptance, the Skakels fancied themselves above such pursuits. While the Kennedys made their broad marks on the world, the Skakels considered their money a ticket to lives of fast-lived pleasure. The Kennedys might laugh, but the Skakels were fun. At Hickory Hill there was always activity, children running helter-skelter, maids coming and going, phones ringing, cars leaving, and Ethel marching through it all.

When Bobby was with Jackie, it was all so different. She was no athlete and had often run off with a book when the Kennedys took to the football field or the tennis courts. Like Bobby, she had much that she had to puzzle out and contemplate. It was one of the great imponderables that she could have loved so profoundly a man who had cheated on her so many times in so many ways, but she did, and her grief was as intense as it was deep. As for Bobby, grief was hardly word enough for how he felt. And so he had Jackie to talk to about *their* Jack, and about much else.

Jackie introduced Bobby to the Greek tragedians. He read them as a young man reads books, underlining passages, memorizing sentences, seeking answers to the eternal "why?" Here he found dark, mournful visages of life, as well as solace in the existential struggle of the heroic individual. "Men are not made for safe havens," he underlined in Aeschylus. "The fullness of life is in the hazards of life. . . . To the heroic, desperate odds fling a challenge." He read deeply and remembered well, and not simply to impress his audiences by studding his speeches with memorable phrases.

In March, Bobby was scheduled to give a St. Patrick's Day speech in Scranton, Pennsylvania. On a sentimental occasion, speaking to a sentimental people, he had only to stand on the podium to evoke all the richly poignant feelings of what had been lost and what lay unfulfilled. He insisted, however, on speaking directly about his brother and quoting a poem about the great Irish freedom fighter Owen Roe O'Neill. Bobby's press aide, Edwin Guthman, warned him that he might not be able to get through the reading without breaking up. He practiced standing in front of the mirror, reading the poem again and again and again. On the day he gave his speech, every time he said the words "President Kennedy," which was often, he spoke with sweet softness, brushing the words with tenderness. At the end of the speech Bobby read the emotionally powerful poem:

> Your troubles are all over
> You're at rest with God on high,
> But we're slaves and we're orphans, Owen!
> Why did you die?
> We're sheep without a shepherd,
> When the snow shuts out the sky—
> Oh! Why did you leave us, Owen?
> Why did you die?

That was the question Bobby asked himself. He would never find the answer, and in standing before this assembly of Irish Americans without crying as he read the words, he was signaling to himself and to the world that he could indeed go on not knowing the answer. "So, on this St. Patrick's evening, let me urge you one final time to recall the heritage of the Irish," he concluded. "Let us show them that we have not forgotten the constancy and the faith and the hope of the Irish."

Bobby realized that it was not only he and his family, not only these people in this hall or those across America, but people throughout the world who saw his brother as a symbol of hope and optimism. As millions of people saw it, the eternal flame burned not only in Arlington National Cemetery but also in the heart and soul of Robert F. Kennedy. Wherever he went, he evoked the deepest and most heartfelt of sentiments.

In June he traveled to Europe, and he saw that flame of hope burning as brilliantly there as it did in America. "I have come to understand the hope President Kennedy kindles is not dead but alive," he told a youthful audience at the Free University of Berlin. He flew to Poland, behind the Iron Curtain, bearing the same message, knowing that his brother's name evoked all that was best and most promising in humankind. He brought with him to Warsaw not only a pregnant Ethel but his three oldest children, Kathleen, Joe, and Bobby Jr., to imbibe this extraordinary spectacle.

In Warsaw, outside the medieval cathedral after mass, he did what would in the next years become almost a campaign routine with him. He jumped out on the roof of the embassy automobile, pulling Ethel and his children up there with him. The Lincoln Continental made its way gingerly through the masses of cheering Poles. The American ambassador feared that the car roof would cave in, but Bobby would not get down. He was not going to hide from these people or from people anywhere. He thrived on these moments of public witnessing, these glorious confrontations when he tapped into wellsprings of discontent and hope. His three oldest children saw this too. They were immensely proud of the daring of their father and this raw new energy that he brought to politics. For these children their father blended into one great transcendent figure, an image that would both inspire and haunt them.

Games of Power

yndon Johnson knew that he was the beneficiary of an immense
emotional response to Jack's death. Even if the new president
backtracked on issues championed by his predecessor, as long as
he did not openly walk away from those ideas and policies, or act in too partisan a manner, he would probably have a strong mandate. He had every
incentive to please the Kennedys. He wooed Jackie with homespun, often
vulgar homilies, talked politics with Ted, called Joe, and tried to better his
rapport with Bobby.

In the first months of the Johnson presidency, Bobby stayed on as attorney general. Most of Kennedy's aides also heeded Johnson's admonition to
remain in the new administration. They remained full of memories of the
fallen president and were often unable to show the slavish obedience to
Johnson's will and whim that he expected of his closest aides. At times the
new president sensed betrayal where there was simply sadness. At other
times he was correct in his suspicions that some of Bobby's minions were
working stealthily to create a groundswell that would carry the attorney
general to the vice presidential nomination at the Democratic Convention in
August 1964.

In February 1964, after a cabinet meeting, Johnson called his attorney
general into the Oval Office to tell him to have his agent back off. It was still
a time of the deepest grief. The martyred president was being memorialized
not simply in the renaming of airports, highways, buildings, schools, and a

space center but in the nation's legislation. The day before the civil rights bill had passed the House of Representatives, sooner and more overwhelmingly (290 to 130) than it would have if Kennedy had lived. Yet at this moment Johnson could not fully savor this triumph any more than Bobby could. The attorney general described the encounter as a "bitter, mean conversation . . . the meanest tone that I've heard."

That was probably more indicative of Bobby's sensitivity to the new realities than a fair-minded analysis of what had transpired. There is a natural deference shown to the president of the United States. Bobby was not willing to bow his head even an instant to a man who squatted where his brother had once sat. "President Kennedy isn't president anymore," Johnson said. "I am." That was nothing more than the truth, but it was a truth that Bobby could hardly bring himself to accept.

The reality was that Bobby was too embedded in sorrow to contemplate realistically his political future. Beyond that, he had made his disdain for Johnson pointedly clear when in 1960 he opposed his nomination as vice president. Bobby knew about the humiliation and disinterest that was the natural plight of a vice president, and he was emotionally the least likely person to hold that office. Still, he wanted Johnson to name him as his running mate. That would make him the almost certain Democratic presidential nominee in 1968.

Johnson was not about to anoint a man who believed the new president was "mean, bitter, vicious—an animal in many ways." In his more paranoid moments, Johnson contemplated not running rather than being saddled with Bobby. "I'm not going to let them put somebody in bed with me that'll murder me," he confided to Governor John Connally of Texas.

Bobby was not the only Kennedy family member who sought the vice presidency. The other prospective candidate was Bobby's brother-in-law, Sargent Shriver, the head of the Peace Corps and the new director of the administration's nascent War on Poverty. In the years since he first went to work for Joe managing the family-owned Merchandise Mart in Chicago and then marrying Eunice Kennedy, Sarge had learned well the arts of deference so essential to a vice president. He represented the most idealistic aspects of the Kennedy administration and was far better suited for the position than Bobby. Sarge had pointedly told the press that he was not running for the office and that his brother-in-law would be a "terrific" vice president, though he may well have doubted that.

"I think that a man who runs for vice president is a very foolish man," Johnson told Sarge. "The man who runs away from it is very wise. And don't you ever be a candidate, and don't let anybody else be a candidate. Tell them, anybody ever runs for it never gets it."

While the Kennedys showed toothy smiles to the world, among themselves they had begun for the first time to squabble jealously over the spoils and prospects. Bobby believed that Johnson was considering Sarge only to avoid choosing his attorney general, an analysis that showed how little he thought not only of Johnson but also of his own brother-in-law. The Kennedy men had never before competed for the same position, but the world was a different place now.

"Sarge really wanted the vice presidency in '64," recalls Harris Wofford, Shriver's longtime associate. "He was very close to being wired with the president's aide Bill Moyers [Sarge's former deputy at the Peace Corps] and Johnson. That's when Kenny O'Donnell [a close Kennedy aide], who had stayed on in the White House, heard Moyers on the phone telling Johnson that the family could not oppose Sarge's nomination. O'Donnell stormed in and said, 'The family's gonna oppose you.' It was an end run around Bobby, and they were chortling over it. And O'Donnell said, 'The hell they won't. We'll fight you to the end. If they're gonna pick any Kennedy, it's gotta be Bobby.' "

T ed did not favor his brother's seeking the vice presidential nomination but thought instead that Bobby should run for a Senate seat from New York State. It was generous advice: he surely understood that if Bobby entered the Senate, he might cast the junior senator from Massachusetts into perennial shadow.

In the wake of the assassination, Ted had returned to work almost immediately, and not merely to sit despairingly at his desk. He found purpose in a senator's life, with its endless hearings, meetings, discussions, and interviews. He was arguably the best natural politician in the family, a back-slapping, jovial, endearing public man whose hands were so enormous that when he reached out and grasped a constituent's hand, it was like being enveloped in a baseball glove.

The year before, when thirty-year-old Ted had arrived in the Senate, he was, by many judgments, a spoiled pup who had ridden a name and unalloyed privilege to victory. He had few credentials beside his name, a reality about which he was not unaware. He was a loyal Democrat with no distinctive issues that he had made his own, and no high public posture. But he had kept his eyes focused downward, learning the ways and whys of the Senate and winning acceptance among his colleagues.

The Senate is often called the most exclusive club in the world, but in the sixties it was two clubs. All senators belonged to the outer club, but there was also an inner club that tapped its members from those who adhered most closely to its rituals and folkways. Ted was a natural man of the inner Senate. He understood and appreciated the institution's elaborate courtesies. Part of his deference to his colleagues was that of youth to age, but part was acknowledgment of this reality. These politicians of the inner Senate debated issues, but they understood that their terms were long and the issues many, and that while they might leave scratches on their opponents, they did not leave scars.

Ted shared with his grandfather, former Boston mayor John "Honey Fitz" Fitzgerald, the sense that politics is a glorious game in which win, lose, or draw, the players return to play again. He was not a man who bore grudges, not a hater, not a malevolent person. Part of his good nature may have stemmed from his own insecurity, but whatever its origins, it was an enviable trait in a politician.

When Ted talked to the president, there was none of the posturing, bickering, and dissembling that had poisoned the relationship between Bobby and Johnson, but that did not mean that they squandered candor on each other.

On March 30, 1964, Johnson called to congratulate Ted on his Sunday morning television appearance on *Meet the Press*. "You can take my job anytime you're ready," Johnson said, "because I thought you survived it wonderfully."

"That's very very kind," Ted said.

"I'm just a trustee that's trying to carry on the best I can," Johnson said with modesty as effusive as it was false. "And I know my problems and limitations better than anybody else."

Ted was a man of instinctive shrewdness, a very measure of it being that he often hid it so well that others did not realize he possessed it. He thanked Johnson for his compliments, but he did not praise his stewardship of JFK's program, a gesture that might have been construed as a betrayal of Bobby.

When Ted was not talking politics with colleagues or working in the Senate, he was an honored emissary of his brother's legacy. Although it was his staff who wrote the eloquent, deeply felt speeches he gave about his brother, it was he who stood before audiences and read the words. It was a wrenchingly difficult role.

The youngest Kennedy brother, who stood on podiums commemorating his older brother's greatness, was a man of the most contradictory qualities. He was self-doubting, forever measuring himself against his big brothers and the fine minds that served them. Yet he was full of willful arrogance, driving his Oldsmobile, for instance, as if traffic laws were for lesser mortals. Sensitive and sentimental about his family, he was often deeply thoughtful, and not only in a politician's calculating manner. Yet he could be crudely aggressive and lustful, especially with a few pops of booze in him.

Jack had been both admiring and jealous of his exuberant, seemingly carefree and healthy youngest brother. He thought that Ted would have a different, freer life, but his kid brother chose the same narrow road as Jack. "Ted, who has all the physical apparatus of being rather easygoing, a rancher or something, you know really taking life easy, he got an ulcer for a while there last year because he didn't know if he was going to be able to stay in law school, because he's not terribly quick," Jack told his biographer James Macgregor Burns in 1959. "He stayed in and has done pretty well, but that demonstrates that's really induced by outside pressures, because he's most physically well-balanced and healthy, but I mean it's gotten to him now so he's doomed to this treadmill too."

His father and mother had wanted Ted to marry and to stop his restless romantic wanderings. On the evening before the wedding in November 1958, he confessed to one of his closest friends that he feared he was making a dreadful mistake. This was not a mark of his selfish disregard but almost the opposite. He hardly knew Joan Bennett, a beautiful graduate of Catholic Manhattanville College. He felt he was being locked into something not of his own making. Just before the wedding ceremony, Jack told his little brother not to worry. His romantic life could go on as before. That exchange had been filmed, and when the couple returned from their Jamaican honeymoon, Joan watched the wedding film and heard the brotherly advice.

Shortly after their marriage in 1958, Ted and his young bride thought of moving to a western state to start a new free life, but Joe vetoed that the moment he learned of it, and Ted set out to live what were largely his father's aspirations.

T he Joan Kennedy who had flown up to Springfield, Massachusetts, ahead of her husband in June 1964 to the commonwealth's Democratic Convention had long since made peace with her marriage. The couple had two little children at home, Kara Anne Kennedy, born in 1960, and Edward Moore Kennedy Jr., born a year and a half later. They had been hoping for a third, but Joan had just miscarried for the second time. She would have preferred to rest at home, but she was expected to get up and play the loyal political wife.

Ted would already have been at the convention, where he would be nominated for his first full Senate term, but his vote was needed in the historic vote to pass the civil rights bill. It was close to eight o'clock by the time he arrived at National Airport to fly a private plane to Springfield.

Senator Birch Bayh of Indiana, who was scheduled to give the keynote address, was there too, with his wife, Marvella Bayh. Ed Moss, Ted's administrative assistant, was also traveling with the party.

Wherever most politicians go, they are always a little late. That way there is added anticipation and the sense that they are the most important one at the event. Few politicians, however, overscheduled themselves as much as Ted did. His days were a constant race from hearing to hearing, interview to interview, meeting to meeting, speech to speech, event to event. It was in the sheer multitude of activities that he found satisfaction, even exhilaration. If that meant that he drove through traffic lights or storms, that was just fine. In the frenetic life he had created for himself, he had minutes to think about issues but rarely hours to reflect about ideas.

For Ted there was never a question about setting out in a plane, even if the forecast was for stormy weather. He was used to flying in private planes in inclement weather. That was one of the unspoken, calculated risks of the political life if a politician wanted to stay in office. As the plane flew north-ward, the pilot, Ed Zimny, moved around thunderstorms and turbulence. By the time the Aero Commander reached Springfield, the local airport was lost under a bank of fog. All Ted had to say was that it was not worth the risk to make an instrument landing, and Zimny would have flown to another air-port. Ted had two siblings who had died in planes, Joe Jr. in 1944, when his plane exploded over England, and Kathleen Agnes Kennedy four years later in a crash in the French Alps, when she and her lover insisted that the pilot fly in conditions not unlike what they were flying in now in western Massa-chusetts. But Ted was as cavalier about the dangers of flying as he was about everything else in his life.

Zimny guided the plane down through the thick fog. Then the bottom of the plane grazed apple trees on a ridge a good two hundred feet above and across the road from the airport, and the Aero Commander nose-dived into the earth. Bayh and his wife managed to stumble out of the wreckage. The Indiana senator returned to pull Ted away and lay him down in the field. "He [Bayh] made the decision to come back to the plane," Ted reflected de-cades later. "It didn't catch on fire, but it could have. It was a clear reflection of the quality of his character and courage."

Ed Zimny, the pilot, was dead, Ed Moss was dying, and Ted was in ter-rible condition. When Ted arrived at the hospital, he had almost no blood

pressure, three of his vertebrae were damaged, one of his lungs had collapsed, nine of his ribs were broken, and his back had been broken in twenty-
six places. The doctors were so worried about other internal injuries that
they used no anesthesia on him. After several days he stabilized, and the
physicians started giving him painkillers. For about a half hour he had a
warm, floating feeling. Then the pain came back, and he lay in agony, count-
ing the time until the nurse came in and gave him another shot.

The doctors were realistic men. When Ted had regained full awareness,
they felt that they had no choice but to tell their patient the truth. One of the
doctors came into his room on a rainy afternoon, the droplets cascading
down the window pane. "You may never walk again, Senator," the lead doctor told him. "You're just going to have to face that."

In any accident, one's initial instinct is to say that one is fine, mumbling
a mantra to oneself as much as to the world. Ted continued to chant this
mantra until it became the truth, or much of the truth. This was in some
ways Ted's and Bobby's greatest strength. No matter what happened, no
matter how deep the wounds or how twisted the wreckage, they were fine.
Maybe they did this only through a willful distortion of reality, but it was no
less extraordinary.

Jack had suffered most of his life with back troubles, and now Ted had
his own crushed vertebrae. The specialists recommended back surgery. Ted
decided to take his chances with time and nature and had the doctors strap
him to a Stryker frame, a medical teeter-totter on which he was rotated every
hour. He was a man who felt secure only when he was in control, driving a
car, leading a floor fight, giving a speech. Strapped to this frame, however,
unable to move, and with nurses hovering over him, he had no control at all.

Ted had an inordinate faith in the wisdom of experts, and a parade of
Harvard professors made their way to Baptist Hospital to educate him on
economics and politics. He paid more attention than he had in his college
classes, largely because he saw the practical efficacies of this instruction. He
was not much of a reader of lengthy tomes but preferred to spend his limited
time perusing memos and essays and speeches or being briefed by the most
knowledgeable experts in the field in question. He was a man not for philo-

sophical musings but for practical ideas that he could pan through, searching for nuggets that he could apply in the Senate. He knew he wasn't as smart as scholarly men like John Kenneth Galbraith, Samuel Beer, and Patrick Moynihan, all of whom visited his hospital room. But he was smart enough during these months to realize that he could use their minds, multiplying his impact on the Senate and on America.

After nearly six months the doctors took him off the board. His muscles had atrophied so completely that he could not even stand up on his own. The aides helped him up and maneuvered him into the swimming pool. He used his shoulders and arms and a weak flutter kick to propel himself back and forth across the water. Finally, after days of that, they strapped him into a corset and gave him a cane, and he began to walk.

Ted discarded the cane as soon as he returned to public life, but the back pain stayed, his occasional companion the rest of his life. It would come upon him unexpectedly, sometimes when he was giving a speech on the Senate floor, or taking someone's testimony in a hearing. He would almost black out and then catch himself and go on. Those who were close to him could see that look in his eye and the slight grimace and knew what was happening, but most people had no idea of the pain that he endured.

CHAPTER FOUR

The Senators Kennedy

here were those in 1964 who thought that it was too soon for
Bobby to run for political office, with his soul still draped in
mourning. "I thought that Bob might develop intellectually more,
a greater sense of history and so forth, a sense of where the hell the country
is going, and not sort of always being so activist that he didn't retreat for
reflection," said Fred Dutton, one of his closest advisers. "If you thought
out your own things, and [are] almost at peace with yourself, there'll be
more of a quiet strength about it."

Bobby might peruse the ancient Greeks, but he was constitutionally
incapable of sitting back and reflecting for an extended period of time.
Instead, when Johnson made it clear that Bobby would not be offered the
vice presidential nomination, he set out to become a United States senator
from New York. Although he had lived with his family in Bronxville as a
boy after the Kennedys moved to New York from Boston in 1928, he had
lived most of his adult life in Washington. He realized that he would be perceived by many as little more than a carpetbagger who had come into New
York with his brother's legacy as his greatest apparent attribute.

Bobby could not afford to fume over Johnson's rejection of him, for he
knew that his and the new president's future were harnessed together, and
soon after he decided to run, he called the president. "If I could work it out
so that I wasn't just landing from Massachusetts to New York, I'd try to do
so," he admitted to Johnson.

That was only part of the problem. "If it looks like I plainly didn't become vice president and then I decided to run in New York because it was the second best thing, I think that looks terrible and I start out with such a burden," Bobby confided to a president he did not trust. Bobby was not a New Yorker. Nor was the Senate seat his first choice. In his soft, underspoken way, he was attempting to solicit Johnson's help, or at least to get him not to undermine him. "It's awfully important beyond myself," he said to Johnson, who was secretly recording his words. "If I lose up there, it's a reflection on the whole family."

T o ensure that there would be no emotional stampede to nominate Bobby, Johnson did not allow Bobby to talk at the Democratic Convention in late August 1964 until he and his vice presidential running mate, Senator Hubert Humphrey of Minnesota, had been nominated. On the last evening in Atlantic City, Bobby finally had his moment. Even before he spoke, the delegates applauded for thirteen minutes while Bobby stood with a shy half-smile on his lips.

In his address, Bobby proudly placed his beloved brother as part of a Democratic Party tradition of stalwart leaders that he traced back to Thomas Jefferson, James Madison, and Franklin Roosevelt. He talked of his fallen brother and of how the late president had wanted to help the mentally ill and the mentally retarded, those not covered by Social Security and receiving less than a fair minimum wage, those in need of decent housing, senior citizens having trouble paying their medical bills, and "our fellow citizens who are not white." These had not been his brother's primary concerns, but Bobby was creating a list of noble endeavors that he hoped to make part of his brother's legacy as well as his own.

This man who stood before the convention audience was far different from the intemperate young attorney who had once worked for Senator Joseph McCarthy, or even the forceful attorney general who had authorized wiretapping against Martin Luther King Jr. and led a secret war against Castro's Cuba. He was becoming the most important progressive politician of his time. With his ideas wrapped in the sacred mantle of his martyred brother, Bobby seemed to his admirers not merely a politician but a priest of politics who would lead his followers and his nation to an exalted place. For the first time he reached out pointedly and profoundly to the youthful generation to claim them as soldiers in the mighty army that would complete what his brother had begun.

"His [JFK's] idea really was that this country, that this world, should be a better place when we turn it over to the next generation than when we inherited it from the last generation," Bobby said. "And that's why he made such an effort and was committed to the young people not only of the United States but to the young people of the world . . . When I think of President Kennedy, I think of what Shakespeare said in *Romeo and Juliet*:

> When he shall die,
> Take him and cut him out in little stars,
> And he will make the face of heaven so fine
> That all the world will be in love with night,
> And pay no worship to the garish sun.

No one there that evening, no matter what they thought of his politics or his family, could hear Bobby's words without being moved. Although thirty-eight-year-old Bobby talked of the golden promise of this agenda, there was a deep sadness to this man that he did not attempt to hide and a foreboding that was rarely heard or seen on a political platform. He ended not with a call for bold action, but with a quotation from Robert Frost that was almost ominous in its implications.

> The woods are lovely, dark and deep.
> But I have promises to keep,
> And miles to go before I sleep,
> And miles to go before I sleep.

During the fall of 1964, in his race against the Republican incumbent Kenneth Keating, time and again Bobby mentioned his late brother. There was a pained immediacy in this invocation that was far different from a politician's ploy of linking himself with a revered name. For one moment, he was the proud bearer of his brother's legacy, repeating his words and his gestures, his mere presence an affirmation of his brother's life. Then a moment later he was awkward, stumbling, withdrawn, sharing his vulnerability. And everywhere Bobby went in the state, he saw the immense emotional pull of his brother's name and the Kennedy mystique.

Despite this extraordinary response, there was a morbidity in Bobby that he could not shake off. He often spoke in a listless monotone, reading words that others had written. He appeared incapable of the schmoozing with party leaders that was so much a part of a politician's life, a weakness that his detractors blamed on arrogance, not on spiritual malaise. He had an unruly mop of hair that was the delight of political cartoonists and a shy, diffident manner. At staff meetings he often gazed out the window as if he were no more than an observer. "Do what you are afraid to do," he had written in his notebook after his brother's death, underlining Emerson's precept. That was one of the axioms he lived by.

B obby was no longer seeing Jackie as often as he had earlier in the year. She had just moved to the Upper East Side of New York City to seek a life unshackled from the omnipresent memories of the assassination. In moving to her five-bedroom, five-bath apartment at 1040 Fifth Avenue, she was not only freeing herself, or part of herself, from living as a national icon, garbed in widow's black, but she was freeing her children. John would forever be the president's son, but here he would have a chance for a real life that he would not have had in Washington.

Bobby decided not to call on his brother's widow to campaign for him, and he debated the idea of using John. There was a cachet and celebrity in the little boy that none of the other Kennedy children had, not even his older sister, Caroline. The campaign staff wanted Bobby to take his three-year-old nephew out to the house where the Kennedys had lived when they first moved to New York State. There Bobby would introduce his nephew to one of his boyhood homes while the campaign's cameras rolled.

Bobby's aides told the *New York Daily News* that "taking the child along on the visit was part of a sincere effort to 'be a father' to the young child," as if this role were best performed before scores of cameras and reporters. Wearing red shorts and a long-sleeved white sweater, John held a stuffed dolphin in his hand and had a ring of chocolate smeared around his mouth.

The next morning Gabe Bayz, the campaign television consultant, happened to ride the elevator up to the campaign office with Bobby. "Bob, by the way, I'm sure you'll be pleased to know that John-John's visit to your home got on 99 percent of the New York stations last night, and most of them ran the footage in its entirety," Bayz remembers saying.

"Well, Gabe, that just proves one thing," Bobby said as he left the elevator. "We're running the wrong candidate."

I n late September, Bobby went to Ferris-Booth Hall at Columbia University to answer questions in an open forum. He considered the two thousand students gathered there part of his natural constituency, and he responded well to their pointed, passionately felt questions.

"The Warren Report was issued this week," one student said. In the immediate aftermath of Kennedy's death, President Johnson had created a commission to investigate the assassination and to issue an authoritative report that would squelch the endless rumors. On the day the lengthy report was issued, stating that Lee Harvey Oswald had acted alone in killing the president, Bobby canceled all his campaign appearances after issuing a short statement saying that he agreed with the conclusions. "Now, do you agree with its findings, or do you think there is any credence in the views of Mark Lane, who has made a statement saying that the report is wrong and that Oswald did not kill President Kennedy by himself?"

This was an inevitable question, and it was asked not in a hectoring tone but as an honest query. "I've made my statement. Is that one yours?" Bobby snapped.

"Well," Bobby tried to begin again, his voice scarcely more than a whisper and tears in his eyes. "When I, ah, spoke to the students in Poland, I said, uh, that the death of my brother was the work of, ah, one individual, and, uh, and I think the Warren Commission Report shows that conclusively."

As the questioner walked back to his seat, another student stood up to ask something else, and as he did, tears washed down Bobby's face. "Ah, just a minute, please," he said, and tried to regain his composure. It was five minutes before he could proceed, five minutes during which the students sat there wondering about the complexities of this man who stood before them.

Moments like this did not hurt Bobby's campaign. Many New Yorkers felt an obligation to the brother protector and thought that, if he wanted their vote, he should have it. There were other Americans who so despised him that they scarcely listened to what he said and only waited for the day when they could vote against him. Part of their aversion could be traced to the image of ruthlessness that had hung over him since he had done the dirty jobs and made the enemies that had helped advance his brother to the White House. It was an image enhanced by the way in which he had elbowed his way into the Empire State, wrenching the nomination away from men who had served the state for decades. Beyond that, the West Side liberals viewed Bobby as a faux progressive. To them he was a dangerously ambitious man who might speak finely wrought words but who sought the support of old-line political bosses such as the Bronx's Charles Buckley. There were others who could not articulate their precise reasons for opposing him, but they loathed Bobby and all others who bore his name.

In a supreme irony, Bobby's greatest attribute as a candidate was that he was part of a Democratic ticket headed by Johnson on the verge of one of the greatest landslides in American history. The new campaign posters read: "Get on the Johnson-Humphrey-Kennedy Team," a team that existed on the billboards and nowhere else. When Johnson came to campaign for Bobby, the two men traded profuse public compliments, but privately they remained wary of each other's sting. Johnson carried New York by over 2.7 million votes, and Bobby won the state by 719,693 votes. Two million Johnson voters could not bring themselves to pull the lever beside Bobby's name, and he

would remain a man who engendered not only love but mistrust, and sometimes hate. That was only part of the bittersweetness of his election. "If my
brother was alive, I wouldn't be here," Bobby said at the victory celebration.
"I'd rather have it that way."

O n the day in January 1965 when Ted arrived back in Congress for the
first time since his accident, he drove over to the Capitol in his blue
Chrysler with Bobby for their swearing-in as senators. Ted was wearing a
stiff brace beneath his suit, similar to the one President Kennedy had worn.
As he moved among his colleagues with painful caution, about as many
Republicans as Democrats came forward to welcome him back and wish
him well. What they had in common was seemingly a genuine regard for
the man.

Ted took pleasure in the minutiae of politics. He had a young man's
view that life was long, and politics eternal. He understood and celebrated a
basic law of politics: nothing happens alone. During his first days in office he
had made a courtesy visit to Senator James O. Eastland. Ted spent two mid-
day hours with the chairman of the Judiciary Committee talking politics and
drinking scotch. Given that Eastland was an unreconstructed segregationist,
it might seem that the latter activity happened more easily than the former.
But the young senator not only enjoyed talking to the powerful Mississippi
senator but learned from him. There were things the two politicians understood without even saying anything. "You've got a lot of Italians up there,"
Eastland said, which was his way of suggesting that Ted would like a seat on
the immigration subcommittee. That was precisely the subcommittee Ted
wanted, and he got it.

Eastland might fume about what he considered his young colleague's
dangerous, facile liberalism, but when in 1965 Ted was trying to push
through an immigration-reform bill that Eastland opposed, the Mississippi
senator allowed him to chair hearings in his subcommittee dealing with the
legislation. That was an example of the courtesies that were a part of the life

of the Senate in the sixties, and one of the reasons Ted looks back on that time with such fondness.

Ted, a fourth-generation American, frequently alluded to his ancestors' difficult journey from Ireland, and he would make immigration one of his major concerns. Here alone he had an impact on American society important enough to merit a place in the history books. Reform is often in the eye of the beholder, and conservatives bemoaned the bill. This controversial legislation ended the quota system based on national origin and ended up turning matters upside down. Within the next two decades the bill diminished European immigration from 50 to 10 percent, while increasing Asian immigration from 8 to 45 percent.

When Ted introduced the immigration bill, he pointedly praised others who had led the long fight. When Bobby got up to support the bill, he noted simply that "the central principle of his bill was first incorporated in a bill drafted in the Department of Justice while I was attorney general."

The Senate was full of prickly, ambitious men who did not look kindly on Bobby's refusal to partake in their customs. He identified with the youth movement of the sixties and its insistence on authenticity. Most of the rituals of political Washington restricted his movements. He was not yet forty years old, but so much of the poetry that he loved was full of impending death and tragedy, and he lived as if he heard time's clock resonating in his ear.

The brooding Bobby was a man of unpredictable moods, a handicap in the collegial world of the Senate. "One day we'd crack jokes for an hour, the next day he'd chop you off," recalled former senator Walter Mondale. "Your relationship would be just as though it never happened. I'm not sure that he realized how some people were hurt by that."

In the Senate, Bobby overshadowed his brother in almost every respect, from the media attention that followed his every word and movement to the dramatic issues that he set forth. At times, however, his little brother was a mentor teaching him lessons he did not intend to learn. "Is this the way I become a good senator—sitting here waiting my turn?" Bobby asked as the two brothers sat side by side on the Labor and Public Welfare Committee. "Yes," Ted said. "How many hours do I have to sit here to be a good senator?" "As long as necessary, Robby."

For Ted it was a fine thing to have sitting next to him a fellow senator whom he trusted completely and could work with on common goals. Bobby learned his way within a few months, and by early 1966 it was Ted asking his brother's advice on a speech about refugees ("Is tomorrow's hearing an appropriate form [sic], or should an address be made on the Senate floor—or both? Do you suggest any changes?"). As difficult as it was politically for Ted to have an older brother who overshadowed him, for a man who trusted almost no one, it was of inestimable value to have Bobby there. Ted sent his brother a newspaper cartoon that showed a little boy standing in front of an enormous dinosaur with the caption: "Bobby only wants to pet it. He won't hurt it."

"Dear Robby," Ted began his note, using his favorite nickname for his brother. "I am sick and tired of hearing about your trip to Africa." But the letter ended with "Your loving brother."

Buried within Ted was that natural frat brother, always ready for a keg and a good time. He simply didn't have his brother's self-absorbed seriousness. Bobby was the guide who led him toward a full emotional commitment to some of the major issues of the time. Ted recalled that Bobby "would suggest that I speak at an early morning assembly at a local high school urging the students to stay in school and continue their studies. He would remind me to give whole-hearted support to a fund-raiser to be held in Boston for Cesar Chavez's farmworkers. He would tell me of a recent trip to the Mississippi Delta, describing the extraordinary conditions of hunger, malnutrition and poverty he saw."

Unlike his little brother, Bobby did not play the long-term legislative game. He rarely nurtured political relationships that might prove useful on some distant day. He hardly ever gently pushed issues toward the forefront. He was not one to barter with his colleagues over votes or favors, an integral part of the whole mosaic of legislative politics. "Bob was always impatient," reflects Dutton. "I would use that as one of the key half-dozen words of his personality. He'd say, 'Cut out the shit and let's get to the point.' He was not one who liked rumination and dragged out questions."

For the most part, Bobby pursued a liberal agenda. His first major bill, cosponsored with New York Republican senator Jacob Javits, extended voting rights to citizens who could not speak or write English but had other academic qualifications. Like the immigration legislation, this was a watershed bill in its philosophical premise as to what defined full American citizenship. Bobby also debated in favor of a bill to require health warnings on cigarette packages, another issue on which he stood in the vanguard. In his first floor speech, he talked about issues that resonated with the concerns of his late brother. He called for negotiations with Russia over a nuclear nonproliferation agreement. He proposed aid to countries that were developing peaceful uses of nuclear energy based on their agreement to the mandatory inspections of their reactors.

I f you sought to work for either Bobby or Ted, you were bright, ambitious, well educated, and ready to work hours that would have seemed insane to anyone but an intern. But there was a difference in the kind of associates with whom each man was comfortable. Bobby brought in as his legislative aides Adam Walinsky and Peter Edelman. They were perfect products of the meritocracy: Walinsky had gone to Cornell and graduated from Yale Law School; Edelman had both his undergraduate and law degrees from Harvard. The two attorneys had worked in the Justice Department before joining Bobby's senatorial campaign.

Walinsky was a stirring speechwriter, as good at capturing the intense, passionate nature of Bobby as Theodore Sorensen had been at capturing the wit and intellect of John F. Kennedy. Walinsky pushed his agenda into his paragraphs and argued his case to Bobby with uncivil earnestness. Walinsky was one of those intemperate, youthful voices of the sixties that usually shouted their truths on the street, not in the corridors of power in political Washington. Bobby not only tolerated such a man but found him indispensable.

Edelman, a year younger than Walinsky, was no less politically passionate, but he was more the legislative craftsman, and he did not shout so
loudly. Both Edelman and Walinsky were uncompromising idealists, but
they also were ambitious men who believed they were working for the most
important figure in American politics. When they strode down the corridors,
it was as if to say that they worked for *the* senator, and their endeavors had
little in common with the tedious routines and platitudinous realities of
many other senatorial aides.

In the early days Ted's staff had been mediocre. Then he was simply
learning and ingratiating himself with his colleagues, not attempting to
make a major impact, and it hardly mattered that they were often inept. Now,
not only was Ted more skilled in procedures, but with his brother in the
Senate, he found it a devastating comparison when he saw Edelman and
Walinsky advancing Bobby's career. Ted rarely expressed his discontent in a
diatribe or in an articulate discussion of how he would like things done. He
just got increasingly nervous.

Ted brought David Burke into his office as his new legislative assistant.
Like Bobby's two new aides, Burke was not a child of privilege, but the son
of a Boston cop. If Walinsky and Edelman were smart beyond their years,
Burke was wise beyond his. He had an Irish sense of politics as the best of all
games, and he had astute insights into his new employer. Though his boss
and he were the same generation, he had a perfect awareness that Ted was
the senator and he was not. As close as Burke came to be to Ted, and as much
as the two men shared, as long as Burke knew him, even long after leaving
his employ, Burke always called him "Senator."

Burke brought in a whole new group of aides—talented, ambitious
young men, most of them attorneys. Each had his issues, his causes, and
they worked with dedication, shrewdness, and political savvy.

Burke was Ted's closest aide, but he was no Sorensen, an alter ego mimicking his speech and manner. As Ted walked across to the Capitol from his
offices in the Old Senate Office Building, he usually had someone in tow,
sometimes a reporter or constituent or friend, but usually an aide, priming

him for the hearing or the meeting of the moment. Although he made
speeches on the Senate floor and took part in debate, he devoted his greatest
energy to his committee work. Committees were supposedly employed to
help Congress write legislation, but they had become entities unto themselves, used for investigations or publicizing issues.

Ted enjoyed people of all kinds, and yet there was a distance between
him and everyone else and some final iota of trust that he gave to no one. "If
you're the brother of an assassinated president and the brother of Robert
Kennedy, and you're a member of the United States Senate and there's a
public adulation around the name and that sort of thing, then someone's
going to most likely take advantage of you," reflects Burke. "And somebody's going to speak about you to journalists, and you hope that it's in your
best interest, but you're sure that's going to happen all the time. And so that
puts a leash upon the extent to which you can give full trust. I've seen him be
himself more than he's told me about himself. That's the difference. I've
seen him do silly things. I've seen him do great things. But he's never, to my
knowledge, pulled me aside and given me an enormous insight close to the
depths of his soul. I don't think he's done that to anybody. I think there are
times when he doesn't even do it to himself."

That reticence began with his own wife. The lesson of his family was
that a man could be an adulterer and remain an exalted family man. Ted had
good reason to believe that he was doing no more than carrying on a family
tradition, but Joan did not see it that way. She was a beautiful, artistic
woman, and she began to doubt herself, wondering if she was unattractive
or had other defects that drove her husband from their home. Ted had a wife
with what in retrospect were the beginnings of a serious drinking problem.
She was his co-conspirator in creating an image of an idyllic marriage, but
the emotional price was heavy, and the deceits large.

Ted had nowhere he could go to be himself. When he was home, he had
a troubled wife with whom he could not speak frankly. When he was outside
the house, his friends might josh him mercilessly, but there was always a subtle deference. They were constantly aware of him and of what he needed or

wanted or seemed to need or want. He appeared much more outgoing than
Bobby, much more comfortable in the public craft of politics, but as Edelman understood, "Ted is actually not that comfortable with people either."
No matter how much Ted drank, there were never boozy confessions and
revelations; there was always a part of him he kept back, circumspect and
cautious.

T ed appeared outgoing but was eternally wary; Bobby seemed much
   more introverted yet was capable of deep candor. By the traditional
criteria that define power in Congress, Bobby was near the bottom of the
hierarchy of authority. Yet he walked with a power and poignancy that
brought him attention wherever he went. He did not need the floor of the
Senate as a podium, for wherever he went the crowds formed and his voice
was heard.

The Johnson White House watched this with increasing discomfort; its
perceptions, while sometimes unkind, were not necessarily without a measure of truth. "He is trying to put himself into a position of leadership
among liberal senators, newspapermen, foundation executives, and the like,"
wrote one of Johnson's top aides, Harry McPherson Jr., to the president in
June 1965.

> *Most of these people mistrusted him in the past, believing him (rightly) to
> be a man of narrow sensibilities and totalitarian instincts. . . . He will pick
> his issues for both immediate and cumulative effect. He will not care
> whether he ever becomes of the Senate. It will be enough for him that he is
> in the Senate and can use it as a platform in his search for power. . . . He
> will become a voice of reason and enlightenment—powerless within the
> Senate, so far as majorities are concerned, but well-regarded in the country
> among the millions of people who believe Senators are self-seeking
> blowhards who don't understand the real issues and who sit around all day
> making deals from which the public never benefits."*

McPherson's final words of advice were as good for Bobby as they were for Johnson. "You have the office, the policies, the personal magnetism, the power to lead and inspire, and above all, the power to put good ideas into effect. An obsession with Robert and with the relationship of your best people to him may, I believe, distort policy and offend the very men you need to attract."

CHAPTER   FIVE

Peaks and Valleys

obby loved his brother Jack beyond all other men, and to love him meant to emulate him, to go forth into the world with the courage that Jack called "the most admirable of human virtues." In early 1965, Bobby was invited to join an expedition attempting the first ascent of a 13,880-foot mountain in the Canadian Yukon named after Jack. Bobby had never climbed a mountain before. Though he was the father of nine children, and a surrogate father to two other children, no one had any doubts about the expedition—not Bobby, who said yes; not his wife, who celebrated his daring; not his sons, who wanted to follow in what they considered his heroic path; not *National Geographic,* which chronicled his ascent; and not the American people, who applauded his intrepid attempt.

This climb to the summit of Mount Kennedy became the prototype for all Bobby's risk-taking, a public assertion of courage in a physical act that took on a spiritual and moral dimension. He was not simply risking his life but making a statement about life. He was risking his life to prove himself worthy.

The Kennedys are such risk-takers that it is tempting to see that trait as part of their genetic makeup. In 1996 a group of Israeli and American researchers announced that they had isolated a gene called D4DR linked to "novelty-taking," a characteristic that the popular press extrapolated to mean risk-taking or adventurous behavior. Few of those with the gene were fast drivers, pilots, skydivers, or mountain climbers, but the gene appeared to increase somewhat the propensity for such behavior. But no responsible researcher would suggest that it could have been the prime factor in the conduct of Bobby and the other Kennedy men. At most, the gene points to a predilection, a tendency that the individual can deal with in a multitude of ways.

Unquestionably, Bobby was in a family of risk-takers, but the risks each one took, and the reasons they took them, were very different. Professor Frank Farley, past president of the American Psychological Association, has coined the term "Type T," or thrill-seeker. He divides thrill-seekers into four groups: those prone to take mental risks, those who take physical risks, those who take intelligent risks, and those who take stupid risks.

Most of these sons of Camelot were Professor Farley's Type T, but one of the great questions of each of their lives was the form that thrill-seeking would take. Would those who took physical risks attempt these activities in the safest way possible, or would they mindlessly up the ante? Would their risks make sense, or would they be silly or futile? And would they choose physical daring to avoid taking on more difficult intellectual or political risks?

As Bobby headed up Mount Kennedy he was not living out a quest that his genetic patterns had doomed him to take. He was an immensely complex man who had grown up in a unique environment, the social, cultural, and historical context of his life. He carried a profound load of heredity—not simply the family blood but what could be called social heredity, all the conduct he had learned in the Kennedy home, the expectations his father had placed on him, and the burden of Jack's death.

Modern science has taught us that we are born with the slate of our lives partially marked, but there is still room for character and choice. It said much about Robert F. Kennedy that he would have agreed that he was responsible for his own life. In the greatest moments of his political career, he tried to get people to assert more sovereignty over their own lives and to make moral choices that they often avoided.

Bobby had little more than a long weekend in late March 1965 to make his ascent. He arrived at a base camp already prepared for his climb. Mal-

colm Taylor, a writer and photographer, observed that Bobby displayed little of the comradeship that was then so much a part of climbing, only a
dogged determination "to get there, get up that mountain, do what he came
to do, get out, and return to civilization."

All of his life Bobby had stood among men who were bigger, stronger,
and more experienced in the ways of the world. He had played football at
Harvard on a broken leg, never outwardly flinching at the pain. He had
stuffed away his self-doubt and fears, and he had stood with his teammates,
both on and off the field. And so he did with everything in his life.

From high camp, Bobby set off roped between six-foot-five-inch James
Whittaker, one of the five Americans who had climbed Everest in 1963, and
Barry Prather, another experienced climber. "He knew that people were
killed on the mountain, on the [Everest] expedition; and that we had a real
difficult time, and that I reached the goal even though it was the most difficult," reflected Whittaker. "And all that excited him. It was really exciting to
him, and I think that's why he liked me essentially."

On the way up the mountain, Bobby fell chest-deep into a crevasse, a
terrifying moment for a novice climber. "Bobby was a physical person,"
reflects Martin Arnold, a *New York Times* reporter who had gone along as
far as base camp. "I think he understated the effort that this was going to
take. It was quite a bit of an ordeal." Bobby slogged onward, and the two
seasoned climbers let Bobby climb ahead the last hundred feet to the summit
to have the honor of being the first man to stand on the peak. As he stood
there, ten or so airplanes and helicopters flew above, filming him.

Whittaker became one of Bobby's playmates who was invited to come
along on the family rafting, skiing, kayaking, and hiking trips. The man was
a great athlete. He was impressed that Bobby could keep up with him. Bobby
wanted and needed to have people around him of apparently superior abilities with whom he could endlessly compete.

That summer of 1965, Whittaker joined the family on two rafting trips.
On the Salmon River the water was swift and cold. That would have been
adventure enough for most people. Twelve-year-old Joe climbed a cliff
twenty feet above the river and prepared to jump. The pool of water below

led into the rushing current. If he jumped too far out, he might be carried downriver. Yet if he did not jump out far enough, he risked falling against the rock face, skittering down into the water. Young Joe flung himself out into space, successfully falling into the pool.

"We better do it or, you know, we won't look good," Bobby told Whittaker. The two men jumped off the cliff too. They then looked up and saw five-year-old Kerry on top of the cliff, held by Joe. Whittaker could see that her knees were shaking, and he thought enough was enough.

"Come on, Kerry, let's go, come on," Bobby and Ethel yelled. And so she flung herself out into space and plummeted into the cool blue water

For Bobby's sons these sojourns in the wilderness were pedagogic exercises of the first order in which they learned how they must live as men. "The people my father surrounded himself with were people who had proven themselves with physical courage," his son Bobby Jr. reflects. "He liked guys who were tough. He took us out into the wilderness to make sure that we had those experiences of physical courage and tested ourselves and that kind of stuff. But we had fun. We'd go on these river trips, and we'd go mountain climbing with him or whatever, and we always just had a ball. We loved it."

"Kennedys never cry," Bobby told his sons. And if they were so weak and irresolute that they indulged in tears, he had the way to shame them. "I can remember him showing me scars on the top of his head that he had received when he was younger and saying, 'Do you think I cried when I got all those scars?'" Joe recalls. "Of course I didn't think so, therefore, I'd have to stop crying."

I n the last week of September 1965, Bobby borrowed the SS *Nehris,* an exquisitely maintained 1917 schooner, from a friend. He set sail from Sag Harbor, Long Island, with Ethel and ten children—six Kennedys and four of their friends. Three other adults went along too, a married couple and Barrett Prettyman, a former associate at the Justice Department.

It is an axiom of shipboard life that in such cramped quarters you have to be neat. The Kennedy children threw their duffel bags, books, snacks, and clothes around below deck. What most people consider a good sail was a tedious bore to Bobby. As soon as the sails were up, he said that he and Ethel wanted to water-ski. "We've come on this trip to have fun," Bobby said, "and we can't just lie around the deck." The dinghy was hardly a speedboat and did not have much gas. There were sharks in the waters off Long Island, but Bobby insisted. When that was over, Bobby ordered that the final sail be hoisted, though the captain was opposed. The captain went below, and the children took turns jumping off the front of the schooner, catching the boat's ladder as it sailed by. This merriment continued until the fin of a shark appeared near the boat.

Prettyman reflected that weekend that "the Kennedys live freely and without restraint." Those who traveled with Bobby did not notice or suspect that he was still at times depressed by his brother's death and might be using these adventures to jump out of a pit of darkness. It was different for his children, and he was teaching them that life must be lived only one way.

That evening the SS *Nehris* took harbor at Newport, Rhode Island. The weather turned dark and menacing, the seas churning up. The next morning the captain looked out on the Atlantic and said that they would not be going out that day. "Oh, it's not that bad," Bobby said, though it was bad indeed. As they headed out, the boat was pounded as it rose and fell in the troughs between the waves. These were dangerously heavy seas for the old sailboat. Prettyman was not a timid man, but he reflected later that he had been "scared to death." The Kennedy children, however, were squealing with delight, as if on a carnival ride. Prettyman had never sailed in such troubled seas, and he feared that if the storm went on, the boat might sink.

Out of the gloom a Coast Guard cutter appeared, close enough that they could see an officer on deck holding a bullhorn. "Senator Kennedy, your daughter Kathleen has had an accident," the voice said, the sound almost lost in the wind and surf. "She's been hurt falling off a horse, and I think you had better come back."

The Coast Guard wanted to bring their boat close enough to the SS *Nehris* so that Bobby might be able to move over to the faster vessel. Bobby had another idea. He prepared to jump overboard, not even wearing a life preserver, which he felt might hinder his swimming to the Coast Guard vessel. "You have to wear a life vest," Ethel yelled. "You have to." Bobby reluctantly agreed and leapt into the swirling seas. For a while the other passengers saw nothing but foam and waves. Then his head bobbed up, and he began swimming fifty yards to the ship.

As the Coast Guard cutter sailed out of sight, Prettyman feared that the SS *Nehris* was going to go down, broken into pieces. The sail was torn. Then the boom spun around, hitting young Joe in the head. Prettyman thought for a moment that the boy had broken his back. The captain managed to sail the boat into Woods Hole harbor, where they learned that Bobby was already with Kathleen, who was not seriously hurt. As Ethel and the children left, the owner returned to claim his sailboat. "You should have seen the guy's face when he saw his ship," Prettyman recalls.

The lesson of each mishap, each close call, was not to take down a sail, or stay closer to shore, but to head out again in the highest seas. When Joe broke his leg careening down the ski slopes of Sun Valley, this was no cautionary tale, no reason to stay off the double diamond runs. The following Christmas vacation Joe was out there again, racing down the mountain, and again he broke his leg. His brother Michael grew into the best skier of the bunch, but one could spot any of the brothers on the mountain, careening downward almost out of control, on the razor's edge of possibility.

There were no mountains on the family estate, so the children invented their own. Hickory Hill was a child's vision of freedom, a laboratory for experience. There were two swimming pools, a tennis court, a great lawn for football games and roughhousing, chicken coops, a paddock, an obstacle course worthy of Marine Corps basic training, and an ever-changing menagerie of horses, rabbits, snakes, ponies, iguanas, lizards, dogs, pigeons, donkeys, sparrows, parakeets, cockatoos, harassed secretaries, beleaguered cooks, and frantic governesses.

The Kennedy boys went out in the woodlands after school and hurled black walnut "grenades" at each other. David and Bobby Jr. built tall walls made of bales of hay and ordered little Michael to jump off. He did so with aplomb, no matter how high they piled the bales. They led the little boy up to the top of the barn roof, and with hardly any prodding, he jumped out into space.

One Saturday morning David came swinging down out of an ancient hickory tree on a rope. "I'm Batman," he yelled as he chased after Bobby Jr. and Michael. Not to be upstaged, Kerry jumped off a shed, using a bush as a natural mattress cushioning her fall. Meanwhile, Courtney floored the pedal of the golf cart she was driving, nearly hitting a gardener, on her way to the tennis court, where Joe was having his regular lesson as the two toddlers, Chris and Max, dumped a basket of tennis balls on the court.

I n November 1965, Bobby flew down to Lima on a three-week visit to Latin America. He sought to reach out to the lives of the poor. He visited *barriadas* where at times the stench was so overwhelming that it choked the uninitiated, and a person wondered how anyone could live there even for a day. He cradled impoverished children in his arms and walked through the garbage-strewn lanes.

"Wouldn't you be a Communist if you had to live like there?" he asked after a visit to one *barriada*. "I think I would." As much as he continued to despise Castro, he expressed great admiration for his romantic cohort Che Guevara, calling him a "revolutionary hero."

Wherever he went in South America, Bobby did not pander, but confronted his audiences. The Johnson administration had stopped all economic aid to Peru until the government finished negotiating a new contract with the International Petroleum Company. Peruvian president Fernando Belaunde Terry was a moderate, but he shared his people's anger at IPC, a subsidiary of Standard Oil of Jersey, as a symbol of foreign exploitation. Bobby did not approve of Washington's bullying techniques. Nor did he countenance the tendency of some in Latin America to blame Uncle Sam for all of their evils. "You are the Peruvian leaders of the future," he told university students in Lima. "If you object to American aid, have the courage to say so. But you are not going to solve your problems by blaming the United States and avoiding your own personal responsibility to do something about them."

Bobby went to Latin America as a witness to the poverty and deprivation and to exhort as much as to learn. Although he met the presidents of the countries he visited, he spent little time with business leaders, middle-class citizens, or progressive elements in the military, all people crucial to meaningful social reforms. Instead, he headed out almost daily to the slums and talked to student groups. These were both crucial elements of what he considered his natural constituency

Several reporters were traveling with the group, and scores of others joined them in each country he visited. If that kind of coverage magnified his truths a millionfold, it also made him an actor, one who needed crowds and poignant scenes and dramas. He was supposedly living within the moment, yet every evening he checked with his press secretary in Washington to learn about the coverage he was getting, and he was unhappy that his trip was not the front-page story back home that it was all over South America.

While Bobby was seeking to identify with the lives of the desperate, anonymous poor, he was traveling with a ten-person entourage that included a secretary, two full-time aides, three other associates, including one with a wife; his own wife and two of her friends. This group traveled with two hundred pieces of luggage.

In Argentina the American ambassador's wife happily fronted money for flowers, scrapbooks, stirrups, and cigars. Storekeepers allowed Bobby and his entourage to take various items with nothing more than a signature. The group left the country owing over $750, including unpaid hotel bills. The debts were not paid until the embassy warned "that the press has gotten wind of the situation and intends to publicize the fact that the Kennedys and their friends owe so much money in Buenos Aires."

Bobby was in Salvador, Bahia, a city in Brazil, on November 22, 1965, the second anniversary of his brother's death. At the Church of São Francisco, he and Ethel sat among poor women for the eight o'clock mass. The Kennedys came forward to take communion. After they took bread and wine, they stayed on their knees even after all the other worshipers had walked back to their pews.

When the great doors of the church opened, Bobby was back in the world of abject poverty. He was seeking to make himself the tribune not only of America's poor but of the world's poor, a unique, indispensable figure. Bobby walked these streets in his suit and tie, and people looked up at him in disbelief. He was used to being greeted almost like a Hollywood star, a celebrity descending onto the mean streets. The stench was so overwhelming that the Brazilian security men returned to their car, but Bobby hurried on. Many of these people did not know who he was, or how this man and his entourage should be greeted.

Mothers held their children up to catch a glimpse of the spectacle. And everywhere Bobby stopped to say a few words, he admonished the children to remain in school, "as a favor to President Kennedy." These children plagued by disease, hopelessness, and poverty were to educate themselves to honor a man who at best was a vague phrase, a promise they had heard, a hope out there somewhere beyond the horizon.

When Bobby flew out of the city that day, he sat by himself with his head buried in his arms. Was he reflecting on his brother's death two years ago to this very day? Or was he thinking of how he had promised to look into these people's woes, and now he was gone, and the dust still lay deep on the streets, the stench was still foul, and the lives were still desperate?

When his formal journey to Brazil was over, Bobby decided to go on a trip up the Amazon. He flew to the legendary city of Manaus. From there he took a paddle wheeler for two days up the river. As he got on a 1939 seaplane to fly farther up the river to a tiny Indian village, he said: "I must be crazy to get on this thing." From the village he traveled in a dugout canoe, at times jumping into the piranha-inhabited water to push the boat along.

When Bobby returned to Washington, a journalist told him that for a man with his obligations, "crazy" was precisely the way to describe his Amazonian excursion. For a moment Bobby said nothing. Then with his lips suggesting a smile, he said: "Perhaps it was not the wisest thing I've done."

A Brother's Challenge

I n 1966 one of Bobby's closest advisers, Frederick G. Dutton, wrote the senator a series of memos that would have sounded mindlessly grandiose to most politicians. Dutton, an attorney and Kennedy administration official, told Bobby that he should not compare himself to politicians like Johnson and Vice President Hubert Humphrey, but to great historical figures such as Mahatma Gandhi and Alexander the Great. "Many tens of millions of people in this country and hundreds of millions abroad look to you as the hope of a world and future built not just on power and the limitations of the present but somehow, and still most inarticulately, in terms of a further reach and fuller measure of human achievement and community. My own guess is that your greatest eventual contribution will not be just as President but in leading toward greater international pulling together —not in a formal organizational sense but in drawing together, through yourself, a very considerable por tion of mankind."

A man whose destiny was of such enormous magnitude had to create a persona far beyond that of a politician. In another memo, Dutton advised Bobby that he should take "at least one major, exciting personal adventure or activity every six months or so, as mountain climbing, river boating, etc, to allow dramatic insights into your own life and interests especially for the younger people for whom you are almost a personal model." Dutton perceived that beyond their personal value, these dramatic sojourns in the wilderness and on the seas were brilliant devices to expand Bobby's image from a politician to a cultural icon.

His brother's close friend, Chuck Spalding, who had known Bobby since he was a little boy, saw the dangerous excursions in a similar way: "I think Bobby wanted very much to engender the kind of excitement that President Kennedy did. He had to find a way of doing it on his own. I think it was compensation. It turned out to be just that, and it began to attract a lot of attention."

Bobby saw the young as the dynamic engines for change and sought to make them his natural constituency. "The same young faces are all over the country and they don't want to know anything about old men or young old men," wrote New York journalist Jimmy Breslin in December 1966. "The only name they know in politics is Robert Kennedy."

The young had a far broader range of political heroes, but it was an immensely flattering idea to be told that you were *the* troubadour of youth. That image was already so strong that when Peter Mansfield, the president of the National Union of South African Students, invited the New York senator to speak in Capetown, he did so because "we thought he represented the younger generation of political leaders and the new ideas of youth."

In accepting the invitation, Bobby was taking a certain risk, not only to his political future but to his very life. South Africa was indeed a different country. Nelson Mandela was not the universally beloved figure and Nobel Peace Prize laureate that he is today, but an advocate of violent revolution who had been imprisoned for life, his name hardly known beyond his continent. As for Bobby, there were many white defenders of apartheid who considered him a radical agitator whose presence in their country was an intolerable provocation.

Bobby had perfected his brilliant political theater, and when he arrived in Johannesburg airport near midnight on June 4, 1966, his advance people had primed a large, pushing crowd, many of them carrying placards. His frenetic four-day visit created immense publicity and shone an intense light into some of the darkest regions of apartheid.

Bobby visited South Africans who were confronting the evils of apartheid, and then in Cape Town he gave a great and powerful speech elucidating many of his major themes. Bobby began not by excoriating white South Africa for its vicious system, but by talking about the racial difficulties in his own country. He saw a fundamental commonality in the human struggle. "Our answer is the world's hope; it is to rely on youth," he said. "The world demands the qualities of youth: not a time of life but a state of mind, a temper of the will, a quality of the imagination, a predominance of courage over timidity, of the appetite for adventure over the love of ease. It is a revolutionary world we live in . . . and it is the young people who must take the lead."

Bobby took on the causes and angers of the young as his own. In the fall he gave a speech at the University of California at Berkeley, the capital of youthful politics and culture. Many of those at Berkeley's Greek Theater had read the Students for a Democratic Society 1962 Port Huron Statement, one of the seminal documents of the emerging New Left. Some of them could recite the opening lines word for word: "We are people of this generation, bred in at least modest comfort, housed now in universities, looking uncomfortably to the world we inherit." They listened in disbelief and anger as Bobby appropriated the very words that they thought defined them. "You are a generation which is coming of age at one of the rarest moments in history—a time when all around us the old order of things is crumbling and a new world society is painfully struggling to take shape," Bobby said.

To his radical critics, Bobby was taking their flowering ideas, uprooting them, and growing them in his own hothouse in neatly cultivated, domesticated rows. He was unquestionably trying to corral much of the energy and creativity of the leftist young and direct it toward his own advance, not simply as a political figure but as a cultural leader of a worldwide movement. And wherever he went, his diagnosis of society's ills was far more radical than his prescription.

T he Robert F. Kennedy known privately to a number of journalists was a different figure from the man on the podium; he was far more candid in his analyses and far more sweeping in his social remedies than in his speeches. Bobby brought all kinds of people home—politicians, athletes, journalists, artists—and mixed them all together. It was endless, boisterous fun, but there was a higher purpose in these frequent evening gatherings: as Bobby was beginning to accumulate a network of influential, well-placed people who would be with him when he made his inevitable race for the presidency.

Bobby had relationships with reporters unlike those of any other politician of his generation. His journalist friends were not testy chroniclers but partners in an adventure. He did not give them what in Washington is usually called "access," a calculated shuttling in and out of a professional life. He invited them out to Hickory Hill, where he initiated them by blocking them viciously in football or just as viciously attacking their ideas. He did not want shills and sycophants salivating at his every syllable. He wanted those who appreciated his life and vigorously argued the questions of the day. The journalists celebrated the glorious pandemonium of Hickory Hill. They were there for breakfasts and when Bobby and Ethel got down on their knees with their children to say prayers before going to bed. Of course, there was calculation in this, but no other Washington politician lived in such a world or invited reporters home as often or as freely as Bobby did.

These journalists sat at dinner among the unruly, vigorous brood. There was such a range of age in the children that while Ethel had given birth to two more sons, Matthew Maxwell Taylor Kennedy in January 1965 and Douglas Harriman Kennedy in March 1967, the oldest children were off to prep schools. Within the family there were the same gender differences in upbringing as in the rest of American society, only in exaggerated form. Neither Bobby nor Ethel envisioned political roles for their daughters and gave them a freedom of conduct their sons did not receive. Kathleen had been allowed to follow her wish to go to the Putney School in Vermont, an ultra-liberal school that was a feast of freedom for their oldest daughter.

As for their eldest son, Joe was being groomed to go down the same path his father was following. He matriculated at the Milton Academy, as his father had done. In October 1966, soon after Joe arrived, the school's athletic director, Herbert G. Stokinger, wrote Bobby: "I am delighted that Joe, III [*sic*] is with us, and his name on the 3rd football roster brings back fond memories of the 1942 and 1943 football seasons." A month later Bobby wrote his fourteen-year-old son; he was sending Joe some news clippings and wanted to make sure that Joe was reading the paper "and not just the funnies." He instructed his second-born on some of the nuances of politics ("The party in power always loses some and with the majorities that the Democrats have in Congress you could expect to lose at least 30 seats").

This was not the typical letter that a father would write to a teenage son. There was a pedagogical imperative here: Joe was to inform himself about the world of politics. Bobby said that he would be coming up to the school to watch his son play football. That was another part of the imperative: the boy had to prove himself on the football field as much as he did in the classroom.

Bobby was an immensely compelling figure to his sons. He was the father they had to emulate in the intensity of their play on the football field and in the earnest endeavors of their lives. He was an impresario of a father, a ringmaster who clapped his hands and life began. Bobby's sons not only profoundly loved but profoundly admired their father.

I n his three and a half years in the Senate, Bobby made no great legislative record, but he did advance the dialogue on the great issues of the day in often prophetic ways. There are some who consider him the quintessential American liberal. One biographer, Arthur Schlesinger Jr., has called him the "tribune of the underclass." Another author, Michael Knox Beran, asserts that Bobby was essentially a conservative, the last American patrician. Bobby was drawing ideas from both sides of the spectrum and had much in common with the New Democrats of the Clinton era: he had a liberal vision of justice and equality but was willing to employ conservative means to obtain them.

Bobby did not have Jack's brilliant conceptual mind, especially with regard to foreign policy. Nor was he like Ted, who was generally comfortable with the conventions of mainstream liberal Democratic politics. Bobby was restlessly seeking out new solutions to old problems, bushwhacking his way through life rather than traveling marked pathways. Most of his colleagues celebrated uncritically at least one element in society, be it business, labor, teachers, the middle class, the minorities, the intellectual elite, but Bobby saw that all sectors of society were susceptible to corruption. He condemned great corporations that "play so small a role in the solution of our vital problems," labor unions "grown sleek and bureaucratic with power," and universities that had become "corporate bureaucracies."

Some of Bobby's criticisms of social programs for the poor could just as easily have been made by the ultra-conservative Senator Barry Goldwater. Bobby argued that "the institutions which affect the poor are huge, complex structures operating far outside their control. They plan programs for the poor, not with them." He was one of the first major figures in American political life to perceive that welfare and social programs had become bureaucracies that were often more concerned with perpetuating themselves than with helping the poor. He was not in favor of what he considered handouts, a guaranteed annual wage, and he instinctively understood the pernicious impact of welfare entrenched in families generation after generation. He knew that life was tough, but his skepticism about human nature set him apart from many liberals.

Bobby's concerns had new urgency after the first massive urban riot of the sixties broke out in Watts in August 1965. Before the fires and violence died down in Los Angeles six days later, thirty-five people had died, nearly a thousand were injured, and there was over $200 million in damage. Like most Americans, Bobby feared that the match might have been set for conflagrations across America. A few weeks later he was having lunch at Hickory Hill by the pool with three of his aides—Joe Dolan, Edelman, and Walinsky. The subject naturally turned to what could be done about the plight of ghetto blacks.

Walinsky set forth ideas for the reconstruction of America's inner cities based in part on an article he had written in *The New Republic* criticizing the poverty program. Bobby was not a social theorist. He operated more on deep feelings about human beings and their ways. Bobby listened intently to Walinsky, then said simply: "Well, keep working on it."

Walinsky and Edelman prepared three speeches outlining a fundamentally different approach to poverty. In January 1966, Bobby gave talks in New York City that dealt with what many considered contradictory goals: integrating America and revitalizing ghetto communities. Bobby proposed the radical idea of considering "special federal aid to suburban schools which take in slum children." Such social engineering went far beyond the parameters of either liberalism or traditional conservatism. His other ideas may not have been so extreme, but they were threatening to the traditional power structure. He wanted to rebuild the ghettos by hiring the unemployed to work on everything from playgrounds to subways, focus public education on rebuilding efforts, and make business a partner with government. In the end, he wanted to create true communities in which citizens chose to live, transforming the areas currently populated by people too poor or troubled to have anywhere else to go.

Bobby's speeches did not set off a dialogue in the *New York Times* or a debate on the floor of Congress; they were largely ignored. He took that not as a failure however, but as a mark of how different these ideas were, how threatening to the establishment, and how crucial it was to try them out not in a social laboratory but in the world. As the community for this extraordinary experiment, Bobby chose Bedford-Stuyvesant in Brooklyn. Although the 400,000 residents made it the largest community of blacks and Puerto Ricans in New York City, Bedford-Stuyvesant was the city's forgotten ghetto, in the shadow of Manhattan's Harlem, the celebrated center of black culture. There was no famous 125th Street, no Apollo Theater, just block upon block of crumbling brownstones, vacant lots, and poverty.

The plan was to be financed in part by a special impact amendment to the poverty bill that Walinsky had drafted. Bobby paid little attention, not even bothering to read the document. On the day of the committee meeting he attempted to wing it, but his abysmal lack of preparation was in essence an insult to his fellow senators. There were suggestions that the meeting be adjourned and the matter brought up later. Afterward, Walinsky walked into the office moaning, "My God, what are we going to do? I can't get him to focus on it." In the end it worked out, and the program received a $7 million appropriation.

Bobby could have limited himself to pushing through enabling legislation, cutting a blue ribbon or two, and standing by observing a social experiment with his name on it. He went far beyond that in attempting to rebuild the blighted area. Bobby had the clout and the far-reaching contacts that few of his political peers could muster, and he used them enlisting America's elite. In one of the early gatherings Bobby managed to get William S. Paley, the chairman of CBS, to hold a meeting at his corporate boardroom. Bobby addressed the group and then proceeded to announce that he had business in Washington and was turning things over to Paley. Few politicians could have gotten away with that so easily, but Bobby could and did.

Bobby wanted the federal and city efforts to be in full partnership with philanthropy and business. He believed that the richest and most privileged Americans could work with the most disadvantaged. He wanted businesses to go into the ghetto not to dispense charity but because they could be socially beneficent and make money. Most sixties liberals were very uncomfortable marching arm in arm with big business, and this eclectic mix of government, business, and philanthropy set Bobby apart from many of his fellow Democrats and from Johnson's War on Poverty.

"It is Bedford-Stuyvesant that is the vanguard," Bobby told a local group at the beginning of the project. "But if this effort—with your community leadership, with the advantage of participation by the government, and with the help of outstanding men in so many fields of American life—if this community fails, then others will falter, and a noble dream of equality and dignity in our cities will be sorely tried."

The Bedford-Stuyvesant model depended on there being scores of Robert Kennedys around America, political leaders willing and able to move beyond their defined roles and to work as catalysts. It depended on the elite acting as Bobby's father had believed privileged Americans should act, giving back something of what their nation had given them, not simply writing off part of their income for charity and feeling good about themselves at a safe distance from the poor and deprived. It depended on politicians working to seed money not to their supporters, or through established conduits, but to the people and institutions most able to create real change, people who might in the end confront the very forces that helped them. It depended on local leaders who were willing to go beyond sloganeering. It depended on the poor themselves having the energy and the initiative to take advantage of the possibilities that flashed before them.

For all Bobby's hopes that the American business establishment would make a firm commitment, only one factory opened in Bedford-Stuyvesant in the next few years. It belonged to IBM, and it generated only 250 jobs. But the concept of community development corporations developed there became a model followed across America, and according to the standard by which most poverty initiatives judged themselves, the experiment changed people's lives. If Bobby had been able to return to Brooklyn in the early years of the twenty-first century, he would have seen a twelve-block historic district of well-maintained brownstones, a shopping center with a drugstore, chain restaurants, gift shops, art galleries, blocks of new townhouses, the Billie Holiday Theatre, and the Restoration Dance Theatre. But the poverty was still there, in block after block of substandard housing and vacant lots, amid the drug dealers and the violence.

Bobby's concern for black Americans had surfaced when he was attorney general, but it was Edelman who first read about the struggle of Mexican migrant laborers in the grape fields of California and about the United Farm Workers strike led by Cesar Chavez. And it was Edelman who vigorously suggested to Bobby that he make a journey to the agricultural heartland of California in Delano for hearings of the migratory labor subcommittee. Bobby was constantly being importuned to lend his credibility and the media entourage that inevitably trailed him to an endless variety of causes and issues. And thus it was understandable why he kept saying no and only reluctantly flew to the West Coast.

On this trip, early in 1966, he gave wide publicity to a national shame—

workers treated little better than farm animals and the Delano sheriff so devoted to his duties that he arrested striking workers before they committed any crimes. For Bobby the great inspiration of this trip was Cesar Chavez. Chavez was a man of deep faith, both in God and in nonviolence, and against enormous odds he was rising up against the injustice that plagued his people. Yet Bobby spent little time with Chavez, then or later. "They were talking about the nature of nonviolence, and they were talking about how difficult it is to make a strike when you have no assistance, no protection under the law," said Edelman. "But curiously enough, I think that was probably the longest discussion that they ever had."

In the spring of 1966 Bobby made another journey of discovery to the Mississippi Delta. He had gone there for hearings held by the Senate Labor Committee's subcommittee on poverty. As was his wont, he went off on his own with a small group, including aides and reporters, to talk to the poor themselves. "We first went to a home where there was a great deal of photography outside the house because the children were all lined up, barefoot, ragged clothes . . . undeniable swollen bellies, and sores that would not heal, and just clearly seriously malnourished," remembered Edelman. "Kennedy remarked to me that these were the worst conditions that he'd ever seen in the United States. Then in the next house there was a small child which either because of a physical handicap or maybe for a nutritionally related problem was perhaps two years old but unable to walk. He just sat on the floor. And Kennedy went into the house by himself. He went and just tried to get this child to respond, for maybe five minutes."

Bobby's admirers thought there was a near-saintliness in the way he wanted and needed to touch and feel the lives of the poorest and the most disheartened. His critics noted that when he cried, there was usually a camera around or a reporter with his notebook. There have been mixed motives, however, behind most of the world's great accomplishments. Sometimes Bobby's idealism and self-interested ambition were laminated in a way that made each one stronger. At other times his brutal realism held down his noble ambitions, limiting how far he reached and how much he dared. With

him it was the eternal question of ends and means, not whether the end justi-
fies the means, but whether the means he was willing to use changed the end
he sought. What must be said, though, is that as mercilessly ambitious as he
was, he never stepped back from what he considered his brother's mandate
for the country they loved.

Bobby took a long trip to the West Coast campaigning for Democratic
congressional candidates in the fall. On his return he had lunch with Edelman at an expensive New York restaurant. "How did it go?" the legislative
aide asked. "What do you think you accomplished?"

"Well, it's really very different from 1960," Bobby said. "All of these
people that looked like they were having a tough time now have cars, they
can get to the rallies. You can schedule the rallies in the parking lot of a
shopping center, and they're all very comfortable. You get the sense talking
to them that the major thing they're worried about is any threat to that comfort. They see the rising black aspirations as a threat to that comfort. They
see inflation as a threat to that comfort."

Bobby was all for helping the poor and the struggling minorities, but he
was woefully unsympathetic to the complaints of the struggling middle class
and of those who feared that if things did not change they would fall into
that class or even lower. Worrying about your monthly mortgage payment
or struggling to come up with the dollars to send your son to summer camp
were not difficulties on the same level as those of a Mississippi Delta African
American family worried over where they would get a meal, but these were
the legitimate concerns of millions of Americans.

"Well, that's rather distressing," Edelman said, after listening to
Bobby's mournful litany of the narrow preoccupations of these Californi-
ans. "What do you think can be done about it?"

It was understandable why he might find it difficult to listen to carping
constituents and others worried primarily about protecting their middle-
class preserves. Yet he was neither a public moralist nor a preacher, but a
democratically elected legislator.

"I think if you could work with them," Bobby said, "if you could stay there and really talk to them—of the ones that I did talk to—you could really turn them around."

Bobby could not take each one of these people and fly with them to the farm fields of central California or down to Cleveland, Mississippi, and show them what he had seen. He could attempt in his speeches and politics to make them concerned about minorities and the poor. This was as much a challenge in domestic politics as ending the war in Vietnam was in foreign affairs. As Bobby mused that afternoon, he wanted it to be a challenge that he would face. Edelman found it both "messianic" and "somewhat egotistical," but he nonetheless thought that if Bobby reached people directly, not on television, not on the podium, but one on one, he could change them.

CHAPTER SEVEN

War in a Distant Clime

ack's Pulitzer Prize–winning book *Profiles in Courage* discusses several different senators who risked their careers standing against majority opinion. They could have compromised and tempered their voices, but they did not, and they often paid for their principles with their careers. In his increasing opposition to the war in Vietnam, Bobby wrestled with some of the same dilemmas. He wanted to be seen as a man willing to stand up for what was right and true, but he was not about to do it at the price of his presidential aspirations. He was not going to allow himself to be marginalized as a ranting dissenter in a country that still favored war. He feared, moreover, that if he spoke out, his words of peace would so outrage Johnson that he would ratchet up even further the machine of war. So Bobby held back and was quiet. His critics suggested that his principles were held in place by expediency, while those who most believed in him felt that he was only waiting for the time and the place.

In the early months of 1967, Bobby decided that the time had come to oppose the war more publicly than he had done before. When he got up on the Senate floor on March 2, he held in his hands a speech that had been written and rewritten by half a dozen aides and advisers. In the address, Bobby stated that the responsibility for the war lay with many people, including himself. He said that he had not come there to focus on "the wisdom and necessity of our cause [or] the valor of the South Vietnamese, but on the horror."

"The Vietnamese war is an event of historic moment, summoning the grandeur and concern of many nations," he said. "But it is also the vacant moments of amazed fear as a mother and child watch death by fire fall from the improbable machine sent by a country they barely comprehend. It is the refugees wandering homeless from villages now obliterated, leaving behind only those who did not live to flee." At the center of his speech was an indictment of the war and its costs that went far beyond the White House to American homes everywhere. "It is we who live in abundance and send our young men to die. It is our chemicals that scorch the children and our bombs that level the villages. We are all participants."

Bobby was speaking words that were rarely heard on the floor of Congress. But when it came to offering solutions, he largely refurbished suggestions that had been made before. He called for a halt to the bombing followed by negotiations that would lead to the South Vietnamese government sitting down with its enemy, the Communist National Liberation Front (NLF), to discuss an open election.

"Politically and diplomatically, it was a fuzzy and maybe even an opportunistic speech, but more than any other Presidential figure he at least dealt with the human agony of the war," wrote James Reston in the *New York Times*.

Reston had written the truth, and when the upset Bobby called him, the columnist expressed himself "puzzled and embarrassed by your irritation." Bobby felt he had taken considerable political risks. Though it hurt to be called "opportunistic," the lasting pain was that Reston had struck at the essential contradiction in the liberal antiwar movement. "I read the speech as the thoughts of a man challenging the President, and as a candidate to succeed him at some point, and frankly, I don't think it measured up to that test," Reston wrote the New York senator. "Your premise was that serious peace talks were blocked because we would not take one single step. Maybe I'm a lousy reporter, but sympathetic as I am to your objective of starting peace negotiations, I was forced reluctantly to the conclusion that Hanoi did not want 'peace' but 'talks' under conditions that would end the military pressure on the North and allow them to maintain their pressure on the South."

There was the conundrum. "Peace" to the North Vietnamese and the NLF meant a united Communist Vietnam. To achieve that, they would fight until they were dead or their revolutionary resolution had been bombed and burned into submission. Johnson wanted "peace" as much as Ho Chi Minh did, but it was not the same peace. The American people did not appear any more ready to accept what the North Vietnamese and the NLF called "peace" than they were to accept the brutal, long-term war that might result in what some would call victory. That was Johnson's reality, as it was the reality of any who would challenge him.

Bobby did not yet foresee that one day Americans would leave South Vietnam in defeat. "He was a fervent anti-Communist himself, and secondly, I think he felt a sense of responsibility in terms of his brother's administration for our involvement," says William vanden Heuvel, a close adviser. "And thirdly, and most importantly perhaps, he was not at all averse to trying to save South Vietnam from Communist domination."

The same Robert F. Kennedy who excoriated the Vietnam War in biblical language got up at a Democratic dinner in June 1967 and praised Johnson for "the breadth of his achievements, the record of his past, the promises of his future."

"How can you say all those things?" journalist Peter Maas asked him afterward.

"If I hadn't said all those things," he replied, "that would give Lyndon Johnson the opportunity to blame everything that was going wrong on that son of a bitch Bobby Kennedy."

That may have been a lame justification to some, but Bobby was rare among politicians in his ability to create an atmosphere in which a journalist could ask bold and even insulting questions. When most people ask for advice, they merely want their judgments and decisions validated. Politicians are no different. When politicians seek to hear varying opinions on an issue, what they generally want is to have each end of the debate snipped off, so that no one presents them with what they consider extreme or outrageous solutions. Bobby entered into dialogue, however, with an unprecedented range of individuals and listened to a cacophony of opinions and deeply felt views. He was comfortable with debate, and yet surely it must have been emotionally exhausting to be so confronted and challenged and battered and pushed.

Edelman and Walinsky offered their own fervent views on the Vietnam War, imploring Bobby to enter the presidential race against Johnson. In the senator's office they were the intemperate surrogates for the youthful generation. But neither Walinsky nor Edelman was present at important meetings where other associates discussed the senator's political future. Bobby was all for reading the powerful passages in Walinsky's speeches and listening to his passionate arguments, but when it came down to making decisions in the real world, he excluded his youthful aides. "Both of us certainly offered our political views, and I suppose that I think he probably listened somewhat more amused than influenced," says Edelman.

"They weren't respected as they hoped to be for their ideas because they were too young," recalls press secretary Frank Mankiewicz. "I don't think Bob ever seriously considered their views. He should have maybe."

Ted was present, however, at all the important discussions about a possible presidential race, and he laid out what he considered his brother's three options. He could enter the primaries and take Johnson on directly. Or he might hang back and wait to see whether Johnson's rising unpopularity would force him to decide against running for reelection. Or if even a grievously wounded Johnson would not retire, the president could be convinced that in order to win he would have to open up the vice presidential nomination, tantamount to asking Bobby to join the ticket.

No one among Bobby's close advisers was as opposed to his candidacy as his own brother. "*My* feeling was that the real issue was the question of the extent to which he'd be able actually to change policies," Ted told his biographer Burton Hersh. "*That* was the test that should be applied, and I think it appeared very primary to me that the issue wouldn't be the war issue but the personality issue, the apparent search for power by Robert Kennedy, which would awaken many of the apparently latent feelings among some people. The opportunity for bringing about new kinds of directions, new departures, appeared to be elsewhere."

Ted believed that his brother would be unable to defeat an incumbent president. He would leave only scorched, smoldering earth between himself and the 1972 nomination, and he would do irreparable harm to the Democratic Party that was the Kennedys' political home. "Ted was much more political than Bob," says Mankiewicz. "Ted really did understand what this fight about the war was doing to the party. Ted was always a much better politician."

Ted could whip off half a dozen reasons why a piece of legislation was crucial to America's future, yet he was inarticulate when talking about personal matters. And nothing was more personal and emotional than his own brother's future. He could not bring himself to speak of the deep fear that he had for his brother—that he might be shot down like Jack. Even decades later, when he talked about it, he veiled the language so that the pain and the memory would stay half hidden. "We weren't that far away from '63," he said, "and that still was very much of a factor."

It would take not just political but physical courage to run in 1968. At one gathering at his apartment at the United Nations Towers, Bobby listened to more of the interminable arguments about whether he should run. "You can't be sure that you'll come back alive from a trip to Dallas," one man said. Bobby turned and walked away to a bedroom. When he returned, he said: "I don't want to be maudlin, but I've thought of all those things."

Inevitably, those who wanted Bobby to challenge Johnson were far more zealous in discourse than those who opposed it. Walinsky implored the senator to stand against a president who was "an ignorant bully," in a country that "is getting used to garbage in the White House." Another adviser, Richard Goodwin, played to the romantic, daring, idealistic side of Bobby. "People can forgive mistakes, ambition, etc., but they never get over distrust," he wrote Bobby. "You could equally argue it would be a disaster [to take on Johnson]. Maybe it would. But a glorious one."

If Bobby did take on Johnson, he had to have faith in the judgment of the American people that despite what the polls said, they would come out for him and against the war. Yet he had deeply ambivalent feelings about the voters and what they were likely to do. In the fall of 1967 he gave a number of speeches in upstate New York to political rallies. Writing the talks was not the most important task in the office. It fell to the newest and youngest speechwriter, Jeff Greenfield, who attempted an eloquent draft. "No, no," Bobby told Greenfield on one occasion. "You have to understand, all they want to do is clap. They're not listening here. Most of them aren't sober. Most of them can't understand these words. You just have to make them clap. Write a children's book."

On October 18, 1967, Bobby called Arthur Schlesinger Jr., a Harvard professor and Kennedy biographer close to the family. Schlesinger, a leading liberal, had just finished talking with Senator George McGovern. The South Dakotan was one of the few antiwar senators. Although Bobby was the peace movement's first choice to run against Johnson, the activists could not sit back hoping that he would end his musings and enter the race. Thus, Schlesinger and other antiwar Democrats had gone to others, including McGovern, to find someone to stand up against Johnson. McGovern was up for reelection in 1968. He told Schlesinger that he felt he could not take the risk, but the peace groups had found a willing candidate: Senator Eugene McCarthy of Minnesota. This was the news that Schlesinger relayed to Bobby.

As soon as Bobby hung up with Schlesinger, he called McGovern to verify the information. "Is that true?" he asked urgently.

"Yes, it is," McGovern replied, as he recalled later.

"He's going to get a lot of support," Bobby said. "I can tell you right now, he'll run very strong in these primaries. He'll run strong in New Hampshire. I'm worried about you and other people making early commitments to him, because it may be hard for all of us later on."

"I can't recall anything that ever so much disturbed him as McCarthy's announcement," McGovern said. "I think he thought, 'Oh Christ, I should have done this. Why didn't I move earlier?'"

By the next day Bobby was far more contained in his evaluations of McCarthy's candidacy, as he surely knew he had to be. His private feelings were different. "He would rather have had Johnson renominated, probably even reelected, than see me as president," McCarthy reflected three decades later. That was a harsh judgment, but given Bobby's own ambition and his view of McCarthy, perhaps not as exaggerated as it may seem.

The two senators were supposedly colleagues in a journey toward peace, but they were very different men. What Bobby condemned as McCarthy's congenital sloth, the Minnesota senator's admirers considered a wise reluctance to indulge in the frenetic posturing of men like the Kennedys. What Bobby's supporters whispered was unseemly dealing with special interests was to McCarthy's defenders a sophisticated awareness of the ways of getting things done. What Bobby's friends condemned as McCarthy's mediocre ADA liberal ratings was to his defenders proof that the man voted his conscience and his mind.

S ince his brother's death, Bobby had wanted to go to Vietnam to see for himself the war that he was condemning so eloquently. He was a man who had to see to feel, and to feel to think, and he needed to go to Southeast Asia to experience those realities himself. Yet he knew that the moment his plane landed in Saigon he would become the focal point and that he would lose control over the issue and be seen as a pernicious influence, undermining morale, stirring up rancor, waging a personal vendetta against Johnson. His brother Ted could go as his surrogate without creating such massive publicity and controversy. Thus, when the senator from Massachusetts flew to Vietnam at the end of 1967, he went ostensibly as the chair of the subcommittee on refugees, but he also went as his brother's emissary.

When Ted arrived in Saigon on New Year's Day 1968, his advance team had already been in Vietnam for two weeks. The group consisted of Ted's administrative assistant, David Burke; Barrett Prettyman and John Nolan, two attorneys who had worked on the Cuban prisoner exchange;

N. Thompson Powers, another prominent D.C. attorney; and John Sommer, who had done volunteer work in the country and spoke Vietnamese.

Ted turned down the briefing that the embassy gave to VIPs. It was not like him to spurn official hospitality, but he was signaling that he was not going to be led on a sanitized journey during the next eleven days. He did not want any reporters around him either, and he was not going to take the obligatory journey to the battlefronts.

Ted had not been a bold proponent of political solutions for the war, but on his subcommittee he had provided a sympathetic forum for those who had witnessed the horrendous human cost of the war. He chose this approach largely because Bobby was confronting the issue politically, and he never took on what he considered his brother's issues. He was more sympathetic toward the dilemma in which the president found himself, but he could hardly have a position different from Bobby's or Johnson would exploit even the smallest of nuances.

It was one thing, however, to sit in a hearing, as he had in October, listening to experts talk about the Vietnamese fleeing not a Communist oppressor but American bombs and herbicides in a war that had already created two million refugees and it was another to actually witness the devastating human cost. Ted had been to Vietnam two years before. His statements about the war upon his return had been so deeply hedged that they had led the debate nowhere.

In South Vietnam, Ted set out to learn the realities that lay beneath the administration's optimistic rhetoric. "Ted was very hardworking," Prettyman says. "He was good. He knew his stuff. I mean, it was almost embarrassing that we'd confront [Ambassador Ellsworth] Bunker or [General Creighton W.] Abrams or one of these people. They'd say, for example, 'Well, you don't have to worry about that refugee camp [Chop Chai in Phu Yen] because they've all got roofs now.' And he'd say, 'That's not true. All those roofs have been stuck for six months down the road here in someplace [Nha Trang].' And he would know because we had worked very hard and gotten him prepared. But he would know his facts. He seemed very focused.

He seemed very, very concerned about Vietnam. He seemed determined to try to find out what was going on."

In large measure, Ted was confirming what he thought was true, but that did not make the journey any less disturbing. "We must have seen every hospital and every refugee camp in the place," Burke recalls. "It was a very extensive trip. We were both emotionally crippled up at times by what we were looking at, in the hospitals seeing children with their bodies burnt horribly.

"Ted wouldn't cry. He'd stop talking. He'd turn and walk away. We were both very affected by some of the stuff we saw, even though we were well warned and well prepared for what we were going to see."

Ted liked to have a problem neatly set out before him, where he could diagnose it, offer his solution, and move on. In Vietnam, however, those Americans who spoke optimistically of their efforts had hardly finished speaking before their words were recanted in another defeat, scandal, or other setback. South Vietnamese refugee officials boasted of their humanitarian concerns while they stole from the most wretched of the poor. The enemy was everywhere, within and without, and even the most peaceful, bucolic scene might erupt into bloody combat.

One evening Ted was supposed to have dinner in Can Tho with David Gitelsen of the International Volunteer Service. Ted, who was forever jiggering his schedule, decided at the last minute to change the meeting to breakfast the next morning. At precisely the time he would have been at Gitelsen's home, a bomb went off on his doorstep that might have killed the famous guest.

After his trip, Ted flew for a little R&R to Hawaii, where he was met by his Harvard classmate Claude Hooton Jr. Like Jack and Bobby, Ted compartmentalized his friendships— Claude was his rousing good-time buddy. They had had many adventures that would never make it into an authorized biography of Edward M. Kennedy, and Claude expected some more raucous evenings. Ted, however, seemed different. He cried as he described what he had seen in Vietnam. Claude had been there before when his friend went through hard times, but he had never seen Ted like this.

T ed returned firmly believing that the war had to end. His name hardly appears in most histories of the Vietnam War, but the work he did with his subcommittee may have had as much impact on changing public opinion as any of the addresses given by his brother or other antiwar senators.

"Ted looked at it like a politician who has an understanding of what it takes for human beings to get together and accomplish something," says David Burke. "If you have a foreign policy objective in Vietnam, our way was not the way to achieve that objective. To have this enormous steam-roller of power rolling over a primitive country, it just wasn't going to happen. His way to influence the electorate here at home was not by using diplomatic words or military jargon that they don't understand. He approached it talking about civilians, talking about casualties, talking about children and refugees, people fleeing, people hungry, people burnt. You have doctors come and testify about the war in Vietnam. You don't have generals. People know doctors, and listen."

Ted's comings and goings in Vietnam had been tracked by government officials who cabled the White House that upon his return Ted "intends [to] concentrate on [government] corruption as well as on refugee and civilian medical programs." That was a worrisome threat to the White House. Larry O'Brien, a Kennedy political aide who had stayed on in the Johnson White House, served as a conduit between the two camps. After talking to Ted, he reported to Johnson that the Massachusetts senator said "that he knew the president held him in high regard and hoped nothing would cause deterioration of that relationship." O'Brien advised Johnson that he should ask Ted to come into the Oval Office to make his recommendations; that would temper the reaction to the speech about his trip that the senator was scheduled to give in Boston.

On January 24, the very day when O'Brien made his suggestion, Ted spent an hour and a half with Johnson, an enormous block of time in the president's day. Ted understood the basic axiom of human behavior that if you want to change people, you praise them first and then offer your studied criticism. Ted told the president that the refugee situation had improved since his previous visit, but that corruption was rampant in the South Vietnamese government. It wasn't enough to shrug it off as an Asian problem, since countries such as Singapore, Malaysia, and the Philippines had managed to curb corruption. He had begun "to wonder whether 'round eyes and white skin' could solve Asia's problems."

After briefing the president, Ted gave his speech in Boston and went on Sunday-morning television. His anger was directed primarily at the South Vietnamese government in Saigon, whom he called "colonists in their own nation," a regime unworthy of continued American support. It was a crowd-pleasing business attacking the lackadaisical, luxury-loving South Vietnamese elite, but it was a pleasure that those in the White House thought not only unfair but also unseemly.

"When he was asked on *Face the Nation* what he would do if the Government of Viet-Nam failed to respond to his demands, his answer trailed off because he will not follow the logic of his position to the point of an American withdrawal," wrote Henry Cabot Lodge, ambassador to Vietnam from 1965 to 1967, in a top secret memo. "He is obviously not clear in his mind on whether the Viet-Nam war is a vital American interest. In one place, he says we are there so as to help the Vietnamese—as though it were not really an American problem. Yet he will not cross the Rubicon of offering an American withdrawal if the Vietnamese fail to do what he wants."

CHAPTER EIGHT

Standing in the Rubicon

he weeks after McCarthy announced were the most excruciatingly difficult of Bobby's political life. He lived for action, yet he sat on the sidelines berated by those who once had praised him. Every day Bobby watched the comings and goings of McCarthy, whose campaign was beginning in New Hampshire not with full arenas and surging crowds but more often with coffee klatches and handshaking along half-empty village streets. By any standard, the Minnesota politician was far from galvanizing the Granite State, though he maintained a wry self-assurance, as if he knew something that his less poetic colleagues did not.

On January 30, 1968, Bobby said that he would not run "under any foreseeable circumstances." Not only was it an unnecessary utterance, but the timing was exquisitely wrong. As he was speaking, the Vietcong were beginning their Tet Offensive, a massive frontal engagement that brought troops into the main cities, even for a while taking over the American embassy in Saigon.

For the first time, many Americans began contemplating that their country might be on the verge of the greatest military loss in its history. For the Johnson administration, that loss was unthinkable, and the Joint Chiefs of Staff called for 206,000 more troops. For the antiwar movement, it was further evidence of the duplicities of the White House and its facile optimism. The protesters made the president a virtual hostage in the White House—he was unable to make a public speech before anything but a controlled audience. As for McCarthy's candidacy, the roads to New Hampshire were beginning to fill with young people going "Clean for Gene," cutting their long hair, donning coats and ties, and knocking on the doors of citizens who greeted them, if not with the promise of a vote, at least with hospitality. For Bobby, who had thought of youth as his natural constituency, it was as if the kids had been stolen away.

Bobby rationalized that he had not run in large measure because "most everyone whom I respect [has] been against my running." His wife, sisters, senatorial aides, journalistic friends, and a multitude of others had pleaded with him to run. He had not listened to their cries but to the sonorous sounds of men like Theodore Sorensen and others from the Kennedy White House, as well as his own brother. He also blamed the professional politicians who said that his running "would bring about the election of Richard Nixon and many other Republican right-wingers because I would so divide and split the party."

Bobby knew that what passed as political wisdom was often nothing more than rationalizations for the status quo. Every time there was a dramatic development in Vietnam, Bobby rustled impatiently. Burke called it the "panic signal," an alert that set off Ted's minions to repeat their arguments about why Bobby should stay out of the race. Then Ted made the same arguments to his brother. This man who thought that courage was the highest of all values would not and could not be seen as a coward. His senior political advisers tried to quell his disquiet, but they could not prevent him from hearing the epithets shouted at him from the street.

Ted understood Bobby's emotional imperatives, and he began backing away from his determined position against his brother's running. His reason, as Burke described it to others, was that "even though politically he should resist these pressures on his brother, for the sake of his brother's personal peace of mind he should let things go along, take their course." Ted, that most political of all the Kennedy men, saw that his brother's psychological needs transcended what Ted considered the political imperatives. It was Bobby's own sense of how a man must act that led inevitably to the difficult, divisive race. As Ted said in a lighthearted moment: "Bobby's therapy is going to cost the family $8 million."

Bobby's defenders made strong, compelling arguments for why their champion was correct in waiting so long to run. But these were political arguments, and this was not ultimately about political judgments. If Bobby had challenged Johnson early on, he would have lived the life he said he wanted to live. But he did not confront Johnson early, and there would always be those like his biographer Ronald Steel who would conclude that he was a man motivated by "self-interest and that we have read into him what we have wanted to read."

It was not only a historian writing years afterward but several of Bobby's closest associates who began to doubt him. The rhetoric that Walinsky wrote in speeches was not just for public consumption. As Bobby continued to hem and haw, Walinsky decided to leave the senator. He knew that if it was not time for Bobby to stand up, it was time for him. Edelman was probably leaving too. The senator had been a vehicle for his two young aides' ambitions and ideals, but they saw little reason to remain.

For Bobby it was time for an excruciatingly difficult reevaluation of his options. He was still hearing a cacophony of opinions, and he knew that Ted thought him "a little nutty" for considering running. Not only was Sarge Shriver heading Johnson's War on Poverty, but Bobby had reason to think that Sarge was quietly working against him. In December, Joseph Califano, a Johnson aide, wrote the president that Sarge had told him that "Bobby Kennedy was getting ready to run against you in 1968" and that Shriver "was interested in Ambassador [Arthur] Goldberg's job at the UN," a job "that would put him in your cabinet."

As he edged toward his decision to run, Bobby had a series of intense meetings. The man who had become attorney general without ever practicing law and had run for the Senate in a state where he was a newly minted resident did not hold off from this race simply because he was afraid he might lose. There were uncertainties and doubts that he could hardly articulate, and yet an inexorable movement pushed him toward announcing. His friends in the press wanted to write about him as a hero, and as they exhorted him, they waited with their pens poised. "The press tried too hard to pin down when he decided to run; my guess is he never entirely decided to run," says Dutton. "He'd take two steps forward and one step backward, and pretty soon he found he was too far in the tunnel to get out."

Political protocol dictated that McCarthy be given the courtesy of hearing about Bobby's intention before he learned about it in the press. Bobby deputized Ted to take care of the unenviable chore of telling McCarthy that he was "probably going to enter the primaries after New Hampshire." Ted did not do what his brother asked him. Ted suggested later that he thought that McCarthy would use the words against his brother, but his larger reason was that he still did not want his brother locked into the race, goaded on by McCarthy's barbs. Any other adviser who had refused his commander's orders would have been rebuked, if not summarily dismissed. As it was, the cost was large. When he did enter the race, Bobby looked even more like an opportunist capitalizing on another's efforts, and McCarthy's barbs had a venomous sting.

On the day of the New Hampshire primary, Bobby and Lyndon Johnson had one thing in common. They both wanted McCarthy to lose. Bobby was hoping that the Minnesota senator would win less than 35 percent of the vote. That way he could come valiantly into the race as the only hope of defeating Johnson. Anything above that would generate problems, and 40 percent or more would create a momentum for McCarthy that would be hard to overcome.

McCarthy received 42 percent of the vote. He had come largely unknown into the towns and byways of New Hampshire. When he left, he was unknown no longer. Bobby had fancied himself the troubadour of youth, but the students marched now to a different beat, a youthful army of the smart and the true singing McCarthy's song. The worst of it and Bobby probably did not envision this—was that he had lost most of them for good, and they would mock him and his words. He could not go into the campaign with an idealist's pride. Nor could he stand back.

B efore he announced, Bobby wanted to see whether he could work out an arrangement with McCarthy so that the two candidates would avoid splitting the antiwar vote by dividing up the remaining primaries. It was an idea that McCarthy's admirers were likely to see as little more than extortion, and Ted had the disagreeable task of proffering it. Ted flew to Green Bay, Wisconsin, to meet McCarthy late at night. "I wasn't that eager to see him," McCarthy says of Ted. "And he had nothing to say, talked about St. Patrick's Day in Boston, and I had nothing to say. There was no point in talking to him."

Ted flew back to Washington immediately and at six in the morning was already out at Hickory Hill. From now on Ted's task was to help Bobby run the best race he could run. "Cut it as close as you can," he told a barber who had arrived to cut Bobby's notoriously long locks. That was only the first of many things that Ted did for his brother during the campaign that Bobby never knew about.

For much of that morning, Ted walked around the grounds at Hickory Hill, plainly apprehensive. He could not tell Bobby his qualms any longer. He could not tell Burke. He did not have well-articulated doubts that could be easily discussed, but an inchoate fear about what was going to happen.

CHAPTER NINE

A Race Against Himself

I n Bobby's first campaign speech at Kansas State University, the students shouted his name and jumped with excitement. They came forward to touch his face and to grab at his cuff links. They tried to take something of him as a tangible souvenir of this day. In the next weeks Bobby flew across the country, talking often to what he considered his natural constituencies, minorities and students. Wherever he went, he bellowed the same message: Johnson must go, and the nation must move toward peace in Vietnam.

At his best, Bobby touched the human heart with a depth and intimacy of which few politicians were capable. On the podium he had neither the wry, ironic style of Jack nor the pounding rhetorical bombast of Ted at his most partisan, but an open emotionalism. He seemed to promise a transformation of human beings into something more enlightened and elevated. He peppered his speeches with phrases such as "I think we can do better" and "I don't think that is satisfactory."

Bobby gave substantive speeches, but there was no systematic critique of American foreign policy, no questioning of the Pentagon budget, only an outraged call for movement toward peace in Vietnam. "In retrospect, I would have difficulty defending him against the charge that he really didn't have any new departure in foreign policy as a campaign matter," admits Edelman. Bobby had innovative, prophetic ideas about welfare reform and new programs to help America's farmers, but he made no serious attempt to make the white middle-class Americans own up to the problems of poverty and race in an idiom that they would understand and not find threatening.

Bobby's campaign was unique in many respects, not least of which was that he was consciously giving up part of the electoral coalition that had been the essence of the Democratic Party since the New Deal. "We have to write off the unions and the South now, and replace them with Negroes, blue-collar whites, and the kids," he told journalist and friend Jack Newfield. "If we can reconcile those two hostile groups, and then add the kids, we can really turn this country around."

Bobby was trying to bring together a rainbow of Americans, including students, the white working class, African Americans, Native Americans, and farmers all shouting his name in ten thousand streets to lift him to the nomination. They would yell so long and so strong that they would become the will of the people, and the Democratic Party would follow meekly behind. He might not have the students in the numbers he felt he deserved, but he believed that sooner or later they would come to him. It was a dangerous, provocative strategy, but given the fact that most states did not have primaries and that the delegates in those nonprimary states had largely already been chosen, it was one of the few options left. This was not revolution but street theater designed to convince party leaders that Bobby could not be denied.

"We're trying to demonstrate who really strikes a response in the American people," said the cerebral Dutton, one of the chief strategists, to the *Wall Street Journal*. "The politicians are going to have to decide who can move the people."

Applause had become the final arbiter of merit and possibility. As Bobby looked out on those audiences, he gauged his success by the decibels and the number of hands that reached out to grab him. "He could have got hurt, they liked him so much," said Jim Tolan, an advance man, after the first day in Kansas.

Bobby had important things to say, but often it was not the words that were remembered but the emotions. He flowed into oceans of people, the waves threatening to pull him into their midst and even drag children along with him, trampling them beneath a thousand feet. In San Jose a mother yelled out when her child was about to be crushed, and Bobby whisked the frightened boy to his shoulders. When he arrived at Los Angeles International Airport, the spectators pushed and jostled one another, and two men ended up in a fistfight. On the way to his speech someone pushed a child up against the rear of Bobby's car. He could not reach out for this child, and by the time the tumult subsided, the one-and-a-half-year-old was bleeding and bruised.

As Jack's friend Chuck Spalding observed the fervor around Bobby, he thought he was mimicking his older brother. All through his political career, Jack had projected a movie star appeal. He had been a handsome, charismatic figure, but he did not consciously promote it the way Bobby did. "I think he [Bobby] thought that was necessary," says Spalding. "When he saw people begin to respond, he developed this crowd appeal on his own. I think he had lost sight that he was going to be the most important figure in the Democratic Party for years to come, and he could have matured. I think he was shortsighted."

Among themselves, the journalists debated whether they were witnessing elements of demagoguery. Bobby wanted the journalists to write about the massive, enthusiastic crowds celebrating his candidacy. But there were harbingers of tragedy in his campaign, or at least of bloody mishap, and several of the journalists wrote stories about the omnipresent danger. "Kennedy Crowds on Coast Unsafe for Small Children," warned the *New York Times*. "Rowdy Teen Fans Get in Bobby's Hair," observed the *Chicago Sun-Times*. "Crowd Madness and Kennedy Strategy," headlined the *Washington Post*. "Mob Scenes and Stirring Words," summed up the *Wall Street Journal*. The emotions that Bobby engendered frightened journalist Stewart Alsop, who in the *Saturday Evening Post* compared the senator's most passionate followers to the Red Guard in China.

One member of the campaign staff, Floyd Boring, said that he had called for the campaign to use more caution. Security could have been increased. The crowds could have been cordoned off instead of letting them break away and run to Bobby. The candidate could have spoken in bigger auditoriums and arenas where the crowds would not overflow. None of that would have impinged on the campaign message, but Bobby told Dutton that he needed these crowds to fill him with energy.

"I think they've decided to risk it," Boring told the *Wall Street Journal.* "It's a gamble you just decide to take."

O n March 31, 1968, in a nationally televised address from the White House, Johnson said that in a move toward peace, he would limit the bombing of North Vietnam. Then, at the end of the speech, Johnson dramatically and totally unexpectedly said that he would not run for reelection. Bobby's advisers momentarily exulted in Johnson's withdrawal, but the more they reflected on the decision the more they realized that it robbed them of a focus. Bobby's crusade had begun as a direct, visceral confrontation with the president. Now Johnson had walked away, and standing there on the field of battle was McCarthy, who had been anointed a saint in the snows of New Hampshire, and behind him the still shadowy figure of Vice President Hubert Humphrey, a liberal icon who would probably now enter the race.

Bobby returned to the campaign trail, but he had lost something of the energy, moral certitude, and direction that had made him such a formidable candidate. McCarthy was much of his problem. Bobby and McCarthy shared common views on the war and the deteriorating quality of American life, but they were opposites in many ways. Bobby was passionate, quick, revengeful, caring, raw, and intimate. McCarthy was dispassionate, languid, intellectual, snobbish, refined, and distant. Bobby read literature as if he were the first one to read it; McCarthy read a book as if he had read it all before. Bobby was criticized for overweening ambition; McCarthy was disparaged for not caring enough about reaching the White House. Bobby wooed blacks, other minorities, and the poor, though they did not a majority make; McCarthy courted students, professors, educated progressives, and upper-middle-class professionals, all of whom, despite their influence on American life, were not numerous enough to take him to the White House.

For the crowds that grasped at Bobby, even tearing away his shoes, his descent into their lives was an emotional moment, but Bobby was still not connecting with the majority of Americans. Tom Johnston, one of his aides, wrote a memo on April 20 in which he worried about "the impression you are making on the middle-class white American whose judgment will ultimately be decisive."

Yet as much as he appreciated communing with the poor, and as comfortable as he was in the haunts of the wealthy, Bobby had no great empathy or concern for the middle class. "White people [are] living better than they ever did before; no matter who they are, having it better: if they had rented before, now they own," he told the journalist David Halberstam. "It is extraordinary how ungenerous they have become. . . . They don't see the poor and they don't want to."

Among many of Bobby's associates and sympathetic journalists, there was an implicit snobbishness toward the majority of the American people. "These colorless shopping centers are the heart of suburbs everywhere and the people from Delaware County who went to their shopping centers to see Bobby Kennedy last night were part of the environment," wrote the proud son of the working class Jimmy Breslin in the *New York Post.* "Colorless and withdrawn people who came out of the rows of cars in the parking lots and watched, listened and left."

Arthur Schlesinger described Oregon as full of many of the same colorless, withdrawn people whom Breslin had seen in New Jersey. The northwestern state was "a pleasant, homogeneous, self-contained state filled with pleasant, homogeneous, self-contained people, overwhelmingly white, Protestant and middle class."

"Even the working class was middle class," Schlesinger wrote, "with boats on the lakes and weekend cabins in the mountains." These workers had once been the stalwarts of the New Deal. Now many of them had started listening to Alabama governor George Wallace's shrill shouts, as he came north to run as an independent, or to Richard Nixon's conservative rhetoric. Others had simply tuned out, feeling that none of the candidates spoke for them, not Johnson, sending their sons to war, not the cerebral McCarthy, lecturing America in words they sometimes barely comprehended, and not

Bobby, who on television was the center of frenetic action and spoke as the tribune of the white poor, the blacks, Native Americans, and Chicanos, people they didn't know and sometimes feared.

On the evening of April 4, when Bobby flew to Indianapolis to deliver a campaign speech, he learned that Martin Luther King had been killed in Memphis, where he was supporting a strike of black garbage workers who sought a union. Several of Bobby's advisers said that he should not go to the event. He thought for a minute, then said, "We're going to the rally."

Bobby sped away from the airport toward 17th and Broadway, in the heart of the black slums of the city. There, on a flatbed truck before a thousand Indianapolis African Americans, stood an improbable figure, a wiry, windblown, shy Irish American scion of wealth, looking even smaller and less significant bundled in his overcoat. Walinsky, who had arrived a few minutes earlier, gave him some hurriedly written words. Bobby stuffed them in his pocket and took out another sheet of paper with a few notes on it. He may have had the finest of speechwriters at his disposal, but he spoke his own words that evening. They were as strong, true, and telling as any words he ever spoke. The very tenor of his gentle voice seemed to plead for understanding. He had the terrible duty of telling his audience that their greatest leader was dead, and then he had to seek some meaning in the tragedy:

> For those of you who are black—considering the evidence there evidently is that there were white people who were responsible—you can be filled with bitterness, with hatred, and a desire for revenge. For those of you who are black and are tempted to be filled with hatred and distrust of the injustice of such an act, against all white people, I would only say that I feel in my own heart the same kind of feeling. I had a member of my family killed, but he was killed by a white man. But we have to make an effort in the United States, we have to make an effort to understand, to go beyond these rather difficult times. Let us dedicate ourselves to what the Greeks wrote so many years ago: to tame the savageness of man and make gentle the life of this world.

Later that night a restless Bobby entered the bedroom where Walinsky and Jeff Greenfield, the newest of the speechwriters, were working on an address. "Well, after all," Bobby shrugged, "it's not the greatest tragedy that ever happened in the life of the United States of America." Although the words startled Greenfield, they were not as cavalier as they sounded, but a mask that Bobby felt he had to wear. He had just seen a leader picked off the same way his own brother had died. He would be going out there in those streets again. And his seeming insensitivity was probably just another way not to flinch.

"You know that fellow Harvey Lee Oswald, whatever his name is, set something loose in this country," Bobby said to the two speechwriters, knowing that it was still loose in the land.

T here were riots across America after King's death, and when Bobby began campaigning again, the ghettos were still smoldering. It is axiomatic that in democratic politics a successful candidate must talk about the issues that matter most to the voters. In Indiana that issue was law and order. No one in American politics had a more legitimate right to talk about the issue than the former attorney general who had fought crime and corruption his entire political career. He alone could do what he began doing in Indiana—balancing his talk of law and order with a call for social justice.

Many American liberals and leftists thought of "law and order" as little more than code words for white racism, and they sought to pry Bobby's fingers off the issue. The *New York Times* wrote of his supposed tilt to the right without any sense of the legitimacy of the issue or awareness that he was saying things he had said before. Newfield believed that the Hoosiers were consumed with the issue because they were a people with "provincial mores." In the liberal *Washington Post*, Mary McGrory wrote that "the people of Indiana are not a compassionate people."

Walinsky, Edelman, and the other young advisers were appalled that their champion of the new politics should resort to what they considered right-wing rhetoric. Every evening Dutton had the unpleasant task of listening to Walinsky's tirades about how Bobby was destroying himself. "I think we probably incorrectly inferred that he was being reached or even semiprogrammed by these other people," Edelman reflected three decades later, "when the fact is that he was perfectly capable of using that terminology completely on his own."

Any presidential campaign is an enervating business, but since Bobby had started so late, there was an even greater urgency. There was the imperative to sound fresh and inspiring wherever he was, whether he was talking to senior citizens in Portland or convention delegates in New York. He found relaxation only in action, and even though he tried to take weekends off at Hickory Hill, he never fully relaxed. His face at times had the sallow look of exhaustion, but he was still full of nervous energy. When James Stevenson of *The New Yorker* talked with Bobby, he felt as if he were sucking the last juices out of the man. "He wouldn't give you a slick answer, and that was so awful," Stevenson said. "He'd sort of wring his face and try to rub away all the weariness and exhaustion, and then he'd slowly come out with an answer that he's really thinking about because he wasn't inclined to just give a premixed answer to anything."

In running the Indiana campaign, Ted's associate Gerard Doherty was having fits dealing with the people around Bobby. There was a haughty, self-important quality among some of these people that once experienced was never forgotten. "Well, they'd go through some town like Richmond, and the next thing you know, they busted the guy's [mimeograph] machine, insulted his secretary, peed on his floor, and left," he recalls. "And we're there trying to pick up the pieces." Doherty considered Bobby's staff "arrogant, overweening, supercilious, whatever the words are." That was only part of the problem, for there was also an extravagant entourage, an eclectic crew of socialites and celebrities who found it fun to hang out with this champion of the poor and the deprived. "I think he had an affinity for more kooks and more nuts than I've ever seen assembled," Doherty reflects. "It used to remind me of a Cecil B. DeMille Crusades movie when you've got dancing girls, jugglers, and dogs."

Bobby had his victory in Indiana, but it was not as decisive as he had hoped, and within its numbers were some troubling signs. He won 42 percent of the vote as opposed to favorite son Governor Roger D. Branigin's 30.7 percent and McCarthy's 27 percent. Despite Bobby's frenzied campaigning and vast advantage in money, he had scarcely risen higher than his polls a month earlier. McCarthy had little money and a disjointed campaign, but his legions of students had worked the state, many of them arriving in buses that the Kennedy campaign had paid to bring in *its* students.

The mythic ideal of Robert F. Kennedy was that he alone could tame the savage earth and bring the races together, that he alone could calm the troubled ghettos and the restless suburbs. However, as much as his supporters tried to say that he had done so in Indiana, he had been unable to slow the racial polarization. He won Lake County, a polyglot industrial area south of Chicago with a mix of black ghettos and ethnic white communities of Poles, Hungarians, Serbs, and others. But he had done so by taking 80 percent of the vote in the black Gary districts and only 34 percent in the white communities. "The lesson of Lake County," wrote vanden Heuvel and Gwirtzman, "was that the more personally involved the white voters were with the racial struggle, the more they identified Kennedy with the black side of it, and turned to his opponents as an outlet for their protest."

T he next big primary was in Oregon at the end of May, and from the day the Kennedy people arrived in the northwestern state, they knew that they were in trouble. Walinsky considered the Oregonians "very morally superior and very disdainful,"—a natural match for McCarthy. "Oregon is just one large chapter of the League of Women Voters," cracked Frank Mankiewicz, the press secretary, remembering the campaign.

Bobby complained that in Oregon "there's nothing for me to get hold of," meaning that there were no ghettos or large minority populations. He went into an oscilloscope factory outside Portland and talked about the litany of urban problems, the desperate ghettos, the wretched poor, the hunger in the hinterland. The employees listened to him as if he were a

National Geographic lecturer talking about an exotic and distant land. Bobby came out afterward muttering about "the strangest workers he had ever met," but these workers were part of an emerging middle class. They didn't worry about union dues and worker solidarity but about crime, health insurance, taking care of elderly parents, inflation—and the war in Vietnam.

The staffers hurried around the state exhibiting what they thought was bold authority and many Oregonians considered eastern arrogance. As they saw it, the traveling staff and the journalists blended into one pushy, surly lot. "We're gonna lead those labor leaders down the primrose path," one of the staffers shouted in a bar in a Eugene hotel, the words heard, remembered, and repeated by locals over and over.

Bobby talked a bold game about new politics, but he was using the oldest of means to help get him to the White House. Pierre Salinger recalls being asked to bring $100,000 in cash to Oregon. "This is something you've got to take to Bobby right away," he was told. There had been lots of cash in the 1960 campaign too, but this was an immense amount of money, to be used for whatever purposes Bobby chose. "I don't know what the money was for," Salinger says. "They just asked me to bring it to him."

Salinger wasn't the only aide carrying money. Joe Dolan has similar recollections. "I would never like to handle money," says Dolan. "Especially if you've read the statutes, you don't want to handle money. We used the money for candidates or whatever it is you don't want to talk about. I was in charge of the airplane. I said to one of the other aides, 'Could you take the money?' So he had a satchel. I said, 'Where's the package?' He said, 'My God, I left it in the police car.' So I said, 'Get off.' We held the plane. And Bobby's in the back with the reporters saying, 'Joe, let's go. Joe, Joe. What's holding us up?' "

When briefcases full of money arrived at the campaign and Bobby surreptitiously distributed a discredited pamphlet, his idealistic young aides were uncomfortable. "It was a style of campaigning which was unattractive," says Edelman. "He's a bridge figure. He's got a foot in the old politics. But I'm probably stuck a little bit in my own youth when I describe it in that way. But he was someone who combined the old political methods with genuine passionate idealism and commitment to issues of social and economic and racial justice that was unique then and is unique today."

Humphrey had entered the race, but he had chosen not to compete in any primaries. Instead, he was working behind the scenes garnering delegate votes. As primary day neared, Ted arrived in Portland as the bearer of bad news. While Bobby and McCarthy had been campaigning day and night, Humphrey had been working the states where there were no primaries. So had Ted and much of his staff, trying to pick up votes piecemeal. There were few speeches and little applause, but endless phone conversations, backslapping, and hand-holding. In Pennsylvania, Ted had outraged some Humphrey supporters by giving a speech to the delegates when he was supposed to say only a few words of greeting. "But what surprised me was almost a total lack of strength there [for RFK]," recalled former vice president Walter Mondale, a Humphrey supporter in 1968. "I thought, you see, it's a strong Catholic state, a strong John Kennedy state, and I thought, boy, we're going to get wiped out here and it wasn't true at all."

Ted was far better at this endeavor than Bobby and his high-strung, proud aides would have been. But much of it had been for naught. The vice president was immensely popular with party activists and labor leaders. In state after state, caucus after caucus, Humphrey was picking up delegates or holding strong.

Bobby's worst mistake was to say that if he lost in Oregon, he would no longer be a "viable candidate." That statement became a problem on May 28, when McCarthy defeated Bobby in what amounted to a landslide, 44.7 percent to 38.8 percent.

CHAPTER TEN

Journey's End

he California primary was only six days away, and the Kennedy people knew that if Bobby lost, the campaign was over. For the first time, Bobby agreed to debate McCarthy. By now the campaign had become nasty and ill-tempered. The McCarthy people, for instance, had run an advertisement accusing Bobby of responsibility for the American invasion of the Dominican Republic in April 1965, when he was no longer part of the administration.

As Walinsky helped prepare Bobby for the debate, the speechwriter believed that every weapon should be used. "I wanted to make sure that there were three or four baseball bats that he could, you know, just take and backhand Gene McCarthy right in the mouth and move him freaked out on the floor," Walinsky said. One of the "baseball bats" concerned the issue of how to deal with blacks in the ghettos of California.

Even years later McCarthy bristles at what he believes Bobby and his people did to him in speeches and in a debate that he believes lost him the primary. "Bobby employed what I would say are the three most dangerous prejudices in this country," McCarthy charges. "One is the antiblack prejudice. So he said I was going to move blacks out to Orange County. Anti-communism is the second prejudice. And he said I was going to negotiate with Communists. And the third one is anti-Semitism. And he did a very dangerous thing. There was an issue of selling fighter jets to Israel. And the accepted practice was to vote in Congress to sell them. The Israelis paid for them with our money. The Israelis wanted it that way. It saved stirring up anti-Semitism. And the State Department wanted it, and we just went along. But Bobby was going around saying he would give the jets to Israel, which was never the issue. And some of Bobby's people I think were sensitive to the falsification of the record and the exploitation of these prejudices. But they kept quiet."

In fairness, Bobby had in no way accused his opponent of anti-Semitism, or of fostering the Communist cause, but McCarthy had a strong point in suggesting that Bobby used antiblack sentiments against him. In the debate, Bobby said that he was in favor of revitalizing the ghettos through private enterprise. McCarthy argued that unless the suburbs were integrated, "we are adopting a kind of apartheid in this country." Bobby believed in integration, but he had serious, principled reasons why he also believed in the economic and social development of black ghettos.

Instead of presenting his own position in fulsome detail, Bobby said that he was "all in favor of moving people out of the ghettos," but that you could not move millions of people. "I mean, when you say you are going to take ten thousand black people and move them into Orange County . . . putting them in the suburbs where they can't afford the housing . . . it's just going to be catastrophic." McCarthy had not proposed anything like that, but the image of ten thousand poor blacks set down in right-wing Orange County was enough to raise anxieties while Bobby dangled the prospect of social progress without social pain.

"What Bobby did was a clever perversion of McCarthy's position," reflects Edelman. "And I was very disappointed in that. But you know, in retrospect, would I have advised him to do it? I wouldn't have advised it then. I wouldn't advise it now."

Bobby was so exhausted by the endless campaigning that he may have done things and said things he would not normally have done or said. Walinsky felt that as tired as the senator had been when the campaign began, he was now in some respects a "zombie." Any candidate going through the days Bobby was having would have been exhausted, but he may have had even more serious health worries.

In August 1966, Senators Mike Mansfield and Everett Dirksen had sent their colleague a letter pleading with him to take better care of himself and get a physical: "Please forgive us if we sound pompous, presumptuous, or paternal, as if we had passed the Medical Board and were practicing surgery. In our bumbling way we are trying to reassure ourselves that you are taking care of yourself, particularly at the end of a long, hot, and debilitating summer."

Although Edelman says that his employer was in perfect health, his autopsy suggests that he may have had serious potential health problems. Bobby's para-aortic lymph node glands were as much as three times normal size. This condition is often found at autopsy in those with latent prostate cancer.

Bobby also had endocrine abnormalities that in Jack had resulted in Addison's disease. Whereas Jack had no adrenal glands at death, Bobby's had atrophied down to a thickness of "little more than one millimeter." That would have made it difficult for his body to produce the quantities of adrenaline needed to cope with stress and difficulties. The autopsy also revealed signs of chronic gastritis that would have often put him in considerable pain.

Bobby may have been suffering from pain and ill health while the world imagined him to be a resiliently healthy man in the prime of his life. Like his older brother, if he had lived, health problems might have become one of the central realities of his life.

S ix of Bobby's children arrived at the end of May to spend the last week of the primary with their father. David, Courtney, Michael, Kerry, Chris, and Max were there not so much to see their parents but as irresistible photo ops to be shuttled in at crucial moments. They had arrived with Bob Galland, a camp counselor hired by Ethel's secretary to run the family's personal summer camp, teaching camping, sailing, and other water skills at Hickory Hill and Hyannis Port. Galland had not even been interviewed by Ethel but found himself sleeping in a bedroom at the McLean estate next to Bobby and Ethel, attempting to direct activities during the day.

Waiting at the Beverly Hills Hotel for the group was twenty-three-year-old Diane Broughton, a typist-secretary in the campaign office. Four hours before the plane arrived, she had been asked to help Galland with the children. On Sunday, three days before the primary, Galland and Broughton dressed the children in matching clothes, complete with campaign buttons and PT-109 pins. The children rode with Bobby and Ethel in a motorcade to a strawberry festival and from there to Disneyland, but were later returned to their two overseers. What struck Broughton was how rarely Bobby and Ethel called their children, not even once a day, though Galland insists that he spoke with Ethel daily.

The children had the kind of fierce, probing intelligence that often suggests either extraordinary creativity or deep troubles ahead. Of the six children, David was the perfect exemplar of this. In recent months his father and mother had both been gone most of the time from Hickory Hill. Seventy-one-year-old Jack Kopson, who lived in a little house that abutted the Kennedy estate, said that a group of boys, including David and several of his brothers, tormented him by attacking his home with firecrackers, exploding cherry bombs in his mailbox, and dumping lime on his lawn. Kopson ran out with his shotgun at the ready, frightening the fleeing youths by firing at the ground. Kopson condemned Ethel for "traipsing all over Indiana" and leaving "only a colored maid over there to watch these children." Several weeks later David and a friend were caught throwing rocks at cars on a highway and breaking one windshield in an episode that could have led to a serious accident. The parents of the other boy showed up at the police station to pick up their son. A governess arrived to pick up David.

Despite all of these difficulties, David had a great sense of wonderment. He had an awareness of the world around him far beyond his twelve years. On the cusp of adolescence, David was old enough to realize that the Kennedy name was a magic wand that he could wave to get what he wanted—and that all was not right with himself and his family. When his father descended on David's life, he gave a rapid-fire quiz on history and current affairs over lunch or sent a football spiraling toward David in an invisible end zone. Everything about David's family was different, even the

way they played football. On every play the end zone changed, and you
could throw forward passes beyond the line of scrimmage. You weren't supposed to lose, and if you got hurt, you weren't supposed to whimper.

David knew that there were things you didn't say or ask. That was
something that his eleven-year-old sister Courtney had not learned. During
one discussion she asked Broughton why her parents were staying in downtown Los Angeles at the Ambassador Hotel and not with them. There was
no motherly call to assuage her, no Ethel sweeping into the bungalow to
make it all right.

B obby spent a long, frenzied final day campaigning before election
day. There were some hysterical crowds, but he was now so tired that
it appeared he might collapse into a stupor. He had been going like this for
close to three months. In San Francisco the crowds were more respectful
than delirious as he and Ethel paraded through the streets sitting up on the
backseat of a convertible. As the car entered Chinatown, shots rang out, and
Ethel fell to the seat, but Bobby continued shaking the hands of kids who
trotted out to greet him. The sounds were nothing but firecrackers, and on
they went as Bobby nodded to an aide to comfort Ethel.

Bobby was the least fatalistic of men when it came to humankind's ability to shape its own destiny, and the most fatalistic of men when it came to
his own life. He dismissed the idea of a phalanx of bodyguards and
omnipresent police protection. He did not attempt to distinguish the sound
of a cherry bomb from the volley of a pistol, or a paranoid's ranting from an
assassin's threat. He was obsessed with not flinching.

This pattern had been going on ever since his brother's death. While
Bobby was on a trip to Montana campaigning for Democratic candidates in
October 1966, the *Butte Standard* had received a phone call saying, "Kennedy dies at four p.m." Bobby was supposed to speak to a crowd of three
thousand at Miners Hall, and when he was told the news, he insisted that the
rally go on. "When we got there, there's a guy there with a gun in the window," Dolan recalls. "Turned out to be a policeman. We always had trouble

with the police wanting to have too much protection and wanting to be in the crowd. You'd see a guy with a loose coat, a double-breasted coat, and he's a different age than the people who are screaming. And right away, you figure, 'That's it.' And we'd always try to get in there between that guy and Bob."

From his hotel in Lansing, Michigan, during the 1968 campaign, Bobby was warned that the police had seen a man carrying a rifle on a nearby rooftop. Not only did Bobby not want the blinds to be drawn, but he was upset when Dutton ordered that his car exit the hotel from the underground garage instead of the hotel entrance. "Don't ever do that," he said. "We always get into the car in public. We're not going to start ducking now." Ducking was not a coward's ploy, and it said much about Bobby's mental state that he seemed to think that it was.

On election eve Bobby and Ethel took over the beachside home of film director John Frankenheimer. The next day, June 4, dawned blustery and overcast, but Bobby nonetheless walked across the beach and plunged into the ocean. He had taught his sons that there were few waters too cold in which to swim, few waves too high not to jump, and few currents too strong not to be overcome. It was cool and overcast, and the surf was high, but he led David into the water.

A great wave rolled in, slapping the father and son down. When the water retreated, Bobby was holding his son, dragging him back onto the beach. Bobby had hit his head in the sand and bruised his forehead. Any other father would have wrapped his child in blankets and warned him away from swimming for the day. David, however, toweled off and joined his siblings frolicking in the pool.

When Bobby and Ethel arrived at suite 511 at the Ambassador Hotel shortly after 7:00 P.M., the living room was full of favored journalists, politicians, and aides. The suite was little more than a holding room. As soon as the results were known, Bobby would walk down to the crowd of supporters in the Embassy Ballroom to thank them for their work. The early returns showed McCarthy leading, but Bobby soon drew ahead, winning 46 percent of the vote. It was a victory, but hardly enough of one to drive the Minnesotan out of the race.

Bobby pulled Goodwin into the bathroom for a private discussion. "I've got to get free of McCarthy," he told the adviser, who, in search of a winner, had moved over from the Minnesota senator's campaign. "While we're fighting each other, Humphrey's running around the country picking up delegates. I don't want to stand on every street corner in New York for the next two weeks [campaigning for votes in the primary]. I've got to spend that time going to the states, talking to delegates, before it's too late. My only chance is to chase Hubert's ass all over the country."

McCarthy had done well in California, and he was not about to end his quest. Bobby was in a close race with McCarthy in his own state of New York. While he went back to campaign there, Humphrey's people would be out across America wooing delegates. Bobby was adept at weaving a promising tale to the journalists who surrounded him and at giving speeches to his supporters that were hymns of hope. But he understood the difficulty of what he faced. "I just think he really did not think he was going to win," Jeff Greenfield reflected. "I really had the sense a lot of times that he just didn't think the country was ready for what he wanted to try to do." That sentiment was echoed by Edelman, who three decades later admitted that his assertion that Bobby would have won the nomination and the election and changed America was his youthful idealism obscuring his rational political judgment.

Bobby may have been more realistic about his prospects than his staff. It was easy enough hearing the applause, the ovations, and the cheers to believe that you were at the center of a vast social and political movement that would change the nature of American society. It was not so easy to accept that Humphrey, who was not even running in the primaries, probably would win the nomination. Nor was it easy to realize that there was another movement to the right favoring Richard Nixon and George Wallace, whose numbers might well prove larger than yours.

Most of those supporters who filled the Embassy Ballroom that evening knew nothing of his doubts, nothing of the backroom politics. They were there to celebrate a clear and certain victory, and they cheered and hooted as Bobby stood before them.

"What I think is quite clear is that we can work together in the last analysis," he said, "and that what has been going on within the United States over a period of the last three years— the division, the violence, the disenchantment with our society, the divisions, whether it's between blacks and whites, between the poor and the more affluent, or between age groups or on the war in Vietnam—is that we can start to work together. We are a great country, an unselfish country, and a compassionate country. I intend to make that my basis for running."

Everything he was saying may have been true, but the results of the primary either belied or challenged much of it. He had won almost all the black and Mexican American vote, but nothing close to a majority of the white votes in the suburbs. He had not been able to fashion a rhetorical appeal that drew together all these peoples. Though he had not preached class warfare or promised endless bounty to the poor, he'd had little impact on a polarization that said far more about the social realities of America in 1968 than it did about the nature of Bobby's campaigning.

There was just one more contested primary left in New York, one more chance to win votes directly. Then it would be all back rooms and televised speeches and perhaps a trip to Europe trumpeting him as a world leader. He was not standing there that evening as he had hoped—as the overwhelming victor of a string of primaries, a triumph so singular that it would have sent McCarthy home and made Humphrey seem a pathetic stalking horse for the discredited policies of the Johnson administration.

There were those that evening who believed that in the end the Democrats could not possibly give the nomination to Humphrey and turn away from the candidate who spoke directly to the pain and anguish and uncertainty of so many of the American people. But the cold-eyed arithmetic of politics probably had another answer.

"So my thanks to you all," Bobby concluded, "and it's on to Chicago, and let's win there."

For the most part, Bobby's campaign days were choreographed by others. He moved along pathways and routes that others had planned out for him well ahead of time. He was used to hurrying through hotel garages, back stairways, restaurant kitchens, and service elevators, rushing along in a cocoon of aides and advisers. Bobby had entered the Embassy Ballroom through the kitchen corridor and pantry area from which servers entered and exited. The plan originally was for him to exit by moving through the crowd and out the main door to a second, smaller reception area in another room. Those well-wishers had seen the address on closed-circuit television, but since it was already after midnight, Dutton cut that event off the schedule, deciding to move the candidate to a press conference in the temporary press room. The quickest way there was to go back the same way they had arrived. "It's been changed," Bill Barry, his bodyguard, told Bobby. "We're going this way."

Bobby moved through a sea of well-wishers, hands grasping, voices shouting, and through double doors into the raw, dark bowels of the hotel. Along this narrow corridor stood many of the kitchen staff and others wanting their glimpse or touch.

Bobby was used to hands reaching out to him. He probably did not even see the hand that held a .22 pistol and fired at his head. Bobby's hands reached instinctively to his face, and then he lurched backward and fell on his back, his eyes open, staring upward. All around him was turmoil. Sirhan Sirhan, a Palestinian American enraged at Bobby's stand on Israel, continued firing toward Bobby. Five people collapsed onto the concrete with their wounds. Others struggled with the assailant, wresting the weapon from him and holding him down. Only Bobby was quiet and still.

T here had been a rally that night full of ominous premonitions, but that had taken place in San Francisco at the municipal auditorium, not in Los Angeles. Ted had been there congratulating the campaign workers in California's second city. Jack had always feared mobs, and there was much to fear as Ted and Burke bulled their way through a smothering onslaught of well-wishers, drunken revelers, and the merely curious while protesters in the balcony screamed, "Free Huey! Free Huey!"—a reference to Black Panther leader Huey P. Newton.

Back at the Fairmont Hotel, Ted and Burke turned on the television. When they saw footage of a milling crowd, Burke thought that something had happened at the San Francisco celebration, just as he had feared it might. The administrative assistant was a man of procedure and process, and he remarked to his boss that from now on the senator would have to have better security.

Ted said nothing but stood in front of the set, staring at the screen. His brother-in-law Steve Smith appeared on the screen imploring people to stay calm and to leave the Ambassador's Embassy Ballroom and go home. Ted stood there, maybe thirty seconds, maybe three minutes, maybe five minutes. Burke was never sure how long it was, but the moment hung there forever. "We have to get down there," Ted said finally, still staring at the screen.

At Hamilton Air Force Base north of San Francisco, Ted talked for a minute by phone to Salinger in Los Angeles. "It's going to be all right," he told Burke, though Salinger had said no such thing. On the flight to Los Angeles in a military jet, Ted said nothing, and neither did anyone else. For months he had felt an unspeakable fear that Bobby's campaign would end not in victory but in death, and that from the moment his brother announced his candidacy someone would be stalking him. They were a family of talkers, but this was not something he could talk about, as if by the very utterance of his fears the dark clouds would begin gathering. He had said everything else he could say to convince Bobby not to run, but there was nothing he could say to prevent his brother from doing what he felt he had to do. And now on this night, through these skies, Ted was flying south to Los Angeles, the City of Angels, where forty-two-year-old Bobby lay in mortal agony.

When the children had returned from seeing their parents for a few minutes on election night at the Ambassador Hotel, all except David went to bed. He sat on the sofa, still in his blue blazer. Like his siblings, he was so hyperkinetic that he had not simply sat there watching his father giving an interview on television. He had reached out trying to grab Mr. Spock, the spider monkey that they had been given earlier in the day in Malibu.

"See that lump," David said, looking at his father on television. "Dad got that when he pulled me out of the ocean." David watched closely as his father made his speech in the Embassy Ballroom. Afterward he asked about the local politicians and others whom his father had mentioned in his speech. The television reporters droned on. Then they started talking with tightly controlled urgency about a shooting, and Broughton did what almost anyone would have done—she jumped up and turned the sound up. "What happened?" David asked, and Broughton said soothingly that nothing had happened. Every image and every word on television belied the woman's assurances, and as the magnitude of the tragedy became clearer and clearer, David sat on the sofa in his grown-up suit trembling, staring at the screen.

Broughton had no idea how long she allowed David to stare at the screen. Finally, she wrapped him in a blanket and hugged him, touching his golden hair. Then Galland took David out of the bungalow and moved him to a room by himself in the hotel.

T ed arrived at Good Samaritan Hospital shortly after 3:00 A.M., while a three-hour operation on his brother was already in progress. "I can't let go," he told one of his aides. "We have a job to do. If I let go, Ethel will let go, and my mother will let go, and all my sisters . . ." All the grand dreams had come down to this: holding on. He held on when he learned that Bobby's brain waves were flat. He stood in the bathroom, the door half open, doubled over.

Much was made in later years of the idea that David had seen terrible things that night and was wounded in ways that would never heal. Much less was ever said of the impact of Bobby's death on his other sons. At Hickory Hill, fourteen-year-old Bobby Jr., the oldest child still at home, woke up early and read about the shooting in the morning paper. He took the paper and cast it into the cold hearth. Later, when sixteen-year-old Kathleen and fifteen-year-old Joe had arrived back at Hickory Hill from their boarding schools after they had been told the tragic news, Bobby Jr. went out and played touch football with them. When the game was over, he walked back and forth across the lawn, his shoulders slumped in despair.

As the eldest son, Joe flew across the country and was driven to Good Samaritan Hospital with the idea that his father might well recover. Bobby was under an oxygen tent with his face distorted almost beyond recognition. He had survived a day, but as Joe stood in that hospital room that long night, his father was pronounced dead on June 6, 1968, at 1:44 A.M. He told his cousin Chris that it had been a "hellish environment," with the adults losing control of themselves. He had been taught that Kennedys don't cry, but his family members were sobbing uncontrollably. His mother could not tell her other children that their father was gone, and so it fell on Joe to tell his younger brothers and sisters that their father was dead.

This was the defining psychological moment in young Joe's life. In this ultracompetitive family, he was already faced with a physically quick, intellectually brilliant, and verbally facile younger brother who outpaced him in everything except for sheer physical strength. And now he was given a crushing duty, to explain the unexplainable, to console siblings when he needed to be consoled himself. If he no longer had a father, in some measure he no longer had a mother either, a mother who could nurture and protect him. In this large family, he was alone, odd man out, first only in age and in responsibility.

Ted was not in any condition to help out with Bobby's children, but he held himself together at Los Angeles International Airport, where he helped lift the casket containing his fallen brother onto the plane. He held himself together on the flight back to New York, sitting with his brother, at times falling asleep resting on the coffin, at other times muttering about what had been done to his brother. "I'm going to show them," he said. "I'm going to show them what they've done, what Bobby meant to this country, what they lost." He held on too when the plane arrived at La Guardia Airport, and when he lifted the casket up, with Bobby's sons Joe and Michael and several of Bobby's closest friends, and carried it to the hearse. Sarge Shriver had flown in from Paris, where he was the American ambassador, and he moved forward to help lift the casket. Bobby's aides pushed him back. He was serving Johnson, and his supposed betrayal was not forgotten.

Bobby's eldest two sons, Joe and Bobby Jr., stood vigil beside the casket as the mourners slowly passed by at St. Patrick's Cathedral. Bobby Jr. helped his grandmother Rose as she left the cathedral. Even the youngest of the sons had their roles to play. Ethel wanted some of the youngest sons to participate as altar boys. "I was an altar boy, and I put on a suit, and I was there to put an eye on them," recalls Galland. "I think it was almost a pageant to them."

Those around Ted felt that he could not possibly deliver the eulogy at the service at St. Patrick's, but he insisted. He did not talk of the Bobby he had known as a youth, a pesky goad nagging him for his failures, or of the ambitious, determined man who had served Jack so well. He talked of what his brother had aspired to be and do in the last years of his life. "My brother need not be idealized or enlarged in death beyond what he was in life," Ted said, his voice breaking. "He should be remembered simply as a good and decent man, who saw wrong and tried to right it, saw suffering and tried to heal it, saw war and tried to stop it." In Hyannis Port, Joe Kennedy lay in bed watching the service on television. Since the shooting, all the old man had done was cry, watch television, and sometimes pray.

To his admirers it seemed that far more than Bobby had died. "When Bobby was killed, my generation was almost critically wounded," reflected Senator Walter Mondale five years later. "As a matter of fact, I don't know, we may never get over it. It's just terrible." So much promise had been swept away. The army of idealistic youth that Bobby had thought he would lead had largely dissipated. The civil rights movement had moved past its integrationist beginnings into a militant separatist phase, from Martin Luther King Jr. and "We Shall Overcome" to H. Rap Brown and the Black Panthers. The Vietnam War would continue for five more years, with fewer American lives lost but largely with American weapons and American intent.

Bobby had tried to be a force of reconciliation, but he had also been a man who polarized Americans. He had praised what was fine and good in human beings and human institutions and condemned what he considered ignoble or unworthy. He cared little about how you got where you wanted to go as long as you got there. He had not solved many problems, but he had acknowledged them, introducing millions of Americans to the realities of poverty in an affluent land and the costs of racial discrimination. Far ahead of his peers, he was a progressive politician who saw that government could hurt as much as it helped and that in its lumbering attacks on social ills government sometimes left damage in its wake. Unlike most of the politicians who later followed that same pathway, he articulated an exalted ideal of what Americans could be and do in the world.

On the slow eight-hour train ride that carried Bobby's body to Washington to be buried at Arlington National Cemetery, Ted stayed with the casket in the last car, elevated so that the thousands along the route could see it as the train passed slowly by. Joe, accompanied by Ethel and wearing one of his father's suits, walked from car to car, thanking the guests for being there ("I'm Joe Kennedy. Thank you for coming"). Joe's voice was as firm as his handshake, and though what he was doing was what any eldest son would have done, to many of those present it seemed that he was taking his first public step as the leading Kennedy of the next generation. His older sister Kathleen recalls that, while she too walked the train, no one wrote of her that day.

David rode with his head out a window. If a friend standing beside him hadn't pulled him back, he might have been decapitated by a steel girder. He put his head out again and saw two people killed as they stood on the tracks when a train passed in the other direction. The youngest of the Kennedy children were there as well, including seven-year-old John Kennedy Jr. and four-year-old Chris. John asked his cousin if his father still headed out to the office each day. "Oh, yes," Chris said. "He is in heaven in the morning and he goes to the office in the afternoon."

For many of the eight hundred mourners on the twenty-car train, the trip seemed to go on forever, suspended outside of time and place. Alongside the tracks stood a million Americans young and old, waiting to pay their respects to the man they called Bobby. There were more poor than rich, and they showed a nation's bereaved face. It was hot on the train. There was a bar, and some people were drunk, or half-drunk. It felt like an Irish wake, with as much laughter as tears. There were senators and congressmen, famous athletes and celebrities, activists and socialites. On and on the train rolled. The liquor ran out, night fell, and the train still moved onward.

One friend along for the journey was Adam Yarmolinsky, who had served at the Pentagon during the Kennedy administration. "As I took a turn standing over the coffin to brace it against the swaying of the train, now slowed almost to a walking pace, and watched the silent crowds lining the tracks, the strongest emotion I felt was wanting the trip not to end. Whatever it was, we knew it was over."

Ports of Call

T ed spent much of the next weeks sailing, but as far from shore as he ventured, he always returned to port. He didn't consider himself the equal of his three fallen brothers, yet he had a burden heavier than any of them had borne. He was a surrogate father to Ethel's ten children, with an eleventh still in her womb. He tried to be a second father to young John and Caroline too, and strength to Jackie, who was so distressed by the violence and so consumed by a desire to obtain her own unassailable fortune that she was planning to marry the Greek tycoon Aristotle Onassis. Ted was heir to all the political aspirations of his family, to the ambitions of those who had served his brothers and the millions who had found in Bobby's campaign a new hope for the nation. He had a wife who had a drinking problem, one made worse by his various infidelities. And of course, he had his responsibilities in the Senate.

Ted sailed up the coast to Maine or over to Nantucket with one group of friends or another. He took John for a sail, sometimes letting the seven-year-old take the helm. He accompanied Joe to Spain, where the young man played the matador to young bulls. He visited Onassis on his island of Skorpios, where he may have had a role in negotiating a prenuptial agreement between Jackie and the shipping magnate, and he was there for their October wedding. He also tried to watch over Ethel, who went to mass each morning and then sat down for an hour or two on the front porch to pen handwritten thank-you notes.

Ted let his beard grow long, and he wandered the compound. He spent time around his children, nephews, and nieces, and sometimes he drank around them too. He was the kind of drinker who talked loudest and laughed longest when he was out on the town, with a drink in his hand and another on the way. With a few good jolts in him, he looked like just another jovial holiday drinker, the sort who made his way down from Boston for a summer weekend and considered every hour he was sober a waste of a good time. The convivial drinkers who pushed their way up to Ted at the bar treated him like a comrade, but unlike him, they were not full of overwhelming sadness and responsibilities too overwhelming to contemplate.

While Ted attempted in his erratic way to keep the family together, his father lay in his bedroom on the second floor in Hyannis Port. The Kennedy patriarch had purchased the sprawling oceanfront house on two and a half acres in 1929. While his children were growing up, Joseph P. Kennedy had sat on the deck on the second floor, surveying the grounds like a general his troops. His children had been expected to move from one activity to the next all day long, playing tennis, swimming, sailing, reading books, discussing current events, every moment full of high purpose. A mere nod of his head was enough to push his children onward from momentary lassitude.

The children had gone off to schools and careers, but they always returned to Hyannis Port. Except for Pat and the institutionalized Rosemary, all of Joe and Rose's sons and daughters had built their own homes there. Jack, Bobby, and Jean all had homes on the street side of the grassy plain from which it was only a short walk to their parents' white clapboard house with green shutters. Eunice had built her large home about a half-mile away, and Ted had constructed his house on Squaw Island, within easy reach of the rest of the family.

There had been such a sense of order and purpose in those family summers. Now there was disorder and indolence, frenetic movement and activities that were at best a caricature of the old disciplined times.

T here was a sixteen-year difference between the oldest and youngest of
Ethel's ten children, and while little Doug did not understand that his
father had died, the four teenagers—Kathleen, Joe, Bobby Jr., and David—
understood very well. Their father had given them a model of what a man
should be. "My father surrounded himself with people who had overcome, had
proven themselves with physical courage," reflects Bobby Jr. "John Glenn, Jim
Whittaker, Rafer Johnson. Our house was filled with Cubans who had fought
at the Bay of Pigs to distinguish themselves. You know, he liked those guys
because they were kind of tough guys and he thought that although war was
horrible and unspeakable and should be avoided at any cost, that it was also the
time when not only the worst but the best of human virtues emerged. He took
us out into the wilderness and places to make sure that we had those experiences of physical courage and tested ourselves and that kind of stuff."

Fourteen-year-old Bobby Jr. combined intelligence, intrepidness, and
boundless inquisitiveness about the world around him. At Hickory Hill he
kept his own personal zoo, everything from a bear to a boa constrictor. In his
bedroom were huge photos of Lenin, Stalin, and Cardinal Spellman, a tri-
umvirate seen there and nowhere else.

Bobby Jr. and his brothers knew that if they were to be true to their
father, they had to lead lives of courage. It was a difficult directive in a country fighting a war their father died opposing, and with politics so full of
moral ambiguities, but if they were to honor his memory, they had to seek
arenas where they could prove themselves not only fearless but heroic.

"James Dickey said that it bothered him that a person can live his own
life in this country and not know whether he's a coward," says Max, the most
deeply read of any of his brothers. "I grew up after Vietnam and before the
United States began involving itself in other conflicts around the world, so
those kinds of conflicts were closed. So what's left? Climbing mountains,
shooting rapids, and kayaking, those kinds of things."

When the sons looked back on their father's life, they did not think of the weeks he had been away, the despair that he had sometimes felt, or the complicated realities of their parents' marriage, but only of Bobby's strength, love, and courage. When they looked back at Hickory Hill, they saw those as blessed years. Young Bobby in particular felt that he and his brothers had had "a little paradise" in the woods and pastures around the house. He and his younger brothers had run free up and down the hills and meadows and streams of Pimmit Run, near Hickory Hill. Then one day the builders had come and flattened the hills, covered the stream, and felled the trees to build the Dolley Madison Highway, followed by homes and a strip mall. He and his brother David had taken some of the stacked pile of culvert pipes and sent them smashing down an embankment, their little sabotage.

For the sons of Robert F. Kennedy, life would forever be marked between the time before their father died and the time after, when nothing would ever be the same. Their mother would not be the same either, not to the world and not to her sons. Ethel had been more fun than a forty-year-old mother had any right or reason to be, her impish joy growing out of a love for Bobby that anchored her securely in the world. It was a measure of that love and its unselfish ardor that despite all that she knew about life and its dangers, no one wanted Bobby to run for president more, or pushed him harder, than Ethel.

Ethel's anchor was gone, and in that first summer she drifted whichever way the currents of her mood took her. She had seen so much death, and not only Jack's and Bobby's. Her parents had died in a plane crash in 1955, and her brother in another plane accident a decade later. She was as much Kennedy in mentality as any of the blood family, and despite being pregnant played games of tennis with fierce competitiveness. She had grown up in a large family, and she could not stand to be alone. Her shingled, rambling house was like a summer camp, with kids and guests running through the rooms.

Even if their father had been alive, Bobby's teenage sons would have begun to confront the dangerous freedoms of their time that summer.

Youths of their age were already experimenting with drugs and sex, and the Kennedy sons were nothing if not precocious. They needed the comforting strength of a father to help guide them away from the dangerous shoals of adolescence, and more than ever they needed a watchful, concerned mother. They were rambunctious, difficult youths full of anger at a world that had grievously wounded them, and they took it out on those around them, testing, taunting, and teasing.

Early in the summer Ted headed off with Ethel and her children on a chartered cruise from Mystic, Connecticut, to Hyannis Port. On the boat, as elsewhere, the youths were noisy and unruly. Ethel spent much of the time far up on the stern of the ship sunning herself, but at one point, according to authors Peter Collier and David Horowitz, she ordered Bobby and David into the cabin, where she beat them with a hairbrush. When the ship reached port, Ethel told them: "You guys have got to get away from here." It was not a bad idea for the youths to spend some weeks away from the emotionally oppressive life on the Cape. They learned that they had to move gingerly around their mother, both psychologically and physically. They were full of a corrosive anger.

The devastating truth is that most of Ethel's sons would become alcoholics or drug addicts. They may well have had a gene for addiction that increased their propensity for such problems, but that hardly accounts for this plague. Their home became a virtual machine for creating addiction, the type of environment written about in any number of psychological textbooks and spoken of with deep, painful truth by addicts in Alcoholics Anonymous meetings.

So many addicts speak of a terrible event in their childhood—a death or a trauma that never quite ended. They need to speak up and speak out, but so much in their families has been walled off emotionally that it is difficult for them to honestly confront the problems that haunt them. In many such homes, the child learns that people around him may explode emotionally and that he can't depend on anything but his own cunning. There is yet another childhood lesson frequently learned in an addictive, compulsive family—always to project the proper image to the outside world. Truth is betrayal. They must never divulge the family's emotional secrets, in some senses not even to themselves. When these circumstances come together, they are often enough to create an addict, devastating his life and the lives of those around him.

J oe and his closest friend, Chuck McDermott, spent much of the summer of 1968 in Spain on the ranch of a bull breeder named Salvador Guardilla. Joe played the matador and danced with a beautiful young gypsy at Ulrera, an old Moorish town. For a fifteen-year-old, it was an incredible adventure, yet Joe's despair over his father's death did not lift.

Even before his father died, he was having a troubled adolescence. He was his father's son in that everything was difficult to him, especially his studies. There were knowing whispers that he might be a Kennedy but he wasn't up to muster. He did not have the upper-class veneer of many of his schoolmates at Milton but rather a rough-hewn, awkward manner. He was a stolid, muscular young man with an intimidating manner that people feared might erupt in violence.

That same summer, Bobby Jr. headed off to Africa with Lem Billings, Jack's best friend from his days at Choate. Lem had befriended young Bobby, whom he believed had the magical qualities of his beloved Jack. As Lem saw it, Bobby was "the best of them all, very much like his uncle, with the brains and spirit and a hell of a personality."

Lem was fifty-two years old, three times Bobby's age, yet he was no aging chaperon deputized to keep Bobby out of mischief, but a friend whom the youth considered a boon companion. In Kenya, the two went on a photo safari memorialized in *Life*. In Egypt they sailed a raft in the Valley of the Kings and with members of the Egyptian Supreme Court watched belly dancers at a club. And Lem talked of Jack and Bobby, and what had been. "The stories he told and the examples he set gave us all a link to our dead fathers and to the generation before us," Bobby reflects. "In many ways Lem was a father to me, and he was the best friend I will ever have."

David had seen terrible things, and if any of the children needed counseling, a mother's tender solace, or a special friend, it was he. "Lem could have chosen any of us," David said later. "I remember the day it happened. Lem appeared and they just sort of walked off together. I thought to myself: Bobby's luck. I wish I had someone."

David was treated very much like his two older brothers—he was sent off on a foreign trip, but without an adult companion. David and his cousin Chris Lawford flew to Austria to attend a ski and tennis camp. David was an avid skier and a fine tennis player. If his father had been alive, he would surely have competed. But the scores did not matter anymore, and the glacier runs did not have the same excitement. A seventeen-year-old woman, who had heard that David was a Kennedy, picked him up. David was only thirteen years old and had scarcely begun to think about girls, but he was soon in bed with the older teenager. In the following weeks he and Chris made the rounds of the camp playing on the sympathy of young women willing to give their all to cheer up the tragic Kennedy cousins.

For the younger children who stayed in the States, the summer meant Hyannis Port, and if anything, the pace of activities was even more intense than in previous years. At Ethel's the day began earlier than in the other homes in the compound. Galland got two of the younger children up, dressed, and sitting in the car with its motor running for Ethel to make the 8:00 A.M. mass at St. Francis Xavier. Later in the morning he took his charges for a sail on the sixty-one-foot *Mira,* including a picnic lunch. Then there was tennis, swimming lessons, excursions to town, and perhaps some touch football.

The other person who was crucial to Ethel's children was her housekeeper, Ena Bernard, who stayed with the family all her life, through an endless parade of cooks, maids, gardeners, and tutors. The Costa Rican woman was a formidable matriarch in her own right who became not only the de facto manager of Hickory Hill but a surrogate mother. She did not have the wild mood swings of Ethel and was always there for the children. Even if there was sometimes unspoken tension between her and Ethel, it was unthinkable that she would ever be fired. When the children were old enough to realize that Bernard was an employee, they did not subtly distance themselves, as most wealthy scions would have done. Instead, they treated Bernard almost like a maiden aunt, albeit one who was endlessly overworked.

That summer of 1968 Bernard had much to keep her busy. Seventeen-year-old Kathleen had hardly gotten back from teaching on an Indian reservation when she fell off a horse and injured herself. Bobby returned in mid-August suffering from inflamed red eyes that had been treated in Kenya just before he returned to the States. David sported a bandaged right hand, hurt while rigging a sailboat. Joe arrived home from Spain telling tales of how he had been knocked down by a frisky calf when he attempted to play a matador. Michael had to be hospitalized to remove a cyst on his neck.

The tennis pro giving lessons to the children at Hickory Hill had the unenviable task of teaching a sport played on a rectangular court with clearly marked lines and standards of decorum that were almost as important as the rules. Kennedys did not lose easily or gracefully, and David had discovered a unique expedient to lead him to victory. He cheated on the line calls. His cheating was so persistent and so outrageous that the pro called Ethel over to watch. Ethel was a fierce competitor herself, and she watched her son.

"You see what David's doing, Mrs. Kennedy," the pro said.

"What do you mean?"

"He's cheating on the line calls."

"I pay you to teach him tennis, not morals," Ethel said, ending that matter.

When the hour-long lesson was over, the court was strewn with tennis balls. "Now, I tell you what I'm going to do," the pro said, as his pupils stood on the side of the court. "Whoever picks up the most balls gets to play a game with me."

"We're Kennedys," one of the boys said, as his brothers nodded agreement. "We pay people to pick up balls."

At Hyannis Port, Jackie lived next door to Ethel's house, so close that she could hear her unruly neighbors. Caroline and John were very different from Ethel's children. Like her cousins, ten-year-old Caroline rode ponies, but she rode at a sedate and gentle pace; she did not gallop. Seven-year-old John loved to sail and swim and run and play with his boisterous cousins. He was not the natural athlete that some of his cousins were, and he was not one of the first ones chosen for childhood games. He tried to act like Ethel's sons, but Jackie was bringing him up with manners he could not discard merely because he was on the playing field.

Those summers in Hyannis Port were almost as important to John's development as they had been to his father. He and his cousins could be themselves on those grounds. John spent much time swimming, boating, and playing games, especially with Willie Smith, who was his schoolmate at St. David's, and Tim Shriver.

John was an irascible child with limitless, nervous energy. He craved the manly comradeship and direction that Jack could have given his son, and he looked for it everywhere. He found it not with his Uncle Ted, who was the strongest male relative in his life, but with several Secret Service agents. Congress had authorized the protection for Jackie and her children, and it was yet another reason why living a normal boy's life was such a challenge.

John also spent time in Newport with Jackie's stepfather, Hugh Auchincloss. The old man took John for walks on the beach while he filled his charge with vivid tales of the pirates who had supposedly buried their treasure somewhere nearby. Jackie asked several men, including the fathers of neighboring boys, to dress up in pirate garb and descend on the children in a longboat during a picnic. John and his summer friends picked up wooden swords and drove the evil interlopers off, capturing the biggest of the pirates and forcing him to walk the plank to his make-believe death. It was all great fun, but John began to cry, even after he recognized that the hulking pirate was six-foot-three-inch John Walsh, a Secret Service agent who had become another father figure to him. "You can't die!" he cried out in despair. "You can't die."

The other boys had recognized that this was nothing but a splendid game, but John's vivid imagination made the scene only too real. Beyond that, there was a deep-seated sense of loss in a little boy in whose world people were constantly disappearing. Nannies. Governesses. Maids. Secret Service agents. Most of this turnover may have been simply the reality of a wealthy life, but for a fatherless boy such losses took on deeper meaning.

Although John aspired to be an athletic youth like his older cousins, he was something of a klutz. All of the boys had their scrapes, bruises, and broken bones, but Ethel's sons earned theirs jumping off roofs, careening down ski slopes, or falling off horses. John's came from stumbling over his shoelaces or tumbling down the staircase.

In 1966 Jackie and her children had been on an overnight camping trip in Hawaii. The group was relaxing when five-year-old John fell into the fire. Walsh rushed forward and pulled him out. John had second-degree burns on his arms and behind, but without the alertness of the Secret Service agent, the results could have been tragic.

Jackie insisted that the Secret Service agents who spent the most time around John be youthful manly types. Many of them were also Irish Catholic. As much as John sought to be just another boy, he could hardly be that when wherever he went Secret Service agents followed him. When his mother walked him to St. David's School on East Eighty-ninth Street, they were shadowed by agents. During his years at the school the agents were either in the back of the classroom or in the basement, within immediate call.

When John was in first grade, the teacher had a show-and-tell in which the children talked about what they had done the previous weekend. "Well, I went fishing with my Uncle Bobby," said John, who was as outgoing and high-spirited as his sister was shy and restrained. "Why didn't you do it with your dad?" asked another boy. "He's in heaven," John said. "He was a Commie," another boy sneered. John jumped out of his seat and rushed the child, and the two boys began throwing punches at each other. The teaching assistant, Patti McGinty, moved to separate them, but the Secret Service agent hurried up from the back of the room and held her arm. "Let them fight," the agent said. "He's going to have to learn."

When McGinty met John's mother at a school event, she praised the young teacher. "I'm so happy you let him fight," she said.

John's mother could have groomed him in her own image, as a perfectly mannered young man, a bit prissy and wan perhaps but always a gentleman. Instead, she wanted her son to have a boy's natural challenges, romping in Central Park, bicycling with his friends, going off on adventures. That said, there remained an inevitable tension between her civilized, philosophically cautious life and her son's desire to be an intrepid man.

Wherever he went, whatever he did, John was reminded of the fact that he was John F. Kennedy Jr. People were always coming up to him, sharing their fondest memories of his father. The older he grew, the more reminders he received that he was different from his peers. This began from first grade on, if not before. Many of his classmates at St. David's were from prominent families. They might go off to California for the holiday. He flew to England to meet the Queen. His friends headed to Vail during the Christmas holidays. He jetted to Gstaad, where his mother was pestered by a brigade of paparazzi. There was an aura of excitement around him that as a boy he hardly comprehended. His friends' parents affected indifference and displayed a studied casualness when he came over to play; when he left, they called their friends to say who had spent the afternoon with them.

So much attention was paid to John, often unconsciously, that he risked becoming an impish little show-off. He was the kind of boy who would poke a friend in the ribs and stand there all innocence while his playmate took the blame for striking the gentlemanly Kennedy boy. He was forever stirring things up the moment the teacher left the room, a quality he never lost even when he no longer had teachers. His idea of play was to tear around wildly until warned to desist. John was a page at the 1966 wedding of his half-aunt, Janet Jennings Auchincloss, in Newport. He wore a velvet coat and silver-buckled shoes, a sissy's outfit by most boys' measure. He may have been decked out as a little prince, but he redeemed himself by attempting to herd ponies into the reception tent while Secret Service agents ran after their charge.

Caroline was a little mother to her brother. She played this role in a loving, secure manner that an older brother probably would not have shown to a younger sister. Unlike the relationship between Joe and Bobby Jr. or between the two oldest Shriver sons, there was no jealousy between John and Caroline.

D espite the patriarchal traditions of the Kennedys, the mothers were the overwhelming psychological forces in the lives of their sons. Their impact was especially evident during those summers in Hyannis Port when their husbands were often elsewhere.

Pat had brought her son Chris for the last part of the summer of 1968, along with his three younger sisters, Sydney, Victoria, and Robin. Pat was the gentlest and most vulnerable of the Kennedy sisters. She had probably suffered more over the loss of Jack than any of them. Not only had she begun drinking heavily, but she bore the considerable onus of the first divorce in the family after she left her movie star husband. Peter Lawford had been a debonair image on the screen and on television. He was not losing his looks but deepening them. If he had not dissipated himself with drugs and liquor, he might have become another Cary Grant. His son did not quite have his father's looks, but at thirteen, Chris was a tall, appealing youth who found in Hyannis Port a home and family that he had nowhere else.

The Smiths were there too. Jean was the youngest daughter. Her husband, Steve, was a charming man who had been adored by Jack and Robert. He aped most of the Kennedy men only too well. He not only had affairs with endless women but maintained a series of mistresses and a life apart from his immediate family. As the youngest sibling, Jean was forever pushing her two sons forward, asserting in every way that Steve Jr. and Willie were at least the equals of their cousins. Despite her efforts, there was a doughy, shapeless quality to the two boys' personalities that set them apart. Joan was present too. Despite her nascent alcoholism, or maybe partially because of it, she clung to her sons and their lives.

The Shrivers were not in Hyannis Port for much of the summer of 1968, since Sarge was ambassador to Paris. Like her brother Jack, Eunice suffered from Addison's disease, and the daily doses of cortisone probably elevated further her nervous nature. Eunice remained an iconic presence, carrying the banner of faith in one hand, the banner of family in the other, and ignoring her wounds and losses as she charged onward on the battlefield of life. Her sons were supposed to follow her. While whatever her sons did was good, it could always be better. Eunice took the most detailed, minute interest in her sons' lives, focusing on whichever one seemed to be faltering and, whether he knew it or not, needed her guidance.

The two oldest Shriver children, Bobby and Maria, were born into a family in which Jack's future dominated their mother's life. Eunice spoke like Jack, thought like Jack, and revered him in a way that she did not revere her own husband. She was gone much of the time, if not physically then emotionally, thinking about Jack's future, his race to the White House, and the success of the Kennedy presidency. At their Maryland country estate, Kennedy family photos and memorabilia were everywhere, and there was little room for pictures of Sarge's distinguished, if diminished, Maryland family.

Sarge had his own bedroom, and if one wanted to know what it meant to be a Shriver, it was here one went. Sarge often appeared to be luxuriating in wealth that was not his, but it was more that he was a lover of beauty, with exquisite taste, and if beautiful things were expensive, so be it. He had fine paintings on the wall, leather-bound books on his table, and tailored suits in his closets. He usually had a religious book next to his bed, not one full of platitudinous bromides but a serious study—Jacques Maritain or Pierre Teilhard de Chardin. His son Bobby liked to come to his room and sensed that here life was different. Like his father, Bobby appeared to be a bold extrovert but in fact was more of a secret introvert who, with his loudness, pushed the world away from knowledge of his sensitivities and the pain of being weighed down by the Kennedy heritage. He loved his father with a love that was simple and true, but his feelings toward Eunice were the complicated sort that boys often have toward their father. It would be the great struggle of Bobby's life to resolve those feelings and to become not just another Kennedy, or even a Shriver, but something of his own making.

Sarge was so enamored of the fact that he had married into the Kennedy family that his own ample accomplishments gave off a dim luster. He called his father-in-law "Mr. Kennedy." He took delight in a wife who could sit with her feet up on a chair, smoking a cigar and speaking in intemperate language, as if she had just come off a hard day's work on the docks. And yet he knew that Eunice could put down her cigar and turn with the most exquisitely sensitive concern to an underprivileged child or a mentally retarded adult. During the Kennedy presidency she started a summer camp for the developmentally disabled at the Maryland estate. These campers were not poster-perfect children shuttled in for photo ops but severely mentally challenged kids. She was often down on her hands and knees with them. Her five children were expected to help out and encouraged to see this not as a tedious exercise in charity but as an exciting opportunity.

In the summer of 1968, Eunice took what she had learned in those summer camps and applied it on a national level. That July, while the rest of the Kennedy family was at Hyannis Port, Eunice was in Chicago at the first Special Olympics games for those with mental retardation. Some saw an embarrassing public spectacle: those who could hardly walk were asked to run, and children who had never been permitted to swim in a public pool were expected to dog-paddle in a shallow, temporary pool. But Eunice saw in them true athletes who were as worthy of celebration for their efforts as any in America. And when others looked out at a thousand participants playing in a largely empty Soldier's Field, Eunice saw the beginnings of games that she believed would one day fill great arenas.

It was a strange juxtaposition that summer in Chicago, those first Special Olympics games and a Democratic Convention in which protesters were clubbed, tear gas wafted through the streets, and the Kennedy name was invoked with sadness and regret.

Ted's Way

O n August 21, at Holy Cross College in Worcester, Massachusetts, Ted gave his first speech since his brother's death. "Like my brother before me," he said, "I pick up a fallen standard. Sustained by the memory of our priceless years together, I shall try to carry forward that special commitment to justice, to excellence, and to courage that distinguished their lives." He was proudly picking up his brothers' fallen banner and announcing that, whatever the cost, he would attempt to carry it forward.

Ted said that he knew "there is no safety in hiding" and that while he would return to the Senate, he would "not run for office this year." The Kennedy name was such a powerful, resonant force in American politics that despite Ted's youth, inexperience, and mourning, some Democrats had implored him to be Humphrey's vice presidential running mate, or even the presidential candidate. He had said no, but only for this year, and most Democrats assumed that one day the last Kennedy brother would return to the White House.

The most important passages of Ted's address were those in which he defined how he intended to carry on Bobby's legacy. "Each of us must take a direct and personal part in solving the great problems of this country," he said. "Each of us must do his individual part to end the suffering, feed the hungry, heal the sick, to strengthen and renew the national spirit." He was pledging to take on Bobby's most liberal agenda as his own. What he did not mention was the other part of his late brother's political analysis, the critique of government social programs and large anonymous institutions—be they government, unions, corporations, or universities—that so dominated American life. He was severing this latter part of Bobby's agenda from his own concerns. It would be another generation before a major Democratic politician could criticize an entrenched welfare bureaucracy and big government, as Bobby had done, without appearing to be a traitor to his party's ideals.

Ted returned to the Senate as an American icon. In his eulogy for his brother Bobby, he had said that his brother "need not be idealized or enlarged in death beyond what he was in life." As exquisite a sentiment as that was, it was also a cautionary note, for in American life, whenever a public figure stood too high and was revered too much, his reputation came tumbling down, often falling far below its justifiable place. That had not yet begun to happen with the Kennedys. The assassinated brothers stood alone in an American pantheon. Millions of Americans remembered them not for what they thought they were but for what they believed might have been and what still could be in the living presence of the last son.

Ted was not merely a stolid bearer of tragedy and the last great hope of the Kennedys, but a working politician, one who was good at his craft and made his own contributions to American life, in his own name. If he had not had this crushing burden of legacy, he might have slowly grown in the strength of his resolve and one day have run for the presidency on his own merits.

In Washington, Ted spent much of his time pursuing issues not in the name of his brothers but in his own. The immediate international issue that most dramatically concerned him was the civil war in Africa, where the oil-rich Ibos of the Nigerian region of Biafra were attempting to secede from their country. Before Biafra was finally subdued, over a million people died from combat, disease, and hunger. The conflict was a harbinger of the vicious civil wars that would so decimate Africa and other regions in the last decades of the twentieth century. Ted took what was his archetypal position on conflicts from Bangladesh to Bosnia, calling for humanitarian intervention and invoking the name of the United States as the world's bright hope. "We hold ourselves out as something different on this globe," he said on December 6, 1968. "Perhaps the starvation of people in Nigeria-Biafra is not in our vital interest. But it is our conscience."

By now, of course, the United States had been sucked into a bloody quagmire in Vietnam. Few Americans were willing to risk their sons' lives in an African conflict in which both sides shot down relief planes and European nations supplied arms while waiting to siphon off oil from the great reserves of Biafra. Yet if an international peacekeeping force had separated the belligerents, not only would hundreds of thousands have lived, but the precedent might in the future have saved countless millions, from the Congo to Eritrea.

Ted signaled his forceful reentry into American political life to his colleagues when in December 1968 he ran for Senate whip, the second-ranking leadership position. Ted defeated the incumbent, Russell Long of Louisiana, largely because his Democratic colleagues knew that Ted's constituents lay not only in Massachusetts but in their states as well.

The victory only increased the talk about Ted's running for president in 1972 against the incumbent Richard Nixon. A troubled Charles Bartlett decided that he should have a talk before things got too far. The political columnist had been one of Jack's closest friends.

"All these young guys are pushing you, but you take this thing and it will be a disaster," Bartlett said. "I'd hate to see the Kennedy name harmed. You're just not as bright as Jack. You don't have the experience. You should lay off and raise your children." Bartlett would never have dared to say such a thing to Jack or Bobby. If he had, he would have been abruptly cast out of their counsels. Ted listened as if he were hearing polite advice. "Charlie, I appreciate what you're saying, and I agree with a lot of it," he said. "There's only one thing."

"What's that, Teddy?"

"The old man taught us, if it's on the table, pick it up."

What Bartlett did not know was just how troubled Ted was becoming, and the chances he was taking with his family name and legacy. He had had any number of short-lived liaisons over the years, but for the first time he had fallen in love. During the last days of Bobby's campaigning, Ted had met an exquisite blond Austrian woman who combined the high style of a European jet-setter with a subtle, cultured mind and was fluent in several languages. Unlike most of the women he had known, she was a challenge to his boyish nonchalance. Her name was Helga Wagner, and when he met her, she was estranged from her oil tycoon husband. Ted called her constantly, wrote her letters, and even introduced her to other family members.

As Ted continued his passionate affair, he tried to be a worthy head of the family. In December, when Ethel gave birth to her and Bobby's eleventh child, Rory Elizabeth, he was in the operating room, wearing a smock and mask, and for a moment almost passed out. And he was with Ethel driving her home from the hospital, stopping first at Arlington National Cemetery where on the cold winter day she stood with her baby before her husband's grave.

When Ted reached his home in McLean, he tried to be a good father to his two sons, Teddy and Patrick, and his daughter, Kara. There was an un-Kennedy-like vulnerability in his children. "If Uncle Jack was shot at, Uncle Bobby was shot, will Daddy be shot?" Teddy asked. The boy was haunted whenever his father left, above all when Ted stayed out overnight.

Ted could not manage to hold everything together, and he turned to his two solaces, his mistress and his liquor. During the winter of 1968–69 in Palm Beach, Ted was staying at the family's oceanfront home, where his parents had wintered since the late twenties and he had spent much time. The home had once been full of Kennedy laughter and shouts, but it was mostly quiet now and full of memories. Ted, who had been drinking, drove home alone in his car late at night along the ocean. A quarter-mile south of the Kennedy house the road makes an almost perpendicular turn to the left. Ted drove straight ahead, crashing the vehicle. He was not hurt, however, and returned to his parents' home. He woke up the chauffeur, Frank Saunders, who went to the accident scene and took care of the matter. Ted went to bed never mentioning the accident again.

In April 1969, Ted flew to Alaska for hearings of the special subcommittee on Indian education. He arrived with a media entourage worthy of a world leader. That was in part because he was considered President Richard Nixon's likely opponent in 1972, and in part because he was the last Kennedy brother and a deathwatch had assembled, to see whether he too would be shot down.

For Ted, drinking was for nights and weekends, bars and buddies, lady friends and parties. That was the saving dichotomy. On this trip, though, as scrutinized as he was, he pulled out the silver flask that had once been Bobby's. "First time I've used it," he told writer Brock Brower of *Life*, as if that were justification enough. Bobby would never have done such a thing in front of a reporter he did not know and trust. "Stars! Stars!" he shouted later in the trip. "First one to find the northern lights gets a beer."

Ted could drink about as much as any man and still appear relatively sober. That was the most dangerous of gifts. But something was different now, and this trip brought him back to thoughts of death and dying. "They're going to shoot my ass off the way they shot Bobby's," he said as the reporters listened and took their private notes. Wanting only to pop a few more drinks, he did not eat at the airport in Fairbanks on the way home. He got on the plane and asked the flight attendant for a drink, and then another. He swaggered up and down the aisle, bouncing a pillow on the head of one of his aides, shouting for him to wake up, and then weaving along shouting, "Eskimo power! Eskimo power!" The journalists listened and noted Ted's sad state, but none of them wrote about it in their publications when they got home.

In early June Ted flew to the Cape in a private plane after giving a speech in Kentucky. That was just a part of his life, a lengthy flight to give a one-hour speech, and then another long flight home. The night was black, and there were storm warnings ahead. Ted still wore a back brace from his plane crash five years before in Massachusetts. He squirmed in his seat in the six-passenger plane, trying to find a position that would relieve the pain. He sat next to Joseph Mohbat, an AP reporter who was traveling with the senator and wrote about the trip.

"Good crowd tonight," the reporter recalls saying. "I guess so," Ted shrugged, as he downed a mixed drink. He could rouse an audience out of their seats, as he had just done with the Kentucky Democrats, and still feel half-dead inside. "What's it all for?" he asked. "I used to love it. But the fun began to go out of it after 1963, and then after the thing with Bobby, well. . . ."

Ted might receive a call from his party to run in 1972, but that call would not be as much for him, Edward M. Kennedy, as for his brothers. He was the convenient receptacle for all the sentiment and ambition of his party, a device to regain the White House. As the plane droned on through the night, he sucked on his drink and mused: "So even if the time is right [to run for president], why should I? You talk about the family obligation, the public service. Is running for this the best way to meet that? Just so many responsibilities. I worry about the kids, I never feel I'm giving enough."

S oon after that flight, Ted, Joan, Ethel and her older children, Chris Lawford, and Jim Whittaker and his brother Lou traveled down the Green River in Utah and Colorado on a whitewater-rafting trip. Bobby had taken his brood on annual summer excursions that had been rituals of risk, bonding, and communion; his sons had gained a spiritual sustenance from them that helped sustain them until the next adventure with their father.

Ted was not Bobby. He might drive with wild abandon, but he did not seek physical challenge the way his older brother did. He would climb no mountain in Bobby's honor, and on those long languid days he was content to float along with his most honored companion—the daiquiri, he held in his hand. Chris Lawford proclaimed Ted "an incredible asshole," an opinion that was shared by several of his cousins.

The young Kennedys saw Jack and Bobby as heroic figures, and in their eagerness to find something of that quality in their Uncle Ted, they were mercilessly unforgiving of his weaknesses. It angered them that he could not be what they thought they wanted and needed and that he made no attempt to be anything but the sadly vulnerable figure who stood before them nursing his drink.

CHAPTER THIRTEEN

The Road Not Taken

On July 18, 1969, Ted flew down from Boston to Martha's Vineyard to race Jack's old sailboat *Victura* in the Edgartown Regatta. He was also there to attend a Friday-evening party afterward honoring Bobby's six "boiler room girls," the young women who had worked for Bobby during his campaign gathering intelligence and helping to win delegates. Joan was not there because she was pregnant.

Ted got nervous when he did not have someone around to take care of the mundane responsibilities of life, leaving him alone to push his way through the world. Thus, Jack Crimmins, his part-time chauffeur, was waiting at the Edgartown airport with the senator's Oldsmobile. When Ted said that he wanted to go for a swim before the race, Crimmins drove him through the quaint old whaling village of Edgartown to the two-car ferry to the tiny island of Chappaquiddick, a mere 170 yards away. From the landing, the chauffeur drove to the nondescript cottage that had been rented for the weekend party. After Ted changed, Crimmins retraced his route on the paved road and made a sharp turn onto Dike Road. The car bounced its way along the secondary dirt road, across a tiny bridge, and down to the beach, where some of the other guests were already swimming.

After his dip in the Atlantic, Ted went back to the two-room weekend cottage to change clothes, then, along with the other guests, headed back to Edgartown for the regatta. The six women watched from a boat rented by one of the other guests, Paul Markham, a Kennedy friend and political

operative who had been a U.S. attorney. Ted and his two-person crew finished a mediocre ninth in their class. At the victory party Ted had three rum
and Cokes. Then he checked into his room at Edgartown's Shiretown Inn.
After a Heineken or two, at around seven, he returned to the Chappaquiddick cottage.

Ted's cousin Joseph Gargan had rented the cottage and put together the
party. Rose and Joe had always had someone around their youngest son to
watch out for him. Gargan's parents had died when he was young, and Joe
and Rose had taken care of the boy. Gargan knew from an early age that his
job was to watch out for his younger cousin and see that he did not get in
trouble. Gargan now had a law degree and was vice president of a Hyannis
bank, but he was still watching out for Ted.

Gargan had invited a number of men who had worked on the campaign.
Many of them had other plans, and in the end the group consisted of six
married men and six single women. As suspect as that may have appeared,
the party was nothing but a sentimental reunion. For the most part these
were Bobby's people, and there was a sense among several of them that this
evening was one reunion too many. Nothing, not booze, not rock 'n' roll on
the radio, not jokes, not sweet recollections, could make anyone forget that
Bobby was gone.

Most of the guests had several drinks, and Crimmins had more than
that. At around eleven, Ted took the Oldsmobile keys from the driver and
left with one of the women, twenty-eight-year-old Mary Jo Kopechne. Ted
had such a reputation as a womanizer that the mere fact that he was alone
with a woman in a car late at night on an isolated island would later lead to
an inevitable conclusion. Ted could never raise the excuse that he was in love
with Helga Wagner, and Mary Jo was unlike the women to whom he was
generally attracted. He liked stunning, sexy women, and that was not
Kopechne. She had the fresh good looks of a high school prom princess,
pretty but not beautiful, educated but not sophisticated. A good Catholic,
she had graduated from a small Catholic college in New Jersey and lived in
the fashion of the archetypal Washington single woman—in a Georgetown
house sharing the rent with a group of secretaries and others. Since Bobby's

death, she had gone to work for Matt Reese and Associates, political consultants, and was on her way to a successful professional career. She did not drink very often, though she had had several drinks that evening, and had a reputation as a demure, chaste young woman. As the two headed off, Ted did not even know Mary Jo's last name.

Ted told Crimmins that he and Mary Jo were going back to Edgartown by themselves. When Ted's mother, Rose, reflected on what happened that evening, she said that Ted's mistake was that he had not followed one of his father's fundamental axioms: when with a woman, always have another man in the car. "Of course, I was surprised that none of the men went with him that night," Rose said in 1971 to Robert Coughlan, with whom she was writing her autobiography, "Mr. Kennedy told me years ago he always had this man, Eddie Moore, with him so that no matter what happened Eddie Moore would be there. And that night it would have been so easy for one of the men to step in the front [seat] of the car with Teddy, and I didn't understand why one of them didn't have the gumption enough—or why he didn't have gumption enough—to think of that. They all were standing around there. I thought that was quite stupid, the way Teddy had been brought up, and all of them knew enough about politics, if you go out with a woman, you or the chauffeur at night and there is an accident, you are sure to be blamed."

Ted steered the 1967 Oldsmobile down the paved road back toward the ferry landing, the headlights cutting a clear path in front of him. Then he made a sharp turn and headed up the dirt road toward the beach where he had swum earlier that day. There have been books written about what happened next and why Ted made that turn, but the best answer may be simply to stand on that isolated road and look both ways. The paved road to the ferry landing is the only route across the island, and the signs to Edgartown are clearly marked. To turn up Dike Road toward the beach, a driver has to make a ninety-degree turn to the right onto the unmarked track instead of following the paved road left to the ferry landing.

Ted said later that he made a mistake and turned onto Dike Road. By his estimation, he traveled toward the beach at twenty miles an hour, a fast speed given the dark night and the dirt road. Driving at that speed, he and his passenger would have been jostled up and down on the rough surface, the next thing to a siren going off to alert him of his error. If he did in fact unknowingly blunder down the wrong road and did not even realize it for the two minutes it takes to reach the bridge, the chances are that he was drunk or otherwise disoriented. The best argument that he was telling the truth and that he had made an honest mistake was that he was equally unaware of what lay ahead or he would have slowed down.

Ted saw a narrow bridge that angled leftward. He tried to put on the brakes, but the car plunged off the right side of the rail-less bridge, overturning and landing roof down, submerged in the water. It was as shattering a moment as the plane crash five years before. One instant he was driving along a country road, and the next he was jolted enough to suffer a mild concussion and find himself lying upside down in a car filling with water. He was a strong, athletic man, far more vigorous than most thirty-seven-year-olds, and he managed to work his way out of the car and to the surface to fall exhausted on the beach. He says that when he realized that Mary Jo had not escaped, he dove down seven or eight times, trying to rescue her from the heavy black currents.

No one besides Ted will ever know for sure why he made that turn, or how hard he tried to save Mary Jo's life. What is known is that Ted walked back to the rented cottage, and for that there is good reason to fault him. "A lot of people think that the one thing Ted should have done is to run to anybody and say, 'For Christ's sake, help me out, this girl's drowning!'" reflected Sarge two years later to Coughlan. "And he didn't. When Ted could not get the girl out of the car, he went down the road to get help. Why didn't he go to the first place where there was a light on? Why didn't he stop and ask for help? Then it begins to look like instead of worrying about the girl he was worrying about himself, and that's what bugs people, that he connived, manipulated to help himself."

When Ted arrived at the cottage wet and shaken, Gargan recalls his urgent question: "Joey, what are we going to do?" That was his plaintive plea, not, "What am I going to do?" or, "What must I do?" but, "What are we going to do?" meaning, "What are *you* going to do?"

Gargan and Markham drove Ted back to the bridge. Gargan stripped his clothes off and dove down. "I grew up here on the water in Hyannis Port," says Gargan. "If a sailboat turns over, you don't call the Coast Guard. You get in the closest, quickest boat to get out to the boat to get the people from underneath the boat. That was my immediate reaction that night. I did everything that I thought I could do under the circumstances. I swam into the car and attempted to find Mary Jo. I had no equipment except my own body. I have the ability to swim and to hold my breath under difficult circumstances, with a current running through there and a car that was totally underwater. I didn't even know whether she was in the car. I never did feel her."

While Gargan continued a rescue attempt, Ted lay on the ground, his hands clasped behind his head, knees up, moaning, "Oh, my God," as Gargan remembered the evening to author Leo Damore. "What am I going to do? I just can't believe this happened! What am I going to do? What can I do?" He did not ask, "What have I done?" or, "What about Mary Jo?" His defenders later would argue that his light concussion made him not responsible for his actions. He had been shaken up, but he was nonetheless able to focus on his own future and on not harming the Kennedy political legacy.

Gargan says that he stood up to Ted. He told him that there was no way out. He had to own up to what had happened. Ted continued to seek a way to get out of responsibility for the accident. Gargan is adamant, however, that Ted did not ask him to take the blame. "If he'd have asked me to take the blame, he would not be my friend now," Gargan says. "He didn't ask me to do that."

The threesome drove to the ferry landing, where the discussion continued. Rationally, there was no way they could hide this. If they said that Mary Jo was driving alone, they would have to involve everyone at the party in their conspiracy. And what if someone had seen them at the bridge? Ted's mother and other Kennedys would soon condemn Gargan for setting up the party and then not having the good sense to jump into Ted's car when he left. In many news stories, he was portrayed as a fumbling buffoon. The larger truth is that if Gargan had agreed to try to cover up Ted's role, it would eventually have been found out, and the Kennedys would probably have been destroyed as a force in American public life.

Ted listened more than he talked. Then, without warning, he jumped from the car, dove into the water, and swam toward Edgartown. Gargan believes that as much as Ted may not have liked the idea of assuming responsibility, he knew that he would have to report the accident. "Markham and I understood that he was going to report the accident and that's why he dove in the water," says Gargan. "And the fact that he did not do that immediately was probably because I did not find her in the car. Whether Ted hoped she would show up, or what other kind of thing might have happened, there was no real clarity on the situation when he dove into the water."

The other shore lay only five hundred feet away. Ted was wearing his pants and shirt, and a back brace. There was a fierce undertow and a tide that could have swept him away at some points. He was halfway across when the undertow grasped him and then inexplicably let him go. He struggled onward, his hands finally touching the Edgartown pier, and he was able to pull himself up. He lurched through the deserted streets to his room at the Shiretown Inn. At around 2:30 A.M., he came out of his room in fresh clothes, asking an employee the time. It may be that he simply wanted to know the hour, but since he did not report the accident, he may also have been establishing evidence that he was in the hotel.

G argan and Markham took the ferry across to Edgartown. The next morning, shortly after 9:00 A.M., they took the ferry back to Chappaquiddick along with Ted so that they could use the isolated public phone on the ferry dock. The first personal call he made after the accident was to Helga Wagner, his mistress. She said later that he called her only to get the phone number for his brother-in-law, Steve Smith, the family consigliore. Helga knew the number because she was supposed to go off vacationing with Steve and his wife, Jean, Ted's sister.

Ted did not call Joan. If he was telling the truth and had not been out with Mary Jo for a romantic interlude, then what an irony it was: for once he was innocent, and much of the world would think him guilty, including probably his own wife.

After talking with advisers, who presumably told him to report the accident and gave him guidance on a statement, he took the ferry back to Edgartown and walked into the police station in the town hall. There was no more formidable figure in the whole state than Senator Kennedy. Standing there in his blue slacks, white polo shirt, and sneakers, he was a man of presence, with nary a blemish on his handsome countenance and a manner that called for attention.

Dominick Arena, the Edgartown police chief, had learned to be deferential to the wealthy summer guests. He allowed Ted to submit an unsigned statement. Neither Ted nor Markham knew how to spell Mary Jo's last name, and that was left blank. Nor did the report mention that Ted, Gargan, and Markham had gone back to the submerged car after the accident; it said only that Ted had "asked for someone to bring me back to Edgartown." Arena did not ask Ted a single question about the accident, and since the senator did not sign his statement, it was not even legally valid. After handing Arena the statement, Ted and Markham left by a back door and immediately took a private jet to Hyannis.

"In looking back, it isn't so much that I think I did things wrong as much as I assumed a lot of things," Arena reflects. "I took so many things for granted. I was convinced that I was really going to get more information from him later. Yes, it looks as though I'm thinking the guy's a senator, I'm letting him out the back door and letting him go home. But I have to honestly admit, and very naively probably, I really thought I'd get a chance to ask more questions."

At Hyannis Port, Ted went immediately to his father's room, where the old man lay in his bed, as he had for so many years. Joe ate only ice cream and baby food. His legs were shriveled up, bones without muscles.

"Dad," Ted said, as Joe's nurse, Rita Dallas, stood nearby. "I'm in some trouble. There's been an accident, and you're going to hear all sorts of things about me from now on. Terrible things. But, Dad, I want you to know that they are not true. It was an accident. I'm telling you the truth, Dad; it was an accident." Joe nodded at his son and shut his eyes.

Ted also had a private conversation with his mother at the flagpole in the middle of the lawn, where nobody could possibly hear them. Ted was Rose's favorite son, a last unexpected blessing, and when she looked back at the tragedy, she blamed not her son but the others. "I do not understand why Joe Gargan or Markham did not report the matter to the police even if Ted did not have sense enough or control enough to do so," she wrote in a private note, "especially when the body of the girl was in the car. That is what seems so unforgivable and brutal to me." There in the words of the Kennedy matriarch was the moral imperative of her family. What was so "unforgivable and brutal" was not Mary Jo's death, or Ted's failure to report the accident, but the fact that Gargan and Markham did not on their own make the accident report themselves.

On one of the first days after the accident, Ted went for a walk on the beach with one of his old friends who had flown to Hyannis. "Goddamn it," he said. "All my life I've done everything they've told me to do. Everything. I'm finished. I'm going to marry Helga and leave the Senate and go off with her."

Joan had stuck with her marriage in part because she identified herself so profoundly as a Kennedy woman, a member of an esteemed, privileged family. But now she was suffering a public shaming. Joan nonetheless traveled with Ted to the courthouse in Edgartown, where he pled guilty to fleeing an accident and received a two-month suspended sentence. She also flew with him to Mary Jo's funeral in Plymouth, Pennsylvania. She held his arm and looked strong, but she was paying a golden price. Her mother, Ginny, had a liquor problem, and when she learned of the accident, she began drinking so heavily that she had to be institutionalized. Joan too found a deeper solace in the bottle now, its magical elixir blotting out the world.

The advisers who arrived to discuss Ted's future were mainly Jack's and Bobby's men. They were there sitting in the living room to protect the Kennedy legacy and the man who had so besmirched it. They were there to try to salvage their own hopes of returning to the White House as well. As Ted wandered the grounds, they debated his future almost as if he were not there. He moved in and out of the discussions, hearing a complicated litany of possibilities and problems.

Sarge was one of the few family advisers who was not there. "Look, Eunice, the worst thing that anybody could do for Teddy would be to have me, the United States Ambassador to France, flying in to Hyannis Port to consult with him about what the hell he should do about Chappaquiddick," Sarge told his wife. "For God's sake the thing to do is not to dramatize the situation but to *cool* it." Sarge considered this onslaught of advisers "the worst possible thing for Teddy. . . . I would have said 'Everyone stay the hell away.' I would write my own statement. I'd go on television and give my own speech. I don't give a shit what anybody thinks. All I want to do is talk to the people of Massachusetts *myself*."

Ted left it to others to decide what he should say, how he should say it, and when he should speak. The speech these advisers wrote for him was the worst major address any Kennedy had ever given. In his fifteen-minute tele-vised address, Mary Jo became someone "all of us tried to help . . . [we] feel that she still had a home with the Kennedy family." He dramatized his own story, though not the death of Mary Jo ("water entered my lungs and I actu-ally felt a sensation of drowning, but somehow I struggled to the surface alive") He played long maudlin notes, saying that during that long night he had mused "whether some awful curse actually did hang over all the Kennedys." In the end, he asked the voters of Massachusetts to decide whether he should resign or stay in office. His speechwriters garnished the speech with a quotation from *Profiles in Courage,* and he concluded: "I pray that I can have the courage to make the right decision."

Among many Massachusetts Democrats, the Kennedys were still largely beyond rebuke. The telephone calls, the cables, and the editorials overwhelmingly asked that he return to Washington. Soon afterward Ted flew to Paris to spend some time with the Shrivers. Dr. Herbert Kramer, a speechwriter and public relations expert, was there too, working with Sarge on a special project. "You'd think that there would be some remorse," says

Kramer. "But no. He walked around the living room naked, drunk most of the time. There was a brilliant guy there, an economist, and we took a walk near Notre Dame, and I said, 'What Teddy ought to do is to resign from the Senate and take a year, two years, to do some good work to redeem himself, to suggest that night at Chappaquiddick had been an aberration.' And the man said, 'You got to be kidding. This isn't a moral animal. It's a political animal. He's not going to leave the Senate.'"

Ted was not the sort of man to do penance on his knees. That his conduct after Chappaquiddick was at times as outrageous as his conduct before did not mean that Ted was unaware of what had been lost at the bridge that night. He was by far the most sentimental of the Kennedy men. Ted saw his brothers as unsullied heroes. He had betrayed the legacy they had given him, staining their names as well as his.

That he talked so little about Chappaquiddick in the years to come, and so inarticulately when he did, was not necessarily the mark of a dissembler. It was a mark of how painful a matter it was to him and always would be. It would surely have been expected that as soon as his young son Patrick was old enough, he would have taken the boy aside and told him about the events before the boy was taunted with the word "Chappaquiddick" on the school playground. Ted did not do so, and he avoided the subject inside the family as much as he did outside it. It was not until Patrick was an adult that Ted talked to his youngest son about the day, and then only because the anniversary was coming up, a time when, in Patrick's words, "the media made it into a media circus."

"He told me there isn't a day that goes by that he doesn't live with it," Patrick recalled his father saying.

CHAPTER FOURTEEN

Boys' Lives

T he summer of 1969 was a time of loss not only for Ted but also for his nephews. They had heard Ted's words about a Kennedy curse in his speech, and many of them wondered if they were indeed a cursed clan, doomed to endless tragedy and mishap. That was a terrifying prospect for an adolescent. It made it even worse that as Kennedys they were not supposed to be afraid. They could not allow themselves to acknowledge their fear, though in doing so they would have diminished its power. If they shared a common dark destiny, then they were not responsible for their lives. No matter what they did, who they hurt, how much they faltered, they were not to blame. They were only playing out their preordained destiny, their knees buckling under the burden of inheritance they bore.

In the end, only those Kennedy sons who grasped that they were responsible for their lives were able to come close to controlling their own destinies without being forever haunted by their family name and legacy. As adults, some of them would find a measure of freedom only by moving across the country, choosing a profession far from politics, or dropping out of the public drama of the Kennedys. Whatever the cost of these choices, it seemed a pittance.

As it was, several of them did whatever they could to obliterate their fears and anxieties about their future as a Kennedy man. Ethel had grown up in a family in which alcohol was treated like bottled water. The Skakels probably had a propensity for alcoholism in their genes, but it was their home that taught lessons that should not have been learned.

The Kennedy patriarch had sensed that alcohol was the most dangerous pleasure of all, part of the stereotype of the Irish American. Joe's father had owned a tavern, a business in which many men failed because they drank up their profits. Joe's father did not drink, and he succeeded as well as anyone in East Boston, going into the wholesale liquor business, selling goods he would not touch. Joe was not a drinker either, even though he sold liquor during Prohibition and afterward became a major importer. He promised his sons a thousand dollars if they would not drink until they were twenty-one. At the Kennedy table, there was only one drink before dinner, and woe betide the guest who asked for more or dared to appear inebriated. If this was the Irish vice, it was not a vice that would touch the Kennedys.

There is no way that Joe, Bobby Jr., David, and the others could have had any sense of the immense personal danger that alcohol and drugs held for them. Rather, they learned that alcohol was as much a part of adult life as driving a car, and it was a natural segue for them to move on to drugs. The youths entered the cultural revolution of the sixties and found in drugs an adventure that took them far beyond the provincial borders of the village. When you drop acid, you do not need to fly to Marrakech or Bangkok to test your limits; the streets of Hyannis Port become labyrinths of danger.

Fifteen-year-old Bobby dressed in black and strode through the streets of Hyannis Port with a hawk on his arm. Bobby, David, Chris Lawford, and other friends challenged the civilities of the quiet village. They heaved water balloons at passing cars and then scurried away. They untied boats from the dock. They ran into the street, slapping the fenders of cars, and falling to the ground while their cohorts screamed, "You've killed a Kennedy." They stuck potatoes into the exhaust pipes of cars at stoplights, and waited for the explosion. They threatened a neighbor girl with a knife at her birthday party, stripped her of her new watch, and ran off. The neighbors gossiped about how the boys had bound up the family cook, hung her on a tree, and terrified the poor woman by saying they were going to set her on fire.

One evening the youths roamed through the quiet streets shooting off their BB guns at the windows of homes and firing pellets at the window of a church. The Kennedys were outsiders to the friendly, communal life of the village. Even if the townspeople did not have much to do with the Kennedys, the residents felt that Ethel was still a mourning widow and deserved forbearance. But this latest assault was beyond the pale, and a group set out to confront Ethel with her sons' assault.

"I am a widow here alone," she said, as if that excused her sons' behavior. Her sons needed to be called to account, to face their accusers, make restitution, and repair the damage they had done. If their father had been alive, that would have happened, but if Bobby had been alive, they would probably not have indulged in such antics.

They had all been pranksters once. Ethel was notorious for throwing guests into the pool at her famous parties, and Ted stole a bus and drove it back to the compound when Jack was getting married. There had been such laughter, such fun, in those pranks, however, not the joylessness of these newest activities.

T ed continued to try to be a surrogate father to Ethel's children. That was at times a thankless task, for the sons were growing up into difficult young men. At Milton, where Joe was a senior, his father had been an outsider too, but his closest friend had been Dave Hackett, a great athlete. Joe was part of an alienated group of outcasts. They were a sad, dispirited version of the Muckers, the infamous group that Jack had help found at Choate.

Ted had no softhearted rationalizations for his nephew's poor grades and unacceptable behavior, and he advised the administration to give him a "good swift kick in the ass." But in the late 1960s it would have taken a full-time father—and an insightful, trusted one at that—indeed to have entered the adolescent world of Bobby's three oldest boys. If he had, Ted would have learned that Joe's major extramural activity was smoking cigarettes and occasionally marijuana in the cemetery across from the school, while Bobby thought it a cool idea to buy LSD and had initiated his little brother David by daring him to take mescaline. David told author David Horowitz that during their drug-taking Bobby had screamed at him: "You're dying, just like Daddy." The words so terrified the youth that he was convulsed with hallucinations.

The two eldest sons competed and fought much the way Joe Jr. and Jack had. Joe, however, was not like Joe Jr., who had been an exuberant youth, as good on the fields of sport as in the classroom. Joe found reading painfully difficult, pointing out words sometimes as if they were a foreign language. He was dyslexic, with learning disabilities—he would have benefited immeasurably from a special school or classes that would have brought out his intelligence. He was not slow, however, in his quickness to anger. He sometimes seemed like a humorless bully, but often this was just an expression of his territorial imperative, knocking people around on the football field, slamming them down in the house, testing them to see if they could take it.

Joe was unlike anyone else in the family, either his or any of the other Kennedy families. Joe was taunted for being stupid, but he was not stupid at all. He trusted almost no one, and that mistrust was a tool he used to understand the world. He was full of anger, but his blows rarely fell, and when they did rarely with his full strength. He had his counseling and the family advisers, but no one grasped the essential fact that he was dyslexic. "I'm sure I had it but they didn't have dyslexia when I was a kid," says Joe. "I had trouble reading but that wasn't the word they used. I've never taken a test but I'm sure I have it."

He couldn't read easily. And if he couldn't read he could hardly play the intellectual games that his siblings played, or if he played he rarely won. If anything, his dyslexia only increased the anger and frustration that boiled within him. Everyone walked warily around Joe, afraid that he would lash out with his uncontrollable temper. He was all bottled up emotionally. He could not express the anger he felt toward a father who had chosen history over family, and had risen into myth, leaving his son alone. His father was the strength of the family, the discipline, the hope, the order, and he had left a wasteland. His mother was a mythic figure too, and he could not express his true feelings about Ethel.

Bobby, for his part, had a quickness of mind and body that was almost mystifying to his older brother. In his interest in animals and nature he had a world of his own that was like a companion that always traveled with him. In his fights with his brother he preferred a taunting phrase to a fist, and his words could strike hard.

For Joe, Bobby, and their brothers, the central lesson of their father's life was that courage was the ultimate value. One of the great psychological struggles of their lives was to separate recklessness from courage, and physical audacity from moral bravery. As the children tested themselves at Hickory Hill, they continued to have accidents serious enough to require treatment at Georgetown Hospital. In the spring of 1969, eleven-year-old Michael lacerated his temple. Joe sprained an ankle. David broke his thumb. Kathleen had to be hospitalized for a cerebral concussion when she fell off a horse. She had company, for at the same time four-year-old Max was also overnight at Georgetown for an asthma attack.

There was less than a three-year difference in age between Joe, Bobby, and David. The three eldest sons shared far more troubled, traumatic upbringings than their younger brothers. Michael, the next youngest son, was nearly three years David's junior. He was the middle son in many ways. He was close enough to his older brothers to appreciate their problems, but he shared his younger siblings' sunnier view of life. The three youngest— Chris, Max, and Doug—had less difficult childhoods and for the most part loving memories of their mother, older brothers, and early years.

"Joe was terrific about providing some structure of family and discipline upon the brothers and sisters," says Chris. "Michael was always a great athlete. Both Michael and David were phenomenal athletes. We went to a grade school which went from prekindergarten through ninth grade. And we would all be involved in sports there, and you would see everybody play sports. And Michael and David were both required to switch teams at halftime in their respective grades because they were such dominant impact players in every sport they played. Max would read until ten or eleven o'clock every night from when he was in fifth grade on. And I was older than Max, but he developed some of his intellectual powers perhaps earlier than I did. And Douglas really had more friends than anybody else. So I think that each of those guys brought an element to the party in terms of an impact on me that was unique and important."

Bobby was a wonderful older brother to little Chris, nine years his junior, opening to him a world of adventure and daring that most children never saw. The area around Hickory Hill in McLean still had field, farm, and forest, and with Bobby as his guide, Chris imagined himself out in some great wilderness.

The teenage Bobby had become a falconer after a lengthy apprenticeship with a master falconer who lived down the road. The birds required constant care, and the ancient sport was not something in which a youth could indulge in a desultory way. Bobby seemed a magical, mysterious figure to his younger brothers and their friends.

Bobby set out with half a dozen boys of various ages, including Chris. The one boy old enough to have a driver's license drove the group south in Ethel's station wagon to the fields where Dulles Airport now stands. For Chris it was an incomparable time, to be with his brother off on this quest. Bobby cast his gaze back and forth until he spied a red-tailed hawk far in the distance. These carnivores are the largest North American hawks, with a three-and-a-half-foot wingspan. In preparation Bobby and the boys had captured several pigeons and stuffed them into pillowcases. Bobby plucked out one of the frightened pigeons and harnessed it in a special vest covered with hundreds of tiny nooses made of fishing line. He tied the pigeon to a ball of string and, holding on to the string, tossed the bird out the window.

When the hawk swooped down on what appeared to be a wounded pigeon, its talons became embedded in the nooses. Bobby rushed forward and threw a hood over the enraged hawk's head and brought it back to the car. Within two or three weeks Bobby had trained the bird to follow him through the forest to chase rabbits, pheasants, and squirrels and return instantly when he called it to his falconer's glove known as a gauntlet.

When little Max became old enough, he went out with his big brother too and eventually ended up a falconer. He and his brother had a sense of a boys' paradise at Hickory Hill, a rural redoubt where a boy could go and do what he liked, roaming the woods and pastures in search of adventures. "It was a rich childhood," Max recalls wistfully. "We had this huge tractor, a full Cub Low-Boy with four-and-a-half-foot tires. Christopher and I would mow the lawn and rake the leaves. It was a childhood filled with lots of fun, manual labor, sports, and I definitely got in lots of trouble but pretty much of my own making."

T he Kennedy sons had two great faiths, a faith in God and a faith in family. "Much of my understanding of [religious] faith comes from neither the doctrine nor the history of the Church, but the prayer and faith you bring that comes from watching my mother in church, watching her on her knees, going to chapel every day," reflects Chris. "It was watching my Aunt Eunice at St. Francis up in Hyannis Port. It was watching my grandmother going to church every day and praying with her Sundays. It was having mass at our home, each Sunday after my grandmother got sick. I can't believe that my children will get the same sense that faith can help you over obstacles, that faith can sustain you as it has sustained me, if they don't see me pray."

Joe had taught his sons and grandsons to have faith in their family and its future. Wherever the young Kennedys went, they knew that their grandfather lay in bed upstairs in the great house. As long as he lived, the family had the sense that his hand was still there guiding them in their advances.

For the nearly eight years since his stroke, Joe had spent much of the time lying in bed watching television or being wheeled downstairs to watch a movie in the screening room. "He was engrossed in television all the time," Rose recalled. "He understood television quite well and when you'd talk about it, he'd raise his head in exclamation and say some terrible expression that they used."

Since Chappaquiddick, Joe had stopped watching television and had largely stopped eating. His eyes were often full of tears. There was no way of knowing whether his tear ducts were malfunctioning or whether his lack of appetite was involuntary. But for a man of his strength of mind, and a body of such resilience, it seemed likely that he sensed that he had nothing left for which to live. Everything he had hoped and worked for was gone. Ted was still there, but he was no longer an heir to the presidency. When Rose reflected on the wreckage of their aspirations spread out before her, she said that maybe they had wanted too much and this was God's judgment.

It was unlikely that Joe sensed any culpability for all the tragedies that had befallen them, or for the sad lives of his grandsons who at times romped on the grounds beneath his window. He was a man who never learned from his major mistakes, because he thought he had made none.

By mid-November, four months after Chappaquiddick, death was near, and Joe's children returned to a half-deserted Hyannis Port, with winter soon upon the land. History would not be kind to the old man lying there. Over the years the portrait of him would darken into a man of merciless ambition, a man who had connived and cheated to reach wealth and power, and who had then pushed his sons to the high places in the world from which they had fallen. That was not how his children saw him. They felt that he had given them everything—money, a name, values, everything. They knew that he was sincere in wanting them to do good work in the world. They did not see the tragic consequences of so much that Joe had wanted for them. They were afraid to look truthfully backward and always looked forward with what some called optimism.

Ted and Jackie spent what would be the last night of Joe's life in his bedroom. On the morning of November 16, 1969, Rose and her children formed a circle around the bed, while the other relatives stepped back. Rose touched a rosary to her eighty-one-year-old husband's lips and placed it in his hand. "Our Father, Who art in Heaven, Hallowed be Thy name," Eunice intoned, beginning the Lord's Prayer. Ted came next, and Jean and Pat and Ethel and Jackie, each speaking a passage from the prayer, and then finally it was Rose's turn. "And deliver us from evil. Amen."

Rose asked for a special white mass in which the priests wear white vestments symbolizing joy. Ted's voice broke when he read the prayer written by his mother, but the most powerful moment of the service came when the one grandchild deputized to be part of the service came forward. Eight-year-old

John strode confidently to the altar and in a strong, certain voice recited the words of the Twenty-third Psalm.

> The Lord is my Shepherd; I shall not want.
> He maketh me to lie down in green pastures;
> He leadeth me beside the still waters;
> He restoreth my soul.
> He leadeth me in the paths of righteousness for His name's sake.

Sailing Beyond the Sunset

or Ted and his sons and almost all the young Kennedys, Hyannis Port was the one place where they believed they could truly be themselves. When the young Kennedys left in the fall to return to their prep schools, they remained within a psychological compound, apart from those around them. They knew that one day they were supposed to move out into the greater world and take hold of power and fame, claiming the family destiny. They did not know who to trust, and they came naturally to feel comfortable primarily among their brothers and cousins. They moved from school to school, romance to romance, acquaintance to acquaintance, adventure to adventure.

But they always returned to Hyannis Port, where the important games of life were played. In the summer of 1970, the eldest Kennedy sons were now in their midteens: Joe, seventeen; Bobby Kennedy, sixteen; David, fifteen; Bobby Shriver, sixteen; and Chris Lawford, fifteen. They were old enough that the competitions were fought even more fiercely, and on fields of play beyond the green grass where their fathers had played their games.

The great competition among the boys was between the seven sons of Robert and Ethel and the four sons of Sarge and Eunice Shriver. The Shrivers had inherited their father's abilities on the tennis court. Rarely did they lose crucial matches, and all the matches were crucial. "I always felt we had to struggle to keep up with each other," says Bobby Shriver. "Beating other people at a sport didn't count. The only thing that mattered was, could you beat your cousin or the brother who was closest to you in age. Am I a better tennis player than Bobby Kennedy? Okay, fine. I can relax."

Tennis was a worthy testing ground, but the football field was the arena of glory for a Kennedy, and here the Robert Kennedy clan gave no quarter to the Shriver youths. The Shrivers were equally ferocious. The boys may only have been playing touch football, but the blocking was brutal.

Bobby Kennedy and Bobby Shriver were the young champions of their clans. They were tall, lean, smart, and handsome. Bobby Shriver did not have a scrapbook of adventures as large as his cousin's, but he had his tales to tell too. He had gone off to live in Paris as the ambassador's son and attended a French-speaking school. He had returned inches taller, his hair down to his shoulders, and wearing exotic French suits. And that was piddling stuff compared to the news he brought from the City of Light: at the age of fourteen he had lost his virginity to a lovely American girl.

Bobby Shriver was fluent in French and affected a European sophistication. He had his father's garrulousness, but unlike with Sarge, much of what Bobby said had an edge to it. Like Bobby Kennedy, he was full of his own deep anger, part of it directed at those family members who considered his father a lesser figure than the blood Kennedys.

Instead of letting their eldest son continue to study in France, the Shrivers insisted that Bobby return to the United States and enter Phillips Exeter Academy in New Hampshire in the fall of 1969. He soon found himself struggling to keep up at the elite prep school and felt isolated from his family. "There was no room at Exeter for glibness or just sort of average smarts," Bobby Shriver recalls, the memories still painful. "That's scary. Particularly if you thought of yourself as having been the smartest kid in the class. That winds you up. You never recover from that. It's terrifying. It still gives me the creeps. But I realized, 'I'm going to fight for my life. I'm going to die here if I don't stay in the library until midnight every night.'"

In the summers Bobby Shriver had to fight for his life too, struggling to beat his cousins, whatever the game, and to have an identity of his own, not just as a Kennedy. In truth, it was almost impossible to compete with Bobby Kennedy, who had his own teenage gang that he led beyond the precincts of

Hyannis Port. Bobby Kennedy was a lithe, stealthy young man who could climb in and out of windows with all the predatory skill of a cat burglar and run loping down the dark streets, no more than a shadow in the night. When his mother threw him out of the house, he broke into the basement of the Newman house across the street and slept there. When he was discovered, he found another hiding place.

Bobby Kennedy moved beyond the circumscribed world of the Kennedys and befriended the kind of young men whom the Kennedy sons had been warned to avoid. One of Bobby's new friends was twenty-one-year-old Doug Spooner, who earned what money he had betting with easy marks on the golf course. Bobby identified with his new friend, and as soon as school was over he set out with Spooner seeking to live by his wits. Bobby and Doug hustled black players in a tough Cambridge pool hall. When they tried to get one of the losers to pay, he came after them with a machete. The assailant struck Bobby with a glancing blow, and the two ran away, just another story to add to the growing legend of Bobby.

One of Bobby's favorite poems was Alfred Lord Tennyson's *Ulysses*, a copy of which his father had given him in the last months of his life, underlining the last stanza. It is perhaps understandable why his father had found such meaning in the Tennyson epic as he set out on his own last fatal quest. But it was hardly a young man's poem, this tale of an old man always roaming with a "hungry heart," setting out a last time "To sail beyond the sunset, and the baths/Of all the western stars, until I die."

When Bobby returned to the Cape in July, he did not even go to his own home but stayed with the Kelleys, a large Hyannis Port family with whom the young Kennedys were especially close. One evening Bobby sat in an abandoned car with John Kelley, writing a letter to Mrs. Kelley as the duo contemplated leaving on a trip. "We feel rejected because of your apparent blaze [sic] attitude to the news we gave to you last night of our departure. . . . There is adventure in the air and the high seas offer quite an enigma for an unsatisfied mind."

Bobby Kennedy also befriended a hip twenty-one-year-old taxi driver, Andy Moes, who was willing to drive Bobby, Bobby Shriver, and their friends to Hyannis and elsewhere. Bobby Kennedy was a studied connoisseur of drugs. Not to be outdone, Bobby Shriver announced that he too had smoked his first marijuana the previous school year at Exeter, though that was not true and he was merely trying to keep up with his cousin. The sixteen-year-old youths at times smoked the weed as Moes drove them around. The young taxi driver said he was interested himself in scoring marijuana. "He was whining that he had this girlfriend," Bobby Shriver says, "and that he had to have a joint to get laid and all that. Finally he offered me ten dollars. I said, 'Shit! Ten dollars for one joint? I'll take it.'" Both Bobby Kennedy and Bobby Shriver had their little stashes, and they paid for their rides by giving Moes a joint or two.

Both Bobbys spent as little time as they could at home with their families, but sometimes they had no choice. In early August, on the evening when Ethel's cook was off, her family had dinner at the Shrivers. Bobby Kennedy had been avoiding his mother, and the evening started out civil enough that it appeared there might be a reconciliation between mother and son.

In the midst of dinner the doorbell rang. Outside stood a group of police and several squad cars. The officer in charge announced to the startled group that Moes was a cop working narcotics undercover, and Bobby Kennedy and Bobby Shriver were being charged with possession of marijuana. Ethel was hysterical. As soon as the police left, she ran after her son, chasing him around the table and hitting him as he attempted to fight back. "You've dragged your family's name through the mud!" she screamed. She did not ask him how this had happened, or whether it was even true.

By all appearances, the two youths had been set up because they were Kennedys, and there were suspicions afterward that Moes had been employed by President Nixon's minions to discredit the Kennedys. In any event, the story was played over the front pages of the nation's papers as if the boys were major criminals. In the end, the youths were sentenced to a year's probation.

Sarge was not in Hyannis Port when the police showed up, but he immediately flew back from California. Three decades later, when Bobby Shriver looked back on his early years, he saw that there was no more crucial moment in his life than this. He was blessed in the father he had, and in the subtle, determined way in which Sarge handled the crisis. He gently pushed his eldest son away from danger and toward a different kind of life. "My father said, 'Look, you're not a bad kid. This is going to work out. I'm going to look after it, don't worry about it. And don't listen to anything anybody else says, because I'm here and I'm your father and I'm going to take care of you. Don't worry about it. And when this is over, we're going to go to California, and you're going to play a lot of tennis and we're going to get out of here,' " Bobby recalls. "And I said, 'Okay.' It was madness there for the first twenty-four hours. We were all getting our hair cut, people were screaming and yelling. And the steadiness, the nonmadness of my father was a big thing. It was a crucial moment in my life. My other cousins did not have a dad there. They had a much more difficult time than I did."

During the Kennedy administration, the children had played together on the expanses of the compound lawn, and it was hard for an outsider to tell from which family each child came. In the years since, the Kennedys had tried to keep the commonality of spirit alive, but the other families felt that it was simply too dangerous allowing their sons to hang around Ethel's children. The Shrivers gently nudged Bobby and their other children away from Ethel's family. The Smiths chose to protect their children by building a summer home on Long Island where Steve Jr. and Willie would not even see their notorious cousins. Jackie's and Ethel's houses were separated only by a tree, but Jackie made sure that her children kept away from their cousins next door.

R ose was there every day in the summers at Hyannis Port, a unique trove of family history and lore whom her grandchildren approached with measured steps. Although in her early eighties, she remained a woman of remarkable vigor and discipline. She still continued to learn French vocabulary and memorize inspirational sayings. She swam out by the breakwater as late as November, when few others entered the cold Atlantic.

Rose's willpower and self-control were the most awesome qualities about the tiny wizened woman. The grandchildren all knew from their parents about the coat hangers with which she had disciplined her children. These tales by their very utterance were often sufficient to whip them into shape. When Rose came down to the Shrivers' for a visit soon after the birth of the fifth Shriver child, Anthony, the older children went around collecting coat hangers and throwing them out, lest Rose decide to demonstrate her legendary technique on their vulnerable skins.

Her grandsons approached lunch at Rose's with trepidation, knowing that they would be lectured to sit ramrod straight at the dining room table. Joe felt that lunch at Grandmother's was like "being back in third grade." When Rose asked Joe how he felt about God, her eldest grandson told her the truth: such matters were not in the forefront of his mind. His sisters had been pushed toward the Church, but he had little interest in Church or God. Rose wanted to send the heathen youth to the priests to be fed some badly needed faith, and then she was back to tales of her martyred sons. "She tells you these stories about my father," Joe reflected, "and you have the feeling that she's telling you this so we can emulate them." That was precisely right, and many of the young Kennedys couldn't wait to be out of there, off to play football, hit on a local chick, or hang out, away from these tedious admonitions about what they should do with their lives. For all their professed love for their revered mother and grandmother, most of Rose's children and grandchildren spent little time with the matriarch.

During the summers in the early 1970s, one Kennedy second cousin, Kerry McCarthy, spent many days at the house with her beloved Aunt Rose. Kerry saw that most of Rose's grandsons had in some measure lost the connection between themselves and the world. Willie Smith came down late to breakfast one morning, upset when the cook refused to prepare breakfast for him. "For God's sake, Willie, go down and put in some toast and fix cereal," admonished Kerry, who was only three years older than her second cousin. "If you missed the scrambled eggs, so be it."

"But I don't know how to do it," Willie said, close to tears. Kerry showed the youth how to make toast and presented him with three boxes of cereals from which to choose, and Willie had his breakfast.

Kerry was present sometimes when Rose had lunch with Ted's youngest son, Patrick. He was a weak, asthmatic, soft-spoken child who appeared overwhelmed by his parents' troubled lives and the burden of his name. Rose wanted the boy to learn about his namesake, his paternal great-grandfather Patrick Joseph Kennedy. Nervous little Patrick listened to his grandmother tell stories of a man whose energy and vigor seemed so far from Patrick's closeted world.

Patrick had a serious asthmatic condition that made him weak and vulnerable. His classmates took prime delight tormenting him. Patrick recalls how his father "said some people were always going to hang around me because I came from a special family, and others were going to heckle me because they are jealous and they think they are missing something from their lives." His father was telling Patrick that he couldn't trust anyone, a judgment that Ted had long since arrived at himself. The heckling, as Ted saw it, was only a mark of Patrick's superiority. You didn't have to be a young Kennedy to be heckled in school or to have your peers ingratiate themselves with you. What Ted took as a tool to understanding was equally a psychological chit that his son could use to dismiss criticisms of his life.

His parents' struggle over Patrick's treatment was in some measure a struggle over the meaning of their lives. Although Joan was proud of being a Kennedy, she was equally proud that she was different. She was what *they* called weak and vulnerable, while she considered herself sensitive with an open awareness of the pain and nuances of life. As much as she could, she sheltered Patrick and in a sense celebrated what she considered his awareness of the price of life among the Kennedys. Ted would have none of that. He too had suffered as a boy, but he had tucked away his whiny weaknesses and become a man. He would not have a son who was any less. One summer in the midseventies Ted hired a German governess. The martinet put Patrick through six hours a day of exercise and studies, a regimen that ended only when Rose called a halt to it.

Ted might have admitted that the German governess had been a bit too much, but he was not going to have Patrick sitting around coddled by his mother. He took Patrick with him to speeches and public meetings, events that would have produced anxieties in the healthiest of little boys. When Patrick returned from these events, he was breathing as if he had just returned from a long run. Joan would give him his drugs and hook him up to the asthma machine. "You're just giving in to him," Joan recalls Ted saying, as Patrick breathed in pure oxygen.

Teddy had an emotional shrewdness that his younger brother did not have. He was endlessly gregarious, and rarely did anyone see his vulnerabilities. On one occasion a group was sitting around talking about their families, candidly portraying their parents "What's it like for you?" one of his friends asked. "I think you're thinking too much about this," Teddy said, shoving such thoughts away.

What neither Joan nor Teddy saw was that their children needed the emotional truths that both their parents had to teach them. If they could mature into adulthood full of the courage that their father espoused and the expressive sensitivity of their mother, they would have qualities that would serve them well in the inevitable struggles of life.

As Kerry McCarthy saw it, the loss of the connection between the word and the deed, the idea and the action, had been suffered not only by the children but by Ted. Kerry had what she called "a full-blown love-hate relationship" with her second cousin. At times Ted was a warm, irresistible presence when he would sweep into the living room and embrace Kerry, thanking her for all the time she was spending with his mother. The next time he saw her he might order her to mix him a drink, as if she were the barmaid.

Of all the emotions that described those summers at the compound, what Kerry experienced again and again was an overwhelming sense of sorrow. The family had what she considered a glorious inheritance, of which money was the least of it, and they were emptying it out like a vintage wine into the gutter. "I've seen their anger and bitterness, but what frightened me was their tremendous sense of sadness," Kerry recalls. "David sobbing at the dinner table and others laughing. Bobby screaming and saying things and then being such a sweetheart to other people. There was a tremendous anger in them. There was no telling the boys. I remember once when Jack's old friend Lem Billings and Bobby Kennedy had been drunk and gotten cut up and Lem said, 'I can't say no to Bobby.' Nobody could say no.

"There were two who struck me as different. Michael was a great kid, and he was the leader of his young brothers, Doug and Chris. And John was different too. Rose would sit the grandsons down and tell them to do something or admonish them about something. And the others would sit there politely, no matter what they intended to do behind her back. But little John would pipe up: 'How come, Grandmother?' He wanted to know why, and he was blessedly unafraid. 'How come, Grandmother?'"

CHAPTER SIXTEEN

Running Free

I n September 1968, Jackie enrolled seven-year-old John in Collegiate, a private school on Manhattan's West Side that had a more
eclectic mix of students than its East Side rivals. He was transferred
either because he was considered too immature to be promoted to the third
grade or because he was "restless," "inattentive," and "disruptive" in class.

John set out each morning wearing the Collegiate blazer with his thicket
of brown hair brushed down across his forehead. There was always some
thing mildly askew about him, a shirttail sticking out of his pants, a sock at
half-mast, a collar unbuttoned. Unlike most of the boys, he did not arrive in
the school bus but was ferried across town in a nondescript Oldsmobile by a
Secret Service agent.

Many of the Collegiate parents were obsessed with their children getting into the best colleges, and these students were often years ahead of those
in public schools. In that context, John stood out for his seeming indifference
and academic mediocrity. He had a hard time sitting still in class. He got
bored. He looked out the window. His mind wandered. He blew exams. His
father had not been that different in his early years, but Jack had shown a
precocious intellectuality that John lacked. He couldn't wait for the time to
head out to play football in Central Park, where he was just another one of
the boys, at least until the Secret Service men herded him into their car while
the other youths got on the Collegiate bus. That was just another way that he
was forced to be different when all he wanted was to be like the other boys.

Academics had become an obstacle course that John didn't like running. There were already those who viewed him as prima facie evidence of the decline of the Kennedys, a self-indulgent rich boy who wouldn't study in the classroom and on the playing field was never the teammate to choose if you were concerned about winning.

John knew he was not stupid. Not until he had run the gauntlet of elementary school, prep school, college, and law school did he come to his own understanding of his problem. He was well into his thirties when he shared a series of books and the revelations within them with a friend who had the same difficulties he did.

The books talked about a condition known as attention deficit disorder, which had not been widely understood when John was in school. He learned that ADD is a neurological condition that manifests itself in a number of symptoms defined by the American Psychiatric Association. "When required to remain seated, a person has difficulty doing so." He had always had that problem. As a child, he squirmed and fidgeted in his seat at Collegiate. As an adult, he got up at dinner parties and roamed around the room. "Stimuli extraneous to the task at hand are easily distracting." Anything could disrupt him, a plane passing outside, a cough, a person passing by. "Holding attention to a single task or play activity is difficult." He was always zoning out, so much so that one of his friends would say, "Calling John, come back, John, where are you, John?"

"Before a question is completely asked, will often interrupt the questioner with an answer." That was him, always cutting to the chase. "Impulsively jumps into physically dangerous activities without weighing the consequences." All his life he reveled in dangerous, physical challenges. "Easily loses things such as pencils, tools, papers, etc." He was forever misplacing passports, wallets, and papers; if it was important, he was bound to lose it.

John read that ADD was not so much a disability as a different way of seeing the world. It was a way of acting that John shared with many of the most creative people in society. This was not a eureka moment in which he felt he suddenly understood himself. It was a useful insight that helped him understand why he acted the way he did. The metaphor that appealed to him most was found in Thom Hartmann's book *Attention Deficit Disorder: A Different Perception.* Hartmann hypothesizes that those with ADD are descended from hunters in an earlier stage of evolution and that they are competing with people whose ancestors were farmers and are thus better fitted for a regulated, bureaucratic world. Hartmann's idea may be more poetry than science, but to John it was a compelling metaphor. At Collegiate, squirming in his seat, his tie askew, tuning out the teacher, John was a little hunter sitting among farmers.

I n the apartment at 1040 Fifth Avenue, there were no photos of President Kennedy in the living room, but in John's bedroom Jackie had placed a number of framed photos. When the boy got up in the morning, or played with his toys, his father's picture was there, and though he did not talk much about Jack, the images of his father became part of him.

John never had an emotional affinity with Aristotle Onassis. The Greek shipping magnate was sixty-some years old, and his precise age was only one of the many mysteries about his past. Not only did John want to be like everyone else, but he wanted a father like everyone else's. He doubtless would have been excited if his mother had married a man closer in age to her thirty-nine years. Instead, here was this kindly old uncle he called "Mr. Onassis."

Mr. Onassis took John by limousine to Shea Stadium to watch the Mets take on the Orioles in the third game of the World Series in October 1969. Onassis knew nothing about baseball and looked on at what to him was a bizarre American ritual. Onassis flew John over to his island of Skorpios. Although it might have sounded exciting to have a whole island to yourself, it was just adults doing their things. When eleven-year-old John invited his classmate Bob Cramer to join him on the island, Cramer pronounced the two weeks "mostly a big bore." John did have moments that most boys would have envied, such as flying over the island in a helicopter piloted not by a professional but by Onassis's own son, Alexander. John did not see much of his mother's new husband since even when Onassis was in New York, he rarely stayed at the apartment, as if that would have violated some right of privacy. The marriage began to unravel almost as soon as it began.

John was watched over primarily by Marta Sgubin, a young Italian woman who became his and Caroline's governess. "I was told that she didn't speak English," John recalled. "That made me immediately suspicious, because I'd heard this before about the other bilingual au pairs who had helped care for us. It was part of my mother's tireless effort to get my sister and me to learn French. But after a few weeks they'd break down and we'd say a tearful good-bye (in English) and wait for the next one to appear. Marta broke down and started speaking English after about twenty-four hours, and thirty years later she is still part of our family."

Sgubin was not the surrogate mother that Ena Bernard became for Ethel's children, but she watched over the details of the lives of her two charges, the clothes they wore and the homework they did. That is not to say that John and Caroline were merely shuttled in and out of their mother's life, monitored by nannies and governesses until they were sent off to private schools. Nonetheless, during John's formative years Jackie was often away from home, most often because she was managing a marriage to Onassis in Greece while her children lived in New York. Her marriage to Onassis gave Jackie and her children wealth beyond that of any of the other Kennedys, but she also knew that her children would have little choice but to live their lives as public figures. She cared for her children beyond anything else in her life and had an uncanny way of sensing when something was wrong with John.

Despite the mannered realities of life as a wealthy, privileged scion, John remained a devilish child. On one occasion he and Caroline traveled to the Florida Everglades with Peter Beard, a photographer and family friend. They were going snake hunting, just the kind of activity that appealed to John. Unfortunately, they were being hounded by reporters and photographers. John understood the game and its challenge. When John saw one of the photographers asleep in the hotel lobby, he ran over and danced his version of the Rolling Stones' "Jumpin' Jack Flash." The man woke up to see a photo opportunity that would have netted him a month's salary, but by the time he got his camera up, the diminutive Jumpin' Jack Flash was gone.

I n February 1971, Jackie returned to the White House at President Nixon's invitation, largely to give her children a sense of their father's life. Nixon and his two daughters waited outside the Oval Office so that Jackie could have her private moments there with thirteen-year-old Caroline and ten-year-old John. At dinner with the Nixons, Jackie worried that John would be his klutzy self and spill his milk. She primed her son to be careful, but true to his nature he knocked over his glass, and Nixon cleaned up milk spilled by a Kennedy.

After flying home on a presidential jet, Jackie put her son to bed and described the photos in his room. "There you are with Daddy right where the President [Nixon] was describing the Great Seal," she said. "There, on the path where the President accompanies us to our car."

Jackie was not only teaching John to cherish his father's memory and place in history but to respect the office of the presidency. Unlike most of the Kennedys, when Jackie said "the President," she generally meant the current holder of that office, not merely her late husband. This was a lesson that John learned well. He grew up having a fundamental reverence for the institution of the presidency and none of the partisan view that the White House was the Democrats' rightful home occasionally visited by Republican interlopers.

In having John write a lengthy, detailed thank-you letter to the Nixons ("I really loved the dogs. They were so funny. As soon as I came home my dogs kept on sniffing me. Maybe they remember the White House"), Jackie was teaching her son another lesson. While others in the family thought they could get away with vague, boilerplate pleasantries, Jackie knew that manners were in the details. Part of the discipline of a good life was to take the considerable time to thank people as they deserved to be thanked.

John's mother was an aristocratic American whose exquisite manners were the cloak she wore in the world. For all her love of her only son, she was not a mother to embrace him spontaneously. In private she could be different, but Jackie was in part employing a royal model in bringing up John and preparing him to live with the scrutiny that would never leave him. But John missed that cuddling emotionality, as his own father had missed it from his mother.

John's sister Caroline was his other guide to the world and its ways. History left her with scars that he did not bear. She remembered the day her father died in a way he did not. She was suspicious where he was curious, reserved where he was expansive, closed where he was open. As they grew up, she became at times almost a second mother to him, delighted with his boyish excesses, but trying to shield him from a world full of disappointment and uncertainty.

When John looked down on Central Park from his apartment high above Fifth Avenue, he saw a vast wilderness, a place of daring and danger. And he came for the first time to equate wilderness with freedom. In adulthood he would find that freedom on remote Alpine slopes, a cold northern sea, places where he could be free of the constraints of civilization, the prying eyes, the nosy pens and cameras. Central Park was a boy's wilderness. He knew it with a guide's knowledge—every twisted secret path and private place.

One day in May 1974, thirteen-year-old John was racing along East Drive on his expensive ten-speed Bianchi bicycle, accompanied by a friend. A man jumped out and, threatening him with a stick, ordered him off the bike. The thief grabbed John's tennis racket and his bike and rode off. The robber turned out to be a drug addict who in his desperate attempt to feed his habit might easily have knocked John to the pavement and seriously hurt him.

The theft made headlines and could have been a psychological watershed in the boy's life: the moment he realized that he could not live like others and had to accept the inevitable parameters of his life as the son of the martyred president. That did not happen, thanks to both the unbridled spirit of young John and his mother's attitude. Jackie might have reacted by insisting that the Secret Service agents never let her son out of their sight.

The agents wanted to tighten up security, but Jackie insisted on the opposite. "No agent is to be in John's pocket," she admonished them. Her son was not to get in the Secret Service car, and the agents were not to walk with him. They reported that, instead of criticizing the agents for laxness, she told them John "must be allowed to experience life and that unless he is allowed freedom, he'll be a vegetable at the age of 16 when we [the Secret Service] leave him."

Every mother faces that anxious moment when her child goes places where she cannot protect him. For Jackie that anxiety was worse than for most mothers. It was not simply that her husband had died on what appeared to be an innocent street, and that her brother-in-law was shot down five years later, but that there continued to be serious threats against her and her children. There was uncontrollable madness out there beyond her view, and she knew it. Yet she insisted that John have more freedom and adventures than his friends' mothers permitted their sons. In doing so, she gave John an emotional inheritance that was as important as anything in his life. There were few things he did as an adult that did not have their beginnings in the boy running free in Central Park.

CHAPTER SEVENTEEN

A Clearing in the Future

senator has two main constituencies, the voters in his state and his fellow senators. In the months after Chappaquiddick, Ted neglected his Democratic colleagues. As the reelection for Senate whip loomed in January 1971, Ted had a challenger, Senator Robert Byrd of West Virginia. Byrd knew as much about poverty as Ted knew about wealth. As a small-town butcher, Byrd had made his way in life in part by ingratiating himself with the powerful. In the Senate he thought of himself as the "waterhole in the forest" where all the senators had to come to drink. He had sweet water for them all—assistance, deals, and myriad courtesies—and when it came time to vote, even some of Ted's liberal colleagues turned against him. Byrd won, 31 to 24.

Ted was incapable of watching out for the concerns of his fellow senators the way Byrd did, and it was just as well that he was no longer part of the formal Senate leadership. He could turn fully to the political concerns that mattered to him. Like Bobby, Ted now had his own coterie of uncompromising, talented, and ambitious aides. One of them, James Flug, was chief counsel of the administrative practice and procedure subcommittee chaired by Ted. Flug was a typical Kennedy hire: Harvard undergraduate and law degrees and a forthright liberal agenda that he worked with immense energy, hard work, and cunning to further.

If one had chanced into Flug's cramped suite of offices in the early 1970s, as often as not he was sitting with his feet up on the desk talking to a senator, a reporter, a leader of some liberal group, or perhaps a Harvard or Columbia professor. Flug might be discussing a bill or working to get someone a job on the Hill, placing him where he could one day call and ask a favor in return. Flug didn't vote on the Senate floor, but he operated not that much differently than many senators, and in those years he had more influence on the nation than did some of them.

Flug worked many of his contacts and sources when Ted fought to deny a seat on the Supreme Court to Clement F. Haynsworth, a conservative North Carolina appeals court judge. Haynsworth may have been a right-winger, but he was a principled man and an honorable jurist. Flug and others portrayed Haynsworth as a southern know-nothing ready to turn the clocks of American justice back fifty years. That was unfair, but the strategy was successful in derailing the nomination. "He [Haynsworth] turned out subsequently to have a very distinguished record," Ted reflected decades later. "I mean you could rethink your vote on that one."

The rejection infuriated President Nixon who sought a measure of revenge by nominating G. Harrold Carswell, a jurist with a segregationist past who made Haynsworth look like Justice Holmes. As Nixon knew perfectly well, it was almost unprecedented to turn back two of the president's Supreme Court nominees in a row. Ted was reluctant to take on this challenge. Flug was not. Ted gave subordinates such as Flug a great deal of leeway, and the aide went ahead pursuing a full-scale investigation of Carswell's past. This rankled Senator James Eastland, the chairman of the Judiciary Committee. Ted had to deal with Eastland every day and could not hold himself hostage to one issue, no matter how important it might seem. "We don't need this," he told his administrative assistant Burke. "So tell Jim that we're sort of slowing down on this stuff." In Bobby's offices, a nod of the head had been a command, but when Burke told Flug, he continued anyway. As the subcommittee counsel, he was Ted's appointee, but he had far more freedom than one of the senator's legislative assistants. Nonetheless, Ted's brothers would probably have fired the man, but Ted accepted that Flug was going ahead. In the end, no single senator, and surely not Ted, had as large a role in the 51-to-45 defeat of the nomination as did Flug.

Ted was still revered by most of his Massachusetts constituents, and appreciated by liberals as one of their authentic leaders, but since Chappaquiddick he was not as much loved by his admirers as he was hated by his enemies. He was in life for the Republicans what Joe McCarthy was in death for the Democrats—a name whose invocation brought campaign contributions and raised partisan fervor.

There was another kind of criticism of Ted common among a number of academics and some journalists. They said that he was little more than a creature of his aides, reading words they had written, proffering ideas they had formulated, and pursuing issues that others championed. That missed the point. Ted had created a machine that spewed out legislation, news stories, speeches, position papers, and interviews. He was using his subordinates, taking their issues and ideas places that without him they could not have traveled. Even when Flug appeared to be going off on his own, Ted probably had his reasons to allow it. Moreover, he didn't have to know all that his aides knew about an issue as long as he trusted their judgment and could quickly prepare to debate the issue on the floor of the Senate, question witnesses in a committee hearing, or discuss the topic with a reporter trotting along beside him as he headed back to the Capitol.

As closely as Ted worked with subordinates such as Flug, Carey Parker, another brilliant and youthful legislative aide, or the much-esteemed Burke, the senator did not trust them or confide in them on the deepest human level. On that level he trusted no one's judgment. No one knew this better than his sister Jean. She was closer to her brother than anyone else in the family. She heard things that no one else heard, but even she realized that her brother had lost that sense of trust in others and that he was alone in the judgments that mattered most. Here was a senator with perhaps more aides, advisers, academic consultants, and think-tank experts than any elected official outside the White House, and he trusted none of them completely. "I think he was very dependent on my father's advice, and his two brothers very much so," reflected Jean. "But as a result of all these things that have happened, he discusses things but there is nobody on whose judgments he relies."

In April 1971, Ted sat drinking a light scotch at the family apartment on Central Park South looking out on Central Park and northern Manhattan. Seated next to him with a stiff bourbon in his hand was James Wechsler, editor of the then-liberal *New York Post*. Ted had met Wechsler a few times before, but they had never talked much. When Wechsler's twenty-six-year-old son Michael committed suicide, Ted wrote an exquisitely sensitive letter. In his reply, Wechsler mentioned his "esteem for the way [Ted] had endured so much private tragedy." Ted was forever writing such letters and attending funerals, few of which came to public notice. Wechsler felt what most others felt when they received such a letter: if Ted could get up and go on, then maybe he could get up and go on too.

The two men were talking about the health issue when Wechsler mentioned that the view reminded him of the one from Bobby's UN Plaza apartment. "He was a very real patriot," Ted said, and set off on a soliloquy about his brother. As the editor noted in a personal memo afterward, Ted told Wechsler that "after Bobby's death he felt his major obligation was to carry on Bobby's fight against the war and to achieve the relationship Bobby had with both the poor and the blue-collar workers."

That led inevitably to a discussion of whether Ted intended to seek the Democratic presidential nomination in 1972. He viewed the prospect without enthusiasm but with a sense of almost tragic inevitability. "I suppose all our kids can see themselves standing in Arlington Cemetery again three years from now," he said. Even though he could understand the dread that often filled his nephews and nieces, he spoke as if he knew that death stalked him but that he had to go on. "For Kennedys the die is cast."

While Wechsler found Ted more interesting and spirited than he had imagined, he also found him "a very troubled, even tormented guy" who perhaps was "masking almost unbearable tensions related to private problems as well as the shattering blows he has taken."

As the editor was leaving, Ted stopped Wechsler at the door and talked about Rose's reaction to her sons' deaths. "This is terribly private," he said, "but it's the only time I've ever heard her question her faith. Somehow she

was able to accept Jack's death, but after Bobby. . . ." Ted's voice hung
there, and Wechsler mentioned Job's suffering in the Bible, an analogy with
which the senator was not unfamiliar.

D uring the Nixon administration, Ted took a major moral stand on the
war in Vietnam, which many Americans thought was largely over.
Nixon believed that the antiwar movement was fueled as much by self-
interest as by youthful idealism. The young did not want to die, he rea-
soned, and once most of the burden of dying had been pushed onto the
Vietnamese, the antiwar movement would lose much of its momentum. Ted,
however, would have none of Nixon's "Vietnamization."

"It is Asian, now, fighting Asian, and they do it for purposes more than
their own," Ted said, asserting that the South Vietnamese soldiers were surrogates for America, pushed forward by American global ambitions, and
dying with American weapons in their hands. "Vietnamization means war
and more war."

Ted was just as fierce in his denunciation of what he considered Nixon's
frontal assault on civil liberties. When Attorney General John Mitchell
threatened mass arrests at an antiwar protest rally in Washington, Ted thun-
dered: "We must destroy the cancer that has been transforming the noble
spirit of our nation in the eyes of our own citizens." The cancer, as he saw it,
was not the protesters but those who sought to arrest them for exercising
their constitutional rights.

Despite his deep concerns over the course of the Nixon administration,
Ted decided that he would not seek the Democratic presidential nomination.
In the spring of 1972, the putative Democratic candidate, George S. McGovern of South Dakota, sought a running mate who would bring votes to the
ticket. The South Dakota senator was thinking of choosing Kevin White, the
popular mayor of Boston, but Ted gave McGovern the impression that if he
chose his fellow Massachusetts politician, he would not campaign for the
ticket. Ted was the most territorial of politicians. He was not about to have
McGovern rummaging around in his backyard.

McGovern's eventual choice, Senator Thomas Eagleton of Missouri, resigned after it was revealed that he had undergone shock treatments for depression. At this juncture, the Democrats knew that if they were to have any chance against Nixon, they had to find a strong new candidate immediately. McGovern decided that he would name Sarge Shriver to the ticket, but Ted for the second time in eight years sought to veto that choice. It may have been unworthy behavior, but it wasn't surprising given the dynamics of the family.

McGovern named Sarge anyway, and the duo went down to devastating defeat, leaving a certain contained bitterness between Ted and the Shrivers. Sarge's oldest son, Bobby, who delayed his freshman year at Yale to work for the ticket, was perhaps the most outraged at his uncle's conduct; it was years before he could muse philosophically, "It's not personal, it's only business," repeating a line from *The Godfather*.

Ted came to evaluate his daily accomplishments largely by the amount of publicity he gained. His aides knew that if he was going to sit through committee hearings, he better have his shot at a quote in the morning *New York Times* or *Boston Globe* or a sound bite on *CBS Evening News*. To his detractors, this was preening, self-congratulatory politics at its worst, but he was also staking out a number of crucial themes and issues that he helped weave into the fabric of American concerns.

In his first years in the Senate he had gotten involved in the Vietnam refugee issue largely because he wanted to do something other than slavishly repeat Robert's political criticisms of the war. Refugees were becoming the flotsam of the last third of the twentieth century. It was a humanitarian issue that tore at the heart, but Ted insisted that the plight of the refugees resulted from political behavior far more squalid than the camps themselves.

The repression in East Pakistan was an issue that most Americans probably would not have followed so closely if not for Ted's loud voice. Pakistan, a bifurcated country created by the partition of India, was America's ally in the cold war. That meant, as National Security Adviser Henry

Kissinger and Nixon saw it, that the military leadership of West Pakistan could take strong measures to hold their country together. In December 1970, the East Pakistanis freely elected Sheik Mujibur Rahman as their president. Three months later the military leadership arrested the nationalistic leader and began a brutal repression, slaughtering his supporters and perhaps a million Bengali Hindus, whose only sin was the happenstance of their birth. In a desperate exodus to survive, millions of other Hindus fled the country, taking up refuge in wretched refugee camps in India. The Nixon administration saw India's seeming beneficence as the cynical mischief of a duplicitous people who were mouthing the words of neutrality and Gandhi-like adages of peace while attempting to sever their enemy.

In Washington a small group of Bengali and former Peace Corps members was working for the end of the brutal repression and for the independence of Bangladesh. They had no money, no prominent support, and no power. While most Senate offices politely showed them the door, Ted's staff welcomed and worked with them.

In August 1971, Ted flew to India to visit the wretched camps. His colleagues were off on junkets too, but most of them sought to educate themselves in places such as Paris and Rome, not in the dust and heat of South Asia.

"It is difficult to erase from our minds a child quivering in fear on a mat in a small tent still in shock from seeing his parents, his brothers, and his sisters executed before his eyes; or the anxiety of a 10-year-old girl out foraging for something to cover the body of her baby brother who had died of cholera a few moments before our arrival," said Ted upon his return. "When I asked one refugee camp director what he would describe as his greatest need, his answer was 'a crematorium.'"

Ted had a crucial role in making Americans aware of what was in effect genocide, and some role in forcing Nixon and Kissinger to retreat at least a few feet from their geopolitical "realism." In the end, whatever pressure Nixon and Kissinger exerted was not enough to prevent India from invading East Pakistan in December 1971 in a twelve-day war that freed the country and installed Sheik Mujibur Rahman as the first president of the newly independent Bangladesh.

Ted and his staff had shown immense, detailed concern over Bangladesh, traveling to the region, writing detailed reports, giving speeches, and lobbying the administration. This effort is largely forgotten by most Americans and by the generally illiterate, desperate people whom he tried to help. It was largely forgotten by Ted too, just another issue on which he made his mark before moving on. Adam Clymer's comprehensive political biography relegates Ted's involvement in the birth of Bangladesh to only three paragraphs; that is a measure not of the author's neglect, but of how many times in his career Ted involved himself in this way in an issue.

Health care, meanwhile, was becoming a key domestic matter. It was an issue on which Ted would arguably have more impact than any of his fellow senators. The world envisioned the Kennedys as a vibrantly healthy tribe. Ted knew how sickly many of his siblings had been growing up, how much Jack had endured, what his father suffered as an invalid, and what his son Patrick was going through with asthma. His family had money for the finest care, but millions did not. He urged comprehensive national health insurance, an idea that Nixon believed made sense. Ted pushed a bill through the Senate to fund health maintenance organizations. Most Americans had not even heard of the term "HMO" and would not have believed that within two decades it would become the primary health delivery system in the nation.

It had been the Nixon administration's idea to make HMOs the way in which most Americans got their health care, but like so many other issues, what distinguished Ted was not so much the originality of his ideas as the persistence with which he pursued them. He negotiated for increased funding in a bill that was finally signed into law in December 1973, two years after the administration first proposed the idea.

A leader is known as much by the quality of his enemies as the quantity of his friends, and Ted Kennedy's greatest enemy was the president of the United States. Nixon feared that the senator from Massachusetts might one day rise from the dark waters of Chappaquiddick to challenge him and his party.

Nixon considered Ted a depraved man and sought to use the senator's sexual encounters as a way to destroy his reputation. In 1970 the president's men secretly photographed Ted dancing with a young woman in Paris and made sure that members of Congress and the press had copies of the candid photos.

"I'd like to get Ted Kennedy taped," Nixon confided to H. R. Haldeman, his chief of staff, in April 1971, suggesting that a wiretap might be helpful. "There's something going on there." Nixon was so obsessed with finding dirt on Ted that he was willing to up the ante with "permanent tails and coverage on Kennedy and other Democrats." He sent a team up to Chappaquiddick to try to dig up more scandal. His operatives shadowed Ted on trips to Hawaii and Pakistan later that year of 1971, but they found nothing useful to destroy the senator.

By July 1971, Nixon felt he needed $2 million in "discretionary money to use if we want to hire somebody to get on Ted Kennedy or some other tactic." "He will never live that [Chappaquiddick] down," Nixon said in the Oval Office in September 1971, and he intended to do whatever he could to ensure that.

In September 1972 Ted requested Secret Service protection because he was worried about threats against his life. Nixon readily agreed so that the agents could spy on the senator. "I want it to be damn clear that he requested it, he requested it because of threats, that son of a bitch, I want to make sure that he is followed," Nixon said. "Just might get lucky and catch this son of a bitch and grill him for '76. He doesn't really know what he is getting into."

Nixon was unable, however, to arrange for the Secret Service agents to spy on Ted, and later that month the senator went off on a sail to Maine on a borrowed yacht with Amanda Burden, a New York socialite. Burden was his latest amour, and it took a certain audacity to sail up the coast of New England, stopping at various ports, with a woman who was not his wife. Nixon thought that Ted had finally offered himself up for public vivisection. "Is the Teddy story being properly kicked around about the woman on the boat and so forth?" Nixon asked his aide Chuck Colson, who answered that a small item had appeared in the *Boston Globe*. "We'll nail the goddamn thing," Nixon exclaimed. Ted was already back from the cruise and out campaigning for McGovern. "And you got a sign out?" Nixon asked urgently. "A man out?" Colson replied: "And somebody got a sign that says: 'Who's Amanda sailing with while you're out here?'"

I n the week after his landslide reelection with 60.7 percent of the popular vote, Nixon had little doubt that the greatest threat to his presidency came from Ted. In June 1972, Washington police had arrested five men breaking into Democratic National Committee headquarters at the Watergate building. One of them was James W. McCord Jr., security director for the Committee to Re-elect the President. Though it seemed little more than a bungled burglary, the crime would lead to the downfall of a president.

Ted had the staff to push an investigation into the administration, but Nixon believed that the senator would back away. "He [Nixon] thinks that Ted and his people will decide that they don't want to take us on," Haldeman noted in his diary on November 11, 1972. "This was based on the basic Machiavellian theory that if you strike a king, you must kill him, and that he can't kill us, therefore he won't strike."

Nixon was not wrong in his analysis, only in his conclusion. Ted considered Nixon an evil aberration, and he became one of the crucial players in bringing the president down. His staff was relentlessly determined. He confronted the Nixon administration as directly and uncompromisingly as anyone in Washington.

From his perch on the Judiciary Committee, Kennedy realized that he was too prominent and partisan a figure to sit on the special committee looking into the break-in at the Watergate complex and other allegations involving the Nixon White House. Others could assume that role, reaching higher and higher up into the administration and the White House, leading to a direct confrontation with the president.

On Saturday, October 20, 1973, Nixon ordered Attorney General Elliot

Richardson to fire Archibald Cox, the special prosecutor looking into the series of matters known as Watergate. The attorney general had pledged to the Senate that he would not interfere with the special prosecutor's work, and instead of firing Cox, he resigned. Deputy Attorney General William Ruckelshaus also tried to resign, but Nixon dismissed him. Then he directed Solicitor General Robert H. Bork to go ahead and fire Cox.

Nixon had made an outrageous assault on an independent judiciary, but he did so because he had been brilliantly manipulated. Kennedy had written the legislation that created an independent special prosecutor, and he had suggested his brother's former solicitor general as his ideal candidate to fill the post. Cox in turn had hired seven former associates of the Kennedys among his eleven special counsels. This was the team whose refusal to back off pushed the president to order Cox's firing rather than let him continue following the trail into the Oval Office. This became known as the "Saturday Night Massacre," and rather than saving Nixon, it probably doomed his presidency. "The Saturday Night Massacre was born in Kennedy's office," Flug says, his words only somewhat hyperbolic.

J oan was the greatest problem in Ted's personal life. He had given his wife an exquisite modern home in McLean, a summer residence on Squaw Island, a French chef to prepare the family meals, a nanny to watch over their children, a maid to clean the home, and a social secretary to handle the details of her life. He had given her money to travel wherever she wanted, to buy whatever clothes she wanted, and to indulge whatever whim struck her fancy. He was no more sexually disloyal to Joan than Joe had been to Rose, and his mother had looked the other way and celebrated family life. His brother-in-law Steve had an endless succession of mistresses and lovers, and his sister Jean put up with it.

But Joan was emotionally and intellectually unwilling and unable to put up with Ted's endless philandering, and she felt that her drinking problem was being spread before the American public to make Ted seem the victim of the piece. "It may be self-evident, but the fact is that Ted Kennedy has held the reins of information," reflected Joan. "When he was in trouble with his womanizing, he fed certain trusted journalist confidants through his press secretary Dick Drayne. And Drayne would say, poor Ted. The irony is that maybe I drank because there was another mistress, one who was in my bed when I was in New York.

"Ted Kennedy would give these leaks, and these little leaks were powerful. They would go out and people started to believe that stuff. My drinking started in our third or fourth year of marriage. Ted wasn't going to point it out. I was too embarrassed to talk about it or I wouldn't talk about it. My story never came out. I was in a man's world. I'm married to a powerful man who has control of the knowledge of the family."

Those close to Ted suggest that Joan's drinking began soon after their marriage, but that does not alter her essential point. She was in a marriage in which she had everything a woman could want except a loyal husband and a partnership in which the couple face together the emotional realities of their lives. Ted had been brought up to think that emotional issues were dangerous, worrisome things that you ran beyond. There were all kinds of things that Ted could not bear to think about his brothers' deaths, his own legitimate fears that he might be assassinated, and now the reality that his marriage was not working. In many ways Joan was crying out for attention, and the louder she cried, the less Ted heard.

Ted could sit downing one stiff drink after the next, and he was merely a senator relaxing, not a man with a drinking problem. But with Joan it was all immensely sad she would stand tottering on the stairs at McLean, accepting her husband's gentle admonition that she go back upstairs rather than come down with the guests. When she was found passed out in the backseat of a car at Hyannis Port, or wandering somewhere with her lipstick smeared and dress askew, she was pointed out as the family drunk, a shameful outsider who could not control her drinking. Ted might drink until he was incoherent, but he was not a drunk, not an alcoholic, not a problem, as long as he could point at Joan.

Joan was insisting that attention be paid to her. She took to wearing the most sexually provocative of outfits to public functions, including a see-through blouse to the Nixon White House. She gave an interview to *Good Housekeeping* in 1972 in which she admitted that she was going to a psychiatrist. This was the first of a number of pieces over the next years in which her triumph over her problems was celebrated, even while she was often drinking again before the magazine came out.

As Joan declined into serious alcoholism, Ted began to assume greater responsibility for their two sons and daughter. He attended school functions and tried to be home more. He played as often as he could with his sons and watched over their homework. On one occasion Teddy was so upset at his father for endlessly going over the homework that he penned a "Declaration of Independence," which he placed outside his father's bedroom door ("You are not asking me questions about the five pages. You are not crecting [*sic*] my home work. It is a free world").

In November 1973, Joan went off to Europe searching for some semblance of a life away from her husband. During her absence, the governess noticed a strange lump below Teddy's right kneecap. The boy was a robust seventh grader at the elite St. Albans School in Washington. He was the fastest runner in his class. He figured he had banged his knee on the football field. That evening the governess told the senator, and Ted immediately set up medical appointments. Within a few days Ted learned that his son's right leg was cancerous and would have to be amputated just above the knee.

Joan flew back immediately to Washington, but Ted was the one who would have to tell Teddy, not his vulnerable wife. Ted said later that as he stood at his son's bed he realized something that was probably as close as anything to the emotional essence of his life: "the brain can only take so much, and the heart can only stand so much." He knew that he had to control his own fear and sorrow and tell this news in a way that would not frighten his son but embolden him.

Ted told his twelve-year-old son that in the morning the surgeons would cut off his right leg. Teddy took it just the way his father had hoped he would. He did not look up toward the heavens and ask why. He did not complain, and he did not get hysterical. Instead, he calmly asked his father just how much of the leg the doctors would take. Then for a while he cried, but when he saw his mother's tortured face, he stopped and asked her to quell her tears.

That evening Ted was scheduled to attend a rehearsal dinner for the wedding of his oldest niece, Kathleen, to David Townsend, a young professor, the next day, November 17, 1973. Kathleen had offered to postpone the wedding, but Ted insisted that it go on. He could easily have opted out of the dinner that night, but he was there. He did not pull the evening down with morbidity and apprehension, but laughed and joked and sang as if he were the most carefree of men. This was Bobby's daughter getting married, and he had to be there with his blessing and optimism.

The next morning Ted was at Georgetown University Hospital when they cut his son's leg off. As soon as the doctors told him that the operation had gone well, he rushed over to Holy Trinity Church to give Kathleen away. The newlyweds had hardly exited the church when Ted hurried off to be in the recovery room when Teddy woke up. As the youth began fitfully to come around, Ted leaned over his son, his forehead touching Teddy's chest. He looked as if he were praying, but also as if he were trying to transfer some of his physical strength to his dormant son. Teddy reached his hands up and wrapped them around his father's head and held them there.

As unlikely as this may seem, the next day Ted wanted Teddy to strap on his artificial leg and start walking. Ted invited Washington Redskins football players to come up to the hospital room, and there were gifts from all over the world, from the great and the obscure, as well as letters and cables and telephone calls.

This all exhausted the recuperating Teddy. After a few days he confided to his mother that he needed some rest from the endless regimen of good cheer. "I'm so tired but I can't tell Dad," he told Joan. It angered her the way she was always being asked to be her children's emissary to her husband. Joan was the place her sons could go to for warmth and protection, a place where their vulnerabilities were seen as marks of sensitivity. "My son was intimidated by his father," Joan says. "With me Teddy Jr. could feel sick, he could cry and fall apart and be a baby, and I would hold him in my arms. But he wouldn't let anyone else know that. He was a Kennedy."

Ted had been told that if his son's leg was cut off from the knee down, he would almost certainly be free of cancer. But when the doctors did further tests, they realized that they had not removed all of the malignancy, and Teddy was left with an often fatal bone cancer.

As Ted was contemplating which of several medical options to take, he privately told Stuart Auerbach, a science reporter at the *Washington Post*, that he feared he wasn't doing enough. He faced a surfeit of choices. His father had tragically agreed to an experimental lobotomy on his daughter Rosemary, in part because he was a wealthy, powerful man and was presented with choices few others received. For the rest of his life Joe mistrusted doctors, and now Ted had to make a similar decision. The advice came in from all across the United States and Europe. In the end, he decided not to go with the then-conventional radiation treatment but the relatively new chemotherapy. At Boston's Children's Hospital, Teddy was only the twenty-second child with whom Dr. Edward Frei III had applied this new approach.

Every third Friday Teddy flew up to Boston with his mother or father for the treatment. His stomach began to turn as he saw the Citgo sign in Kenmore Square. That meant that he was only six minutes from the hospital and the six-hour drip that left him dazed and incoherent. From the moment he entered his hospital room he began to throw up. His father often stayed with him overnight during the eighteen months of treatment and gave him the antidote injections, holding his head when he vomited.

When young Teddy returned to Washington on Sunday evenings, he was not treated by his family like a tender child going through a regimen that would have felled even the most stoic of adults. The word "cancer" was not whispered behind his back but spoken boldly to his face. His "stump" was a "stump." He didn't get out of taking his turn bringing in the paper either. When Ted took the big needle into his son's room for his shot, his brother and sister didn't cringe but shouted: "Here comes the mad scientist." And when Teddy started to walk, Patrick and Kara took bets as to how soon their big brother would fall down. That March, when he was still in the midst of chemotherapy, Teddy was out skiing on the slopes of Aspen.

For Ted politics was personal, and nothing was more personal than overseeing young Teddy's treatment. "Having a child with cancer reaches to the very depths of your soul," Ted reflected three decades later. "Particularly because there is so little you can do, yet certainly more that you can do now than when Teddy had it. We were fortunate to have access to good health care. Secondly, fortunate to have health insurance. Many of the parents I met at the hospital had children who were taking a similar treatment. That treatment was to last for two years. Some parents sold their houses to pay for it. Some could only afford twelve or fourteen months of the treatment. They were asking the doctors: 'What percentage does that reduce my child's chances of being able to survive?' So you ask me why I'm for health care. I didn't need a reason before, but that's a reason I'll never forget."

When young Teddy looked back on the operation and the months of chemotherapy, he could say boldly that God had been bad to him in all that he had been asked to suffer, and good to him in providing him with a family and a home that could take care of him. He had suffered much, but he had learned one important thing: "The great lesson that I have learned from my family is that no matter how terrible things might appear at one point or another, there is a clearing in the future."

Outcasts

fter his arrest for marijuana possession in the summer of 1970, sixteen-year-old Bobby felt that he was being driven from his home. He bought an old Ford Falcon and headed west to Los Angeles with a friend, Conrad Lowery. Always before when the Kennedy men had taken intrepid journeys, they had done so to enhance themselves in the political and social worlds in which they lived. Bobby set off not to enrich his résumé as a Kennedy but to leave everything about it behind.

In California, Bobby sold the car, and he and his friend took off hitchhiking and riding the rails. They panhandled for meals and took odd jobs. Along the route Bobby met other youth of his generation who had dropped out to live in hippie communes or to hang out along Telegraph Avenue or on Sunset Boulevard, but he did not meet teenagers living the way he lived. At one point he rode for three straight days in an open freight car before finally tottering off, black and exhausted. "I was riding around with the bums," he said. "It was good: I could be one of them and not be a Kennedy."

Bobby had been thrown out of two prep schools in three years; after four months of wandering, he returned east and entered Pomfret School in Connecticut, where it did not take long for him to be thrown out again. He brought his girlfriend, Kim Kelley, into his room for the night, and during the day he took her down with him to the basement, where he cooked meals on a hotplate.

After being expelled for his indiscretions yet again, seventeen-year-old Bobby returned to Hyannis Port to spend the summer of 1971. One evening in nearby Barnstable a policeman told Bobby to move on when his car was blocking traffic on Main Street. The officer arrested Bobby after he took a bite of his ice cream cone and, according to the officer, "spat a bit of the ice cream cone in my face." Bobby had the attitude that others were always to blame. He told the judge, "This is all a bunch of made-up lies." When he refused to call his mother, he ended up spending the night in jail. He followed the dubious advice of one of the other prisoners and pled no contest. As soon as the proceedings finished, he left the village to spend the rest of the summer at Lem's apartment in New York City.

Fifteen-year-old David sought to emulate his big brother, seeking experiences that took him beyond the pale. Even more than Bobby, he could not envision himself traveling up the arduous road that his father and uncles had traveled. He sought some other road where no Kennedy had ever traveled, and he would not be judged by the standards his forebears had set.

In the summer of 1970, David and fifteen-year-old Chris Lawford set out hitchhiking to New York to spend a few days away from home. They panhandled passengers at Grand Central Station. "It was great being just ordinary people and not Kennedys," David said, though hustling passersby for change was a curious concept of ordinary. The two youths took their money to Central Park, where they copped bags of heroin.

When David returned to Hyannis Port, Ethel did not hug or scold him, but acted as if he were no more than a casual boarder who had been away for a while. David decided to sleep much of the rest of the summer in neighbors' houses or wherever he could sack out. David and his brothers must have thought that there was something terribly wrong in a mother throwing her teenage sons out of the house or emotionally disengaging from them. It was this seeming indifference that was the most devastating to them and had an incalculably large impact on their psyches. They were somehow supposed to gather the strength to go out in the world and be great men carrying forth a noble legacy, but at home their mother acted as if they had already soiled the family name so much that they scarcely deserved to bear it. In their most trying times, apparently Ethel had little to give her older teenage sons except complaints and indifference.

In the fall, David went to Middlesex School in Concord, Massachusetts, where he smoked marijuana with his breakfast coffee and dropped acid regularly. His hair was long, he was willfully sloppy, and his mother rarely called him.

Several of the sons began seeing Robert Coles, the distinguished Harvard psychiatrist and author, though it was not said, even among the family members themselves, that they were in therapy. To them therapy suggested weakness, a disgorging of the truths of their lives to a stranger, even if he was a close family friend. Over the next decade Coles was called in to "work on" Bobby, David, or Joe. The esteemed psychiatrist was treated by the family like a small-town doctor who passed out sugar pills, said a few encouraging words, and sent the relieved patient on his way. "Bobby has gone to a psychiatrist who has helped tremendously who is also a friend and admirer of Bobby's father," Rose said in January 1972. "And he is tremendously interested. And I think the boy's been straightened out now. He's back in school."

As a young doctor, Coles had met Bobby Kennedy and traveled with him to visit the poor and the hungry in Mississippi. The two men had talked of many things, from the nuances of politics to the works of Shakespeare. When Coles wrote his book *Lives of Moral Leadership*, the first two chapters were on the late senator. Coles was devoted to the memory of Robert Kennedy and so emotionally ensnarled in the family that he may have found it difficult to lead the youths to any semblance of psychological freedom.

If these young men were ever to have their own true lives, they had to come to terms not only with themselves but with their family's past, and they needed a guide willing to shine a lantern into the dark crannies. The family was constitutionally incapable of realizing the enormous emotional strain with which these young men were living. For two generations the young Kennedys had been told that no matter how wounded they might feel, how tortured in spirit or sick in body, they were to go out and take on life. If they pretended long enough and hard enough that they were well, then they would be well. That attitude was based on a flawed optimism, for it required that they never look at the darkest realities but instead pretend that they did not exist or were manufactured by their enemies.

As a young man, Ted had stomach problems that signaled his emotional distress, but that was not to be discussed or admitted. So now did young Joe, and his stomach problem was treated in much the same way. "Joe came home the other day from Africa, and they thought he had an ulcer," Rose said. "It must be a nervous tummy."

J oe was always going to be troubled by school. Nonetheless, he matriculated at MIT in the fall of 1972, one of the most intellectually rigorous schools in America. "I was nominally at MIT," Joe says. "I was there until I had to take an MIT course and I backed out." He could not keep up with the fierce pace, and even before the first semester was over he dropped out. Soon after that Diane Clemens, a young professor at Berkeley, received a call from an old friend, Dr. Joseph Brenner, a psychiatrist closely associated with Dr. Coles. Dr. Brenner asked the professor if she would let Joe live with her while he studied at Berkeley.

Clemens, who had a young daughter and was separated from her husband, had just begun her new position, but she reluctantly agreed. "Joe called me and we talked," Clemens says. "He said he didn't want to come here. He didn't want to do this. Coles and his colleagues were suggesting this. He was defiant. I said, 'If you don't want to come, don't come.' As far as I was concerned, there wasn't going to be a negotiation."

Joe arrived at the celebrated university and entered as a freshman. Like Bobby and David, Joe sought the anonymity that would allow him to build what was truly his own life. But he was a scion of the most celebrated family in America, and he could hardly walk the campus without people knowing who he was. That was even true of one of Clemens's colleagues, who told her he wanted to guide young Joe in his reading, since, as a Kennedy, the young man needed a unique education, and he was the one to direct it. Clemens vetoed that idea, telling the professor that such an approach might drive Joe away from the university.

"Joe was kind of tentative about what he was going to be next in life, and that made the vulnerability," Clemens says. "He wanted all the input he could have from others, but he was very careful about it if someone came on too strong. Without having a father, he felt lots of pressure to make the kinds of decisions that a twenty-year-old has to make, and his mother certainly put a lot of pressure on him."

Clemens often heard Joe screaming at his mother over the phone; by the measure of the sound bellowing up from downstairs, it was their preferred method of communication. "One of the lines that kept being repeated on the telephone was, 'I'm going to throw you out of the family.' For a young person who has lost his father and feels that he cannot really be part of the family, it's not easy. He never said to me that he was finding it difficult being part of this family, but I saw it. I was keenly aware of what was going on, the context of the situation."

Joe struggled to contain his anger and turn that emotional energy into something positive. One afternoon he became so irritated with Clemens's eleven-year-old daughter Lani that he chased her through the house. She shut the door leading upstairs and bolted it shut, leaving Joe unable to enter the upper stories of the home. Joe screamed at her. When she refused to open the door, he pushed so hard trying to get out that he tore the bolt from its hinges. The little girl was so frightened that she ran into her bedroom and climbed out the window into a tree.

When Clemens returned home, a chastened Joe was sitting on the front stoop waiting for her. "Ah, hey, Prof, there's something I gotta tell you," he said nervously. "I did something really bad." He had a sheepish look on his face, as if he had to confess and put the matter behind him. "I could hardly keep from laughing," says Clemens.

On another occasion Clemens's German shepherd bit her leg. She screamed and Joe ran up the stairs to help. By then the dog was simply standing next to its wounded mistress. Joe picked up the massive animal and threw it down the stairs. It was a gesture of help, and a brave one at that, but it was hardly what should have been done.

As difficult as he was at times, Clemens had empathy for Joe and his struggles. "Messages are incredible things in the family because things aren't said explicitly, so there's all this ambiguity," she reflects. "Just how do you get the impression that you're to behave one way or another? Joe was not a very self-revealing person. He didn't want to stand out. He didn't want to attract people who would use what he was. He was like Siddhartha, wandering around, figuring it all out. The person who interrupts that process is really demonic."

After several months Joe showed up in Clemens's office with a determined grimace. "I don't want to study anymore," he said, speaking in the same confrontational style that he had affected the first time the professor talked with him on the phone.

Joe worked for a short while in Mayor Joseph Alioto's office in San Francisco but gave that up too. He purchased a Toyota pickup and headed off to the Northwest with his friend Wilbur James. A few days later he crashed the new truck, shattering the windshield so badly that it left shards of glass. Joe did not stop to repair the truck but limped back to Berkeley. There he had a second accident in his already damaged Toyota that left him in the hospital overnight.

Clemens was appalled that Joe would contemplate driving back to the East Coast without fixing the windshield that he had broken in his first accident. She offered to replace the glass herself. Joe would not listen and drove away in the vehicle.

Two accidents in as many weeks would have chastened most drivers, but Joe seemed only emboldened. If Clemens was correct in viewing her young charge as Siddhartha off on a quest for meaning, that journey had come to an end both spiritually and physically. He arrived back in Hyannis Port in

August 1973 prepared to enter the University of Massachusetts in the fall. Before school started in September, Joe figured he would have one last great blast at a rented house in Nantucket. David was there with his new girl- friend, Pam Kelley, a member of the Hyannis Port family so much in tandem with the young Kennedys.

There was a soft vulnerability to David that drew people to him but caused him endless pain. Even for David, things were going better now. He had gone out west with Pam for the summer and worked on a ranch outside Boulder, Colorado. He had grown up with Pam, and though they had started this trip as cronies, they were lovers now. He finally had a companion he could talk with. His mother didn't like Pam, but his mother didn't like lots of things, and for the first time in a long time he felt good about the life that lay ahead.

David loved his big brother, though he wished Joe had more time for him. In that way too this week was special. At the end of the stay Joe set off to drive David and Pam back to the ferry landing. For all the desire of Joe and his brothers for anonymity, they were at times shameless showboaters whose actions shouted, "We are Kennedys, and you are not." They were notorious for breaking into ski lines in the winter, and this day Joe gunned the Jeep through the woods, as David and Pam stood up in the vehicle. Joe sent the car careening one way and then twisted the wheel back the other way as they laughed wildly.

The Jeep lurched onward, as guffaws and shouts spilled into the air. When Joe came to the highway, he did not look to see a station wagon com- ing down the road. He bolted out onto the highway, and when he saw the vehicle bearing down on him, he swerved and the Jeep bounced into the ditch, rolling over several times.

Joe was fine, but David had fractured vertebrae, and Pam's injuries would leave her paralyzed for life. The Kennedys had become adept at get- ting out of scrapes that would have destroyed the public careers of most others. Pam was taken care of on generous enough terms that she neither sued nor talked to the media. As for David, he was given morphine in the hospital. It was like someone with a sweet tooth tasting sugar for the first time. No matter how much the nurses gave him, it could not assuage the pain that he felt.

Joe had his license suspended, but he was able to get on with his life. By all appearances, Joe treated the accident as Ted had treated Chappaquiddick, never talking about it and keeping whatever pain or guilt he felt to himself. The accident did serve, however, to quiet him down. He kept a low profile at the University of Massachusetts; after graduating, he managed his uncle's 1976 reelection campaign. That was precisely how his father had gotten his first serious political training.

Twenty-four-year-old Joe had some of his father's virtues, such as a zealous intensity and endless energy, as well as some of his father's faults, including a nasty temper and a certain lack of empathy for campaign work ers. What he lacked was his father's high vision of politics and the ambition to pursue that vision at any cost. His uncle was so popular in the common-wealth that he hardly had to campaign; this was hardly the tough test that Joe's father had passed in steering Jack toward his first Senate victory, but it gave the young man a taste of politics that he found inviting.

In the mid-1970s the Kennedy name was still magical in Massachusetts politics, and Joe was soon approached to run for statewide office, either trea-surer or lieutenant governor. Joe was used to being pushed forward solely because of his name. He had a weakness often shared by those with the kind of uncontrollable temper that plagued him: a vulnerability to flattery. And the suggestion that he was ready for statewide office was nothing more than the most shameful flattery.

Now that her husband was gone, Joe's grandmother Rose saw herself as the final repository for her family's political values. She feared that her grandchildren would end up the spoiled children of entitlement, and one afternoon she invited the two eldest grandchildren, Joe and his older sister, Kathleen, to her Hyannis Port home. "I want you to learn something impor-tant about politics," the wizened matriarch said. Rose took out a scrapbook and opened it to a picture of her father flying in a primitive airplane, and another of Rose and Mayor Fitzgerald on the cover of the *Boston Post*.

Joe figured he was going to have to sit there as Rose reminisced about the turn-of-the-century world. Instead, his grandmother turned to the back of the scrapbook where there were newspaper want ads. "I want you to look at what's written there," Rose said pointing at the page. "See those words: 'No Irish Need Apply.' Lots of people suffered mightily to get you in the circumstances you have. Never take it for granted."

The Shriver Table

unice decided that during the summer of 1974 fourteen-year-old Tim would have a series of life lessons. His mother wanted her sons to have close friends, and she arranged it so that three of Tim's closest buddies could go along. The boys began by flying out to the Pine Ridge Indian reservation in South Dakota to spend a month working at various tasks. It was probably the least welcoming place in America for four privileged youths to learn about Indian life. The year before a group of several hundred armed Native Americans had occupied Wounded Knee, the scene in 1890 of the last major battle between federal troops and Native Americans. The descendants of these warriors had been protesting against the alleged complicity of the Oglala Sioux leadership with paternalistic Bureau of Indian Affairs officials. Two Indians died during the seventy-one day siege, and hundreds were arrested. A year later, the reservation still boiled with recrimination and violence.

Tim saw that there was nothing romantic or mystical about the alcoholism, despair, and anomie in the ravaged community. "I had this idea of Indians out there with tears in their eyes because the rivers were polluted," he says. "What we saw was real life for that particular community, serious problems of poverty, alcoholism, pollution, racial issues, and that people didn't like us."

This was not the "adventure travel" of staying in first-class accommodations, and venturing out each day to explore another culture. "We were

living on peanut butter sandwiches in this desolate place," says Danny Melrod, one of the friends. "One day we got in a fight over a jar of peanut butter, until the glass broke, and we had nothing. We worked in a moccasin
factory, and then on a ranching operation. We were riding horses that we
barely knew how to ride. We went off on a cattle roundup and branded cattle. We're all saying, 'Why are we here? If my mother only knew.' "

The youths survived that challenge and were excited when they learned
that they were next to fly to Washington State to spend time with Jim Whittaker, the first American to reach the summit of Mount Everest. When they
got to the Northwest, they learned that they were joining a group of adults
climbing Mount Rainier, among the most challenging ascents in the continental United States. The other climbers were at least twice their age and had
previous climbing experience. The only other climber close to their own age
was Tim's big brother, twenty-year-old Bobby, a student at Yale University.

Tim and Bobby had an intense, often bickering relationship. As so often
happens with the oldest child, Bobby was the battering ram who opened up
easier lives for his younger siblings. He had been not only the most troubled
of the Shriver sons but by far the most sensitive to the dynamics of the family. He was struggling to build a life for himself out of something other than
the discarded timbers of his past.

His big brother's misconduct had served as warning to Tim, but unlike
Bobby, he was shrewd with his parents. He had a way of appearing the dutiful son when he also had a full measure of the anger and self-doubt that
plagued his big brother. There was a half-decade difference in their ages, but
the brothers competed against each other as if they had been born within a
day of each other and there was only enough food on the table for one of
them. The fact that Bobby was along for the climb made it almost inevitable
that Tim would follow in his steps, no matter how high up the mountain
they went.

By climbing Rainier, Tim and Bobby Shriver were doing something
that the sons of Bobby Kennedy Sr. might have done, or that Jackie would
have signed John on to do, but the Shrivers' philosophical and psychological
reasons were very different. The Bobby Kennedys considered the mere act

of physical risk salutary. As for John, he would have climbed that mountain too, but it would have been part of his journey into manhood, an assertion of himself that he might not have done in such a pronounced way if his father had lived. For the Shriver sons, however, these were authenticating experiences that gave them a sense of confidence that they had accomplished something useful and important.

After a few days spent learning basic mountaineering skills, the group set out at one in the morning from base camp. "It's a blizzard like I'll never see again," remembers Melrod. "We're tied onto a rope, and you cannot see more than two feet in front of you. Your goggles are freezing up. They told us, 'If you can't make it, we're going to put you in a spot and get you when we come back from the ascent.' Everyone said, 'There ain't no way I'm going to be sitting there in a sleeping bag.' The blizzard got even worse. They shouldn't have continued, but we reached the summit, and we fought our way down. We were proud we had done it. Fear drove us to that summit."

Because his brother had set out to the summit, Tim had been more willing to head up the mountain, but he was equally relieved when they finally came down from the icy precincts of Rainier. Tim's friends now thought there would be time for a deliciously leisurely end to the summer, a few weeks on the beach or some gentle sailing. They were stunned when they learned that for the third act of the summer Eunice had arranged for the youths to join up with Ringling Brothers and Barnum & Bailey Circus, where the legendary Gunther Gebel-Williams would teach them lion-taming. Tim's friends had had enough. They flew back east, glad not to have been shot or starved in South Dakota or frozen to death on the high reaches of Rainier.

Tim set out alone to join the circus in Texas. "One thing people don't know about the circus is that these people work their tails off," Tim recalls fondly. "I mean, you're setting it up, three shows on Saturday, two shows most other days of the week, and the show is three and a half hours long. Gunther was developing a leopard act at the time. Nobody had ever trained leopards. So at eleven o'clock at night Gunther did the leopard thing. He'd bring out six leopards into that cage and work them for two hours. I did the elephants, and I begged him to let me in the cage, but he wouldn't do it. I lived in this little back bunk in his car. He probably thought it was crazy having me come along. But it was fun."

E very summer the Shriver children also spent time helping their mother with her camp for the developmentally disabled. The mentally retarded had been pushed so deeply into the closet of American life that many people were uncomfortable around them. They might mutter a few words of conciliatory greeting or give a pat on the back or a gentle exchange, but that was about it. "When I was a boy, you'd see people my age or younger, when they saw a person with Down syndrome, they didn't know how to act with them," says Mark. "You know a Down syndrome adult can come up and give you a big hug and some people step back. For us, that happened since I was a little kid."

Their social concerns did not begin and end with mental retardation. "My sweat suit had the Peace Corps written on it," reflects Tim. "I got the message from my parents that there was a value to be placed on social justice and empowering the unempowered. I grew up being dragged around to speeches where my dad spoke about peace and poverty in the Third World. These things were part of my life. The same way some family might be a swimming family or a baseball family. This was the embryonic fluid of my development. This was what we ate, slept, and drank."

When the Shrivers sat down for dinner, it was a gabfest, and outsiders had to bull their way into the conversation as the family argued over everything from the Baltimore Orioles' pennant chances to the prospects for President Richard Nixon's impeachment. Sarge and Eunice did not seem like parents but like just two other voices shouting for attention. "It was a great joy at the table," recalls Neal Nordlinger, who spent many meals there as Bobby's closest friend. "You never heard more laughter, or had more fun, more jokes. People would stand up and do little skits. It usually entailed giving somebody a really hard time. Like making fun of somebody, all in fun, but everybody I mean, Sarge, Euni, everybody was a victim. Everybody."

Sarge and Eunice may have been philosophically and religiously modest, but they were proud of what they were doing with their lives. They also
had the compulsive sense that whatever they were doing, no matter how
valuable, it wasn't enough and was only a way station on their life's journey.
There was a restless, driven quality to both parents that they imparted to
their five children. Every moment had to be filled. If you were not learning,
you were doing, and if you were doing, it had to be something from which
you were learning. The sons picked up on this natural adrenaline and were
constantly moving from activity to activity. They were so hyper that if their
parents had not been much the same, the children would probably have been
candidates for suburban America's newest drug of choice, Ritalin, used to
subdue the hyperactive.

For all that Sarge and Eunice wanted their sons to experience the
tough realities of the outer world, they did not want them to face the difficult, ambivalent truths of the Kennedy past that they were unwilling to
face themselves. The older generation of Kennedys had gloriously positive
images of their family. The younger Kennedys, however, were living in an
age of political disillusionment and were confronted with countertruths
and revisionist history that challenged almost everything they had been
taught.

That whole process began in earnest in the mid-1970s with the Church
Committee investigations revealing that during the Kennedy administration
there had been assassination attempts against Fidel Castro and also that
the president was having an affair with a woman who at the same time was
having a liaison with a mob chieftain. The woman, Judith Campbell Exner,
subsequently held a press conference and wrote a book detailing her relationship with Kennedy.

The rumors and revelations reached such a pitch that Eunice asked her
husband if there was anything to this business of Jack's fooling around in
the White House. "Well, I can't imagine it," Sarge replied. "Jack's a careful
person. I can't imagine he'd do something like that. On the other hand, when
he was a bachelor, I don't assume he led a pure life. I just can't believe that
after he married Jackie, and then he was running for president, that he would

be doing these things. But the only one now alive who really knows would be Ted. Why don't you talk to him?"

Eunice asked her brother, who told her that "there's nothing to any of it." Ted knew the truth about Jack's private life, but he considered this a man's business. He was not going to talk about it with his sister or a brother-in-law whose pristine morality set him apart from the other men in the family. For all Ted's disregard for his brother-in-law, there remained a serious question for Sarge and subsequently for his sons. Could a man who could not even look at the darkness within his own family possibly lead a world full of evil and good?

T he Shrivers were trying to nuture in their sons the strength and vitality of their ancestors. To do so, Eunice and Sarge were trying to create an atmosphere in which their sons and daughters would be sorely tested and would not use their name as a chit to get them through life.

Bobby was just enough older than Tim to realize that these were not mere games they were playing when they climbed mountains or fought ferociously on the tennis court, but preparation for what his family considered the grand battles of life. Bobby had the deepest insights of any of the brothers into his family, in part because he was the oldest and had confronted most directly the realities of life as a Kennedy grandson, and in part because of his nature. By the time he headed off for college, he sensed that if he was to have his own separate identity, not as a Kennedy son or a Shriver but as someone unique called Bobby Shriver, the time had come to begin to assert his own separate identity.

For Bobby's eighteenth birthday, in April 1972, Sarge planned to give his eldest son a Volvo that he could take off to college with him. Most youths would have grabbed at those keys, seeing the vehicle as a symbol that they were independent and fully a man. Bobby turned the gift down. He could not see himself as an independent man if his father was giving him a car.

"When you're a kid, it's who your friends are, do any girls like you, are you a good athlete," reflects Bobby. "Then at eighteen, nineteen, or twenty you start thinking, 'Uh-oh, now I've got to grow up and get a job and what am I going to do and oh my God, look what my parents did. Oh my God, I better do something important.' "

When Bobby entered Yale in the fall of 1972, his freshman roommate was impressed with the generosity of his new friend, who had chosen the top bunk in their dorm room. The next morning at seven when the phone rang, and many mornings from then on, the roommate realized why Bobby had given up the lower bunk. "It's Mrs. Shriver," the caller said. "Is Bobby there?"

"He's asleep, Mrs. Shriver."

"Tell him to get up and seize the day. You hear. Seize the day!"

Bobby had all the credentials for social success at an Ivy League college. He was the son of famous alumni, a graduate of a prestigious prep school, and a handsome, tall, outgoing, bilingual, sophisticated young man. He was tapped for membership in Scroll and Key, the second most prestigious secret society at Yale, as was his father before him. In his senior year he was the editorial page editor of the *Yale Daily News*. He had the natural skills of an editorial writer, making his judgments with certainty. He called for students to boycott a speech by the Nobel Prize–winning scientist William Shockley, whose appearance was being sponsored by the ultraconservative Young Americans for Freedom. Yet he defended the man's right to present his eugenics theory that blacks were intrinsically inferior. He wrote that "the arrogantly insensitive YAF invitation, the pseudoscientific genetic quackery of William Shockley, the morally repugnant nature of the debate's tacit assumptions merit our total disgust and indignation."

Bobby sometimes appeared overwrought. A young man whose voice swaggered as much as the rest of him, he crushed his victims with his verbal assaults. His intentions were probably no worse than those of his adversaries, but he was almost too good at it, and sometimes his wounds left scars that did not go away.

"In college Bobby had a polarizing personality," recalls Lloyd Grove, who worked alongside him on the *Yale Daily News* and is now a gossip columnist for the *New York Daily News*. "People either liked him or were deeply irritated by him." Bobby was acting no differently than he did at home, but the world was not the Shriver dinner table writ large. Bobby was not so much arrogant as full of a kind of manic energy that exhausted all those who were not up to matching him idea for idea and quip for quip. Although his detractors assumed that he was acting with Kennedy entitlement, his struggle was precisely the opposite: to become a person who was not making his way through the happenstance of birth and name.

Bobby took a step toward that goal the next summer when he worked for *Rolling Stone* in San Francisco. He needed a car to get around the city, and he found one through an ad in the *San Francisco Chronicle*. The Pontiac convertible had ninety thousand miles on it and looked as if every mile had been a hard one. Bobby had only $200 saved from his job. He borrowed the other $300 that he needed to purchase the car, then drove it to the house where he was staying. "I sat down on the curb in front and looked at this tattered car, and I had a just overwhelming emotion about it," Bobby remembers. "It was weird. I had gotten it myself. I was lucky that I didn't accept the car from my dad because I wouldn't have had that emotion just getting my crummy old car for myself."

Despite Bobby's need to make his own way, when Sarge announced that he was running for the Democratic presidential nomination in 1976, Bobby helped his father and could do so because he had graduated a semester early from Yale. He was proud to be part of the futile campaign. There was a seamless quality between Sarge's personal and political life, and though he did not come close to winning any primaries, unlike his opponents, he talked regularly in his speeches about the meaning of family and government's role in fostering families. Sarge had suffered the courtier's fate of always waiting, and he was able to announce his candidacy only because Ted had decided not to run. Ted was all for a Kennedy reaching the White House, as long as he was the one. He did almost nothing to advance Sarge's campaign. "Teddy didn't do shit for my father," Bobby said afterward.

Bobby had helped his father out, but unlike his cousins, no one expected him to spend his life fulfilling his father's destiny. He wanted to be a journalist, and he started out at the *Evening Capitol* in Annapolis, Maryland, in the fall of 1976, then moved from there to the City News Bureau in Chicago and then to the *Chicago Daily News*.

Bobby arrived at the end of the Chicago newspaper era immortalized in *The Front Page*, whose coauthor Ben Hecht had worked for the *Daily News*. Bobby liked the raw realism of Chicago journalism. For a man of his intense aggressiveness, nothing could have been more thrilling than covering the crime beat in a newspaper war in which every day was a battle against the two other dailies, the *Chicago Sun-Times* and the *Chicago Tribune*.

There was no sweeter moment in life than showing up at the Billy Goat Tavern on North Michigan Avenue when you had a story on the front page. The bar, neatly situated underground between the *Sun-Times* and *Tribune* buildings, was the kind of place where it was always two in the morning and no matter how many you had, the bartenders never cut the drinks. One evening Bobby left the cheap hotel where he had rented a room to have a meal at the bar before starting work on the night shift.

Mike Royko, the gritty philosopher of the Chicago streets and at that time a columnist for the *Chicago Daily News*, stood at the bar wreathed in cigar smoke. "Hey, kid," Royko said as Bobby sidled past him in the crowded room. "That was a good piece today."

Mike was standing next to a longtime flack, the kind of man who always paid for the drinks even if he wasn't drinking. "Mike, I can't believe you said that!" Royko turned to the man as if he could not believe what he had just heard. "You're sucking up to the Kennedys, man. I can't believe it."

Royko pulled his fist back and hit the man square in the face, then threw a drink at him. "Oh, it was an unbelievable thing," Bobby recalls with immense relish. "I always respected Royko. He was The Man. The guy insulted him. Mike Royko would never suck up. That's ridiculous. Ridiculous."

It is one of the endearing ironies of American journalism that newspapers are often more exciting, more innovative, and more sheer pleasure to read and to work for when they are dying than when they are successful. Bobby had both the good and the bad fortune to show up at the *Daily News* during its death throes, and when it expired, he was out of a job. He moved to Los Angeles to work for the *Herald Examiner*, which in a professional sense was like moving between hospices. Shortly before that paper folded too, he decided that he needed greater tools to understand and affect the world, and he sought them back at the Yale Law School.

Bobby had a fierce intelligence, verbal agility, a fine memory, and a competitive drive, but he was still immensely insecure. For his first opportunity to argue a case in moot court, he had two women as his opponents. These students were brilliant and resourceful, and he sensed that they would take special pleasure in tearing him into easily digestible morsels. Bobby asked Tim, a freshman at Yale that year, and his brother's friend to act as witnesses in the trial. Tim was in fact a double witness, testifying on Bobby's behalf in the court and later standing witness to how he had done. When Bobby argued, be it a moot court or elsewhere, he lost all self-consciousness and was merciless in his attacks.

This was not just a class exercise but a family event, followed by his parents as closely as a match between their sons on the family tennis court. Afterward, Eunice called Tim to ask him how his brother had done. "It was hysterical," Tim said. "Bobby just acted like he does at the dinner table, and he won."

John's Song

 everal times a day in the summers, the passenger ship *Prudence* sailed past the compound in Hyannis Port on its hour-long cruises, the tourists hanging over the rails trying to catch a glimpse of one of the Kennedys. The sightseers would happily have settled for any of them, but the most cherished sighting of all was Jackie and her children, Caroline or especially the boy they called John-John, a childhood nickname that John found deeply irritating.

On one hot August afternoon in 1974, the passengers paid no attention to the three boys crowded onto a twelve-foot-long Sunfish in the water just beneath them. The youths sailed their tiny sailboat directly at the 150-passenger replica of a Maine coastal steamer, then dove off at the last moment just as their boat veered away before crashing into the side.

If the tourists had stopped peering so intently at the shore and glanced at the water beneath them, they might have realized that the laughing youth paddling back toward the little boat was John himself. Next to him, shouting gleefully, was Tim Shriver. And the third youth, whose shouts rang out above all the others, was William S. "Billy" Noonan, the final participant in this game of chicken, seeing who would be the last to dive off before the boat hit the tourist ship full square.

Like almost any youth entering his teenage years, John's great desire was to be like his friends, to not stick out as different. He did everything he could to camouflage himself, to dress and behave as his peers, but it was not that simple. Even now, as he was playing this game with his friends, Secret Service agents in a motor launch watched through binoculars.

Jackie nudged John toward companions who she thought might inspire him to develop a more serious attitude. In April 1975, she arranged for her son to fly to Russia with the Shrivers. Tim was a year older than John. They had first become close during the summer of 1970 when Jackie invited Tim and Maria Shriver to fly to Skorpios to spend much of the summer with John and Caroline. On this trip to the Soviet Union, John experienced the intense world of the Shrivers, moving from factories to ballet, from receptions with ranking officials to attempts to learn what the average Russian felt and thought. For Tim it was not easy being around this younger boy who was the constant center of attention, celebrated for his mere presence, and the two youths became competitive. In Tim, though, John had found someone who would be one of his closest friends.

D uring the summer of 1975, a carload of agents followed Billy Noonan as he drove fourteen-year-old John and his sixteen-year-old cousin Anthony Radziwill around the Cape. Billy was two years older than John, and turning sixteen meant that he could drive. He was well over six feet tall, with a blunt, open face and the muscular build of the tackle that he played in high school. He looked more like one of the young Secret Service agents than like John's contemporary. Billy's father had served as head of the Small Business Administration in Boston during the Kennedy administration and had been a candidate for ambassador to Ireland. His father had died when Billy was thirteen. He and John didn't have to talk about that to realize what they had in common.

In some ways Billy was a throwback to the small legion of Irish Americans who had stood as a phalanx around Jack. Irish history is so full of treachery that loyalty is prized above all other virtues. It was out of Billy's sense of loyalty that all his other virtues grew. Even as a teenager, John understood and valued that quality almost as much as Billy did, and he gave as much as he received.

John was also like Billy in that he did not share the common American belief that the best way to be psychologically healthy is to vomit up everything about your life to your close friends and sometimes even to strangers. John did not spill his emotional innards to anyone. He appeared to be candid to his friends, but there were parts of him that nobody saw. He did not judge the closeness of a friendship by the number of secrets he told a person.

John may not have seen Anthony Radziwill as often as some of his other friends, but Anthony was the closest thing to a brother John would ever have. Anthony's mother, Lee, was Jackie's younger sister, and the two Bouvier women had been competitive with each other since they were girls. Jackie had married a senator, while Lee had married a Polish prince, Stanislaus Radziwill. Lee had a romantic involvement with Aristotle Onassis; Jackie married the man. Anthony still had a living father, but he was a child of divorce, brought up in London. The two young boys had identical long haircuts and looked more like brothers than cousins. They played together in the summers and could talk about their time on Onassis's yacht the way nobody else could. Onassis had died in March, with his estranged wife elsewhere. The event had little emotional impact on John.

John and Anthony were considered putative royalty when they only wanted to be boys. Anthony was back in America now, living with his mother and trying to drop his mannered London ways and become an American teenager. Anthony had the curious idea that summer that if he couldn't yet speak with an American accent, the next best thing was to affect a cockney voice so perfect that it would have gotten him safely through London's East End.

All of the young Kennedys were adept at the minor manly art of the putdown, but John and his friends carried it to a place far beyond the others. They didn't go at each other with wry ripostes but with brutal punches to the emotional solar plexus. If you couldn't take it, you didn't belong. Billy could seem as cruel in his remarks as any of the boys, but some of it was calculated. He fancied that it was his pedagogical imperative to knock John down when he got to swaggering. "When he got too full of himself, he'd get a noogie," Billy recalls fondly. "That's why our relationship lasted so long. I think he didn't have any other chance. He was all boy. He was leaving his coat, forgetting his hat, tripping over his untied shoes. And we had a lot of fun at his expense. That's how it worked."

It was harsh stuff, but Billy's barbs were love taps compared to the way John and Anthony treated each other. They went at each other as if they were playing a scene in *Who's Afraid of Virginia Woolf*, so brutal and mean that even their friends tiptoed away from the worst of it. Both Anthony and John seemed different from Billy and the other Kennedy youths; they were a bit precious, with a delicacy of manner unknown to most Americans. They tried to lose that, though, as they tore each other down, hung out at the beach, attended the dances at the Wianno Club, checked out the local girls, scored beer, and cruised country roads singing along to David Bowie.

J ackie did not want John to spend his vacation time only aimlessly amusing himself. In the summer of 1976, fifteen-year-old John traveled with his sixteen-year-old cousin Tim down to Guatemala to help out on a Peace Corps relief project. An earthquake had struck the Central American country, leaving about twenty thousand dead and wreaking havoc over most of the country.

Jackie would not have sent her son off with any of his other cousins, but she felt that with the Shriver son, John would experience the most idealistic and noblest part of his father's legacy. Their times together were good for both teenagers. Tim was almost too serious, and he benefited from his younger cousin's gaiety and seeming nonchalance. As for John, he experienced the Kennedy ideal of giving back to society—and in the form of an adventure with Tim.

Guatemala was in the midst of a bloody civil war, and the army, brutally attempting to put down a guerrilla insurrection, had killed tens of thousands, most of whom were little more than bystanders. The boys did not speak Spanish and would have been hard put to find their way in a countryside full of rebels and the army brigades that chased them. They were well protected since, as always, John had his Secret Service agents in tow, but he and Tim lived in a tent and ate the local food. All day long they rebuilt homes and dug outhouses, work that, in Tim's words, was "brutal and tough." "We were both sick the entire time," Tim recalls.

It was an adventure to be savored mainly in memory, but they stuck it out, and John learned things that summer that he would never forget. He saw the excitement in doing good and helping others. If there was a continuum of virtue, he learned, the highest level was to do things not seeking applause or rewards but for the intrinsic merit and reward in the act itself.

I n the fall of 1976, Jackie sent John to Phillips Academy in Andover, Massachusetts, an hour north of Boston. The prep school, known as Andover, had long since faced the melancholy realization that a gentleman's "C" was no longer enough to get a gentleman into Harvard and Yale. Andover had rigorous academic standards and expected long hours of study. In the meritocracy, the children of wealth and privilege had to run the same gauntlet as everyone else. It made for what some students felt was a tense, almost merciless environment; they leavened their long hours of study with drugs, booze, and sex.

The wealthiest, most self-consciously sophisticated students kept to themselves, while scholarship students and the academically ambitious trudged between school and library. Most students lived somewhere between the two poles. John entered the school his junior year, long after friendships and cliques had been formed. New arrivals were usually looked down on as intruders illiterate in the language of the place, but John was different, of course, arriving with a cachet that would have brought him immediate entrée with Andover's "beautiful people."

John could so easily have slipped into that world, but he consciously chose to stay away. He did not become part of that group or any other. He had shoulder-length, unkempt hair and dressed with what even by the loosened standards of the seventies was determined slovenliness. He was a lean, good-looking young man, but still an unformed adolescent largely indistinguishable from his classmates.

John was one of the worst students in his class; it did not help that because of his November birth date he was close to a year younger than some of his classmates. In the end, he achieved a distinction that he shared with his grandfather Joe at Boston Latin School: he almost flunked out and had to take an extra year to graduate. That didn't seem to bother John, since he was primarily determined to have a good time. Drugs would have a devastating impact on several of his cousins, and he was fortunate that marijuana was his drug of choice, not heroin or cocaine. His cousins pursued these more serious drugs as a journey into a netherworld. John's smoking of marijuana was hardly an ersatz trip into the unknown. In fact, it became a lifelong indulgence, doing for him what a martini before dinner did for many in his father's generation.

When John arrived at Andover, he brought with him a coterie of Secret Service agents who hung out at the Andover Inn. When John flew back to Boston after the Thanksgiving holiday, he came alone. He had just celebrated his sixteenth birthday, and that meant the end of Secret Service protection. Up until now, the agents had driven him to Andover from Logan Airport. As he picked up his bags, he realized that he had no idea how to get back to school. He called his mother, who hired a car to take him the twenty-five miles to school.

Jackie was, of course, the dominating influence of John's life. Growing up with Jackie made him comfortable with sophisticated, intriguing women. It was natural then that when he fell in love for the first time, it was not simply with one of the real beauties of the school, but with one of the more interesting and complex of his fellow students. Jenny Christian was a young woman who intimidated many of his classmates, in part because of her keen, often devastating perceptions about her contemporaries.

John swam into this relationship with Jenny head-on, appreciating both the sensuality and the psychological intimacy. His father had been incapable of treating women as intellectual equals. Whenever important political decisions were to be made, no woman was in the room. Unlike his father, John could have women as lovers and as friends, and that a woman was his lover did not mean she was less a friend.

John didn't feel a need to have a sexual relationship with every woman he knew, and he enjoyed the company of women generally. That set him apart not only from his father but from almost all of his male cousins—and much of the adolescent male population, for that matter.

During that first year he befriended Sasha Chermayeff, who became one of his two closest lifelong female friends. Sasha came from a family of artists. When John talked about dining with senators as a boy, she told of meeting Salvador Dalí when she was ten years old. Like John, she entered the elite school her junior year. Some of her new classmates considered her weird. She had a poetic, bohemian soul rarely seen at Andover. She did not care about *things*, power, money, scrambling. She later became an artist in her own right and pursued a country life in which family, friends, and nature inspired her art.

What drew the two together originally was not being kindred souls, but the two worst math students in their class. No one wanted to squander time studying with them, but it was an open question what they could learn from each other. In their unsuccessful attempt to unlock the mysteries of math, they started talking about life, and thus began their friendship. John was a great friend in part because he had a profound sense of empathy. It was a rare trait in someone brought up in such a privileged, protected manner. John could place himself in another person's world and appreciate his lot, an immense attribute in any profession he might pursue, from acting to journalism to politics.

"He had this beautiful, sweet, kind, loving side," Chermayeff says. "He was someone I could go to and he was always going to be warm and honest. How many people in your life do you have that you can always go to for a clearheaded view of things? He had a good, objective, intelligent opinion on everything, a warm way of helping you see through your strengths and weaknesses. He wasn't a liar. He wasn't deceptive. He wasn't manipulative. He didn't try to use people."

These were extraordinary qualities in one of the most famous teenagers in the world, brought up in a world of endless entitlement. If some of these qualities came from his mother and sister, then part must be credited to John himself.

John met Ed Hill, his closest male friend at Andover, the day Hill roared into the dining room and tossed a Frisbee across the tables. This was forbidden, and everyone in the room simply sat there—except John, who jumped up, grabbed the Frisbee, and threw it back. That led to them both being ordered out of the dining room and the beginning of a lifelong friendship. Hill was a popular student and had been elected school president. He was an excellent athlete, a letterman in cross-country and track and field, and his physical daring verged on recklessness, all qualities that John admired.

Hill might have seemed to have little in common with Sasha, but he had his dissident side as well, and not just when it came to winging Frisbees across the dining hall. He was later thrown out of Andover for more serious misconduct. "John's friends were all orphans and misfits, although I am not too comfortable with the term 'misfit,' except for its literal meaning," says Hill. Jenny, Sasha, Ed, and most of the friends John made later in life were just a little off-kilter from the way most of their peers acted or lived. They were never quite one of the group. Neither was John or his mother.

"The trouble with me is I'm an outsider," Jackie admitted in one of her most telling remarks. "And that's a very hard thing to be in American life." She was an outsider to the girls at her private school, to the Kennedys, to political wives, and to the White House. She was loved by Americans in part because she was so different. Despite all the surface appearances, John too was something of an outsider, an orphan comfortable with those of his own kind.

John would have deeply loved to be a great athlete. Since he could never play tennis, football was his chosen sport, even if it was just for pickup games on the broad fields of Andover. For all his nervous energy, however, he was hopeless. He threw the ball with a girlish sidearm, stumbled downfield pigeon-toed, and lost the ball to opponents cutting around him or slapping the ball away when he reached for a pass. He was an elegant swimmer, though, and he developed into a fine skier.

In the summers Jackie sent him off for outdoor adventures that did not depend on the ability to throw a spiral pass or sidestep a determined foe, and here he found something that he loved and was good at. In June 1977, he spent twenty-six days on an Outward Bound program off the coast of Maine.

The following summer Jackie sent seventeen-year-old John west to work on a 22,000-acre Wyoming cattle ranch. The Barlow ranch was run by thirty-one-year-old John Perry Barlow, a lyricist for the band the Grateful Dead. On their album released that spring, he had penned the words to "The Music Never Stops"—a precise statement of the way he lived life. He was a scion of the flower child sixties who hung out with the Dead and had dropped acid with Timothy Leary. His was hardly a résumé that suggested he was a proper guide for a young man being groomed for an exalted place in the American establishment. It said much about Jackie that she would even think of sending John to Barlow's ranch, and it said much about John that Barlow became a lifelong friend. Barlow was quirky, witty, dynamic, irascible, and blessed with an acute sense of language, all qualities that John admired.

"There are some people you meet, and two minutes later you know they're going to be a lifelong friend," Barlow says. "He was rambunctious and like a lot of seventeen-year-old boys, with more energy than sense. He was bouncing around a lot, but get him wound down on something—tell him to dig a posthole through six feet of glacier rubble—and he'd go after it like he was killing snakes."

At Andover, John found a niche for himself as an actor. He appeared in several productions and starred as the "bull goose looney" McMurphy in *One Flew over the Cuckoo's Nest*. Ken Kesey's dark, comedic masterpiece was hardly typical high school fare. John had something of McMurphy's irrepressible thirst for life and his sense that at times irresponsibility is the highest responsibility.

Few came to the theater to watch John because he was the son of a martyred president, and the applause at the end was well earned. When he flew to New York to be with his mother, he saw what celebrity truly was, and most of the time it had little to do with him. Billy was in town one evening, and Jackie decided to take her son and his friend out to dinner. Jackie had not made reservations at Patsy's, an Italian restaurant on West Fifty-sixth Street where she had gone with Onassis. She stood outside, not sure that this was the place. The maître d' hurried out, greeted Jackie effusively, showed her to the best table in the house, and insisted on ordering the dinner himself. John was used to the impact his mother had, but for Billy it was new, and the two boys watched delightedly when the waiter brought a bottle of wine and poured it for them as well.

"We were having a glass," Billy says, fondly recalling the illegal drink. "Then this woman enters the place. Real New York. She's got the gait, the fake nose, hair coiffed, the whole bit. She comes flying in and looks around at who's looking at her, and then, oh, my God, she sees Jackie, and she loses her composure and her whole Paramus, New Jersey, upbringing comes out. She thought she was the belle of the ball, but she loses it, and we're chuckling and laughing."

This was one of the last times that John could stand at the edge of the spotlight. On November 26, 1978, for his eighteenth birthday and Caroline's twenty-first, Jackie took over a private club with a French motif. The whole family was there at Le Club when Ted gave the welcoming toast. "I shouldn't be doing this tonight," he said. "By rights, it should have been the father of these two children. Young John and Caroline bring new life to the family."

The older Kennedys departed by about one, leaving the club largely to John's friends from Collegiate and Andover. Then the young began to party, passing around a joint or two, dancing to rock 'n' roll, and necking in the corners.

Around four in the morning John and his friends came reeling out of the place to be confronted by a swarm of paparazzi. Billy stood at John's side. Billy was not a great admirer of New York or what he considered the overprivileged wimps at the party, but he was John's friend. It sickened him to see these parasitic photographers pushing their way into things that were none of their frigging business.

In the chaos John was pushed down, and Billy stood there in a pugilistic pose worthy of John L. Sullivan himself, protecting his friend from this unseemly onslaught. It was a tense, nasty moment, immortalized a week later in a full-page photo in *Life* magazine. John had been the subject of the media since he was a little boy, but most of the time as ancillary to his mother as he held her hand or arm when she was pursued by the media. In that birthday glare of flashbulbs, eighteen-year-old John saw what his public life would be like from now on.

J ohn had another hint of the public life of a Kennedy the next fall when he went up to Boston to join his family for the dedication of the John F. Kennedy Library on October 20, 1979. The original plans were to build the library in Cambridge and make it not merely a repository of documents but a vibrant center of learning. That grandiose scheme had not worked out, and instead the starkly sweeping I. M. Pei building was built at Columbia Point in Dorchester. The library would prove even more isolated intellectually than physically, never becoming an important part of the cultural life of Boston.

This bright fall day, however, began full of promise in the presence of twenty-eight of the twenty-nine Kennedy grandchildren. Despite the fact that the cousins were sitting together, they brought with them their squabbles and tensions. Joe and several of his siblings seemed jealous of John and Caroline, who were treated with a degree of deference the others did not receive. The Bobby Kennedy children were convinced that their father represented the most authentic part of the family tradition. Joe and his brothers and sisters were irritated that so little of this day and of the library itself was devoted to their father.

Almost the entire Kennedy family was there that day, along with President Jimmy Carter and other dignitaries. Within a few days Ted planned to enter the presidential race to challenge Carter, but this was a day to put that aside and commemorate the Kennedy presidency.

John read Stephen Spender's poem "I Think," a great poem keenly felt and well delivered. There was an understated elegance in President Kennedy's only son that was first on public view that day.

Then Joe got up. He had worked for a while in the Community Services

Administration, where he decided that these agencies "were bureaucracies that needed poverty to stay in business." That was a perception his father had had over a decade before, and little had been done to change the reality. Joe spoke the words he believed his father would have spoken if he had been there to address these seven thousand people. "My father died waging a struggle," he said. "As I have grown up, I have come to appreciate what that struggle was about." He then talked of the groups that his father had helped—migrant workers, Chicanos, blacks, and Eskimos. "One day it may be different here," he shouted, "but only if we acquire what my father called 'moral courage.' "

Joe stood facing the president of the United States, both his finger and his words pointing at him. Carter was a far better man than he was a president, and he could have told Joe how difficult it is to change the world. "I hope and pray that my generation will work to bring about the decent and just world he [RFK] so much wanted to see," Joe said dramatically. "We all know inflation bears down hardest on the poor. Well, what about the standard of living of the people on the boards of the giant oil companies that are squeezing us so hard all the time, and who is stopping them?"

Joe's unspoken answer was that he would soon be the one. He was starting a company, Citizens Energy, to provide cheap housing oil to Massachusetts poor. This speech was not only an idealistic hymn but a calculated political coming-out. His brothers and sisters and cousins applauded with fervor, as if he were their spokesman too. But first there was the matter of Uncle Ted's campaign to wrest the presidency from Jimmy Carter.

Keeping the Faith

T he Edward Moore Kennedy who was planning to run for the 1980 Democratic presidential nomination against Jimmy Carter was a man whose public accomplishments were as great as his private life was troubled. In the years since Chappaquiddick, Ted had grown not only into one of the most powerful senators and the titular leader of American progressives, but a subtle legislator with political gifts acknowledged by senators on both sides of the aisle. Unlike many of his liberal colleagues, Ted did not simply settle for the forceful, emotionally satisfying profession of his ideals but remained a practical politician focused on subtly pushing through his agenda.

Some observers like to categorize senators as either workhorses or show horses. Ted was a workhorse who understood that almost nothing important happened in the Senate without hours and hours of detailed work. While he did not spend days researching issues, preparing testimony, and interviewing experts, he took all that his subordinates had done and used it.

But Ted was equally a show horse. He knew and loved and was immensely talented at the spectacle of politics, the soaring rhetoric of a speech, the glad-handing and the backslapping, the pitching of his activities so that he would see his face on the evening news and read his name in tomorrow's papers.

The Democrats had taken control of the Senate after the 1976 elections, and when Ted assumed the chair of the Judiciary Committee, he had close to a hundred people reporting to him. He was at the center of a machine whose work was to further his legislative agenda. Ted stamped his name on the work of scores of aides and researchers. Each night he left his office with a briefcase full of dozens of memos, directives, and letters. He almost unfailingly returned the next morning with the letters read, the memos digested, the decisions made.

Ted was inevitably one of the most vocal proponents of raising the minimum wage, broadening health insurance, expanding medical research, and providing more aid to the nation's schools. To his conservative critics, he was a Santa Claus who was up the chimney and away when the bills started arriving. He was a political child of the Johnson era as much as the Kennedy era, when the Democrats thought that they could deliver guns and butter, massive social programs and fiscal responsibility.

In some of his most important actions Ted stepped beyond traditional liberalism to support programs and issues that even the most stalwart of conservatives would have applauded. In a classic example of how Ted worked, early in 1974 he had dinner with Harvard Law professor Stephen G. Breyer, who bemoaned the excessive, stifling regulation in the airline industry that kept fares artificially high. The senator convinced Breyer to take a leave and come down to work on the subcommittee on administrative practice and procedure that Kennedy chaired. Breyer spent months developing sources, researching the issues, and prying information out of the reluctant Civil Aeronautics Board (CAB).

When the first hearings were held, Ted had a stellar witness in Freddy Laker, the flamboyant British millionaire who wanted to start $125 New York–London charter flights on Laker Airlines, which would severely undercut the scheduled carriers. It was populist drama, and Ted played it brilliantly, as he did the next day when he took CAB chairman Robert D. Timm to task for preventing Americans from flying by refusing to set fares they could afford.

The CAB did not overnight go out of business, smashing its regulations as it left, but Ted had set in public motion an increasingly irresistible idea.

T*wo-year-old John Kennedy Jr. watching his father leave by helicopter from the White House lawn.* (JOHN F. KENNEDY PRESIDENTIAL LIBRARY)

F*or the children of Camelot nothing was more exciting than a ride with President Kennedy in an electric cart at Hyannis Port.* (JOHN F. KENNEDY PRESIDENTIAL LIBRARY)

P*resident Kennedy with the next generation of the family on August 3, 1963.* (JOHN F. KENNEDY PRESIDENTIAL LIBRARY)

Robert F. Kennedy had a special feeling for his third son, David. (BURTON BERINSKY)

In 1964, Robert F. Kennedy campaigns for the Senate with his sons (left to right) *Robert Jr., Joseph II, and David.* (BURTON BERINSKY)

At a memorial service for Senator Robert Kennedy, his brother Senator Edward Kennedy stands behind (left to right) *Michael and David Kennedy and Christopher Lawford.* (TETI/MILLER)

R*obert Kennedy's widow, Ethel, and his children kneel at Robert Kennedy's graveside.*
Left to right: *Courtney, Ethel, Michael, Joe, Douglas, Christopher, Max, and Robert Jr.*
(half visible). (TETI/MILLER)

*J*ohn F. Kennedy Jr. and
Edward M. Kennedy Jr. at
the Robert F. Kennedy tennis
tournament. (TETI/MILLER)

*T*hree of the younger sons of Camelot: (left to right) *Anthony Shriver,
Douglas Kennedy, and Patrick Kennedy.* (TETI/MILLER)

A *youthful Timothy Shriver, the second son of Eunice and Sargent Shriver.* (TETI/MILLER)

R*obert Kennedy Jr.* (left) *and Joseph Kennedy II, brotherly competitors.* (TETI/MILLER)

Christopher Kennedy (left) *in earnest conversation with his big brother Joseph Kennedy II.* (TETI/MILLER)

Three Kennedy teenagers: (left to right) *Robert Kennedy Jr., Caroline Kennedy, and Robert Shriver III.* (TETI/MILLER)

Patrick (left) *and Edward Kennedy Jr., brothers and friends.* (Teti/Miller)

R*obert Shriver III* (left), *Robert Kennedy Jr., and* (front) *Michael Kennedy.* (Teti/Miller)

Three boys looking for fun in 1972: (left to right) *John F. Kennedy Jr., William Kennedy Smith, and Edward Kennedy Jr.* (TETI/MILLER)

The young Patrick Kennedy was asthmatic and often sick. (NEW YORK NEWS SERVICE INC.)

The teenage Edward Kennedy Jr. did not let his amputated leg slow him for a moment. (NEW YORK NEWS SERVICE INC.)

Robert Shriver III with his father, Sargent Shriver. (TETI/MILLER)

Mark Kennedy Shriver and Jeannie Ripps on their 1992 wedding day. (DENIS REGGIE)

D*r. William Kennedy Smith, who brought notoriety to the Kennedys when he was tried unsuccessfully for rape in Palm Beach, Florida.* (NEW YORK NEWS SERVICE INC.)

M*ichael Kennedy, a great athlete, kayaking with his two children.* (NEW YORK NEWS SERVICE INC.)

Maxwell Kennedy and his wife, Vicki Strauss Kennedy, with their newborn. (NEW YORK NEWS SERVICE, INC.)

David Kennedy and a friend shortly before his death in 1984. (NEW YORK NEWS SERVICE INC.)

Joseph P. Kennedy II off with his twin sons, Joe Jr. and Matt.
(New York News Service Inc.)

Senator Edward Kennedy and Mrs. Vicki Reggie Kennedy at the Cape. (NEW YORK NEWS SERVICE INC.)

As Senator Edward M. Kennedy grew older, he looked more and more like his grandfather John "Honey Fitz" Fitzgerald. (NEW YORK NEWS SERVICE INC.)

He came back to it again and again, not only in pithy sound bites that made it to television, but with formidable documentation to back up his case until the airlines were deregulated.

During the 1968 Indiana primary Ted had watched how appalled Bobby's aides had been when he had spoken about law and order, considering it a code name for all that was dark in America. But Ted picked up the issue and sought—with the help of none other than Senator Strom Thurmond of South Carolina—to rewrite and in many places to toughen up the federal criminal code. Conservatives were fed up with judges meting out light sentences, and the legislation narrowed judicial discretion. The 297-page bill passed the Judiciary Committee 12 to 2 in November 1977, and then passed in the Senate. A revised version of the bill passed in the House and into law finally in 1983, with measures that were anathema to many progressives.

Ted began to take on an increasing number of issues in the foreign arena. On a trip to Russia in 1974 he met with Chairman Leonid Brezhnev and at night visited with a small group of dissidents in a Moscow apartment. He pushed the Soviets to allow Jewish "refuseniks" to immigrate to Israel. Ted's importuning of the Soviet leadership was an important factor in that government's eventual decision to allow hundreds of thousands of them to leave.

Ted worried publicly and tried to stem the massive flow of arms out of the United States whether it was to the repressive Pinochet regime in Chile or to the Middle East. During a trip to China he reiterated his theme that human rights were a natural right of all peoples. It was a lecture heard many places in many ways, but rarely by the Chinese Communist leadership.

Some Americans would have found laughable the idea that Ted was making a record that would mark one of the great legislative careers of the twentieth century. They read in the tabloids about another Senator Kennedy. Ted was not out carousing night after night, but when he was, he at times made a spectacle of himself that did no honor to his name or his contributions to American life. But the next day he would be back in the Senate hard at work.

There was such a dichotomy between these two Ted Kennedys that it was like a jigsaw puzzle that no matter how much you played with the various pieces, you could never quite put it together. The Democrats and progressives heard the stories of Ted driving his convertible recklessly to National Airport at the last minute, running up on the curb or in the wrong lane. They whispered about him getting drunk at parties, propositioning guests, and sometimes not even remembering where he was. They heard that the police in Palm Beach had what they called the "Ted watch": they would gently haul the teetering senator out of one watering hole or another and guide him back to the family estate on North Ocean Boulevard.

These were not just trivial indulgences. Burton Hersh, a respected author with considerable access to Ted and his associates, writes in his 1997 book *The Shadow President* that some of the senator's aides "found themselves bird-dogging a live prospect across a crowded reception, under orders to recommend a drink in the back of the senator's limousine, or even—there were verified claims—a line of coke."

Ted's admirers heard these stories, but to them this was not Ted Kennedy. Ted Kennedy was the magnificent political leader who took their ideals and values to places that without him they might not have reached. Ted Kennedy was a man carrying such overwhelming burdens that it was no wonder he occasionally indulged himself. The senator had an alcoholic wife and troubled teenage children and nephews, all of whom needed him to be a strong, powerful symbol of how a Kennedy man should live. As they saw it, this man did so much good that he should be left alone, his private life cordoned off from public view. What they rarely asked was, what kind of a leader might he have become if he had faced his demons and purged them from his system?

Ted's enemies heard much the same stories, and others that were even worse. They felt that this was the real Ted Kennedy, an emotionally corrupt, adulterous drunk who pushed so-called progressive legislation that was the perfect product of such a dissipated mind. To them he epitomized everything fraudulent in American liberalism.

In December 1979 Ted's sexual conduct was exposed for the first time in a respected publication. In "Kennedy's Woman Problem, Women's Kennedy Problem," published in the *Washington Monthly*, Suzannah Lessard found Ted's behavior startlingly incongruent, so different from the sensitive public man who stood on the side of the angels on issues of women's equality and rights. "Kennedy's womanizing is widely known to the many women who have been approached themselves and to reporters and others who have been around Kennedy and have seen the pattern in action," Lessard wrote. "The type of womanizing that Kennedy is associated with is a series of short involvements—if they can be called that—after which he drops the lady. It suggests an old-fashioned, male chauvinist, exploitative view of women. . . . It gives me the creeps."

Richard Burke, Ted's administrative assistant from 1978 to 1981, developed a serious cocaine habit and after a breakdown left the office. During his time working for the senator, he was astounded at how Ted managed everything. "He'd take home briefcases full of work, and he would do it even if he had a party going on or he was going to be out late with the kids or he was going to be out late partying, he would end up finishing it," recalls Burke. "Even if it meant coming over at five o'clock in the morning and going through papers and notes and everything. If we gave him a packet that had to be done that night, it was done."

And yet Burke had the feeling that Ted could not go on forever without changing his conduct. "I can remember when I had my episode, and I was like saying, 'Not only do I need help, but the senator needs help,'" says Burke. "Larry Horowitz [a physician, top health aide, and later Ted's administrative assistant] came to me and said, 'Rick, there's so much scar tissue on that man, don't even attempt to try to ever think that he's ever going to change or that anybody is going to be able to help him.'"

I n mid-1979, Carter was so weak and unpopular that Democrats feared he not only would be defeated in 1980 but would bring down scores of senators, representatives, governors, and other elected officials in the process. Ted was the most viable challenger to Carter. His was the liberal

agenda favored by most party loyalists, and he was leading the moderate
president in the polls by an unprecedented three-to-one margin. "Kennedy
right now can decide on his own volition whether he wants to be President,"
said Mervin Field, a leading pollster in August 1979. "Ulysses Grant and
George Washington are the only others I can think of who found themselves
in that position."

Ted's brothers had never had such affirmation when they contemplated
running, and Ted found it an irresistible opportunity. "I think it has been
advantageous to be a Kennedy," he said as he announced his intention to
embark on a quest that had been there for him since 1968. "My father, you
know, set a high standard that, you know, coming in first was important.
And I think the advantage was the association with my brothers and my
father. The disadvantage is the constant measurement to their very high standards. I hope I can run a campaign for the presidency that will measure up."

Ted's major problem was probably not his progressive politics in an
increasingly conservative age but his inability to deal with the emotional
truths of his own life. Joan had moved to Boston in the fall of 1977 to live
what she said was her own life. The couple could have announced that they
had separated, but that was unthinkable. Ted flew up occasionally to have
dinner with Joan; these evenings were so unlike a husband and a wife going
out together that Joan considered them "almost like having a date."

When Ted was about to announce his candidacy, Joan flew down to discuss her role. It was a matter between a husband and a wife, but at the meeting Joan was confronted with a panel that included a psychiatrist, three other
doctors, Eunice, Ethel, and Ted. This was her and Ted's marriage they were
talking about, but a conclave of experts and family members had been sum-
moned to determine whether Joan was up to campaigning. It said so much
about the emotional realities of Ted's own life that he could not sit down
with his wife and discuss honestly whether there were enough shards of feeling left between the two of them to justify going on this journey together.
The question asked was whether Joan was up to helping Ted and the family
to achieve their destiny. Joan said yes, and if in part this was a generous act,

it was also a way to stop thinking about her alcoholism, to reconnect with her husband, and to have the public role to which she aspired.

T ed was alone at the end of September 1979 at his house on Squaw Island when Roger Mudd flew up to Hyannis to interview the senator for an hour-long *CBS Reports*. Ted liked and respected Mudd, who had a reputation in Washington as sympathetic to the Kennedys and their causes. He was a natural choice for Ted's first important interview of the campaign. His staff worried about the way the senator occasionally appeared in interviews. There were times when Ted mumbled, mouthing inane non sequiturs, stumbling along from sentence to sentence. He often did this to avoid questions he found impossible to answer, but with Mudd he would get a gentle test-run for what he would face later.

Ted was immediately taken aback and confounded by the tone and content of Mudd's questions. Any politician has responses ready to the inevitable tough questions, but Ted did not have them this day. When Mudd asked him about the status of his marriage, he had no stock answer worked out beforehand.

"Are you separated or are you just . . . what?" Mudd asked.

"Well, I don't know whether there's a single word that should . . . have a description for it," he said, sounding like a shoplifter caught with the goods. "Joan's involved in a continuing program to deal with the problems of . . . of alcoholism, and . . . and she's going magnificently well, and I'm immensely proud of the fact that she's faced up to it and made the progress that she's made. And I'm—but that progress continues, and that . . . it's the type of disease that one has to . . . to work."

Ted was no more articulate talking about Chappaquiddick. As Mudd pressed him, Ted showed his irritation by giving the journalist only tiny scraps of information, holding himself tightly controlled. He didn't even give a strong answer as to why he was running for president.

The *CBS Reports* interview aired on November 4, 1979, three days before Ted formally announced his candidacy, and the reaction was devastating—an inarticulate, uncertain Ted seemed to become imprinted on the minds of the media and much of the public. That same day Iranian students overran the American embassy in Tehran and took sixty Americans hostage, a move that dramatically changed the tenor of Carter's presidency and of those who would speak out against him.

As Ted set out on his futile quest, several of the young Kennedys were afraid for their own lives as well as for their uncle's. Ted's three children were the most immediately affected. During the summer of 1979, when he was still debating whether to enter the race, his aide Larry Horowitz talked often to Teddy, Patrick, and Kara to convince them that the violent tide had receded and that their father would be well protected by the Secret Service.

Ted himself was afraid in a way that his brothers had not been, and he had good reasons to be. He was haunted by memories they did not have, and stalked by dangers of which they had not been so immediately aware. He was the focal point of so much rancor and madness. He received about ten letters a day so full of hate that they were considered possible death threats. There had been crazies in his house in McLean, one discovered hiding under a living room chair. There were the direct and immediate death threats, many of which he was not even told about. Some of these were hushed up in part for fear that they might inspire others. As he went out across the country, there was sometimes someone with a weapon at the edge of the crowd. On the campaign trail he often wore a bulletproof vest, but that did not assuage his fears that there might be someone out there inspired to complete the deadly trilogy by killing the last of the Kennedy brothers.

Ted believed that he stood there on those podiums representing not only himself but Jack and Bobby, the family and the ideals for which the Kennedys stood. Thus, as he lost primary after primary in the first months of 1980, the results seemed not merely a loss but a repudiation of both the candidate and his family. He had brought family members out to Iowa in droves, including even his eighty-nine-year-old mother, but in January 1980, Carter won almost twice the votes Ted received. Ted moved on from there to the first primary in New Hampshire. He was in the most exuberant of moods as he introduced his family in Manchester. "I just want to introduce you to a few members of the family," he shouted. "Heeeeere's my son Patrick! Say hello, Patrick." His twelve-year-old son could hardly bring himself to stand up and acknowledge the applause. Of Ted's three kids, he was the most frightened at the possibility of losing his father. When the boy was not with his father on the campaign trail, Ted had to remember to call him every night so Patrick would know that his father had lived another day.

"And there's my daughter Kara! All right, Kara! Then my niece, Maria! Okay, Maria! And here's one who has been around quite a lot—my nephew Joe! Now, just because he ran my Senate campaign in 1976 and is doing a lot of work for me now, Joe certainly doesn't want to run for public office—in Massachusetts or anywhere else. Right, Joe? There are a lot of Kennedys, and you'll see most of them before we finish in New Hampshire. And you'll be seeing them all long after that! There are more Kennedys coming!"

Ted's words suggested the inevitability of a new generation of Kennedys entering public life, this glorious campaign being their entry point into the family profession. New Hampshire was the next door neighbor, and losing the Granite State was like the Georgian Carter losing South Carolina, but lose it Ted did, another harbinger of what was to come.

When Ted had been contemplating running, Carter said that if he entered, "I'll whip his ass." It was not the language expected of a born-again Christian, but it was an accurate forecast of what happened. There are few more dispiriting experiences than a political campaign that starts out with immense promise and then slowly dissipates. For most of the young generation of Kennedy men, this was their first political campaign. For many of them it was a disillusioning experience. They had been brought up with the idea that their family was uniquely revered across America, and they were startled that their name frequently evoked hate, anger, and studied indifference.

The young Kennedy men did not all act the same way. Tim Shriver cringed at being hauled up on stages as a Kennedy, as if his own identity meant nothing, while his younger brother Mark found it all matchless fun. Other young Kennedys found excitement in the campaign drama, especially

Joe, who was ever ready to defend his uncle's honor with a quick retort, and Bobby Kennedy, a natural campaigner in the most hostile of venues, traveling Carter's South looking for Kennedy votes. "I took a great deal of time off from high school to campaign," says Chris Kennedy. "You go out to Iowa, fifteen or sixteen years old, go knock on doors, areas of the country that we never understood existed. You have no idea. 'Hi, my name's Chris Kennedy, and I'm hoping that you'll think about supporting my Uncle Teddy in the next. . . . ' And you're scared out of your mind."

Ted made a few gaffes in the campaign, most notoriously telling a San Francisco television journalist that the shah of Iran "ran one of the most violent regimes in the history of mankind—in the form of terrorism and the basic and fundamental violations of human rights, in the most cruel circumstances, to his own people." There were elements of truth in what he said, but it was hardly the kind of remark to make while the anti-shah militants held Americans captive in the Tehran embassy. It was not an occasional malapropism or unguarded sound bite that doomed Ted, however, but a reputation that appeared to be as much personal as political. In the last weeks of the campaign he won important races in California and New Jersey, but those victories did not prevent Carter from garnering enough delegates to guarantee his nomination.

At the Democratic Convention in New York City in August, Ted gave a winner's speech, not for the nomination, but for the liberal ideals that he espoused. He said that he was there that evening "not to argue as a candidate but to affirm a cause . . . the cause of the common man and the common woman."

"Someday, long after this convention, long after the signs come down and the bands stop playing, may it be said of our campaign that we kept the faith," Ted continued. "May it be said of our party in 1980 that we found our faith again. . . . For me, a few hours ago, this campaign came to an end. For all those whose cares have been our concern, the work goes on, the cause endures, the hope still lives, and the dream shall never die."

In his speech Ted evoked the dormant emotional soul of the Democratic Party. He had done for his party what Ronald Reagan had done for the

Republicans four years before in his speech at the Republican Convention after losing the nomination to President Gerald Ford. The delegates applauded and screamed and stomped for a good half-hour, and Ted left first in the hearts of the Democratic stalwarts, if second in their votes.

On the final night of the convention, Ted returned to Madison Square Garden to stand on the podium with Carter and other leading delegates after the president's acceptance speech. The two politicians were ready to make up publicly, but not in their hearts and minds. Before arriving at the arena, Ted had practiced the stereotyped political ritual of lifting his arms together with the victorious Carter, but when he walked out on the stage, he limited his congratulations to a handshake.

For eight months Ted had largely set aside his personal problems and marched ceaselessly along the campaign trail. Joan had been there for many important occasions, and she had served him well. But he had rarely mustered the obligatory affection that a husband was expected to show a wife, a kiss on the cheek, a hug, a moment walking hand in hand, solicitous help into a limousine.

Joan thought it was time to have a serious discussion about their future together. The couple flew back to Washington, where they gave a party for campaign workers and the Secret Service agents who had protected Ted. In the midst of the melancholy event, Ted whispered to his wife that afterward they would be flying up to Hyannis by private plane. On the flight Ted kept busy going through papers and documents and said almost nothing to Joan until the plane landed at an airport unknown to Joan. Ted told his wife that they were at Montauk Point on Long Island. He said that he would be getting off there, and the plane would then fly her on to Hyannis. He deplaned, and Joan sat there by herself, and though she did not have the conversation that she so much wanted, she knew that their marriage was over.

CHAPTER TWENTY-TWO

Bobby's Games

obby was living at Winthrop House, where his father and three uncles had roomed, but his years at Harvard were unlike theirs. He didn't try to distinguish himself in any accepted way, but he was one of the best-known students on campus. He had a sheer physical presence that by itself called attention. He was six feet two, with a twenty-eight-inch waist and eyes of searing intensity. He loved to dance along the precipice of danger—literally. He once stood on the roof of a six-story building and jumped to the next building ten feet away. He kept a fanged rattlesnake in his room. When he was caught speeding on a ski trip to New Hampshire in March 1973, he was going almost double the thirty-five-mile-per-hour speed limit.

Bobby rarely traveled alone but with an entourage. Over the next years Bobby and his brothers, cousins, and friends became familiar figures on the hip club scene in downtown Boston. Chris Lawford, a student at Tufts, was usually part of this group, as was Bobby's younger brother Michael as soon as he was old enough to brush past the bouncers. They arrived in a pack, standing together, looking out at the dancers and drinkers. Bobby was so hyper, so intense, that he could hardly stand long enough to chug a drink before he looked nervously toward the entrance. Then they headed out, a phalanx of Kennedys, into the Boston night.

Joe may have been the oldest of the Kennedy children, but Bobby was the anointed one, the one with intelligence, wit, and daring. In August 1977,

Parade quoted a "Kennedy insider" as saying that "while Joe would make an excellent public servant, he doesn't have the intellect or the temperament that Bobby has. Joe has quite a temper. Bobby is the one to watch. He is bright as a star, a damn good speaker and writer, and he is definitely going places."

Other students might smoke marijuana or drop acid a few times, fancying that they had traveled to the other realms of experience, but Bobby and his friends did heroin. "I was shocked at the number of people using heroin," recalls David Humphreville, who became one of Bobby's closest friends. "Stunned. And shit, these guys took me up to Harlem ten times over the years, and we climbed the stairwells with no stairs in them and all that. And just the people I would see up there, the guys in suits, you know, we ran into people we knew. Shocked the shit out of me. But it was trendy, and people wanted to fit in and show that they were cool and they were part of things. There was a lot of peer pressure. It was the thing to do to stand out in those circles. They were all trying to top each other."

Bobby was no junkie sitting in a squalid room shooting up. Drugs were just another trip. He could take it further than anybody else and always arrive back home. While studying for a law degree at the University of Virginia, he managed to write a biography of the esteemed Alabama judge Frank M. Johnson Jr.—while he continued to do drugs. And sex was yet another arena to experience. "Those guys tell stories about women," recalls Humphreville, who lived with Bobby in Charlottesville, Virginia. "I saw it. Someone came knocking on our door one morning at four. Some girl was there. She wanted to see Bobby. You know, you just think, shit, that's certainly nice in some respects when you're in school, but it's also hugely debilitating."

"He had such a great sense of fun and adventure," says Chris Bartle, another close friend and Bobby's roommate at the University of Virginia Law School, "that if you're a young man in your twenties, it's the easiest thing in the world to gravitate to him around whom a lot of interesting fun things happened. They happened because of who he was. They happened because of what he did. They just happened because he made them happen."

At law school Bobby often headed out with his friends in the afternoon to go horseback riding. These were not trots along horse trails but gallops across pastures and unmarked terrain. The rides challenged even his friend R. Kent Correll, who had grown up on horses and technically was a far better rider than Bobby.

When Bobby's dear older friend Lem Billings visited, he was equally fearless, jumping fences alongside Bobby, urging his steed up the mountainous terrain. Jack's Choate classmate was obsessed with Bobby. Lem had begun mentoring the fifteen-year-old youth in 1968, but by the midseventies Bobby was doing the mentoring. Lem grew his hair long and took the same sacrament of drugs. And wherever Bobby led, Lem followed. On one rain-swept day Bobby took the group across a stream and up a slippery bank. Lem's horse stumbled, fell, and rolled over on Lem, who lay there apparently either dead or unconscious. "Then Lem said in his distinctive voice, 'Gee, Bobby, this is a lot of fun,'" recalls Correll. "And he just got up, got back on his horse, no complaints, and we went on with the ride."

I n the summer of 1974, Bobby led a rafting expedition on the Apurí-mac in Peru that included David, Chris Lawford, Doug Spooner, Harvey Fleetwood, a journalist who had come to write about the adventure, and of course Lem. The Apurímac was a wild, uncharted river, yet Bobby and his friends set out with little of the equipment, and taking few of the safeguards, that any seasoned explorer would have recommended. Bobby's father had taken the family on rafting trips in the Grand Canyon and else-where, but compared to what Bobby was attempting, that was like riding on the swan boats in the Boston Public Garden. As for his Harvard classmates, they may have gone on their tours of the Continent or sailed up the coast of Maine, but none of them had gone where Bobby had gone, and they knew it as much as he did.

Bobby set out again in the following years for other great adventures, culminating in a journey in 1980 to Venezuela. As usual, Bobby was the impresario, charging a fee from the other participants, gathering supplies, and figuring out logistics. This rafting trip began at the headwaters of the remote, largely unexplored Caroni River in the jungles of Venezuela. The group of about twenty included Bobby, his brothers Michael and Max, his cousin Chris Lawford, Lem, and Bartle. Lem was sixty-three years old, at least twice the age of the others. By most assessments he should not have been there, but his life was Bobby and his brothers. "How could you *not* have let Lem go on that trip is the real question," says Bartle. "I mean he would have never forgiven Bobby."

Michael first observed the Caroni from a Venezuelan Air Force helicopter flying above the river. He reported back that it looked tough but doable. But once a military transport plane dropped the group and their several tons of equipment in a landing near the headwaters, they realized just how difficult rafting down the Caroni would be. At least they were starting out early in the morning, for, as Bartle recalls, "if anybody was going to get drunk or high, it was before anybody was apt to do that."

"This isn't a rapid, this is a waterfall!" Lem shouted at the first bend of the river. Ahead lay rapids that in a couple of hundred yards dropped a good seventy feet. The rafts plunged into the boiling waters and were immediately drawn into a souse hole, which is a place where the water creates a back wave, or backroller, sucking in anything that enters its purview, holding on to it, and not letting it move on. Everyone was thrown into the swirling waters. Before Lem and Max were thrown overboard, they banged heads so hard that Max suffered a concussion. When Lem was finally pulled out of the churning river, his face was blue and he was in the midst of a serious asthma attack. Tim Haydock, a young doctor, ministered to Lem and patched and sewed up several others.

Many intrepid rafters would have halted right there and called for help, but they headed down the river again, not even planning to stop if they came upon another especially dangerous stretch of rapids. It was a sunny day, and rafting down the Caroni was like riding a spirited horse, alive to the touch. Then they heard a roar, and as they rounded a bend they faced what Michael called "the nastiest and mightiest hole I've ever seen in a river." Bobby's craft plunged into a swirling mass of water that held the raft for forty-five minutes while it cast out the crew. Lem's raft rose up on top of a great wave, dumped its passengers into the water, and then fell back.

Almost everyone—sixteen or seventeen people in all—was tossed into the turbulent waters. They all made it through the rapids and lay exhausted downstream on the riverbank regaining their composure. Some of them were hurt badly enough that they should have gone to a hospital. But they were days from any help. To make things worse, most of the supplies had been lost, and the radio no longer worked.

As they waited for rescue in a tent with rain beating down, Lem regaled the group with heroic tales of Jack and Bobby Sr. "The stories he told and the examples he set gave us all a link to our dead fathers and to the generation before us," Bobby reflected. "The titans became men who we should not fear to emulate. With his encouragement we lost our fear of cutting a path of our own."

No matter how close to death they came, how reckless the risks, they returned again. "During the several journeys we made I saw people facing the possibility of their own destruction," said Michael. Only in this dance of death did they believe they could learn to walk like men.

There was ultimately something silly about their pursuit of danger and their belief that in doing so they were following in the pathways of the "titans" who had gone before. Lem knew that this was not what Jack and his brothers had done. As young men, they had chosen other arenas in which to prove themselves. Joe Jr. had gone out for Harvard football for four years without winning a letter, and Jack had gone out for football too until he hurt himself. Bobby, despite his small size and limited abilities, won his Harvard *H*, and Ted won his football letter too. But none of these young Kennedys played anything but club sports. Lem spun his tales, but even he began to realize that something was wrong. Lem had grown somewhat disillusioned about the journey that Bobby and the others were taking. This was the last trip that he ever made, and the next year he died.

Bartle closely observed what was happening to Bobby and his brothers. "Everybody in the world felt sorry for them," Bartle reflects. "Everybody in the world was willing to bend over backwards to do something for them, for whatever purpose. So it's easy to imagine how a sixteen-year-old might think that drugs are okay. It might fill an emotional or spiritual need to block out those terrible feelings of doom and guilt and all the horrible stuff they went through after those horrible assassinations. They felt doomed in a way. It's like, 'All right, fuck it.' That's partly it. And then a lot of people were saying, 'Don't worry. No consequences.' And then another group of people were saying, 'You deserve this. You're a member of the elite, and elite people can do things that other people can't.' And that all becomes a lifestyle and a habit that's difficult to give up."

One day in 1980 Bobby received an intriguing phone call in Charlottesville. The caller, David Horowitz, and his co-author, Peter Collier, had written *The Rockefellers*, a highly regarded, best-selling account of a family that was in some respects the Republican counterpart to the Kennedys. The best part of the book dealt sympathetically with the young generation of Rockefellers and how they faced the burden of immense wealth and fame. Horowitz said that he and his partner wanted to write a similar book on the Kennedys, and Bobby invited him to visit.

Horowitz was a left-wing Berkeley intellectual, proud of his associations with the Black Panthers and SDS stalwarts. He wore his hair long, dressed with a calculated casualness, and spoke in the genial language of youth. He was a man of subtle shrewdness. Horowitz told Bobby that he admired his father. In this new book he would inhabit their lives. "I'm conscious of the burdens that are on you because of this name, because of the expectations put on you," he recalls telling Bobby. "And I'm going to do your portrait as an individual and show you in your own achievement."

Bobby agreed to cooperate, and in exchange the author promised not to write about the Kennedys' drug use except if it became part of the public record. Bobby invited Horowitz to Hickory Hill and then to Cambridge, where, as Horowitz remembers, his host laid out a line of coke on the table. The author was a generation older than Bobby and his friends. He asserts that he never took drugs that day or any other. Several of those around the family say that he did and that on one occasion he bought drugs, a claim that he also denies. That day in Boston Horowitz says that Michael came into the room when Bobby was snorting coke. "Oh, by the way, this is David Horowitz," Bobby said. "He's writing a book on our family." Michael did not know the deal that Bobby had struck with the author, and he stood there startled at such foolhardiness.

Horowitz traveled back to Virginia and headed off with Bobby and his entourage on a swing around Alabama campaigning for his Uncle Ted's run for the 1980 Democratic presidential nomination. Bobby was running a high fever, but that hardly slowed his steps. Horowitz says that he became frightened when the second car in the entourage pulled up alongside—as they sped across the state at eighty miles an hour they passed marijuana joints between the cars. Bobby had a religious young woman on his lap whom he was unsuccessfully trying to get to sleep with him. In Montgomery there was another woman who did not have such scruples. As soon as she left his friends invited another woman to the house.

"You know, the guy was, by this point, looking like he was ready for hospitalization," Horowitz says. "He says, 'Bring her over.' And he took a bath. I'm thinking to myself, if I were that sick, I wouldn't get any enjoyment out of any of this. I mean, it's just ridiculous. And that also seemed bizarre to me. We had four other men or five, whatever. We're all sitting around watching this guy do his stuff. And I'm thinking, 'Oh, well, I guess this is the Kennedy experience.'"

Bobby led Horowitz to other young Kennedys who did not have Bobby's awesome ability to juggle drugs and daring. Chris Lawford was almost overwhelmed by life as a young Kennedy. When Jack died, his mother, Pat, started seriously drinking. As for his actor father, he became such a devotee of cocaine that on Chris's twenty-first birthday, he gave his son a gift of the narcotic. Three years later, in January 1980, Chris Lawford was busted in Aspen trying to obtain Darvon with a fake prescription. In those years Aspen was full of snow that did not fall from the heavens, and it took a certain foolhardiness or plain stupidity to get busted. Later that year he was arrested again in Massachusetts while trying to score heroin in the Roxbury ghetto.

Chris sought to establish his identity as a Kennedy grandchild. He signed his letters "Christopher Kennedy Lawford" and affixed the name to the top of his résumé in October 1981 when, despite his drug problems, he won admittance to Boston College Law School. His father was slowly dying and had little left but his name, but it was still a celebrated name that Peter Lawford thought was inheritance enough for any son. When Chris returned to L.A. with his Kennedy T shirts and Christopher Kennedy Lawford résumés, it hurt his father immeasurably.

Chris was nonetheless beginning to rethink what it meant to be a member of the Kennedy family. "I used to think that what my family did was involved with a great desire to serve the country," he told Horowitz. "Now I keep asking myself what was it in my grandfather that made him push the family so hard and cause us all such tragedy?"

B obby was David's model in life, and he shared some moral responsibility for what his younger brother was becoming. "Bobby would shoot up David, but it was part of what David wanted," says a close Kennedy friend. "His hand was too shaky to do it himself, so his brother would do it for him. Of course, he introduced David to it. He was the leader of the pack, and David was emulating him."

David would have been much better off if he had not joined Bobby at Harvard in the fall of 1973. If David was going to compete with his big brother at Harvard, there was only one direction in which to go—that was filling his days with drugs and women. He was a handsome young man with deep, narrow-set eyes that seemed to advertise his vulnerability. His neediness and candor appealed to many women, though his temper flared at the most importune of times. David sought to tear away the Kennedy nameplate from in front of his life, and yet he took advantage of all the enviable perks of his name. He had no real sense of his fundamental human value, and drugs were a way to obliterate that reality.

David had become the young generation's version of his Aunt Joan: a convenient emotional receptacle for the rest of the family. His weaknesses were pointed out and gloated over so loudly that the alcohol and drug addictions of the others seemed not to exist. His big brother Joe saw him one day slouched over and stoned. "Look at him," he said. "He was a varsity halfback as a sophomore [in high school]—something none of the rest of us could do. He was terrific. And now he can barely put one leg in front of the other."

Early in 1976 two Kennedy friends found David lying half-dead in his bed at Harvard. The doctors at Massachusetts General Hospital diagnosed his condition as bacterial endocarditis, a sometimes fatal affliction associated with using dirty hypodermic needles. The diagnosis was deeply embarrassing to the Kennedys, and that summer David was sent on a trip to Pakistan, where he hunted boar with General Zia-ul-Haq, the future president, and met the Libyan dictator Mu'ammar Gadhafi, who was also visiting the country.

When David returned to the United States, he dropped out of Harvard and directed his attention to the one thing that interested him: heroin. David started seeing Dr. Lee Macht, a psychiatrist who, while still in his thirties, had been the Massachusetts commissioner of mental health. Dr. Macht tried to wean his patient off street drugs by prescribing Percodan, Dilaudid, and Quaaludes, without informing the state authorities. He may have been worried that news of David's problem would otherwise leak out to the press, but what he did was illegal; the doctor was found guilty of a misdemeanor and fined $1,000.

David found in drugs a solace he did not find in family, faith, or friends. It was dangerous enough to shoot yourself up with one drug, but David had started mixing drugs. In April 1978, he put a syringe full of cocaine and Dilaudid into his arm. He OD'd and was rushed to the hospital emergency room. This time the family and friends decided that drastic action had to be taken. David was admitted to McLean Hospital in Belmont, Massachusetts. His sister Kathleen came to see him and cried that her little brother should not be kept almost a prisoner in such confinement. She was an intense, idealistic young woman who vowed to get him out immediately. Instead, three days later she called her brother and told him, "The others say not to trust you—that you'll say anything to get out."

David left McLean after only three weeks, but the reality was that the other family members did not trust him. They wanted him cured, but they also wanted him elsewhere. He was sent to Sussex, England, for two months of "neuro-electric therapy": he was hooked up with a strange headset that supposedly sent signals to his brain to end his addiction. The signals apparently never reached their target, for after two months he returned to the United States, hungry to score heroin out in the dark streets.

For a number of years he had been making the jaunt up to Harlem in his BMW, where he was known by the nickname "White James." It was there in a junkie's hotel that he was robbed in September 1979, and for the first time the Kennedy drug problems came to public attention.

Full of an addict's extreme selfishness, David was reluctant to accept any strong measures. He was essentially exiled to California, where he went to live with Don Juhl, a Sacramento therapist who put him through what David called "therapy by humiliation." After six months David moved out of Juhl's home and tried to regain his life. He was desperate, disoriented. He had hallucinations. He thought he had murdered his own father, and then that Bobby Sr. was trying to kill him, drowning his son in the Malibu surf. He was a shamed, shunned outcast from his family, the light of drugs and liquor the only beacon in his night.

CHAPTER TWENTY-THREE

A Life to Be Stepped Around

n the spring of 1983 Horowitz drove over the San Rafael Bridge from San Francisco to Marin County, where he met David at Victoria Station, a British theme restaurant. David had started drinking again. His face was red, and he was trembling. When Horowitz looked into David's eyes, he thought they looked full of anger and fear. "I want to tell your story, David, from your point of view," said Horowitz, who lived nearby in Berkeley. "You've been scapegoated in this family, and I mean you're the one who takes the blame for everything. And that's really unfair."

Horowitz wanted very much to interview David, but he realized that he was in such bad shape that it was next to impossible, with David sadly slurring a potpourri of words that signified very little. The author backed off talking to David that evening but arranged to see him again. Horowitz was about the only person in David's world ready to listen to him with a sympathetic ear, and once he began to talk he could hardly stop.

"I feel they should have done something earlier," he told the author. "My mother, although in a sense she wasn't really competent. But even more Teddy, Steve Smith, and the rest of the group who were always figuring out ways to keep Joe, Bobby, and Chris from having to pay the piper, but who just let me go. When they finally did do something, it seemed like it was more to keep me from OD'ing in the street and causing a problem for Teddy's campaigning, than anything else. The thought crossed my mind that if my grandfather was alive the same thing could have happened to me that happened to her [Rosemary]. She was an embarrassment; I am an embarrassment. She was a hindrance; I am a hindrance."

Soon after David began talking to Horowitz, he returned to the East Coast to begin an internship at the *Atlantic Monthly*. His was the most routine of jobs, but he appeared lost, shuffling along, unable to focus. He served out his six months as if it were more a sentence than a job and reentered Harvard in the summer of 1983.

When they began talking, Horowitz beseeched David for interviews and intimate details of his life. In the midst of that process the roles reversed. David became the supplicant, worried about what he had said and what would appear. Horowitz had learned from other sources that Lem had been doing drugs with the young Kennedys. Lem had just died, and Horowitz knew that when the story of his drug use appeared in the book, the family would conclude that David had blabbed to the author.

"I'm going to tell you something I know because I know you're going to be blamed for it," Horowitz said over the phone from Berkeley, telling David that he knew from other sources about Lem's drug use and planned to write about it in his book.

David saw that he might be blamed for the revelations, and that if he was, his banishment from the family would be extended. David called Bobby to alert him to what would appear in the book. David then called Horowitz and told him that Bobby swore he would never talk to the author again if he put a word about drug use in the book and that Bobby was terrified by the prospect of such revelations. Horowitz sensed that Bobby suspected rightly that he knew about his heroin use and was worried that the story might appear in the book.

While David dropped out of Harvard again after the fall semester, everything seemed to be going right for Bobby. Lem had willed Bobby his fancy East Side digs, and he had taken up residence there. He had married Emily Black, a charmingly innocent young woman from Indiana, and brought her into his world. He was working as an assistant DA, and the fact that he had failed the law boards only made him seem a little human.

Those who envied Bobby's blessed life didn't see him driving up to Harlem with his hat pulled down to score heroin. He was making the same dangerous trips in search of heroin that his spaced-out junkie brother was making. But Bobby could have it all. Whenever drugs reached out to try to pull him down, he had the strength and will to walk away—or so he fancied. Only years later did he reflect that his "ability to put drugs aside for long periods and my strong willpower in every other area of my life fueled my denial." David Humphreville, one of his few close friends who did not do drugs, observed the pattern: "Bobby always thought, 'I can quit.' And he would quit for a little while, and then he'd go back."

Bobby's friends were trying to keep matters contained, and there was fear among them that their own drug use might soon be revealed. In May 1983, two of Bobby's close friends, Eric Breindel and Winston Prude, were busted buying heroin in Washington from an undercover agent posing as a street dealer.

Prude had gone on one of their South American expeditions. "Bobby and I were sitting there drinking this Jim Beam whiskey, and Winston came out of his tent," recalls Humphreville. "All of his drugs, pills, everything had gotten wet, it was like soup. He'd eaten some of that, and he was covered with sweat and had chills. His eyes were huge, and he dumped his head in this water, and we're just looking, thinking, 'This is pathetic.' This guy was getting high on drug soup." In April 1985, Prude shot up some especially pure heroin and later was discovered dead in his apartment.

H orowitz had promised to keep Bobby's private drug use out of the book. That promise became moot, however, in September 1983 when Bobby became sick on a Republic Airlines plane flying from Minneapolis to Rapid City, South Dakota. He was traveling to Deadwood to be with a friend, John Walsh, a former priest with whom he could run across the prairies and seek some measure of spiritual relief. When Bobby refused to get into the ambulance that was waiting on the tarmac, the police opened his bag and found about a fifth of a gram of heroin.

When David heard about his brother's arrest, he felt a measure of relief. He was no longer the only public junkie in the family. But the family rushed to Bobby's support, and when David saw the contrast in the ways in which he and his brother were treated, he became despondent.

It may have seemed terribly unfair to David, but it was Bobby's first public crisis, not his umpteenth, and the family had learned from the desultory way in which they had handled David's problems. Beyond that, and most important, Bobby faced his arrest and addiction square on. The arrest had ripped the veneer off Bobby's life. He grasped that. He did not seek to hide any longer. He entered a rehabilitation facility in New Jersey and stayed there five months.

Looking back on it years later, Bobby tried to understand why he had not sought help sooner. "My life experience with public scrutiny and an aggressive and nosy press made me wary of attending public twelve-step programs, where I might otherwise have found recovery," he said. That was probably true, but this was also a favorite Kennedy rationalization. There was always someone else to blame for their problems, some excuse that they merited because of the demands of their name and heritage. Despite such fears, Bobby now became the first Kennedy to regularly attend meetings of Alcoholics Anonymous, an organization that in the following years would become— aside from the Church—the young Kennedys' most important source of spiritual solace.

AA is about admitting and understanding the human flaws that lead one to seek in liquor and drugs what the movement believes can be found only in faith in a higher power. AA is about telling those truths to your fellow alcoholics. When you walk into an AA meeting room, it does not matter whether you are a Harlem wino or an East Side lady who drinks—you're all the same. You know that no matter what you say, no matter how pathetic, or terrible your words are, they do not go beyond the room in which they are spoken.

Bobby had good reason to fear that he did not have such a room. He had seen what happened to his alcoholic Aunt Joan. For years she had wanted to go to AA, and for years Ted and his advisers had talked her out of it. They dreaded her talking about her problems in her breathless way in some meeting. In 1976 Joan had admitted herself to the Smithers Rehabilitation Center. The New York institution followed many of the tenets of AA, including total frankness about one's condition. Two of the other patients, including one who called himself a freelance writer, sold a story to the *National Enquirer*. And it was not over for Joan. While Bobby was seeking a way out of his own addiction, a fellow alcoholic with whom Joan attended AA meetings and whom she had hired as an assistant was secretly writing a book about Joan's life and tribulations.

Despite his justifiable worries, Bobby began attending AA meetings. No one ever abused the sanctity of those meetings, and the experience helped to lift him out of the depths. Bobby said he was sorry to those he had hurt, got up early in the morning and made his bed, drove as if the speed limits were for him as well as for others, and doted on his pregnant wife. But it was not easy, and those who knew him well realized that he remained a young man full of private demons.

D avid dropped out of Harvard once again after the fall semester, and in March 1984 he entered St. Mary's Rehabilitation Center in Minneapolis under an assumed name. Sitting in the chill of Minnesota, David had little to show for his twenty-eight years except that he was a connoisseur of rehab. He had tried them all—the star-studded celebrity ones that the Beach Boys and the Rolling Stones had used, the experimental ones, and the ritzy boutiques where the scions of wealth were shelved away.

The Kennedys was about to come out, and Horowitz wasn't calling anymore. *Playboy* excerpted the most sensational material on the young generation, promoting the issue with radio ads talking of "sexual misconduct [and] drug dealing." The lengthy article, illustrated with a picture of a Kennedy shooting up, was a devastatingly revealing take primarily on Bobby, David, and Chris. The most brutal insights came from David and Chris themselves.

The three young Kennedys read the prepublication excerpt filled with a sense of betrayal and searing rage. Spread out on these pages was an object lesson in how the world worked. Horowitz had promised a transcendent work that would help Bobby, David, Chris, and the others come to terms with themselves and their lives. Instead, he and Collier had merely uncovered the three young Kennedys, using many of their own words and leaving them shivering in the cold glare of the public eye.

Joe felt equally betrayed by the book's portrait of him. He called Diane Clemens, the professor he had lived with in Berkeley, who was on a sabbatical in Berlin. "So, Prof, that's what you thought of me," Joe said. When Clemens finally saw the article and the book, she felt that much was distorted and taken out of context. Clemens had thought of Horowitz as a friend, and she felt violated, but it was nothing like Joe's sense of betrayal.

David was unprepared for the massive onslaught of publicity hurtling toward him. As David saw it, Bobby had introduced him to drugs, setting him out on this despairing road. Bobby had plunged to the depths too, but he was already reaching up again, his transformation making David seem even more pathetically self-indulgent and weak. Bobby had introduced Horowitz to the young Kennedys. But Bobby had not said the kinds of things that David had said, and the wrath that was sure to come would not come down on Bobby's head. David called Horowitz and told the author that he was being blamed for exposing his family, shaming them before the world.

David left St. Mary's just before Easter and flew down to Palm Beach. For his father's generation, the Florida estate had been a place where the Kennedys communed as a family under the eternally watchful eyes of Joe. Now the grandchildren made their desultory visits to the estate. They spent the obligatory time with Rose, but for most of them Palm Beach was primarily a convenient hangout among the privileged.

Rose had had a stroke on Easter Sunday and was expected to die in a few days. David knew that this was probably the last time he would see his grandmother. Rose could be a tedious nag, insisting that he and his brothers sit up at the table, use proper grammar and the right fork, and avoid the slovenly slang of their generation. She was a woman who at times mistook manners for morals, but there was an awesome strength and discipline in her carriage that was one of the spiritual assets of the family. Losing her meant losing much of that strength.

David was so much an outcast in the family, so mistrusted, and so much a source of rancor and unpleasantness, that he was not even allowed to stay at the estate. He took a room at the Brazilian Court, a first-class hotel. The day after the stroke David, along with his youngest brother, seventeen-year-old Douglas, came out to their grandmother's sprawling Mizner-designed home. Rose was a tiny, wizened form lying deathlike in her bed. David burst into tears and ran out of the room. Anyone who saw him there would have found it impossible not to feel for him and to find his banishment cruel and misguided. But soon after he left, the nurses discovered that a bottle of Demerol prescribed for Rose had disappeared from their stand. Although the nurses felt that David had probably lifted the medicine on his short visit, a family witness says that it was two of his cousins who took the Demerol and delivered it to David because he had said he needed the painkiller.

The next few days in Palm Beach were no different from many periods in David's life—a haze of drugs and booze, of low-life companions and dimly lit bars. He appeared a disheveled, wandering derelict, shorts stained, hair unkempt, moving from bar to bar, scarfing vodka and scoring drugs. The only thing that marked him as a Kennedy was his American Express card. At one point he took a taxi out to the estate, but the guard knew enough not to admit this relic of a young man, and he headed back to town.

Caroline Kennedy and Sydney Lawford arrived at the Brazilian Court in the middle of the morning on April 25, 1984, to look for their cousin. David did not answer the knocks on the door. After they left, two members of the hotel staff ignored the DO NOT DISTURB sign and entered the room. They found David wearing his shorts and shoes, lying dead between the sheets on one of the twin beds. He had died from a combination of cocaine, Demerol, and the tranquilizer Mellaril, some or all of which had been injected into his groin.

There were some in the family who thought that David had committed suicide, driven to his death by the *Playboy* excerpt and the book that was soon to follow. But twenty-eight-year-old David had been committing suicide for a decade, and if he had plunged the needle in his leg hoping to find oblivion and avoid further shame, this was only the final act.

The Palm Beach County state attorney concluded that after David died, someone had come into his room and cleaned it up, possibly pulling the needle from his leg and disposing of the drugs and paraphernalia down a toilet that afterward showed traces of cocaine. Caroline and Sydney were the obvious suspects. Although they denied that they had entered the room and were exonerated, it would hardly have been unseemly to leave their cousin with a modicum of dignity before the police and the press got there.

Other things died there that day, among them a kind of trust. Never again would any of David's brothers truly take a journalist into his confidence; instead, they would often affect a confrontational pose that said they could not be fooled by sweet promises. "When David died, Bobby was just feeling really low," says Humphreville. "He said, 'Yeah, maybe it's my fault,' in the sense that if he'd cleaned himself earlier and looked after David, it might have been different. He was just sort of beating himself up because that's the way he is."

The family had memorialized all the previous Kennedy men who died in privately printed books. No one put together a book memorializing David. He was buried in Holyhood Cemetery in Brookline, Massachusetts, near his grandfather, and there he lies, his life lessons purely cautionary ones. He was in death what he always had been, a life to be stepped around.

CHAPTER TWENTY - FOUR

The Games of Men

t was clear with David's death and Bobby's addiction that the Kennedy sons had an inheritance and a mind-set that if left unchecked led to alcoholism, drug use, and foolish physical risk-taking. If these young men could walk around these swirling rapids rather than attempting to run them, they would carry with them the most positive aspects of the family legacy. If they did not, they were likely to pull their family down, their legacy to be written about not in political tomes but in the tabloids and scandal sheets. In David they had a warning, and an immensely powerful lesson. For the most part, however, they were not notably intro-spective and found their answers in pushing onward, not in looking within. That was another part of the family legacy, and in many ways the most dan-gerous of all, for it allowed the most negative aspects full rein.

The Kennedy sons remained a competitive lot constantly looking over at one another, trying to decide just where they stood. They still played their occasional games of football on the grass at Hyannis Port, but the field of battle had moved into the greater world.

Of all the surviving grandsons, none seemed less able to compete than Ted and Joan's youngest son, Patrick. He had an asthmatic condition that, if not psychosomatic, was clearly exacerbated by the tensions within his fam-ily. He had a father who he feared might be assassinated and a mother who moved in and out of an alcoholic haze, in and out of rehab, and in and out of emotional constancy. At Potomac School, which he attended with various other Kennedys and Shrivers, he was the kind of boy whom others would pick on. Patrick's father had had a trying childhood: shuttled from school to school, Ted had been a fat, often friendless little boy. Patrick was being shuffled from place to place too, and he was equally friendless and equally sad.

In November 1982, Patrick asserted himself in the most unlikely of ways. His fifty-year-old father was contemplating whether he should run for the presidency in two years, and he wanted his family to help him decide. He had reason to think that this might be his last chance to grab at the golden ring. He had already hired several key people, including Pat Caddell, Carter's pollster, and Bob Farmer, a fund raiser. His staff had just leased office space on Capitol Hill for campaign headquarters. President Ronald Reagan seemed a perfect adversary, and Ted's rhetoric already rang out boldly: "Ronald Reagan must love poor people, because he is creating so many more of them."

In previous years when deciding whether to run, Ted had gone through the perfunctory ritual of feeling out family members. But until now his children, sisters, nephews, and nieces would not have dared to oppose what was the family's highest ambition. This year Ted already knew how his children felt, especially fifteen-year-old Patrick who did not want his father to run.

On the Friday after Thanksgiving the family got together for lunch at the dining room table in Hyannis Port where Ted had sat so often as a little boy. In that earlier time he had been placed at the side, where he watched full of trepidation for his father, reverence for his mother, and awe for his big brothers. Now he sat at the head of the table. Patrick was there along with his two siblings, Teddy Jr. and Kara, some of his cousins, and Steve Smith. The only outsider was journalist Dotson Rader, one of Pat Lawford's closest friends.

The reporter had just traveled the country interviewing people about the Massachusetts senator for an article in *Parade*. Ted turned to the writer to ask his take on the potential race. Rader spoke eloquently of Ted's destiny to run. Steve Smith spoke next. He had become a surrogate brother to Ted, the man he trusted more than anyone else in the family. And Steve neatly demolished every positive reason Rader had given. Smith looked Ted in the eyes as he dismissed his prospects. No one else dared to look at Ted straight on. "Why don't we take a vote?" Smith asked. Rader raised his hand in favor of the race while everyone else voted against it. Afterward, Smith raised his glass to toast Ted and what he considered his noble career, while the object of these effusions stared off into the distance.

Whatever Ted's melancholic reflections, he had done something extraordinary. He had decided to stay out of the race based largely not on the realities of politics in America but on the desires of the young generation of Kennedys, most dramatically his own youngest son. Some of his nephews still considered their Uncle Ted a figure of mild derision and embarrassment, but from now on the young generation began to see Ted differently— as a man making sacrifices for his family that few public men would have made. "The family seemed intensely loyal to one another and the dominant feeling was optimism, the sense of a whole new generation coming up to new achievements and possibilities," Rader recalled of the atmosphere that afternoon.

The young Kennedys had done Ted an immense service. If he had won the Democratic nomination, he would probably have squandered much of his political capital and energy trying to defeat the immensely popular Reagan.

A s a young teenager, Patrick moved up to Boston to live with his mother. Joan was not so gauche as to trash her former husband, but she trashed the political life that he lived, portraying it as rude and unsavory, a world far from her pursuit of music and art and sensitive discourse.

In the fall of 1983, sixteen-year-old Patrick entered tenth grade at Phillips Academy in Andover, Massachusetts, where his cousin John had graduated four years before. When he was not being hazed, he was ignored. His more sophisticated classmates avoided him. He had asthma attacks and was a mediocre student who despite his famous name was largely anonymous.

Patrick had taken to fishing, standing alone on a Hyannis beach surf-casting. He was so unlike his big brother. Teddy, a student at Wesleyan College, had lost a leg, but he appeared a boisterous free spirit who could out-match his little brother in every sport or game. Young Teddy traveled with their father to famine-ravaged Ethiopia, but Patrick did not because his father feared he might have a serious asthma attack.

At Andover at that time drugs were a main recreational activity not listed in the student handbook. While John had limited himself largely to marijuana, Patrick developed a taste for both liquor and drugs. Despite their own abuse problems and those of other Kennedys, Ted and Joan had no idea of what Patrick was doing until he had a serious problem.

During spring break of his senior year, Patrick was admitted to Spofford Hall, an upscale rehabilitation center in New Hampshire. It was not until six years later, in 1991, that the public became aware of Patrick's problem when one of his fellow patients, Bob Remy, sold his story to the *National Enquirer*. Patrick attributed his personal crisis to the pressures of growing up, particularly as a Kennedy. "I felt terrible about myself," he told the *Boston Globe* in June 1999. "I wasn't living in my own skin. I was trying to escape that feeling, to run away from it. It was me hitting the wall and coming to grips with a lot of unanswered questions about who I am, where I came from, and how I was going to be accepted. I didn't feel like I measured up to my family and who I was supposed to be."

His parents were of little help. Patrick was desperate for someone to reel him in. "Honestly, neither one of my parents was around to set the limits I needed," he said. "They were both so consumed by their own issues."

Ted tried to be there for Patrick, but he was always hurrying off to someplace else. He was not running toward anything but running away—away from intimacy, away from his past, away from contemplation. Then there was Patrick's mother, whose drinking troubled both her life and the lives of her children. She was a witness to a different kind of life, where vulnerability was not weakness but strength, tears were not a betrayal of manhood but a sign of sensitivity, and art and music, love and laughter were not relegated to times of play but resided at the core of life.

When Patrick got out of Spofford, he began seeing a psychiatrist who put him on antidepressants. Since Joe Jr. and Jack were little boys in the early twenties, Kennedy children had been driven out of their sickbeds, their lives crammed so full, and so orchestrated, that there was supposedly no time for depression. Although the family axiom was "Kennedys don't cry," what they also meant was that Kennedys don't get depressed.

Patrick was a deeply mistrustful young man. He saw what had happened to his mother, whose trusted assistant and supposed friend had written a devastating book about her personal life and problems. "It's not always easy for me to spill the beans, if you will, with someone that I have to worry about telling other folks," Patrick admits. "So for me, it really helps to have someone who is protected by professional confidence. If I started telling my friends what I was worried about and what I was thinking about this and that, it may end up in a tabloid newspaper."

At Andover, Patrick shared one dubious accomplishment with his cousin John. He did not graduate with his class but had to take makeup classes at Georgetown University before entering the Jesuit college in September 1986. Wherever he went there was at least one family member who outshone him. At Andover, John's bright spirit was still fondly remembered. At Georgetown, his cousin Maria Shriver was on the board of regents, while her brother Anthony was an undergraduate and a major presence.

After only two weeks on campus Patrick dropped out. Ted was upset that his youngest son wasn't sticking it out. Patrick wouldn't budge. It was the first time he had ever stood up so boldly to his father. He was perfectly aware of the ambivalent nature of the family heritage. "There is a fine line between being consumed by it and wrapping yourself around it and using it and being positive about it," he told M. Charles Bakst of the *Providence Journal* in October 1994. "There's a fine line between having it become a negative albatross on your back and having it become a lift and support to back you up. It could have been a destructive factor in my life as it was a positive factor. In fact, in various periods in my life, it's been both." Patrick was so consumed with his family heritage that whether he embraced or denied it, his life was defined by it.

J oe had as many problems gaining a separate identity as Patrick did, but by the time of his brother's death in 1984 he had moved beyond many of his earlier difficulties and begun what was in many respects a salutary life. Much of that he owed to his marriage. After almost a decade of steady dating, twenty-six-year-old Joe married twenty-nine-year-old Sheila Brewster Rauch in February 1979. Unlike his mother, who could practically have been cloned from the Kennedy DNA, Joe was marrying a woman who was very different.

Sheila was three and a half years older than Joe. When the couple met, she had been a twenty-one-year-old college senior finishing up her honors' thesis at Wheaton College, before entering graduate school at Harvard's School of Design. Joe was eighteen years old, still dealing with his father's death two years before and seemingly unable to make it in college. Sheila became an anchor of support through many of his difficulties, including the Jeep accident that left a young woman paralyzed. She had maturity and patience beyond her years. Sheila was from a wealthy old Philadelphia family that took their Episcopalianism as seriously as the Kennedys took their Catholicism. She did not fancy that she was marrying up but at best marrying sideways—or in the value system of Old Line Protestant Republican Philadelphia, making a decided move downward.

For the most part, spouses married into the Kennedy family, but the Kennedys did not marry into other families. In this case, Sheila no more gave up her family identity than did Joe. She was a serious, highly principled young woman, yet she had a practical demeanor that said, *Bring life on, I'm ready for it*. The couple married in a Catholic church, but Sheila's Episcopalian minister assisted, and the newlyweds vowed to let their children learn about both faiths and make their own choices. "Our relationship had been tested more than some marriages are during a lifetime," Sheila wrote later. "Whatever mistakes we were to make, they were not because we didn't know each other."

Joe and Sheila bonded through their very different strengths. Sheila had the academic smarts and social skills that had passed Joe by. She had a graduate degree from the Harvard School of Design and was working in housing policy for the city of Boston. She was quiet and restrained, while he was boisterous and outgoing. She enjoyed moments of solitude, while he could not stand to be alone. She was well read, while he hardly read at all. He trusted her mind, and she trusted his instincts.

One of Joe's strengths was a booming, exuberant optimism. He was endlessly enthusiastic about his bride and their prospects, doubly so when Sheila became pregnant. He had been traveling in France when he learned the news. He was the firstborn son who was supposed to become president one day, but he wrote his wife that having a child was the greatest thing that "ever could happen in my life." He told her that the birth "is more important to me than anything that had ever happened" and that since his marriage he had "just a happier and more meaningful life because I have you to love and look forward to. I just love you so much."

In October 1980, Sheila gave birth to twin sons, Joseph Patrick and Matthew Rauch. The newlyweds had been living in a pleasant bungalow in a modest neighborhood in Brighton. With the birth of the twins, the couple moved out to a sprawling nineteenth-century farmhouse on the South Shore. It was a good place to raise boys and dogs, and Joe was as much a loving father as Sheila was a loving mother.

Joe's personal and professional lives had come together in a sound way, and he was as happy with one as the other. When Joe went into the city, it was to the new offices of his organization, Citizens Energy, in the Russia Wharf Building on the Boston waterfront. Joe had gotten the idea for the energy program sitting around chatting with his father's former associate Richard Goodwin. "We were sitting watching a news broadcast and they scrolled the profits of oil companies," Joe said. "And Dick said, 'I wonder what would happen if you took all those profits to reduce the cost of oil.' And I said, 'That's a really cool idea.'" Joe was excited enough that he developed a new idea for a nonprofit company for the next six months with Steven Rothstein, who had just lost a race for the Massachusetts House from the

Boston suburb of Brookline. The idea was both simple and innovative. Most homes in New England were heated with oil. It was so expensive that the poorest residents often had to choose between keeping warm and eating decently. Citizens Energy would buy oil on the world market at low prices, have it refined at another low price, and then have it delivered by retailers charging minimal costs as well. They not only needed the oil-producing countries to cooperate but required shipping, refining capacity, financing, terminal storage, and distribution.

"I wrote with Steve's help letters to the OPEC ministers," Joe says. "The first one to respond was the Venezuelan. He wrote me a letter that arrived Friday, saying be there Monday morning. I went down there and spent a few days and signed a contract with them."

It took Joe with his celebrated name to be able to convince the Venezuelans to sell oil at the official OPEC prices and not at the higher spot oil price that most customers were paying. "I assume the Kennedy name had a very significant amount to do with it," admits Joe. For the Caracas officials it was a way to ingratiate themselves with the nephew of the powerful Senator Kennedy who yet might be president of the United States.

Joe asked one of his closest friends, William E. "Wilbur" James, to move from Seattle to help run Citizens. Joe used his political connections with refineries and storage facilities, pushing them to work with his fledgling company at little or no profit. Then he went to the retail dealers and through their associations pressed them to deliver the heating oil at a price far below what they normally charged. Many of the retailers had no desire to deliver fuel oil for what some of them said was less than it cost to run their trucks. But the dealers were already being blamed by their customers for the high cost of oil, and they did not want to risk Joe's wrath.

In October 1981, twenty-nine-year-old Joe stood on the Boston docks alongside Governor Edward J. King to welcome a giant barge arriving with 125,000 gallons of oil. The TV cameras rolled when Joe and the governor turned a valve to start the oil flowing into a waiting truck. The beauty of it for Joe and his associates was that they did all this out of their new offices without having physical plants or involvement beyond making the deals. In the refining process, only about 20 percent of the oil ended up as heating oil. Joe and his partners were left with the other 80 percent to sell and trade, plowing the profits back into cutting the fuel prices, investing in worthy ventures, and paying salaries and overhead.

Most successful politicians learn that they are smart not only to thank others but at times to exaggerate the help that others have contributed. That Joe did not do so was perhaps as much a mark of his insecurities as his egotism. Instead of bringing a couple of retail dealers into the photo op and running down a litany of people who had made this enterprise work, he gave the impression that he personally had drilled the oil, refined it, trucked it to the homes of the poor, and delivered it with a humble tug at his cap.

At the beginning it was an incredible bonanza, each tanker representing $10 million in profits. Joe may have had grandiose ideas of changing the whole nature of the oil industry, but the publicity he received was out of proportion to the results as the profit margins narrowed. In the end it proved difficult to use the market economy against itself. He talked to the oil minister of Saudi Arabia, but the Saudis were not interested in cutting a special deal for Joe, in part for business reasons and in part, he believed, because of his Uncle Ted's strong support of Israel.

"The net number of people who actually found some significant relief was pretty small," said one of the major oil dealers. "By working our suppliers and getting dealers in the chain, there was very little direct expense against Citizens Energy. It's a little intimidating when they come to you, but what the hell is the difference if a couple of cargo ships go to them, it keeps their goodwill. But it boggled my mind. What the hell are we doing here? I think that individually and collectively, we as community dealers do an awful lot more good in total than they do."

The retail oil business in Massachusetts is run by hundreds of small dealers. Even many of those who did well selling at the regulated price felt intimidated by Joe. He simply did not understand the impact he had on the businessmen. "Joe likes having a tough-guy image," says Fred Slifka, president of Global Petroleum Company, one of his admirers in the business.

"He gets hot. He gets a little mad and he flies off the handle easy. I'm not sure he'd hurt a fly, but he'd like to make out he would. He puffs out that chest of his."

As fuel oil prices went down on their own and Citizens Energy was unable to buy large amounts of oil at cheap prices, Joe and his associates moved into a number of other areas. For what was still a small operation with no more than fifty employees, the company had complexities worthy of a major conglomerate. There was a bewildering array of profit-making and nonprofit subsidiaries: Citizens Conservation, Citizens Resources Corporation, Citizens Heat and Power Corporation, and Citizens Health Corporation. At their conception, the "profit-making" companies were almost a misnomer. As a nonprofit, Citizens could not legally compete with a wide range of legitimate businesses. Therefore, by creating profit-making companies, Citizens was able to move out into a vast array of enterprises, always supposedly with the idea of taking any profits and using them for social beneficence. Joe's younger brother Michael had joined Citizens Energy too, with his freshly minted law degree from the University of Virginia. He was an outgoing, exuberant young man who especially on the foreign trips was superb at ingratiating himself with businesspeople and government officials.

By 1986 Citizens was doing close to $1 billion in business, but 70 percent of that ($731 million) was contributed by one profit-making subsidiary, Citizens Resources. They were trading in oil futures, commodity speculation that was in part inspired guesswork, in part political and economic conjecture. They weren't that good at it and made only tiny profits. As *Boston Business* reported in June 1989, "The [Kennedy] magic failed to work in the tough business of oil futures trading, which is governed by a complex of forces beyond the influence even of the Kennedys."

When Joe and Michael traveled to Africa and the Soviet Union looking for deals, it was the sons of the legendary Robert F. Kennedy seeking not to make personal profit for themselves or their company but to follow in their father's and uncle's exalted footsteps. Joe, Michael, and their associates made deals that Citizens could not have gotten if they had merely been hungry young men looking for an angle.

J oe's salary in 1985 as president of Citizens Energy Corp was $56,628. While that was relatively modest, the salary was less than a quarter of his $225,659 income. He also received $50,216 in interest and dividends, along with $118,815 from rents and trusts. The Kennedys' primary business property, and its largest asset, was the Merchandise Mart, which Joe's grandfather had purchased in 1946 when the giant venue for trade shows and wholesale sales was the biggest office building in the world.

Forbes estimated in 1991 that the Kennedy family fortune was $350 million, "most of it . . . in a single, illiquid real estate property with no great future. Shirtsleeves to shirtsleeves in four generations? Close to that. After taxes, operating expenses and improvements, the Mart complex spews out about $20 million free cash flow a year for the family members to divide and spend." The fortune paid out about $70,000 to $100,000 a year to Joe and each of the others of his generation. It was not enough money for them to live like entitled princes, but it was enough to make them immune from the mundane monetary struggles of most of their peers. They had their personal family concierge in the offices of Joseph P. Kennedy Enterprises, a secretive establishment housed at various addresses in Manhattan. Here they sent their bills and sundry expenses, wrote checks on accounts they never had to balance, and received annual money from their trust funds. They had it easy, and having it easy is sometimes not easy at all when you rarely know the satisfaction of achieving something entirely on your own.

As the operations at Citizens Energy grew larger and the company competed against profit-making companies, there were hard questions about how much executives should be paid and what kind of incentives were legitimate. Joe's associates did not have his family wealth. Nor could they parlay the goodwill achieved through the company into a political career or advancement elsewhere the way Joe could. There were no books Joe could look at to give him these answers, no neat calibrations. But if he did not keep

Citizens true to its original premise, it risked becoming little more than a cynical shield for gaining entrée because of its nonprofit persona and then enriching itself more than it helped the poor and the needy.

"I think Wilbur James was doing things that Joe didn't understand," said one close observer and Kennedy friend. "Joe typically doesn't pay attention to what other people are doing. The world surrounds Joe. I've never met anybody more narcissistic to a fault. So it's one thing if you're not that bright and you know it so you can surround yourself with smart people and learn and sort of make the most of situations. He was so pigheaded and insecure about his intelligence that people around him had their own agendas and he wasn't even aware of it because he was so self-indulged."

Joe is adamant that he was so much the center of Citizens Energy that nothing important happened without his knowledge. When Joe was in a good mood, he had great natural warmth and the ability to inspire his subordinates. When he was in a bad mood, James and the other executives had to get used to Joe screaming at them. People at Citizens learned at times to work around Joe, not necessarily with Machiavellian intent but because that was the easiest way to get things moving ahead. They had to get used to handling all the details of their dealings. Joe's associates learned not to present him with endless memos and lengthy briefings but to keep it short and then get out of there.

At a little over six feet tall and about two hundred pounds, Joe was more physically imposing than his mere height and weight would have suggested. Even if he had had a haircut that morning, he managed to look almost shaggy. No matter what jacket or shirt he wore, his body appeared too large for it. His body and personality seemed to be bursting out of their surroundings.

Joe had a forceful, direct way of speaking that could sound threatening and confrontational. That was the way Joe and his brothers were brought up, and for the most part he did not mean to be hostile at all. Those who could accept that and who gave as good as they got often worked easily with Joe. But others, especially women, found it difficult to be around him.

Joe and his associates were full of intriguing, inventive ideas, and fuel oil was only the beginning. Citizens Conservation was allowed to act almost as an arm of government in an energy-saving program for low-income apartments. The company went into an apartment building, assessed how the individual tenants might cut their energy use, and then gave a low-interest loan to the property owner to do the necessary improvements, which usually included purchasing a new fuel burner. The savings were so large that the landlord's annual payments for both fuel and the loan were no higher than his fuel bill had been. This was not a frontal attack on poverty, but it was another immensely useful, incremental measure. And as always, it benefited immensely from the strategic use of the Kennedy name.

Joe and his associates were developing programs that seemed on the verge of becoming national models. In the Reagan years, when deregulation was the model and the welfare state a bloated villain, these programs appealed to the right as well as to the left. Joe's achievements at Citizens were being lauded in everything from *Forbes* to *Mother Jones*. "Horrors! A profit!" headlined *Forbes*. "Shades of Old Joe Kennedy—has some of the patriarch's business acumen rubbed off on his grandson?" The answer was apparently yes, and a call for celebration.

Money trumps idealism most of the time. Although no one had come to Citizens to become wealthy, millions of dollars were floating through the offices, and several of the employees wanted to stick their beaks into the flow. "Wilbur was very focused on making money," says Joe, carefully measuring his words. "Wilbur began to say I came back to help you for a few months and I've been here for a long time, a couple years, you've got to put into place a whole series of financial packages if people are going to work here." Joe was facing one of the merciless conundrums of modern capitalism. He wanted to sail forth on a high tide of altruism, but Joe could not hold on to top business people if he did not pay them commensurate salaries, and he constructed a salary and reward system in which he was among the worst paid. It was a makeshift system held in place primarily by Joe's strong hand.

Joe noticed the way the elderly poor complained about outrageous prescription costs, and he sought to form a partnership with Medco, a young company that sought to cut the costs of prescription benefits by buying in bulk and delivering through mail order. The first two people Joe brought in to run the company didn't work out. When Joe ran into John Doran, an acquaintance from the Cape, on a flight from New York to Boston, he approached the executive to take over the fledgling business. Doran was game as long as he was paid the salary he would have received at another top position and twenty-five percent of any increase in value created under his management. "I thought it was a rich deal, but I recognized that my contract with Medco was up in two years, and we'd see if it worked," Joe says.

Once again it was the Kennedy name that was Citizens primary asset. The company would use its inside relationships with political, labor, and business communities to bring new business to Medco. Doran was a good salesman with a good product. When he signed on major clients including unions, telephone companies, and Blue Cross and Blue Shield of Massachusetts, he was selling more than a program to cut drug costs. He was selling a prestigious tie-in with the Kennedys, not only Joe and Michael but by implication Ted as well. And he was also selling the moral satisfaction that Citizens share of the profits would be poured back into the community.

J oe had strong entrepreneurial skills and the ability to create what had not been there before. Although he was often compared to his grandfather, Joseph P. Kennedy had created nothing lasting in his life but his family. Joe's grandfather had pushed his sons into national politics because he felt that in their lifetimes Washington would be the center of power and action. In the Reagan vision of America, that was no longer necessarily true. There were all sorts of moral conundrums and complexities ahead for Joe, but he did not have to look yearningly at Washington to make a major contribution.

Thus, when Speaker of the House Thomas "Tip" O'Neill announced in March 1984 that he was retiring from his House seat in the same eighth congressional district in which Jack had served, Joe did not rush to announce his candidacy. One of those most opposed to the race was one of Joe's oldest and most trusted friends, Chuck McDermott. Joe had known McDermott intimately since they had spent the summer together in Spain the year of

RFK's death. McDermott had dropped out of Yale to play with a country-rock band, Wheatstraw, and had recently joined Citizens as program director. McDermott was neither a buttoned-down executive nor one of the ambitious young political men who often gravitated to the Kennedys. He saw the immense, exciting possibilities of Citizens and felt that this was no time for Joe to move on.

The eighth is a legendary district in American politics, known not only as the birthplace of JFK's career but as the home of that homespun philosopher of politics, Tip O'Neill. The district stretches from East Boston, where Joe's great-great-grandfather Patrick Kennedy arrived in 1849, to parts of the South End enclave of Irish Americans, to the working-class neighborhoods of Somerville, and from the Cambridge academic ghetto of Harvard and MIT to the suburbs of Waltham and Arlington. It is rare to find a congressional district with such a diversity of cultures and classes. If, as O'Neill famously stated, all politics is local, then politics was nowhere more local than in these forty-two square miles where each constituency had to be wooed in its own way and time. The eighth was a Democratic bastion, and there were at least half a dozen candidates pawing the ground outside the starting gates, most of them well qualified to represent these constituents in Washington.

In January 1985, Joe told Chris Black of the *Boston Globe* that he had no interest in running. "It just is not in me to do it," he said with cut-to-the-chase candor. "It's such a crummy system. It just seems like a bog. I really wonder if it isn't better to go out and do something [rather] than fight this ball of molasses." He predicted what would happen if he got his feet stuck in the Washington morass. "I'd just lose my temper," Joe said with an astute sense of his own weakness.

Soon after Joe's interview, twenty-three-year-old Teddy flew off on a ten-day trip to the famine-plagued reaches of Ethiopia with his father and sister, Kara. With Joe opting out, Teddy had begun thinking about entering the primary. If he won, he would barely be twenty-five by the time of his swearing-in, but his aspiration was not as quixotic as it might appear. Teddy may have lost a leg to cancer, but he treated it as no more than a quirk or idiosyncrasy. He had lived a boy's life and a young man's life. He skied, water-skied, sailed, and drove fast. At Wesleyan, he seemed like just another party-loving guy, but his mates did not visit hospitals to talk to those going through what he had gone through. He was a hefty six-footer with an infectious laugh. He looked like his father's son, but he had an open, gregarious manner and none of Ted's suspicion about the outside world. Teddy had spoken powerfully at the 1984 Democratic Convention about the handicapped in America. In Massachusetts he had started a foundation, Facing the Challenge, to promote the cause of the disabled. His story was celebrated enough that it was being made into a TV movie.

Teddy was almost overwhelmed by the starving, desperate people who surrounded the group wherever they went in Ethiopia. At one point an aid worker stopped Teddy from taking off his shirt to give to one boy.

On their flight back home, Ted had a conversation with his son about his political future. "It's there and you have to think about whether you want to do it," Ted said. He had been brought up to believe that when a political opportunity appeared, you went for it. "What do you think about it? Do you want to do it?"

Ted did not for a moment suggest that his son was perhaps too young to run for Congress. As a young man, Ted had tried to get away from his endless family obligations by moving with Joan to the West and a measure of freedom. He had run for the Senate when he had been just barely old enough to assume the seat. He wanted his son to do the same, though by sailing into Washington on little beyond his name, Ted had exacerbated insecurities that he never lost. When you did things yourself, you stood on firm ground, and Ted still stood on shifting sands. It was as if he needed his son to run for Congress to justify his own life and the massive personal sacrifices he had made.

Young Teddy did not have a heroic martyred father or brother as an inspiration, but a flesh-and-blood dad who drank too much, ran off on his mother, and yet, as his son saw it, was one of the most inspirational, positive figures in public life. While Joe was worried most about what *he* would accomplish in Washington, Teddy was aware of the costs. He also saw other ways he could advance his overwhelming concern for the disabled. Even if he decided to run, he knew that Joe and his family felt that the eighth congressional district was *theirs* and Teddy had no business even thinking of perching there. Ethel had brought up the children with a proprietary sense of politics, and even if Joe was not running, he could hardly abide the idea that his young cousin should do so.

Teddy remained interested enough in running that he moved into Somerville to establish residency and make his political reconnaissance. His father brought in several of his associates to advise his son on the race. Teddy was overwhelmed by the prospect of entering public life and possibly disappointing his father. It was disturbing news too when Joe changed his mind and said that he intended to run.

In July 1985, Teddy announced that he would not be running. "It boiled down to a personal decision," he told the *Globe*'s Bob Healy, a close family friend. "If it had been a political decision I would have been in it. Every indication showed that I could do well, if not win the thing." He would not have anyone say that he had backed out because he was afraid he would lose. He told his father that he was opting to go to graduate school in environmental studies. He chose to make it clear too that he bore no ill will toward his cousin. "There is no grief between Joe and me," Teddy said, though there might have been if he had announced his candidacy.

CHAPTER TWENTY-FIVE

Left Out in the Cold

D uring the last half of 1985, Joe's face stared out of fifty public-interest billboards celebrating Citizens Energy ("No one should be left out in the cold. Write: Joe Kennedy, Citizens Energy Corp."), and his voice was heard on public service ads on radio and television. In the fall, Joe planned a statewide tour to highlight the arrival of the first fuel tanker of the season. Judging from Joe's public presence, it appeared that fleets were sailing into Boston with fuel oil that would soon flow into the oil tanks of the poor from one end of the Commonwealth to the other. The reality, however, was that the program was supplying only 2,300 to 5,000 homes, about 2 or 3 percent of those served by the state's fuel assistance program. Joe bristles at the idea that Citizens oversold itself, and believes that one should look at what Citizens did, not what it did not do. "The truth is that thousands of people under any measurement stayed a little warmer thanks to Citizens," Joe says.

For several months, Joe had been planning to run for Congress. But he waited until he had received months of publicity that would not have been his as a formal candidate before announcing on January 19, 1986. Although he was well financed, with $1.7 million to spend in one of the most expensive primary races ever, several of his opponents were amply funded too. The Kennedy name set him apart from the rest of the field and made him the front-runner. The Kennedy name also brought him unwanted scrutiny. "A distracting covey of fringe candidates, who range from pitiful to absurd, cannot indefinitely deflect notice away from the way Kennedy blows his lines and stumbles over typewritten copy," wrote Chris Black in the *Boston Globe*, referring to several marginal candidates.

Richard Goodwin wrote speeches that sounded as if Joe were running for president, and Joe read them as if reading the words for the first time. He was expected to be articulate and forthcoming on a myriad of issues, but he sounded more programmed than passionate. Joe was at his best when he spoke from his own convictions, and if the syntax was not always as straight as the sentiment, he connected. He was not a classic liberal; he questioned whether the nation could afford national health insurance and remained harshly critical of many aspects of the poverty program.

Joe was not the only Kennedy running for Congress that year. His older sister Kathleen was campaigning for a seat in Congress from a highly conservative district in Maryland. She had moved to the state when her professor husband, David Townsend, took a position at St. John's, the celebrated Great Books college in Annapolis. Joe and Kathleen showed a common front to the world, but they were extremely competitive.

Kathleen had the right to feel slighted since, if she had been the eldest son and not the eldest daughter, the eighth district would have been hers and she would not have had to travel to the politically barren reaches of rural Maryland to run for office. She was not tainted with the stories of indulgent behavior and tragic accidents that were prominent on her brother's résumé, but she had little charisma. She was a plodding, issues-oriented candidate who rang doorbells in her running shoes.

As the campaign progressed, Joe improved in his public presentations and continued at one-on-one politicking. He ran down the corridors at a senior citizens' home, embracing residents. He tore back and forth across the street in a parade. He charged up to commuters during rush hour to grab their hands and shout a hurried "Hi!" He would never be the urbane, subtle, eloquent, issue-parsing candidate of the Cambridge liberals, but he spoke to the hearts and heads of those who most needed government services, be they seniors or immigrants.

In the September primary, Joe won 52.5 percent of the vote, a formidable achievement given the field that he faced. In the overwhelmingly Democratic district, the primary was the crucial election. As Joe prepared to go to Washington, he dismissed the idea that he was picking up a torch from his late father and uncle, calling the idea "a crock of baloney"—little more than a journalistic conceit. "I'm not carrying around all that baggage," he asserted, though he was surrounded by the past. He downplayed the idea in part probably because it was so overwhelming, and so much the theme of his childhood.

In order to run, Joe and his family had given up their country home and moved back to the district to a home on the same Brighton street where they had previously lived. "We were never able to recapture the magic we had shared in our first Brighton home," Sheila reflected later. "The pressures of the campaign and Joe's new congressional life began to take their toll and our marriage became trapped in a downward spiral."

Even the most levelheaded of candidates often end up with an exaggerated sense of themselves. For months they are not merely the center of the universe, they *are* the universe. Joe was used to having others spend their time promoting him, but his ambition was mixed with anxiety. He drank endless Cokes during the day, and he didn't sleep well at night. Sheila had been a crucial adviser, but she wasn't any longer. She sensed that this man she had married had become something different. Joe's friends had a somewhat different perspective. They felt that the marriage was already in trouble—its problems exacerbated by Joe's new career. Whatever the difficulties, Joe and Sheila were good parents devoted to the twins.

J oe headed down to Washington intending to have a powerful impact on his country. He knew that there would be scrutiny of him that few freshman members receive, and he acted as if that were only his due. Most freshmen get in line when they enter the House, knowing that they will rise through seniority, their advance fostered by the relationships that they will build. They know that their enemies will do ten times more harm than friends will do good, and so they spend time cultivating their colleagues.

Joe had rarely stood in a line in his life, and he had never been good at ingratiating himself with others. Even before he was sworn in, he pushed to be named to the Energy and Commerce Committee. Joe went to see Republican John D. Dingell, the powerful, savvy committee chairman. As Joe recalled, Dingell didn't know who the freshman Congressman was. "You like clean air up there?" Dingell asked when he learned the young man was from Boston. Joe nodded in the affirmative. "Well, I'm from Detroit and I don't, and there's no chance you'll ever get on that committee."

That was not exactly the best approach to get Joe to back down. "I said, 'Mr. Chairman, you've just guaranteed that I'm going to try.'" Joe said, "I think I had sixteen people shaking my hand and saying they'd vote for me but a lot of them voted against me and I lost." As a freshman, he should hardly have expected such a plum assignment, but he was disappointed and had to settle for his third choice, the Banking, Finance, and Urban Affairs Committee, an area in which he professed to know nothing. "I could have raised hell with the energy industry," Joe told the *Globe*. "I didn't know a thing about banking. But I'm learning, and when I know enough, I'll raise hell with the banking industry."

That was just the kind of incendiary statement that pleased some of his constituents but made many of his colleagues uneasy. Joe had a natural abrasiveness that he was unwilling to modulate. Joe was all for making a difference, but the daily life of a member could wear one down with its tedium, long delays for votes, droning committee hearings, interminable fundraisers and cocktail parties, routines built upon routines. "I mean, I know I love the job but I don't know that it is right for me or my family, and I don't know that I am the most effective person in it, and I am just trying to figure it out," Joe told the *Boston Globe* after a year in the House. "The lows are the personal side, and that is just horrible, it just is. You know, I have two little boys and I just love being with them. I have been home one night in two weeks and that is really hard."

In March 1987, Joe, Sheila, the homeless advocate Mitch Snyder, several other congressmen, and movie stars Martin Sheen and Dennis Quaid spent a night on a heating grate at the Library of Congress. Snyder was the controversial impresario of homelessness who through his hunger strikes, confrontations, and publicity had raised the consciousness of Americans about a hidden shame. It was thirty-seven degrees with a chilling wind when Joe arrived after a dinner for freshman legislators at the White House. Joe was planning to protest by joining several of his colleagues sleeping out overnight. Joe was only trying to do some good, shining attention on a serious social malady, but his little attempt ended in disaster. As he sought to find a spot on the heating grate, reporters hurried up to the most celebrated member of the freshman class. Joe could tell that one of his new congressional colleagues, who had spent years working largely anonymously on the problem, was irritated that Joe was muscling in on the publicity. Joe got the sound bites that evening, and also a congressional enemy for the rest of his career.

Even the homeless were not universal in their praise. "That's the celebrity grate, buddy, we don't rate that one," yelled one homeless man, upset that Joe and the others had taken over the warmest spot for the evening. *The New Republic*, a journal that might have been sympathetic, turned on Joe and his companions: "The homelessness crisis has remained stubbornly resistant to quick policy fixes, but it has proved advantageous to politicians with a fondness for theatrics."

S nyder lived at the Community for Creative Non-Violence homeless shelter along with Carol Fennelly, his longtime companion. Snyder had coerced and shamed the federal government into giving over a building on Capitol Hill to house 1,400 homeless, by far the largest such shelter in America. Through those portals came a diverse array of humanity. On a rain-swept day late in 1986, a bedraggled African American and a barefoot, scraggly, thin teenager soaked through from the cold rain pouring down outside came to the front desk. "My name is Dougie Kennedy," said the disheveled form in front of them. "And I was out in Virginia, and this guy was homeless, and I found him and I brought him here. Do you have a place for him?"

Of course, they had a place for him. "My father would have liked what you do here," the youth said as he turned and hurried away. Realizing that the teenager might be the son of Robert Kennedy, Fennelly ran out into the rain, where their erstwhile visitor was getting into an ancient American car. "What are you doing out here in this glass-covered street with no shoes on?" she asked. "Can I take you someplace?" The young man shrugged his answer. "Say, are you a Kennedy?" "Yes," he answered, and drove away.

That was the beginning of an intense relationship between nineteen-year-old Douglas Kennedy and Snyder. "He has had more of an effect on my life than anyone," reflects Doug. The youngest son of Robert and Ethel was in some ways the forgotten Kennedy. He had been born prematurely with congested lungs and had almost died in his first seventy-two hours of life. His younger sister Rory had been born six months after their father's death, and that gave her a certain poignant aura and made her seem special. He didn't have that, and if at times he seemed overlooked in the spirited drama of his family, he was left largely free to figure out his own life in his own way. A freshman at Boston College, he was puzzling out what he should do and become.

That winter, to dramatize his cause, Snyder slept most nights on the grate outside the Library of Congress, and for several weeks during his Christmas vacation Doug slept there beside him. For about half a year, whenever he was back from school, he spent most of his time around Snyder, learning and listening. Fennelly had a sense of a deeply sensitive, inward-looking young man. She felt that he was almost desperately trying to recapture something of the excitement and drama of the activist sixties, and in doing so to recapture his father's life. Doug was so moved by the plight of the homeless and the savage inequalities that he saw that he contemplated devoting himself to this cause. "I almost gave away my whole life to become a homeless advocate, to quit school and live on the street," Doug reflects. "I mean, I came very close to, like, giving away all my money."

The more Fennelly observed him, the more she realized that as childlike as he often appeared, Doug had the perceptions of someone who had spent years observing and trying to figure things out. "We were working for the first homeless legislation ever passed by Congress, and it would have been very useful for us to be able to say, 'Douglas Kennedy, youngest son of Robert F., is joining us in this fight, living on a heat grate too.' But he would never let us do that. He didn't want the media focus at all."

"What I learned from Dougie and from Mitch is that celebrity costs everything," Fennelly says, meaning that literally in the case of Snyder, who committed suicide in 1990. "You trade in your life for it. Anybody who wants that kind of celebrity and power ought not to have it, because they don't understand the nature of it and how corrupting it is and how much it costs. And Dougie understood that instinctively even though he was very young. And he wanted none of it."

O f the young Kennedy men, no one was more aware of the curses of celebrity than Patrick. When he was only twelve, a reporter for NBC News came to the house and asked him what he thought about his father's extramarital affairs. Ted had been rightfully infuriated and gotten the tape destroyed, but young Patrick was full of his father's instinctive mistrust of journalists. His big brother, Teddy, had an easy confidence when reporters approached him, but Patrick was always nervous and wary with outsiders, be they reporters or anyone else.

When Patrick left Georgetown College in September 1986, he had gone up to Boston to spend a few days working in Joe's campaign. What impressed Patrick, as he told his biographer Darrell M. West in 1995, was that "there wasn't any kind of family resource that wasn't marshaled in a concerted effort to see him elected. I just remember the size of it and financially the commitment made was huge in terms of putting it together." As Patrick observed the frenzy of the final campaign days, he realized that "it was time for the torch to pass to our generation," though it was far more than a torch that was being handed on.

When twenty-year-old Patrick entered Providence College in the fall of 1987, he was one of the oldest first-year students. By this age, almost all of his cousins had had experiences that had taken them beyond the narrow, confining parameters of family. Patrick had gone nowhere on his own and appeared to seek no great adventure or intellectual challenge. If he had stayed at Georgetown or enrolled at Boston College, the other great Catholic institution that several of his cousins attended, he might have had professors ready to wrestle with dangerous, heretical ideas. The Dominicans at Providence were more often largely concerned with imparting the received wisdom of their faith. Patrick sat in the back of his classes literally and figuratively, receiving mediocre grades. His fellow students were at first intrigued that a Kennedy was in their midst, but once the initial excitement died away, Patrick was just another student shambling toward his degree.

One of his classmates, Scott Avedisian, described Patrick as "timid like a rabbit." He was thin and appeared undernourished, as if lacking some fundamental nutrient. He did not hang out much with his fellow students. At one party a number of the guests were playing a Trivial Pursuits drinking game in which an inordinate number of the questions had to do with various Kennedy scandals. Most of those in attendance did not realize that the diffident young man standing there was Patrick. He did not flinch or walk away, but stood there as if they were talking about somebody else's family.

One of the assumed perks of being a Kennedy was women, but even here he generally came up short. "I had a very pretty friend whose father was big in Boston politics and Patrick pursued her," recalls Kimberly Stender, a fellow student. "He was kind of comical. He was just coming on to her like you would hit on someone at a bar. He wasn't, like, overly drunk or anything. He just wanted to talk to her, and he was like, 'We should go out one night,' and she kind of laughed at him and walked away. Most of the mainstream women at Providence just could not have cared less about him. I mean, if it was his cousin John, maybe we'd be interested, but not Patrick."

Early in April 1988, Patrick started having bad headaches and sharp pain in his back. All his life he had soldiered through in his weak, vulnerable body. At first he tried doing just that, but before long he had difficulty walking. At Massachusetts General Hospital in Boston, he was diagnosed with a tumor on his spinal cord. His brother's cancerous leg had been amputated fifteen years earlier. Now Patrick lay in a hospital bed not far from the institution where young Teddy had gone through a horrific regimen of chemotherapy. It took five hours to remove the tumor because it was far larger than expected (2 centimeters by 1.25 centimeters), so large in fact that if it had not been removed, Patrick might have become a quadriplegic. The tumor was benign, and a few days later he left the hospital wearing a neck brace and walking with the help of a cane.

Patrick had no prosthetic leg to learn to use, and his convalescence was as much psychological as physical. At the age of twenty, Patrick said that he had discovered that "every day counts." After less than a year in Providence, and with limited life experience, Patrick decided to run for the Rhode Island House of Delegates. "I don't think I'm the kind of person that wanted to just stand still and sort of let myself enjoy being a young person," he told the *Boston Globe* in 1999. He didn't enjoy being a young person and sought an adult identity. "I wanted desperately to be part of that [Kennedy family] legacy in both a personal and public way," he said. "It was a way for me to feel my life had a purpose and was worthy."

Patrick's immediate problem was that to get the seat he wanted in the Mount Pleasant section of Providence he would have to take on Jack Skeffington, the well-regarded Democratic incumbent. Skeffington was an undertaker who had buried four thousand of his neighbors. As deputy majority leader in the Rhode Island House, Skeffington had doubtless reached the high point of his career. He had served his constituents well enough and, if not for Patrick, probably would have given up his seat a few years later, his retirement marked by sentimental tributes from friends and neighbors. The district was almost as Democratic as Dublin is Catholic, and the Democratic primary was the true election.

Ted was worried that Patrick might lose, and that if he won he would sacrifice his education. His youngest son convinced him that he was wrong on both counts. His father authorized the family arsenal to be opened, and the weapons pointed at the now-vulnerable figure of Jack Skeffington. When Patrick began his campaign, he introduced himself to voters by saying, "Hello, I'm Patrick Kenn-e-dy?" as if he were not sure who he was. A speech consultant fixed that, but he was an excruciatingly bad speaker; his admirers argued that his Uncle Jack had been just as bad when he first ran for office in 1946.

Skeffington was the unanimous choice of the state Democratic Party convention, and the five-term incumbent fought back with a campaign far beyond anything he had ever mustered. But he was overwhelmed by the shrewd maneuvering and the last name of his twenty-one-year-old opponent. Not only did Patrick have more money, but he knocked on at least three thousand doors talking to Democratic voters.

On primary day Skeffington saw another measure of what he had been up against. At each polling place the incumbent visited, there stood a Kennedy—Joan, Joe, Ted, Teddy, or Kara—next to an advance man and a photographer with a Polaroid camera ready to immortalize the voter's moment with a Kennedy. The district was full of Irish and Italian immigrants who would have been only slightly happier if the new pope, John Paul II, had been standing there soliciting their votes.

When Skeffington arrived at yet another polling place, he saw the instantly recognizable figure of John F. Kennedy Jr., ready to be photographed with the next voter. John was six years older than Patrick, and when people asked him about running for office, he always said he was not ready yet and did not want people to vote for him merely because of his name. After a while John came up to Skeffington. "Jack, I'm going to tell you something," he confided. "I don't like being here. I don't think it's fair for me to be here. I want you to know the only reason I'm here is for my cousin. But I don't believe in it. I don't think it's fair. This is your neighborhood."

Patrick won overwhelmingly, 57 to 43 percent (1,324 to 1,009), at a record cost of $73 a vote. "None of the victories I have ever had in my political life has meant so much as this one tonight," Ted said, as if the victory justified his life as much as his youngest son's. Patrick had what his father thought was the greatest of victories, but John was not the only one of his cousins to believe that he had won in the wrong place at the wrong time in the wrong way.

I n the House of Representatives, Joe did not spend much time plugging away creating legislation but took dramatic public stances on a multitude of issues. His legislative accomplishments were those of a politician who had learned crucial lessons at Citizens Energy about the workings of American capitalism. He wrote legislation that allowed nonprofits to take private money to build affordable housing. In 1989 he made probably his most important contribution when he authored and pushed through legislation that obligated banks to report their loans not only by geographical area but by race and income level. This was a way to expose banks that were redlining—refusing to make loans to poor or minority applicants—and by so doing to open up home mortgages to those who had systematically been denied them. He also worked to enact legislation to allow low-income families to buy homes owned by failed savings and loans.

Joe was a Clinton Democrat before Clinton, a mainstream populist who approved of the death penalty, because he believed in it, and was all for a bill to balance the budget. On some issues he was closer to some far right conservatives than to Democrats with special interests to protect. "A number of very conservative Republicans supported Joe on these banking measures because they agreed with his message about making sure capitalism works for all people and not just for the rich and all that," reflects one of his former aides.

Joe was immensely popular in Massachusetts and was touted for governor almost from the day he arrived in Washington, but the congressman from the eighth district was less popular among his colleagues.

Joe received a lesson in his popularity when for his second term he and Representative Chester Atkins of Massachusetts both sought a seat on the House Appropriations Committee. Joe brought his Uncle Ted in to lobby for him, a step that broke House etiquette. The New England Democratic caucus had a secret vote in December 1988 to decide between the two

Massachusetts congressmen, and Joe lost 9 to 8 when one member who had been pledged to Joe secretly changed his vote.

If there was any major negative take on Joe, it was simply that he couldn't get along with his colleagues and had an uncontrollable temper. In December 1988, he and another driver came close to an accident in Brighton. Joe jumped out of his car, and seventeen-year-old Victor Avila claimed that the congressman rushed up to him and roughed him up. Avila later dropped the charges, but it was hardly the kind of action expected from the most promising young politician in the state.

Joe's aides had learned to deal with Joe's temper. It wasn't just that he got upset over seemingly minor matters, but that he exploded like a tornado out of a summer sky. The subordinates not only excused Joe but for the most part considered suffering his outbursts only a moderate price to pay to be around a man who cared deeply about the issues that they cared about.

When Joe and an Irish priest traveled to Northern Ireland in April 1988, behind them came a truckful of news media and behind that still others. The British soldiers at a checkpoint were presumably not unaware of their famous visitor and his entourage and they gave Joe a unique welcome. They ordered the priest out of the car, and beyond the hearing of the reporters employed language that is not usually used in front of a cleric. "You cannot do that to a priest," Joe said, though the whole history of Northern Ireland was that they could and did. "Sod off," the soldier said, taking Joe by the arm. "Take your hands off me!" Joe exploded. "Since when do you tell any-body what to do?"

Joe was a partisan Irish American, and it rankled him to see how his religious brethren were treated in the largely Protestant country. On a visit to Scores, an airplane manufacturer with large American defense contracts, he was appalled to see how few Catholic workers were employed, and how they were harassed, and he vowed to do something upon his return. He cared, of that there was no doubt, and yet he was intemperate, so undiplo-matic that he brought the emotions of war to the causes of peace.

Upon his return Joe worked in the House to push through an amend-ment to the Defense Bill that would make it illegal for overseas firms with

Defense Department contracts to discriminate in their hiring practices. There would seem to be no argument against the measure, but the Reagan administration opposed it because it would add complexity to relations abroad. The amendment passed overwhelmingly in the House, though not in the Senate, and Joe went with his Uncle Ted to confer with Senator Sam Nunn, chair of the Armed Services Committee and the most knowledgeable senator on matters of defense. Joe invoked everything from Nunn's own Irish Catholic wife to the civil rights movement, trying to get Nunn to support the amendment and pull the Scores contract. Most politicians have one form of rhetoric for the public and another for their peers, but Joe preached as if he had an audience of thousands. "Senator Nunn indicated that he had had enough," Joe recalls. "He said he didn't want to pull the contract. I made another pitch, and Nunn said, 'I'll drop the amendment.' It was a lesson learned and I was quiet."

Afterward, the *Boston Globe* reported that an observer who had been in the meeting said that "outside the office, the aggressive congressman was cornered and admonished by his outraged uncle. 'Don't ever, ever talk to Sam Nunn like that.'"

Joe has a different view. "It wasn't that Teddy chewed me out," he says. "Teddy was in a friendly way trying to commiserate with me, talking about the dynamics." The problem was, however, as so often in Joe's life, others had a different perspective of what was going on, and what they thought was outrage was to Joe mere dialogue.

W hen Joe was having emotional difficulties and often when he just wanted a good comradely discussion, he went to see Dr. Robert Coles, who was as close to a father figure as Joe had. Coles had a sterling reputation not only as a fine teacher and distinguished author but as a humanitarian. It is hard to believe anyone would view him with dismay, but one person close to Joe thought that Coles had been too emotionally close to the Kennedy family to offer perspective. Nonetheless, this same person says, "But who knows if Joe would have gotten as far as he has without Robert Coles?"

Joe's marriage to Sheila was slowly unraveling, and he was unable to grasp just how close his wife was to leaving. He came from a large family, and he wanted to have a whole brood of children running around the house. Sheila said no. Her days were full of mothering with only two children. She would not hand the twins off to nurses, nannies, and tutors the way Ethel had. But her refusal had to do with far more than mothering style. As a woman in the Kennedy family, she had few ways to express her discontent. One of them was to refuse to bear any more children. Sheila loved being a mother more than anything in her life, and Joe did not understand what a profound statement his wife was making.

There was an emotional and physical disconnect in Joe's life. In Washington he had a small apartment on Capitol Hill where he spent the week when Congress was in session. Sheila never once stayed overnight in his Washington apartment, and he had what in effect were two lives, a marriage to Sheila in Boston and a marriage to politics in Washington.

CHAPTER TWENTY-SIX

Joe Jones in New Haven

hen Tim was at Georgetown Prep, he had a number of nicknames, several of them scatological, but the one that maddened him was "the Priest." He had thought about entering a seminary, and although he had given that up, he continued to attend mass regularly and study religious books. "Tim's extremely religious," says Brad Blank, a close friend. "I mean, if you go in his car and you look at what's strewn on the floor, it's like eight little newsletters on Christ, and rosary beads are falling out of the glove compartment."

Tim Shriver still wanted to do good, but he didn't want to end up as "some do-gooding Charlie who wants to run around and help people and you don't know how to do a goddamn thing." In the summer of 1977, seventeen-year-old Tim taught inmates at Lorton, the Washington, D.C., prison. When he arrived as a freshman at Yale that fall, he developed a reputation as an academic roustabout. He was a mediocre, desultory student who preferred discharging fire extinguishers in the dormitory to cracking books. He enhanced that reputation by living with several friends off-campus in a raw neighborhood where few Yalies ventured.

Tim did not make it into the prestigious secret society Scroll and Key, a rejection that would have been far more painful to his older brother Bobby, who did become a member. For the most part, though, Tim was an immensely competitive young man, prone to petty jealousies. He wasn't catching on academically, and he decided to go into Yale's teacher prep program.

The academically snobbish considered education a repository for those who weren't up to the great challenges in the social sciences, the sciences, or the arts. That is a brutally unfair attitude in a democracy where public education is one of society's great glories, but that's the way it is. Tim did not give such matters consideration but went into his education classes with the sense that they could provide a necessary tool for him to lead a socially useful life.

While still an undergraduate, Tim worked as a student teacher at Lee High School in the New Haven ghetto. When he graduated in 1981, almost all of his classmates left New Haven to explore larger worlds. Tim's brother Bobby was an investment banker in New York making lots of money, and his cousin Joe had just started an innovative nonprofit company. Tim made what appeared to be an almost embarrassingly mundane decision to stay in town and work for Upward Bound, a federally funded program to prepare disadvantaged high school students for higher education. "That was the most intensive work I ever did," Tim recalls. "I was actually in housing projects with kids, responsible for after-school programs, running career workshops."

Despite the Kennedy family's long political championing of the advancement of African Americans and civil rights, Tim was the first member of his family to actually spend lengthy periods working with a minority population. Most social do-gooders and progressive theorists do not personally test their ideas against the gritty realities of life. Tim was doing so, and what he was learning was immensely unsettling. "I needed to understand those kids," Tim reflects. "I needed to understand their love, their anger, their fear, their language, their culture, their way of doing, their oppositional nature, their comforting nature. Maybe I had grown up with too naive a concept of what it meant to care. Maybe I thought if you showed up and offered to help, that you would be greeted by a smile and you could lend a hand. But I got in there and it was like, 'Whoa.' Changing the world requires solidarity and understanding and a self-emptying. It is as much about moving the barriers created by distance as lending a hand from across the divide. Nobody wanted my help. Everybody gives you the bird. It was like a shocking new world that I couldn't understand."

Tim worked endless hours. He became obsessed with learning why many of the children didn't like him, and why even those who did like him rarely saw their lives bettered by his intervention. After three challenging, frustrating years, Tim went to discuss the situation with Dr. James P. Comer at the Yale University School of Medicine's Child Study Center. The African American psychiatrist is one of the leading school reformers in America, and he listened as Tim told a troubled tale of his years at Upward Bound.

"Look, I'm working with these kids, and I don't think I'm doing it right," Tim said. Comer saw that Tim was floundering. He helped arrange a one-year fellowship from the Field Foundation so that Tim could study at the Child Study Center and try to figure out how he could truly change these young people's lives.

One day later in 1984 Comer took Tim with him on a visit to Hillhouse High School. For three years Tim had been visiting the high school with Upward Bound, but he had made little impact on the students at the almost exclusively African American school.

"This is Tim Shriver," Comer said, introducing him to the principal. "I think he can help you here."

Salvatore "Red" Verderame had a vague recollection of seeing Tim running around the school wearing a trench coat, but he was hardly impressed. The salty, homespun New Englander had been a basketball coach for most of his career. A few years back he had given up coaching and settled in as assistant principal at Hillhouse, but he had been terribly reluctant to become the white administrator of a black school. He had taken the principal's job largely because the new superintendent, John Dow Jr., implored him to do it. Dow had come from Grand Rapids, Michigan, where he had a reputation as a firebrand. When Dow arrived in New Haven in January 1984, he became the first African American superintendent of schools in the city's history. He faced a challenge changing a school district in which 80 percent of the students came from minority populations and 72 percent from families on state assistance. All of the social problems coalesced at Hillhouse High School; Dow called it "the worst school I have ever seen."

Verderame was willing to take whatever help he could get, but the principal had a visceral distaste for almost anything to do with Yale. Here was an elite university within a mile of a hopelessly poor, disadvantaged black community rife with drugs, prostitution, and violence. Most of the professors rarely looked out beyond the ivy groves into the desperate world just outside the campus. As the principal saw it, the few Yale social scientists who did venture out considered Hillhouse a convenient laboratory where they could go to have questionnaires answered and observe the peculiarities of the genus *African American*. Verderame had a higher opinion of Comer than he did of most of his Yale colleagues, but that did not mean he trusted everything Comer said. He thought it was highly unlikely that this Yale graduate before him would be at all helpful.

But Verderame was not about to do to Tim what he believed the Yalies did to his students—treat them like anthropological specimens devoid of individuality. He decided that he had better at least check the teacher out. The first thing he learned was that Tim was living in a single room above a church on Dixwell Avenue, as blighted a neighborhood as there was in New Haven. Verderame figured that if Tim was living there, he might not be quite the stereotype the principal imagined him to be.

Verderame was nonetheless justified in being suspicious of his new teacher who was supposed to implement some of Comer's ideas. "He was a real Yale student, if you know what I mean," says Verderame. "Tim was just not street-smart." With an ample measure of arrogance, Tim knew precisely what should be done at the school. On one of his first days Tim approached a teenager who told him: "Fuck off, asshole."

"Trying to teach was the hardest, mind-bogglingly difficult thing," Tim says. "Poverty produces chaos. Chaos produces tremendous pressure on kids. Kids transferring schools three times a year. Kids losing their textbook and not having any money to buy the new one. The second week I was there a kid got shot right in front of me. He had blood all over his hands. So it didn't take a rocket scientist to figure out you needed to think differently about how to educate those kids. And it wasn't just the new curriculum. It had to do with the human element."

Hillhouse was being held hostage by some of the older students, who roamed the corridors largely at will. These renegades were as old as twenty-two and twenty-three. They had turned Hillhouse into their private club, a place to hang out, grab a free lunch, and meet girls. There was often random violence in the halls, and the atmosphere intimidated those who came to Hillhouse to learn. The school was not unlike the Lorton Reformatory where Tim had taught, a place run largely by inmates who intimidated anyone who challenged them.

Tim went into Verderame's office late one afternoon to discuss the problem. "We have a lot of experts on child development and conflict at Yale, and we would like to propose that we conduct an analysis of why kids are fighting in the school," Tim told Verderame.

The principal was a tall man with an intimidating physicality increased by the cigar that he was smoking and the silence that greeted Tim's suggestion. "Oh, is that right?" Verderame said finally. "We would be very interested in your solving the fighting problem."

Tim believed that there was a direct and easy line between the academic theories he had read about at Yale and the realities of daily life at Hillhouse. He was ready to prepare what he later called "some nice humanitarian systemic way to address violence prevention." Tim's liberal philosophy told him that he could reason with these difficult youths. He could solve things by giving them counseling and special classes. Tim tried some of the measures he had learned in his classes at Yale, but he was working with teenagers who had no real parents, had lived all their lives on welfare, and saw violence that they carried into Hillhouse. They already had the education many of them thought they needed—they had learned their lessons in the street, not in the classroom. Tim did the best he could do, but the best he could do was no more than a shout in the night.

Verderame and Dow had not gone to Yale, and they had a different way to approach the violence and disregard that was turning Hillhouse into little more than a holding cell. Verderame signed on Tim and two other beefy teachers to his little SWAT team, including one six-foot-four, three-hundred-pounder known as "the Crusher." The four men patrolled the corridors during classes. Anyone they found roaming the halls they kicked out. Some of the students were the size of professional football players. A few times Tim and his colleagues had to call the cops to get the most belligerent of the students off school grounds.

Tim was surprisingly good at intimidating and muscling these teenagers, a skill he had learned neither at home nor at Yale. By the time they had booted 85 of the 1,500 students out for good, the corridors began to grow quiet, and there was a new discipline in the classes as well.

Hillhouse had been so plagued by violence that the school had given up having evening dances, theater programs, and night football games. The school could have hired guards and off-duty cops and ringed the events with security. There was no money for that, however, and even if there had been, this was a school, not a prison. Tim had learned enough in his first months at the high school to propose an idea that never would have occurred to the professors at Yale. He suggested that they go to the gang members in the school who were responsible for so much of the violence and ask them to take over the policing duties. It was a dangerous approach. The gang members could have turned into an intimidating goon squad. But the young men were proud to be asked to perform such a service, and Hillhouse came to life at night once again.

There hadn't been a play at Hillhouse for a decade, but with evening activities starting again, the school decided to put on a major production of *Man of La Mancha*. Some students practiced their parts while others helped build the sets in the woodshop. "Here were these kids standing onstage singing 'The Impossible Dream,'" Tim says. "I mean, it was a classic, if only someone could have taped it."

Tim would receive a Ph.D. in education from the University of Connecticut in 1997, but his real doctorate was earned at Hillhouse High School. "Tim came into New Haven and really earned his way," says Dow. "He wasn't given nothing. As a matter of fact, you would never know he was from that family. The way he acted, the way people treated him, the way he interacted with folks, he wasn't afraid. I'm telling you, he would go right into the lion's den, man, and people trusted him. And he really earned his spurs. He had become a teacher. He had earned that."

I n August 1983, Bobby Shriver attended the Special Olympics World Games in Baton Rouge, Louisiana. As a little boy, he had toddled around the Shriver farm among the developmentally disabled who came there each summer for his mother's camp. As he grew older, he helped out in the camp and took much away from his experience. "I think of myself as having a disability," Bobby says. "I think of everyone as having some disability or another." His mother envisioned Special Olympics as a movement that while enriching the lives of the Special Olympians would also teach others a new humanity. Every four years the games had grown, but like the athletes, they remained marginalized in American life.

The opening ceremonies took place on a plywood stage with a sound system borrowed from the local AFL-CIO. One of the celebrity hosts, John Schneider of *The Dukes of Hazzard*, told the audience that they could call an 800 number to buy his photo. Few spectators attended most events, and some parents felt that something was missing. Why wasn't the president of the United States there? And where were the television cameras and the newspaper reporters recording their children's triumphs?

Bobby shared their feelings and vowed that at the next international games in 1987 things would be different. After graduating from law school in June 1981, Bobby moved to Los Angeles to clerk a year for Judge Stephen Reinhardt of the Ninth Circuit Court of Appeals, one of the most liberal jurists in America. Then he moved to New York City to work in investment banking with James D. Wolfensohn, Inc.

The Kennedy family ideal at its best was that everyone should do their full complement of pro bono work, whatever their field of endeavor or profession. Bobby went to Brandon Stoddard, the president of ABC. Bobby knew the television executive because he had married Bobby's former editor at the *Herald-Examiner*, Mary Ann Dolan. Bobby convinced the ABC executive that the network should run a two-hour-long prime-time special. This had never been done before, and Stoddard took a considerable risk, authorizing an expenditure of $2 million.

Bobby knew that millions of viewers were not going to tune in simply to watch the mentally disabled running around a track or swimming in a pool. They would watch if several of America's greatest stars sang their biggest hits and there was a gigantic display of fireworks. There was no hotter artist in America at the time than Whitney Houston. When Bobby went to ask her to headline the opening ceremonies, her father, John Houston, talked with him first. Houston had worked in New Jersey for the Office of Economic Opportunity that Sarge had created, and talked admiringly of Bobby's father. When Whitney sidled into the room, she listened to the two men talking and announced that she would be there in Notre Dame Stadium in August 1987.

Whitney and other artists, including John Denver, did not charge any fees, but the show still had to pay for all the paraphernalia of stardom: entourages, guitar tuners, hairstylists, airplanes, and the like. As the unpaid executive producer, Bobby knew that the sound system had to be superb, the stage first-rate, and the banners dramatic enough for television. For weeks Bobby made phone calls, implored, argued, and pushed. He could be irritating in his single-minded pursuit. If he wanted something from you and you doubted that you wanted to give it, his was a call that you wanted to avoid taking.

Forty-eight hours before the opening ceremonies, Bobby showed up at Notre Dame, where forty carpenters were working on the great stage and banners were being hoisted. As he looked out on that panorama, it was in some measure the most powerful, most fulfilling moment of his life. He began to cry. He had helped to lift the vision of the Special Olympics to the place where he believed it belonged. He felt an immense affinity with the 4,700 Special Olympians who would in two days parade through this historic stadium. He sensed that he had suffered immense emotional difficulties because of his name and family, but in this one moment the scale balanced. He understood in a way he never had before what his mother had been saying for years: to understand the humanity of these Special Olympians was to be blessed with a sacred vision of life itself.

That year Bobby worked not only on the 1987 Special Olympics World Games but on the first Special Olympics Christmas record. The idea came from Jimmy Iovine, a top record producer, who was married to Vicki McCarty, one of Bobby's closest friends. Iovine wanted to memorialize his late longshoreman father. Vicki pushed her husband to do so in a record for the Special Olympics. It was a complicated effort to get the top recording artists in the world, including U-2, Sting, Madonna, Run-D.M.C., and Bruce Springsteen, to record Christmas songs. Iovine had the connections and the clout, and when *A Very Special Christmas* came out in November, Iovine's efforts were rightfully celebrated. But without Bobby, the producer would have gone to another charity, and the record would not have been promoted with the same panache. In the end the record raised more money than any previous charity LP.

In 1988 Bobby decided to give up his career in investment banking. For the next four years he worked producing another Christmas record and a Special Olympics television special. In 1992 he moved to Los Angeles, where he felt he could build a life for himself without always having a hyphenated name, Bobby Kennedy-Shriver, and his mother constantly prodding him. He continued producing a series of Special Olympics Christmas records. He was given a $120,000 salary out of the profits, and a record label gave him office space. He spent the rest of the time breaking into the film industry. That was not as difficult as it would have been for most newcomers, since in 1986 his sister, Maria, had married Arnold Schwarzenegger, who was becoming one of the biggest stars in Hollywood.

T im sought to get away too. He took the 1985–86 academic year off from teaching at Hillhouse to work on a master's degree in religion and religious education at Catholic University in Washington. During that time twenty-six-year-old Tim married twenty-nine-year-old Linda S. Potter, an attorney at a Washington law firm. Tim's bride was what his Yale classmate Dan Samson called "a babe," one of the basic prerequisites in the marriages of most of the group, but that was only the beginning of her qualities. Not many women of her background would have given up a law practice and in the fall of 1986 set up house in the ghetto of New Haven.

Tim moved into a bigger place with his bride, but the neighborhood frightened the unflappable Verderame. "You don't know what the hell is going to happen there every night," the principal said. "I mean, a lot of people get mugged around there. They had whores hanging around. But he thought he could get the neighbors working together cleaning it up. They started having kids, and Linda had to come home with the babies and the groceries. Then he moved to a nicer place, and he was happier and the family was happier."

Tim had wanted to continue to live in the community in which he worked, but he was realizing the compromises that had to be made. He could compromise either the safety and security of his family or his social ideals, and he chose the latter. Tim had by any measure his own life. He was doing his part as a father to his three young children, Sophia Rose, Timothy Potter, and Samuel Kennedy. He showed up on the public tennis courts to play a tough game with whoever showed up. He went fly fishing with Verderame, which was pure joy even if he had to listen for the hundredth time as Verderame recalled the morning Tim dumped a box of worms in the trout stream, ruining the fishing for the day. In his spare time Tim hung out with his fellow teachers.

Freud said famously that happiness is nothing more than love and work, and Tim had it in a straight suit, love and work, work and love—a sweet passionate life. He knew that he was doubly blessed, not only in doing work that was intellectually and emotionally satisfying but in living what was truly his own life, a life in which his last name bore only what honor or shame he brought to it. "I was very happy being Joe Jones in New Haven," he says. "I was very, very happy not to be defined by other people."

His associates were more aware of his Shriver/Kennedy connection than Tim realized, but it was unquestionably true that his achievements were his own.

Tim would have stayed on teaching at Hillhouse, but soon after he arrived back from his year's sabbatical, Dow named him to be the coordinator for an important new program in social and emotional development. The superintendent believed that in a school system in which children arrived wounded and broken, radical steps were imperative. Two-thirds of New Haven public school students lived in single-parent homes. Like it or not, Dow argued, it fell upon the schools to teach these children how to make moral choices, how to work together, and how to understand the dangers of drugs, truancy, pregnancy, and violence.

This new program was seeking to expand the students' "emotional intelligence," a term popularized in Daniel Goleman's important 1995 book by the same title. Intellectual intelligence is only part of a person's true IQ. There is a psychological or emotional intelligence as well, and it includes the ability to relate to and work with people, skills at working out problems and seeing difficulties through, and moral centering. Among the programs that Goleman enthusiastically chronicles is the program in New Haven, where he sat in on classes and talked with Tim and others.

Tim worked closely with Dr. Roger Weissberg, a young professor at Yale, who had developed an innovative curriculum that attempted to provide aspects of this social and emotional development. Dow wasn't about to have a few of his classes used as guinea pigs, and he insisted that they take the experimental program and make it systemwide, from kindergarten through grade twelve.

Despite Tim's often irritating audacity, he was a natural collaborator who listened to teachers and truly worked with them. Tim's team worked out of a large room at Hillhouse. His office was a tiny, windowless cube that previously had been a closet. Even when Tim was promoted to overseeing the entire social development program for the district, he agreed to do so only if he could stay at Hillhouse and not be confined in the downtown administrative offices. "I loved that place," Tim recalls of Hillhouse. "No windows. Kids everywhere. Yeah, I loved it. I couldn't have been happier."

Tim and his team sold their ideas first to wary principals and other administrators, then to often harassed teachers, and finally through them to the students themselves. Tim and the others worked exceedingly long hours. When the students poured out of the school on Friday afternoons, the team stayed on to analyze their week, like a football coaching staff running tapes of the last game.

They were in part trying to change a culture of negativity and help give the children of New Haven a renewed sense of their own worth. Second graders learned, for instance, not to "put down" but to "put up," to praise the qualities in others. The students discussed the dangers in life and how they could skirt some of them. Third graders heard a story about children disappearing when they went off with strangers. Ninth graders discussed what teen pregnancy does to a young life. There was nothing miraculous in this, no catechism being set out, but the students learned to see themselves and the world around them in a new way. They learned new ways to face many of the myriad difficulties of their lives and how to value themselves and those around them. The program reached beyond the students and teachers to cafeteria workers and others in the school who could be important mentors, and parents were also brought into the process.

Tim understood that often the best leadership does not come from the front of the room or from behind a podium, but from the back of the room. "He was never taking credit," says Karen DeFalco, a fifth grade teacher who became a crucial member of his team. "If we were doing a program for parents or something like that, he would never get up to the microphone and do the introductions. We would be doing it. He was never in the spotlight. He empowered us, made us feel comfortable to be with him, that he had so much faith in us."

Part of the continued pleasure of Tim's life was his contact with the teachers at Hillhouse. When Verderame was taken sick with suspected brain cancer, Tim went almost every day to the hospital. And when the principal learned that he had been misdiagnosed and had a mysterious brain fungus, Tim called specialists at the National Institutes of Health (NIH) outside Washington and arranged for a doctor to fly up in a private ambulance plane to examine Verderame and take him back as a patient.

Verderame was at NIH for months. When his wife and daughter came down to visit, they stayed at the Shrivers' house. When Tim was in town, he brought Verderame back to the house in a wheelchair and set him beside the pool. The principal had become blind, but he could feel his way into the water. After many weeks he began to recover. He could see again and was flown back to New Haven. "After that I got progressively better," Verderame says. "But Tim was responsible for the whole thing. I would not be here if Tim didn't step in. I mean, we're not family, we're just fishing buddies. And I said, 'You know, I'm Italian. Now you've got me around your neck for the rest of your life.'"

D ow had fought the good fight, but he had made too many enemies, especially when he condemned a parsimonious city administration and a state government that he said shortchanged the poorest school districts. Dow was pushed out early in 1992, and with him went much of the initiative for change. Later that same year Tim's other crucial partner, Weissberg, moved to the University of Illinois at Chicago when Yale did not offer him tenure.

Tim would have stayed on longer, but in the spring of 1993 he was confronted with a difficult dilemma. He had tried to distance himself from his family, physically and psychologically, and immensely appreciated living in New Haven, a world away from the political and social world of the Shrivers and Kennedys. He went to the Special Olympics World Games every four years and was on the board of Connecticut Special Olympics, but he had never taken a major role in what was essentially a family charity run by his parents.

The 1995 Special Olympics World Games were originally scheduled for Dallas, where the local establishment had pledged tens of millions of dollars. None of the Kennedys had visited the city since November 1963, though they had all traveled to Los Angeles, where RFK had been assassinated. Dallas itself had played no role in Jack's death, however, and the games would be a marvelous way to reconcile the Kennedy family and the Texas city. But Ted could not bring himself to go to Dallas, and he would not agree to the Special Olympics going there either.

Ted had nothing to do with Special Olympics, and other than in his role on the board of the Kennedy Foundation, he had no right to have any say about the organization his older sister had founded. It may have been selfish to put his emotional frailties above every other consideration, but he had rarely been so adamant on a family matter, so much so that he risked a break with Eunice. In the end, in February 1992, the decision to turn Dallas down was announced and blamed on the controversial new film *JFK*, which supposedly had rekindled questions about the assassination. But that was obviously a fallacious reason: the games would not be taking place for three and a half years, when the publicity over the film would be long gone.

The Shrivers scrambled to find an alternative. Connecticut governor Lowell Weicker Jr. had a Down syndrome son, Sonny, and was a strong supporter of the Special Olympics. The governor came up with a $20 million loan guarantee to hold the event in New Haven. The presidency of these games was usually given to a major local business or political figure who could draw on decades of contacts and experience. No one came forward in New Haven to take on that role, and as the weeks and months passed, thoughts turned to the one Shriver family member with an intimate knowledge of the small city.

"There was this gaping problem, and I was getting dragged into it more and more because my name was associated with it, and I didn't think there was any alternative," Tim says. "But I can honestly say that my mother or father never called and said, 'You have to do this,' or, 'It's time,' or 'Wouldn't you consider. . . . ' You know, there's never been like a pull. Maybe it's been a bit more clever. Gotta give them a little credit."

In the summer of 1993, Tim left the school district to become president of the games. As he was leaving, he invited the whole Hillhouse team to his house for a final party. Beforehand he found a special quotation that he thought exemplified each person. After dinner he handed out the slips of paper to each individual and asked them to read the words. Karol DeFalco read her quote from Martin Luther King Jr.: "We must accept finite disappointment, but we must never lose infinite hope."

Tim's appointment appeared to some to be the latest example of

Kennedy nepotism—schoolteacher Tim Shriver elevated because of his name. What few understood was the pain that he felt in leaving Hillhouse, the love he felt for those with whom he had worked, and the sense of obligation he felt toward his mother and the organization she had founded.

Tim did not have the millions of dollars that the Texans had promised, and as 1993 ended he was $18 million short of the $28 million needed to put on the games. "One minute I'm euphoric and the next I have to sit down and quell the panic," Tim told the *New York Times* in May 1994.

Tim managed to get most of the necessary money, and when the 7,200 athletes marched into the Yale Bowl in July 1995 for the opening ceremonies, there were few among the spectators who were not uplifted by this paean to possibility. One of those there that evening was President Bill Clinton; it was the first time an American president had ever attended the games. That scene would be seen by millions more in the television special that Bobby produced. The games attracted about half the spectators that some had estimated, and merchants were left with unsold souvenirs and other stock, but by any equitable yardstick, Tim's achievement was great.

CHAPTER TWENTY-SEVEN

John at Brown

hen John arrived as a freshman at Brown University in September 1979, he found for the first time in his life a place where he truly belonged. It was not Harvard, where every time he walked across the Yard, trudging along in the family tradition, he would have been reminded of his father, uncles, and grandfather. Nor was Brown the socially esteemed Princeton or Yale, where his Kennedy identity would have endlessly pursued him. Brown was known as the school you went to if you couldn't make it into the prestigious top tier; at least you could say you were going to an Ivy League college, even if it was in Providence, Rhode Island.

John had picked what *USA Today* called in 1984 "one of the USA's hottest colleges—maybe the hottest." Brown was the place to be, capturing the ethos of its time the way the University of Chicago had captured it in the thirties or Berkeley in the sixties.

A great deal of fun was made of Brown's "New Curriculum" in which you took whatever courses you wanted to take. And you took them for an A, B, C, or no credit, with no Ds and Fs set down in your record. Or if that was too onerous, you opted to go for a "satisfactory"; if you didn't merit even that, there was no indication in your record that you had even attempted the course. You did not have to take math or science or a philosophy seminar. Instead, you could sit through courses such as "Rock 'n' Roll Is Here to Stay" or "Semiotics 66: Introduction to Cinematic Coding and Narrativity," where you spent time watching Hollywood films. It was all so media chic.

"All the ink has made Brown media-neurotic," Phillip Weiss wrote in *The New Republic.* "Every quarter the administration publishes a tabloid called *Brown in the News* packed with the latest, including stories about student stars. A recent cover highlighted the visit by the *Today* show. There was a picture of President Howard R. Swearer getting comfy with Jane Pauley in front of space heaters on the green. It was hard to tell which one was the TV personality."

The students at Brown learned from each other as much as they did from their professors, and though one could graduate without studying philosophy or knowing American history, the synergistic mix produced students who had a powerful impact on the media-centric world of the future. Graduates directed high-tech investment on Wall Street, wrote situation comedies for television, made fortunes on Internet and dot-com start-ups, became leading journalists, and wrote the hippest of novels. John was a natural addition to this Brown world. He wasn't much of a student, and he could learn while not seeming to learn. He had his own intuitive sense of the media and popular culture and a unique awareness of politics unlike that of any of his peers.

John met most of his lifelong friends at Brown, one of them during orientation week. It was typical of John that he wasn't going to hang around a dorm room when he didn't have to, and he was off on the beach at Newport. There he chanced across another freshman, Robert "Robby" Littell. Robby was a six-foot-two-inch lacrosse player who had been brought up in the country club suburb of Greenwich, Connecticut. Although he dressed with preppy aplomb and spoke a rich boy's lingo, since his parents' divorce he had not lived a wealthy life. His father had committed suicide when Robby was seventeen, just a few months before he entered Brown. For all Robby's flashy bluster and laughter, his father's death hung over him and was something he had in common with John.

Both young men pretended to be far less introspective and sensitive than they were. "I was having to deal with my dad's legacy, and so was John with his, and we talked about this a bit," says Littell. "We didn't have our dads there to say they were proud of us. Also, they weren't there to make us feel bad. They weren't there to give us pressure to do well or better than they did. We were actually proud of our sensitivity toward women, because we were brought up by she-wolves, a mother and a sister. But in terms of not having a dad, we were tough guys. Not gonna talk about it."

At Robby's core was an impregnable self-confidence. He just knew that he was not only good but also just a little better than the other fellow, including his new friend. "Boy, it was just sort of one of those things where you look over and you go, 'Who's that? Who do they think they are? I must win,'" Robby says. "And that's a great dialectic. I'm probably dead wrong, but I always thought I was better. Still do."

John played up the part of his personality that meshed with the person he was around. Robby was a rowdy, rambunctious guy's guy. He had never seen a party he did not want to crash, a bar he did not want to enter, a football he did not want to throw. Nothing would ever change Robby, and nothing would ever change John's relationship with him.

R obby might think he could compete evenly with John, but he was the only one who thought so. From that very first week at Brown, John had a presence unlike anyone in the history of the school. On the one hand, he was the most famous college freshman in America, and yet he was also just another kid whose mother drove him up to Providence and unloaded the car. Most parents gave their sons and daughters the clothes and other upscale paraphernalia that they deemed suitable for an Ivy League education. Jackie hauled into John's dorm room at Littlefield Hall on George Street a motley wardrobe, a strange statue, and other objects that looked like they had come out of an attic and deserved to be moved from there immediately into the garbage.

But that hardly affected John's celebrity and the cachet he brought to the school. Everyone knew that they had a famous classmate who liked to be called John, not the cloying childhood "John-John." Even those classmates inured to celebrity could hardly help noticing the photographers who dogged his steps just when he was trying to blend in. When there was commotion in the registration line, one of his classmates, Matt Gillis, turned back to see John surrounded by a cluster of photographers and reporters; he had a look on his face that Gillis thought was "weary and disgusted, but also somehow apologetic to the people around him for the disturbance."

Robert Reichley, Brown's executive vice president, handled public relations for the university. Before John arrived on campus, Jackie's secretary called and told him that all media requests should be forwarded to John's mother. "I asked John to drop by my office, and it was very clear that John was very capable of managing himself," Reichley says. "It was understood between us that I would block the obvious and let him know about those that weren't so obvious. It ran to everything, including making a paid appearance at a local strip joint. This producer from ABC's *Good Morning America* wanted John to come on the show, and she hadn't been able to reach him. She got really huffy, and I called John to my office to talk about it. He came over, and I dialed her phone number and said, 'Here's Johnny.' I only heard John's side of the conversation. 'Yes, ma'am . . . yes, ma'am . . . well, ma'am, I'm honored by your invitation, and I want to come on your show . . . but I don't have anything to say, and someday I may have something to say, and I hope that you'll remember me and ask me back.' Which was just about the best kiss-off I ever heard, and there was nothing to say after that."

John handled the massive attention with aplomb. Freshman year he hung out a lot at "the Ratty," the Sharpe Refectory at the center of the Brown University campus. His hair was long, and he was dressed in an ever-changing collection of fraying shorts, Hawaiian shirts, tank tops, T-shirts, and jeans, all of which managed to show off his physique. He was six feet one inch tall and 165 pounds, and he had already developed the startling handsomeness that even without his name would have been enough to make people stare at him.

John's hair was as thick as fur. "I used to kid him about the density of his hair," says Jim Alden, one of his Brown friends. "One time he said, 'Jim, I gotta ask you a question. Does your hair ever just hurt?' Apparently there just was so much hair growing out of his scalp, there was not enough room for all of it, and his head would ache. So I had to explain to him that, no, I never really experienced that, and that was his problem and his alone."

John sat hunched over his food, his right hand hooked around his tray while his left hand pushed the food into his mouth, like a stoker feeding a ship's furnace. As he worked his way through the dishes, never missing a word of the conversation, his fork darted out, speared a French fry, a green bean, or a strawberry from his neighbors' plates, popped the items into his mouth, and moved back to another stab at his own plate. He didn't care if a person was sitting down with him for the first time. His or her meal was as much fair game as anyone else's.

"That's him, that's him," whispered a freshman at the next table as John got up to snag a second slice of pizza. That was the kind of thing that was always happening to John. He just ignored it, grabbed the pizza, and walked back to sit among his friends.

J ohn could not avoid his celebrity even when he sat in classes. That first semester John took a course with Professor Edward Beiser in political theory. Beiser appeared so aware of his famous student that as John sat slumped at his seat among a hundred of his classmates, some of his fellow students had the feeling that the professor was teaching to John alone.

Beiser might have been richly flattered to have John taking his popular class, but that did not mean that he would let him get by without work. When John handed in his first assignment comparing John Locke and Karl Marx, he received a "no credit." Brown had come up with benign euphemisms for failure, but "no credit" was still an F. Beiser had been willing to forgive lapses in grammar, spelling, and philosophical nuance, but it was hard to give a passing grade to a student who had no idea what distinguished Locke from Marx.

"What am I going to tell my mother?" asked John in the professor's office. Although Beiser tried to reassure him that he was not about to be thrown out of college, John was more worried that Jackie would find out. "Oh, my mother's interested," John said. "She knows about everything."

John found it hard to concentrate on a book. He might have a flash of insight, and then his mind went somewhere else. He found it difficult sitting through classes, where he couldn't move around or even twitch. He knew that he was not stupid, even if he could not regurgitate great quantities of material the way many of his classmates could.

In the spring of 1980, John signed up for a seminar on "The American Experience in Vietnam." A student needed Professor Charles Neu's permission to take the class, and most of the students were upperclassmen. The historian thought that even though John was a freshman, he might bring something unique to the seminar. On the first day John told his classmates that they should not hold back on their criticism of his father's role, but from then on, instead of entering into the spirited discussions, he said almost nothing.

The professor concluded that John was content to live off political gossip he picked up at his mother's dinner table instead of cracking his books, coming to class prepared, and entering the debate. "He had a kind of annoying way of speaking," Neu remembers. "As if to suggest that he had not been properly educated. He did not enunciate all of his words. It had a kind of thickness to it, almost as if to pretend he was a slightly dim-witted athlete."

John was supposed to come in to see Neu to discuss his term paper, but John never did. Toward the end of the term Neu ran into John on the Green. "Oh, by the way, I ought to talk to you about that paper," John said. Most students at least came up with some excuse, but John didn't do that. "Forget it, it's too late," Neu said.

Neu gave John a "no credit" and was no doubt glad to be rid of what he considered another one of the spoiled scions whose self-indulgent behavior marred Brown's academic life. Neu was correct in what John had done in the seminar, but perhaps not in his understanding. John had never talked in a serious way to any of his friends about his father, and it was unlikely he would do so before a roomful of strangers. In the seminar he had been a freshman among upperclassmen, expected to defend his father or to admit that he thought JFK had been wrong. He was no longer seeking a father figure the way he had as a boy, but he was still trying to understand Jack. That was the great intellectual struggle of his life: to comprehend the meaning of his father's life and in so doing to understand himself and the world around him.

J ohn and Robby joined Phi Psi, a fraternity known for its good-
looking athletic types who were not too serious, but decent. If you
were a Phi Psi, you played sports well, but you didn't eat at the jocks' table.
You were smart but not overly studious. You were well read but not pedan-
tic. You were perhaps a ladies' man but not a predator. You were proud to be
a Phi Psi, but you were not obsessed with fraternity life.

The fraternity was such a handsome group because those were precisely
the men selected for Phi Psi. One fraternity brother, Richard Wiese, modeled for *GQ*. Whether good looks help build confidence in a man or people
assume that spirit and mind go hand in hand with handsomeness, handsome
people often tend to do better in life than their efforts alone merit. John
gravitated toward such men, though in this he was in a category all his own.
He was working out, toning up his body. He had classic Greek proportions
and a face that with maturity was becoming fully defined. One day he
showed the sixteen-year-old actress Brooke Shields around the campus. She
was six feet tall, and as beautiful as John was handsome, and as they walked
across the Green, students and faculty stopped in their tracks. It was as if
some other species of humanity was strolling there among mortals.

J ohn's room was a shambles, with layers of clothes and books strewn
around it, a fledgling anthropological site where you could date a
sock or a notebook by where it fell in the pile. He and Robby added to their
considerable reputation for the studiously bizarre by keeping a pig in the fraternity house basement.

One time when Jackie came to visit, John was off delivering a paper to a
professor and his mother decided she had time to make a phone call.
"Where's the phone?" Jackie asked as she looked across the cluttered room.
She got down on her hands and knees to follow a mysterious cord that

popped up among the papers and books and clothes. She followed its twisting journey until she came to—the stereo. Jackie got up, giving up the attempt to find the mysterious telephone.

Sometimes young women who were not students hung around outside the frat house. When John sat down in his classes or for lunch at the Ratty, women just happened to sit around him, hoping to merit something more intimate than his famous smile. He was always polite, but he turned down romantic opportunities that most of his peers would not have rejected.

John tended to have long-term relationships and walked away from women who imagined that their physical attractiveness alone made them fascinating. His girlfriend for his last two years at Brown was Sally Munro. She came from a tweedy old New England family whose forebears would have had nothing to do with the likes of a Kennedy. She stood along the sidelines when he played football and waited for him when he showed up late for dates, as he often did. When you went out with John, you ended up not just his lover but his de facto manager, trying to keep him in check, straightened up, on time, and on his game.

J ohn distinguished himself at Brown primarily as an actor. He began in the spring of his freshman year playing in Ben Jonson's classic seventeenth-century drama *Volpone*. Professor John Emigh, who directed the play, remembers John as "a wonderful actor, secure in his body, a good presence on stage, able to make large gestures and to be humorous about it." Although the student critic for the *Brown Daily Herald* initially praised John's performance, in a later column he retracted his admiration, saying that he had been "sitting next to his [John's] mother on opening night, and I guess I was dazzled." When no longer dazzled, he wrote that "John doesn't move well. He's very inhibited and self-conscious on stage. And his voice is off-putting. He sounds like a New York preppie." The youthful critic was suggesting not only that John could not act, but that he gave off the unmistakable aura of the spoiled rich boy he tried so hard not to be.

As an actor at Brown, John did not want to play heroes, idealized versions of himself, the way many Hollywood stars do. He chose roles that took him far away from himself. He played crude, maggoty Big Al in David Rabe's searing *In the Boom Boom Room*. The play is the dispiriting journey of Chrissy, a go-go dancer with aspirations to ballet, and Big Al, her violent lover. John entered into the role with aplomb. To prepare, he went downtown to a barbershop and returned with a rude crew cut. Then he started hanging out at the toughest Providence bars, practicing playing a character like the foulmouthed Big Al to a knowing audience. On opening night he dry-humped his girlfriend and spewed racist words across the stage while his sister, Caroline, sat watching her brother's performance.

John played a similar low-life character in Miguel Piñero's brutal all-male prison drama *Short Eyes*. Most of the prisoners are minorities, and the African American director scoured the campus to find a cast of blacks and Latinos, most of whom had never appeared onstage before. John's character, Longshore, was the only prisoner with the *cojones* to kill another inmate, a child molester, by slitting his throat. On opening night the audience found itself enveloped in a terrifying world where the darkest impulses of the human psyche held sway. And strutting through it all, the king of darkness, the prince of despair, was John, looking, in the admiring words of the Providence *New Paper* critic, "like a Southie punk strung out on downers and reeking of venom and sweat."

Afterward, Jackie came up to the director. "What have you done to my boy?" she asked. The director took it as a compliment.

J ohn had an absolute fascination with acting. It was maddening to him that he felt he could not make it his life. Part of the problem was his mother. Jackie wanted what any mother wants for a son—a worthwhile, happy life. She didn't believe he would find it on Broadway or in Hollywood. She did not want to see her children exploited for their names. The worst thing about his mother's advice, as John realized, was that she was telling the truth. Robby believed that John didn't consider acting masculine enough. "I think he probably perceived it as a little too soft himself," Robby says. "We referred to California as the sperm-free zone." That may have been part of his reasoning, but it was more complicated. John loved acting, yet he knew it wasn't enough, it wasn't right. He would have to turn away from the one career he desired and try to find something else about which he cared ardently.

John took out his frustration on one of his fraternity brothers, Mark Rafael Truitt. In college drama departments there is usually one actor on whom most of the professors rest their hopes, a supposedly prodigiously talented young man or woman who inevitably will rise to the heights of Broadway or Hollywood. Or so it is believed. At Brown, Truitt was the anointed one, and he had already proudly announced his intention of becoming a professional actor.

"John couldn't fucking stand him because he was going to go be an actor," recalls Robby. John had a chance for a little revenge when he played Longshore in *Short Eyes* and Truitt played the child molester whom Longshore tortures and murders. In the scene where John spat in Mark's face, he kept in his mouth bits of a bologna sandwich he had eaten earlier in the play, to spit up on Mark. In a later scene he banged Truitt's head down the toilet with such pleasure that the director asked him to tone it down.

One evening a group of frat brothers were sitting around John's room when he jumped up, grabbed Mark, and started bouncing him down the stairs. Truitt, whose head was hitting the stairs, was laughing, but John had a serious look on his face, and Robby broke it up before Truitt was hurt.

CHAPTER TWENTY-EIGHT

An Actor's Life

John was full of jokes and pranks, and then something would happen and it was clear that there was another part of him that he chose almost never to show. On March 30, 1981, John was rehearsing for *The Tempest*. President Reagan had just been shot and was at George Washington Hospital in uncertain condition. As many at Brown saw it, Reagan was a mediocre actor in the wrong job. The attack was yet another excuse for some of the student actors to make fun of old Ronnie. John kept out of it. Then he looked up from a newspaper he was reading. "You know this is fucked up," he said, and the others grew quiet.

Two weeks later one of John's friends, Christian "Chris" Oberbeck, was in John's room when the spacecraft *Columbia* touched down at Edwards Air Force Base. "Oh, this is a great country," John said, and Chris knew that he meant it.

John could have spent his vacation after his sophomore year wandering around Europe or hanging out at the beach, but as a Kennedy son, it was almost obligatory that he do something more than that. When John spent the summer of 1981 as an intern at the Center for Democratic Studies in Washington, he stayed with Tim in Maryland.

The following spring Tim came to John and told him about ConPEP (Connecticut Pre-Collegiate Enrichment Program), where he was an assistant to the director. He told him that the summer program would have seventeen tutors working with disadvantaged teenagers. "You should do it," Tim said, and John agreed.

"Are you the real John Kennedy?" one of the other tutors asked at dinner the first evening.

"No, he's the fake," said Dan Samson, a Yalie whom Tim had also talked into working for the program. Dan had an in-your-face insouciance and was not intimidated by John. It was a character trait that would help the two become friends. Dan's nickname for John was "Dick." He was referring not to the diminutive of "Richard" but to a part of his anatomy. John always had nicknames for his friends, and if Dan was going to call him Dick, then Dan would be Dick too, even if that made the friends sound at times like an Abbott and Costello routine.

Dan was oversized in body and spirit, a boisterous presence with a booming voice so loud that it was as if he traveled with his own loudspeaker. The first time he visited John in New York, he was in his friend's bedroom, far from Jackie's part of the apartment, but he woke her up. Dan played Mr. Happy-Go-Lucky. He rarely let people know that his father, a distinguished Seattle doctor, was a Holocaust survivor.

Neither John nor Dan was especially enamored with having Tim as a supervisor who took a mite too much pleasure in bossing his subordinates. They treated him as they would any boss, sneaking out at night from watching over their charges to score thirty-two-ounce Slush Puppies at a Wawa convenience store.

During the day both young men worked hard. John tutored the students in math, a subject with which he had long struggled himself. He taught English too, but the best thing he did that summer was to put on a play. These ghetto kids had been conned enough in their lives to know when someone was pandering to them, but John was an enthusiastic, empathetic coach and director.

J ohn almost never lost a true friend or an old girlfriend. Sasha, who had shared with him a total ignorance of the mysteries of math at Andover, was at the University of Vermont. He drove up there several weekends, and she came down to visit him as well. On one occasion they were

headed to New York City in Sasha's new Volkswagen Beetle when she realized that she had forgotten her wallet. As usual, John had no money. He did
not believe in retracing his steps in love or travel. "We'll just blast through
the tollbooths!" he exclaimed, happy to turn this all-too-familiar 180-mile
route into an escapade.

This was great fun until they came upon the Triborough Bridge. The
tollbooth looked as impregnable as the Great Wall of China. "Well, we'll
just have to play the Kennedy card," John said just before he talked the tollbooth attendant into letting them through. They laughed and whooped all
the way down the FDR Drive. Then the car started gasping, and at 158th
Street, in the midst of Spanish Harlem in the dead of night, the VW stopped,
out of gas.

Sasha had grown up in New York City. She knew that by the time they
got back her new car would be stripped. "Okay, well, you stay with the car,"
John said solicitously. "Just lock yourself in and hide in the car and I'll go.
I'll go and come back with gas." John headed off with neither money nor
any identification except his face. An hour and a half later, he returned with
a red gas can and a smile. He had played the Kennedy card once again, talking a taxi driver into taking him to his Fifth Avenue home, where he hit his
mother up for some money.

Billy Noonan was another old friend with whom John stayed close. He
had gone off to spend a year studying in Dublin before matriculating at
Boston College, the proper finishing school for a true son of Erin. Billy
didn't have the preppy veneer of most of John's frat mates, and college had
hardly dented his tough exterior.

In December 1981, Billy was told he had testicular cancer. His father and
several uncles and aunts had died of cancer. Lying in his bed in the cancer
ward at Massachusetts General Hospital, twenty-two-year-old Billy was
learning not only how he dealt with illness, but how his friends dealt with it.
Some of his friends treated his cancer as if he were contagious; they hadn't
even shown up at the hospital.

John and Tim had called during the week. That meant a lot to Billy, but the phone was not ringing now. It was evening, a week before Christmas, he was alone, and tomorrow morning he faced surgery. He knew there was a good chance that even if he survived cancer, he never would be able to have children, he, Billy Noonan, who thought that if you were a man and you were not a priest, then you got married and had a slew of kids.

As he sat looking at the flowers Jackie had sent and thinking about the surgery, a nurse came into the room and told him that he had a guest in the waiting room. It was nine o'clock, well after visiting hours. Billy walked down the hall and there stood John. He had a crew cut that was so close to the bone that Billy thought he was bald, and he was wearing a multicolored sash that he had brought back from Africa. In Billy's eyes, he was looking weird.

"I couldn't sit down in Providence and have you facing this thing," John said. "I know you're reluctant to do this." As worried as Billy was, he was taken aback at how John looked like a multicultural thug who would as soon mug you as greet you. John explained that the crew cut was for his role as Big Al in *In the Room Room Room*. To Billy the skinned head was just right for John's funny riff on the chemotherapy that Billy possibly faced. He said he too might have a hairless look soon, and he and John could be brothers from foot to head.

The people who had visited Billy that week had all tiptoed around his illness, avoiding the dread "c" word. John said it bold and clear. Cancer. Now on the eve of the operation, this was the first time that anyone had tried to face up to what might lie ahead for Billy. And it was John, whom he thought of as his little brother, who was doing so. As the two friends talked, they were held together by their sense of manhood, their friendship, their Catholic faith, and their family traditions. They discussed the odds, the likely outcome, and the recovery. They discussed death too, for that also figured in the odds.

The nurse came into the waiting room at midnight and said that it was time for John to go. At seven in the morning, the surgeons operated on Billy. It went well, and after a long recovery he was fine. One thing was different from then on. Billy no longer thought of John as a younger brother.

ackie was there for her son, talking about Billy's condition and offering help, attending opening nights at Brown, and watching over his
life the way parents rarely do once a child reaches college age. As much as
Jackie wanted John to have a serious, purposeful life, she understood that
first of all he had to become a man. She was the most formidable mother in
America, and if she had put on her imperious air, John's friends would have
been uncomfortable being around her. But she did just the opposite, welcoming John's friends to her homes and making it clear that they were to be
themselves.

These invitations included weekends at Jackie's new summer residence
on Martha's Vineyard. The 425-acre estate had a thirteen-room main house,
but John and his friends hung out more in the barn that served as a six-
bedroom guesthouse, including John's private preserve, a bedroom with a
witty heart-shaped bed. Although by any measure this was a major estate, it
had an intimate feel to it, and as the years passed, it became as much John's
true home as any place on earth.

When Robby showed up at John's Fifth Avenue home for dinner, he was
not about to eat what he considered the gussied-up food that was served
there. Jackie understood the culinary preferences of her son's friend. Even if
everyone else at the table was eating the most rarified of French cuisine,
Robby was presented with his preferred meal: a hamburger so well done that
it approached the rigidity of plywood and rice so white it looked as if it had
been bleached. Jackie might indulge the culinary idiosyncrasies of John's
friends, but she did not change the guest list merely because he was home
from Brown. There were often leading political and cultural figures at the
dinner table, and young John had an easy familiarity with the American
elite.

Despite all of his friendships, John's mother and sister occupied by far
the most intimate place in his life. When Caroline married Edwin Arthur
Schlossberg in 1986, John toasted the newlyweds: "It's been the three of us

alone for so long, and now we've got a fourth." As close as he was to his sister, however, it was hard to imagine two siblings who saw life more differently. His sister was a disciplined, reticent young woman who became both an attorney and an author. He was a largely undisciplined, outgoing young man who was trying on different roles, mimicking his associates, seeing life as a feast of personalities. She led an ordered life, while he flitted from subject to subject and from place to place.

John had an immensely caring mother, but it bothered her son sometimes when people made her out to be almost perfect. He had experienced her mood swings, how she at times appeared distant or preoccupied with her work or social affairs. These were not major complaints when stacked against the kind of loving mother she had been, but it said something of how unreal the world's perception was of Jackie that when John confided such matters to his close friends, they were startled.

There was an inevitable dichotomy between the public Jackie, the eternal First Lady, the American Queen, and the mom at home, and what was extraordinary was how little the public figure characterized the woman at home. "This is a real mom," says Robby Littell. "She was America's first lady, and she didn't give that up. And John may have felt estranged at times from her because she had that public life. She had that very high-level experience which has very little oxygen at that level. She was a person hard for people to hug. But I think John's feelings were just a function of basic human insecurities that we all would have. I mean, he was a mama's boy. That's what I always called him. A mama's boy."

John sought to get away from what he considered the overcivilized East Side world in which he had been brought up. One of the ways he did that during his college years was by taking flying lessons. He was constantly testing the parameters of his world, trying to push outward, looking for space and freedom.

John had freedom that his forebears had not had, and the enviable opportunity to do pretty much what he wanted to do with his life. In a letter to his wife, Abigail, John Adams famously wrote: "I must study politics and war that my sons may have liberty to study mathematics and philosophy, geography, natural history, naval architecture, navigation, commerce, and agriculture in order to give their children a right to study painting, poetry, music, architecture, statuary, tapestry, and porcelain."

The first generation fought on the manly plains of war and politics, the second stuck to traditional masculine pursuits such as commerce and farming, while the third took on the aesthetic, more feminine world of the arts. Jackie was seeking to have children who would nobly benefit from the toils of their forebears. That was fine in Europe, where upper-class inheritors were celebrated for their good fortune and style, but most Americans had little use for nonproductive dilettantes, no matter how generous their largesse, how deep their erudition, or how rich their artistic connoisseurship.

Making money was the thing, not spending it. "Looking at the usual result of enormous sums conferred upon legatees, the thoughtful man must shortly say, 'I would as soon leave to my son a curse as the almighty dollar,' " wrote Andrew Carnegie, after he had built one of the great American fortunes. Two of the wealthiest men of John's time, Warren Buffett and Bill Gates, would gladly have seconded that statement—each plans to give his heirs only a relative pittance out of his immense fortune.

J ohn probably did not even know it, but Jackie was in frequent communication with his professors and the Brown administration, trying to excuse her academically challenged son. She talked to Beiser with whom John took three courses, and told him that John was dyslexic. John had learning disabilities that he later would define as attention deficit disorder, but whatever the name, John had a problem reading and studying that was something more than self-indulgent disregard. Robby knew about John's problem, but his friend wasn't running around using dyslexia as an excuse. "He saw shit backwards," says Robby, "and he had to compensate for it and he did, and he was a smart cookie."

The following July, Beiser wrote a devastating letter to John: "I remind you that you have never passed four courses during a semester since you came to Brown. If you do return to school next fall, you must do so with the absolute determination that you are going to pass eight decent courses, with no safety net to fall back on."

John was in Africa, and Jackie opened the letter, something that many mothers would not have done. She called Beiser, who told her that on his most recent "no credit" he had given his students an extension. All John had to do was to finish a paper satisfactorily and he would have a credit. But the underlying theme did not change. For the most part, as Beiser saw it, John was a lackadaisical student content to slide by on charm and a smile.

John's mother had a new beau, Maurice Tempelsman, a Jewish American businessman with extensive holdings in African diamonds. Tempelsman was a sophisticated man who could talk as easily about Greek myth or modern European poetry as he could about the continued imprisonment of Nelson Mandela in South Africa or the upcoming congressional election. Tempelsman was still married, though living with Jackie at 1040 Fifth Avenue. It was a relationship unthinkable for a lady of stature even a generation before, but it was acceptable now, not only to New York society but also to John and Caroline.

John spent the summer of 1980 working for Tempelsman's company in South Africa. When he returned, he helped found a student group working to educate the academic community about the realities of apartheid. This was his first attempt at political activism, and he helped set up a major conference on the issue that included as speakers UN ambassador Andrew Young and Helen Suzman, the legendary white South African crusader against apartheid. "Almost nobody knew John was behind this," recalls Brown administrator Reichley. "The conference was not as well covered by the media as it should have been. John knew he could put his name on it and it would have been well covered, but it was just like him to stay in the background and get things done right."

John used his unique access to attract such prominent figures to campus, but he was very careful to not become part of the story and to keep the conference focused on the nature of apartheid in South Africa. "Bringing Suzman to Brown was a clear message from John at a time when there was a battle on campuses across the country to make America join in imposing sanctions against South Africa's apartheid regime," recalls Christiane Amanpour, a friend of John from the University of Rhode Island. "The morning after her electrifying speech, John invited Helen and a few of his friends to breakfast at the local diner we called Beef 'n' Bun. I was impressed that John had had the gumption to pull this and the whole lecture series together."

John could easily have become one of the major progressive leaders on campus, but he backed away from further involvement. He did not say why, but he may simply have not wanted to assume a role as a public political figure, even if it went no further than the Brown campus.

I n the fall of his junior year, John moved into a Victorian house at 155 Benefit Street that the owners had studiously restored. It was not simply the furnishings, which were unlike most student settings, but the middleclass atmosphere that was unlike anything John had known before. John had developed his own sensitive detector to ward away those who wanted to be around him to live within the penumbra of his celebrity. This was *his* house, and he orchestrated matters so that he was living among people he considered authentic human beings.

One of those he asked to live there was Christiane Amanpour, whom he had met freshman year. Christiane was three years older, his sister Caroline's age, and a young woman of serious political concerns with whom John could forthrightly discuss all kinds of issues and ideas. "I have no brilliant explanation why we became close friends other than there was a 'simpatico' between us (his word)," Amanpour reflects. "There was never this conscious or subconscious one-upmanship between us. There was never a romantic twist, nor ever the hint of one. I think he saw me as solid. English, sensible, and reliable. I always gave him the truth and the most honest appraisal I could muster."

John may have been a man's man, but he was most comfortable exposing not only his emotional side but also his deepest intellectual side around women. With them it wasn't always a competition, and he wasn't exposing vulnerabilities that could be turned against him. That year John roomed with three other Brown students with whom he felt comfortable: his fraternity brother Chris Oberbeck, the only other man, and two women, Christina Haag, an aspiring actress, and Lynne Weinstein.

John may have had a reputation for studious sloppiness, but he did his share, swabbing down the bathrooms, picking up his clothes, buying groceries. Most evenings the group sat down to dinner together. They laughed and they joked, but they also had serious conversations. Much of the real learning at Brown happened like this—while a group of friends sat around and talked forming their own sense of life. "We had the most intense dinner table conversations about whatever was happening," Amanpour recalls. "Reagan was president, and we endlessly debated his policies, from 'Star Wars' to nuclear war. I remember Chris Oberbeck was very Republican, and I had just discovered the anti-nuke campaign, and so I was full of earnestness and righteous indignation at the thought this man might blow up the world. It all got very boisterous, and John, whose politics were clear, was also an effective devil's advocate."

They were part of a generation in which liberty ruled and hormones raged free. Yet what had they done with their freedom but create their own little middle-class family? Christiane prepared exotic Iranian fare, Lynne brought her own quirky tofu dishes to the table, Chris broiled a mean steak, and John was inordinately proud the evening he cooked his first meal of hamburgers.

Another evening Christiane was cooking fish and she needed lemon. John tossed her the keys to his old Honda. The set of keys was the kind of complicated mechanism with a chain that the driver of a sixteen-wheeler might have carried. At the store she was trying to get the key out of the ignition when it sprung back and hit her in the mouth, cracking a front tooth. Christiane's dream was to be a television correspondent. The fact that she had a heavy accent and lacked an anchor's blow-dried looks had not set back her ambitions, but this would about do it. She hurried back to the house in tears. "Oh, that's okay, Kissy," John said, using her nickname. "It just means you'll never have a career in television. It's all over." Christiane started crying again.

C hris Oberbeck had his own full introduction to the boyish John when he went to stay with him in New York City. The two Brown students stood on either side of Fifth Avenue at eleven o'clock at night, flinging a Frisbee back and forth across the wide thoroughfare. That was John, taking a simple game and turning it into something new, exotic, and always open to the unexpected—the Frisbee whizzing six inches from the head of a pedestrian, for instance, or landing on the windshield of a taxi driven by a cabbie cruising for a fight as much as a fare.

"Obes, come check this out," John said in the midst of their game, pointing out a foot-wide ledge around the Metropolitan Museum of Art that, as he knew from his childhood, ran around the entire building. The two men jumped out on top of the concrete, and in the dark Oberbeck set out gingerly behind John. The ledge was not particularly dangerous, though if they had fallen they could have broken an arm or a leg, but it was typical John, turning an ordinary evening into an escapade.

Besides his own sense of derring-do, Oberbeck had another quality common in John's friends. He had a gift for language. He was not only knocking his buddy down in richly inventive put-downs as he inched his way around the Met but offering richly witty observations and savage riffs on the world they inhabited. As they continued edging their way around the massive building, oblivious to the possibility of guards or new obstacles, they played their verbal games until they arrived back where they had begun.

John was no longer a teenager, but he found freedom in a boy's play. He may have been in his twenties, but he was not about to put away his Frisbees, footballs, skates, bikes, and kayaks. He still wasn't much of an athlete, but he couldn't get his fill of sports at Brown so he joined a racquetball club outside Providence, where he went with a new friend, Gary Ginsberg, to have a go at it. Gary was in the class behind John, but the two young men shared an abiding interest in politics and sports. Gary had grown up in Buffalo, New York, playing competitive tennis and was naturally more gifted at racket sports. It was racquetball they played, not squash, which John considered almost a sissy sport. John usually failed to beat up on Gary on the court, but he tried to win a few clear victories when they debated politics. Their shared interests and competitiveness in sports and politics would develop into a close lifelong friendship.

Reagan had just been reelected, and like most Brown students, Gary was a proud liberal who thought that the president was a plague on the nation, a reactionary spouting right-wing bromides. Gary was into his Reagan rant as the two drove to the club.

"You know, everyone misses the point about Ronald Reagan," John interjected. "He is an extremely effective president because he stands for very clear principles, he's consistent in his messages, you know where he stands. He's a strong leader. He's . . ."

"Oh, come on," Gary said, hardly believing what he was hearing. "How can you like Ronald Reagan? He's tearing down the welfare state. He doesn't care about the poor. He stands for the rich. His wife is buying expensive china."

"No, no, no, you're missing the point," John insisted. "You're missing the point on him."

Gary let it drop, thinking John was simply trying to be provocative. Only years later did Gary look back on conversations like that one and reflect that he had missed something about his friend. John could be assertive when the subject was sports, but he was often tentative in the way he talked about politics. It was easy to conclude that John didn't care much about politics. He had a wariness about anything that touched too intimately on his father's past or his future as his father's son and inheritor. He kept his own counsel, and few realized just how interested in politics he was or how astute his judgments were.

"Obes, I can't believe it," John said once to Oberbeck as he sat in his father's rocking chair in his bedroom reading a book of his speeches. "Look at this! Can you believe he gave these twenty speeches in, like, two weeks? He wrote and delivered! He did all this stuff."

Chris turned and left his friend alone, sitting there "communing with his father's spirit and history and who he was."

CHAPTER TWENTY-NINE

Saving Grace

Bobby Kennedy had come out of rehab, but his life was still at ground zero. His brother David's death continued to be profoundly unsettling to him, and he was struggling out of the darkness of heroin addiction. He had been the gilded Kennedy son. Now he was tabloid fodder, Bobby the junkie Kennedy, Bobby the spoiled, troubled, rich young man, only he wasn't so young anymore. He was almost thirty years old, an age when his father was already chief counsel and staff director of the Senate permanent subcommittee on investigations that probed the conduct of Senator Joseph McCarthy. Ahead of him, Bobby had two years' probation and mandatory community service.

Bobby set out to "retrace my steps to that point where I had started off on the wrong path." That meant going back to before the pain of his father's death to reconnect with a little boy who loved nature and wanted to explore the woodlands and the rivers. In January 1984, Bobby began doing his community service for the Natural Resources Defense Council (NRDC), a respected environmental group. Initially he directed a group of prisoners turning a family estate along the Hudson River into the Castle Rock Field Center. That was how he met John Cronin, who would be housed in one of the buildings as "the Hudson Riverkeeper," dedicated to protecting the river.

Cronin had grown up on the banks of the river in Yonkers, New York. He was not the stereotypical environmentalist, an upper-class gentleman concerned more with aesthetics than jobs. Cronin was proud of his blue-collar roots. A committed, righteous advocate, he was in some ways a throwback to the union organizers of the thirties.

Through Cronin, Bobby met Robert H. Boyle, who had helped found the Hudson River Fishermen's Association, the organization that in 1983 had hired Cronin as its first full-time Riverkeeper. As a reporter for *Sports Illustrated*, Boyle had begun writing about the environment in the late fifties, when almost no one was addressing the subject in popular magazines. He was a proud curmudgeon in a profession that still tolerated most of the diversities of human conduct, be it drunks, scoundrels, or a brash man spouting about the confounded interstate highway program and the dangers of draining wetlands in California. "*Sports Illustrated* has greatly changed," Boyle says. "Initially it was half-participant and half-spectator. Now it's for the dumb ass sitting in the frat house watching the tube."

In 1969 Boyle wrote his seminal book, *The Hudson River: A Natural and Unnatural History,* a biography of a 350-mile-long river that to him was "the most beautiful, messed up, productive, ignored and surprising piece of water on the face of the earth." It was a historian's book and a poet's book, a sad lament, and a spirited call to arms. Boyle posited two visions of the future—the Hudson as "clean and wholesome, useful for both navigation and drinking water, a river toward which millions of people can turn with pride and expectation," or "a gutted ditch, an aquatic Appalachia, a squalid monument to greed." As Boyle saw it, history stood in the balance. The Hudson and every great river in America needed a steward "out on the river nailing polluters on the spot, telling highwaymen they cannot build the road here but over there, talking to schoolchildren, telling anglers where the stripers or sturgeon are running."

Boyle considered Cronin the personification of the Riverkeeper. As for Bobby, Boyle was not impressed. Boyle cared little about the circumstance of name or position, especially not when it was attached to this thin, sad specimen of a human being who somehow had come by the name Kennedy. "Bobby struck me as pretty much a lost soul, very much withdrawn, and inside," says Boyle. "I think what gave meaning to his life was his growing realization that there was a lot to do for the Hudson River and the world. I saw almost an awakening. Look what I can do. That's my sense of what it could do for him. It turned him around."

The vehicle of Bobby's deliverance was the river and the people around it, especially those in the town and city of Newburgh, sixty miles north of New York City. When Bobby first arrived in Newburgh, he had the impression that the community could have been shipped up from the slums of New York City and dumped there. Industrial squalor along the banks walled off the river from the people of Newburgh. A sewage plant and a junkyard blocked part of the waterfront. The few pleasure boats that cruised the area had to navigate around old piers, stumps out of the water, and submerged and abandoned boat and barges.

Even though many residents had long ago grown numb to what they saw and felt, one man was shouting in an outraged voice. Joe Augustine, a pawnbroker, jeweler, junk dealer, flea market impresario, and silver market plunger, was an odd choice for a bedeviling hero. He was a full-time nag who had taken out full-page ads in the local papers attacking state environmental officials who had done so little to protect his city.

Augustine beseeched the Hudson River Fishermen's Association for help. Bobby responded, driving over to Newburgh to meet the man. Bobby had grown up in a political world where leaders talked endlessly about their love of "the people," though they rarely touched them with anything more intimate than a handshake at a speech. Here was one of "the people," a cantankerous argumentative man sketching a dark portrait of collusion between politicians and business leaders while state officials offered little but neglect and indifference. This was nothing but a little town along a great river, a tiny problem flowing into the realities of state and national dilemmas, but it was a stunning microcosm, as much an education to Bobby as any course he had taken at Harvard or the University of Virginia.

"Newburgh was a revelation about government officials," Bobby reflects. "I don't think there's anything wrong with being a little bit parochial, but there is in being utterly self-interested. Newburgh had great potential, and how did that wealth get squandered? It got squandered because public officials were stuffing their pockets and not really caring about the public welfare. I mean, I think human beings are basically corruptible. And the institutions are created and run by human beings. So they're going to end up serving the most powerful interests."

Bobby and Cronin realized that they did not have the power or the organization to confront the whole problem. Instead, they decided to focus on Quassaick Creek, just one of the many sources of pollution. To them it appeared a microcosm of what was going on throughout Newburgh and up and down the Hudson River.

Bobby, Cronin, and two volunteers set out to investigate the stream from its beginning at Chadwick Lake to its end seven miles later where it flows into the Hudson. The upper reaches of the stream still had bucolic stretches, but for the most part Quassaick Creek had become a receptacle for industrial waste. The activists thought that they would find only a few polluters along the creek, among the others mostly trying to obey the laws. They discovered that everyone was dumping waste, from the tiniest auto body shop to facilities run by the town and city of Newburgh. Many of these people did not act as if they were committing criminal acts; they were just business operators trying to get an edge, doing what their neighbors were doing, figuring that the creek would wash it all away down the Hudson and into the Atlantic Ocean.

Cronin and Bobby and the others got down into the creek to trace the foul discharges. Some of the pollutants in the water were so powerful that putting a hand in the water would make it fester immediately. Cronin lowered Bobby on a safety rope to a culvert where he took samples of the liquid spewing from a broken pipe. It proved to be naphthalene, a dangerous chemical that could be traced back to any one of many companies along the creek. They waded in waters full of fecal material that made them retch as they continued up the waterway. Bobby and Cronin donned wet suits and plunged into a pond in the middle of the night to swim to a discharge pipe. They secretly climbed up on the old roofs of the dilapidated buildings of an old textile company to check matters out further. Sometimes along the creek, they sat at an industrial pipe around the clock, waiting for a discharge to start running.

All his life Bobby had sought to emulate his father and uncles, who valued courage above all. For the most part he had merely affected a posturing, self-conscious recklessness. This was different. What he was doing now was truly courageous, even if slogging up polluted Quassaick Creek did not have quite the panache of running the Caroni River in Venezuela. Now he was using almost all the skills that he had developed over his life. His athletic prowess served him well as he sailed on the Hudson or hung over culverts. His deeply felt spirituality came alive and gave even more meaning to his stewardship of the river. He was argumentative, a natural trial lawyer, and was preparing to file lawsuits to end the impunity of the polluters. In trying to give a rebirth to Quassaick Creek and the Hudson River, he was giving rebirth to himself, finding in the great river blessings that even Boyle did not realize were there.

When it became clear that the Hudson River Fishermen's Association was going to sue the town of Newburgh, Linda Fehrs, a part-time reporter for the *Sentinel,* Newburgh's weekly paper, decided to give Bobby a call. "I don't give interviews," Bobby said, slamming down the phone. Fehrs was shy; if she had not been so determined to get this story, she would have given up. She might have given up too if she had realized that Bobby was one of *the* Kennedys. A few minutes later she called again, and he hung up again. She called a third time, and as soon as Bobby heard her voice, he set the phone down. Fehrs knew that Bobby was the lawyer for the organization. It made no sense that he would refuse to see a reporter who was writing about the lawsuit he was about to file. She could hardly believe the arrogance of the voice on the other end of the phone. Finally, on the fourth try, an irate Fehrs got Bobby to agree to see her.

When Fehrs arrived at the riverside headquarters, Bobby was pulling out of the driveway in his blue Toyota truck. "Are you Linda?" he asked, leaning out the truck window. "I'm Bobby Kennedy, and I know we have an appointment, but I can't make it." When Linda saw Bobby's distinctive Kennedy features and realized who he was, she was even more irritated that he would back out like this. "Go talk to John," he said. Cronin was endlessly articulate, and in the midst of her long interview Bobby returned and sat listening, eating a fast-food hamburger while Fehrs pointedly ignored him.

Shortly afterward, Bobby called the young reporter. "If you want an interview, you can come down and talk to me," he said, "and you can go through the files and all that." She returned and did so as Bobby sat surrounded by a mound of files, papers, books, and clippings. Bobby ran off lists of figures, names, and cases, so intricate and detailed that she was sure a third of it had to be wrong, but when she checked it out later, she found no errors. After the interview, Bobby made the strange stipulation that she was not to use his name in anything that she wrote.

Bobby never said why he made such a request, but Fehrs concluded that he didn't want to distract attention away from the important story. "I got no publicity for five years," Bobby reflects. "I talked to Fehrs off the record because she was doing substantive work, but in terms of having my own name in the paper, talking about myself, I had made a commitment to myself I wasn't going to do that."

As the reporter hung out at the office going through files, she overheard Bobby talking on the phone. Listening to him, she realized that maybe Bobby didn't want her to write about him, but he was perfectly willing to use his celebrity as a weapon when other means failed. "If you can't invite him over, or talk to him, or get him to go to see the sewer plant, then take him to 21," Bobby advised a caller, mentioning one of the most exclusive restaurants in New York. Fehrs thought that Bobby didn't want to display himself at the famous restaurant that Fehrs had only heard about, but was insisting that someone else go.

Fehrs wrote a series of articles that helped to create a groundswell of anger against the superintendent and the town board. Where once there had been only a few quiet observers at town meetings, now two hundred or more citizens showed up and asked tough questions. These community members held their own meetings as well, to figure out how to take back their town from these officials.

These aroused citizens wanted Bobby to come to town to lead them. He was reluctant even to be seen in Newburgh, but he agreed finally to address one of their evening meetings. The townspeople rustled nervously in their chairs, impressed that this Kennedy had come to help them. Bobby was becoming a truly inspired speaker, but it was strange the way he almost always started off stumbling around, inarticulate, before his mind and emotions began working in tandem. He did not tell them what they had come to hear. He told them that they had to lead themselves. He told them that they had families, jobs, and responsibilities. All the other side had to do was to wait them out. They had to work together, to organize themselves. These were powerful inspirational words, but they were not what some of these people wanted, and as they filed out there was as much disappointment as enthusiasm. But in the ensuing weeks some of the people of Newburgh did what Bobby suggested: they formed their own groups and began to lead themselves.

The elected leaders tried to make Bobby the scapegoat. The officials said he was "a dope fiend." Robert Kirkpatrick, the town supervisor, charged Bobby and the Hudson River Fishermen's Association with committing "legalized extortion." He claimed that the activists had not come to Newburgh to help the community but through lawsuits and threats to shake the town down for money.

Kirkpatrick had struck upon a note that would be heard again. He and others often pointed to the seventeen-year-long environmental battle against Con Ed's building of the Storm King power plant on the river. In 1980 the giant utility gave up its plans and agreed to contribute $12 million to the Hudson River Foundation. This was the beginning of a pattern of litigation that helped finance the environmental movement. Much of the money funding the Riverkeeper came from a $500,000 settlement with Exxon for pumping fresh Hudson water into its tankers and exporting it to Aruba. Out of such actions the term "environmental lawyer" was just starting to be employed. These activists punished malefactors by making them cough up large settlements that went in part to environmental groups, which often used the funds as seed money to initiate new suits.

F or a man whose office appeared disorganized, even slovenly, there was precision in Bobby's briefs and strength in his evidence. He presented twenty solid cases against Quassaick Creek polluters to the U.S.

Attorney's Office, which picked four of them to prosecute. That left sixteen cases that he took up to Albany to show to officials at the New York State Department of Environmental Conservation (DEC). Bobby thought that he would find allies there, but he sensed suspicion and jealousy at the DEC, and perhaps embarrassment. A few lawyers at the agency understood what a service was being rendered, and in the end the DEC joined with the Hudson River Fishermen's Association to file suits against a dozen more polluters. In the end every one of the polluters settled before going to court. The fines and fees included $200,000 for a Quassaick Creek cleanup fund.

I n May 1989, five years after his arrest, Bobby held a $75-a-head Riverkeeper fund-raiser at his home in Bedford, thirty-five miles north of New York City. He grilled fish for a large group that included his mother, his cousins Caroline and John, and movie stars Ed Begley Jr. and Jill Clayburgh. He appeared the confident host, but his friends knew that his life was still supported by his almost daily attendance at mass and AA. He was so nervous that his legs shook sometimes. He chewed on pencils as if they were pretzels, twisted paper clips with his fingers, and snapped rubber bands. Bobby himself felt that he and by extension his brothers and sisters had addiction in their genes and that it could be said that he was "born alcoholic," that he didn't "become" an alcoholic. That was probably true, and it gave an added poignancy to his ceaseless struggle.

Bobby was helped immensely by his wife Emily, the mother of their two young children. She was a lawyer who had given up her career to watch over Bobby and their children. Emily was a woman of the American heartland who could have been overwhelmed by Bobby's drug addiction, and other problems, but as Bobby's old friend Chris Bartle reflected, "She is his rock. Bobby needed the kind of assurance and support only she could provide."

While he was working for the Riverkeeper, Bobby went to night school at Pace University in nearby White Plains, receiving a master's degree in environmental law in the spring of 1987. That fall he became the head of a new Pace legal clinic in which students filed lawsuits in cases brought to them by the Hudson River Fishermen's Association and the Hudson River-keeper.

Bobby was doing nothing more than trying to make the system work by making government and private industry obey the laws. One person who shared that concern was Ronald Gatto, an officer for New York City's Department of Environmental Protection (DEP). Gatto was supposed to protect the reservoirs and water supplies that fed the metropolis. He was a brash, squat weight lifter with an accent that suggested he had rarely traveled north of Queens. Gatto couldn't understand why a teenager who held up a convenience store for fifty bucks went to jail when executives who fostered the pollution of the drinking water used by millions were let go with at best a measly fine.

In the winter of 1990–91, Gatto caught prison workers from the Bedford Hills Correctional Facility pouring raw fecal material from sewage tanks into a trout stream. He had also come upon workers from the Putnam County Hospital Center dumping medical and human waste into the reservoir. He handed out tickets, but he said later that his bosses had ripped up the tickets and told him that he didn't understand.

Gatto had a big mouth and a bold manner. He went to everyone from state senators and assemblymen to the internal affairs department of the New York police, and wherever he went he got the same message. He didn't understand the politics. He was a cop, and he better be quiet and go back to his job and leave things alone that were beyond his comprehension.

As a last resort, Gatto decided to call Bobby. When the cop told his story, Bobby invited him to Pace, and Gatto drove over to see Bobby immediately. The cop had the feeling that he had known this lean, intense man for most of his life. When Gatto finished his tale, Bobby laid out plans to have the officer tell his story at a hearing chaired by New York City council president Andrew Stein. Gatto's 1991 testimony was widely covered on television and in the local press, and when it was over, two officials were pushed out of their jobs.

Gatto continued his relentless crusade. What some of his superiors considered endless grandstanding was to his mind nothing more than an attempt to do what he was mandated to do. Time and again he got in trouble, and time and again Bobby was there, the threat of litigation in one hand and the hammer of publicity in the other. In resurfacing as a public figure, he was using the Kennedy name as a bludgeon. "We didn't choose to make publicity a tool, but it is a tool, and if you ignore that, you'll go nowhere as an advocate," Bobby reflects. "That's one of the currencies of power."

The two men talked almost every day. It could be about the kids or fishing trips as easily as about pollution or corruption. "When I've run into problems, he's told me to make sure you always pray to God," Gatto said. "And he says that's what he does and that's what keeps him strong." It was one of those friendships in which Bobby could call Ron at two in the morning and ask him to be there, and he'd be there. And Ron knew that Bobby would do the same for him.

Peter Pan on Rollerblades

F or several of his brothers and cousins, Bobby had become a model for how a Kennedy man should behave, an adventurous idealist who had overcome great difficulties to make his mark on the world. John admired Bobby and later in his life became closer to him than to most of his other Kennedy cousins, but John wasn't compelled to stake out his own kingdom of virtue.

Since graduating from Brown in 1983, John had been living with Robby Littell, his Brown roommate, on West Eighty-sixth Street. As John saw it, New York was the only place to live, but he refused to be a denizen of the Upper East Side, which he considered a neighborhood of pretense and veneer. He favored the gritty West Side. John zoomed down Central Park West on his bicycle, weaving in and out of lanes, running lights, and skirting around pedestrians. His bike was lost or stolen so many times that he might have saved money taking cabs.

John ran in and out of parties from Columbia University to SoHo. He devoured hamburgers at Jackson Hole on Columbus Avenue. He came stoned out of Xenon at two in the morning. He jogged around the reservoir, played football with the guys on Saturday afternoon, and then went out for Sunday breakfast somewhere—breakfast, mind you, not brunch. He belonged to three different health clubs and tuned his body up as if he were working on a Ferrari. When he was going to a Knicks basketball game and the traffic was heavy, he didn't call for a town car or grab a taxi; he took the subway. He skied in the Canadian Rockies, kayaked in New York Harbor, sailed in the Atlantic, and swam off Martha's Vineyard.

"People have all these expectations about me and my father," he told Billy. "My father would have wanted me to do whatever the hell I wanted." It was indeed the life Jack might have lived if World War II had not forced on him a new seriousness.

After graduating from Brown, John went off to India for six months. He did not even bring a suit with him, and when he visited Prime Minister Indira Gandhi, he had to borrow an ill-fitting jacket. In Calcutta he had an audience with Mother Teresa. To see her, he traveled through corridors of the dying and the wretched, and he was moved in ways that he could hardly articulate. No other twenty-two-year-old private citizen could have come to India and met with Indira Gandhi and Mother Teresa, but John was not going to limit himself to the powerful and the famous, be they saints or sinners. He was like any other vagabond American knocking around India, trying to figure things out.

During his months abroad, John wrote Christiane, saying that he missed reading about current events in American papers; as a fledgling reporter at CNN, Amanpour had access to endless newspapers. She sent him so many clips that he wrote back to say she'd overwhelmed him and to cut it back.

When John returned to New York, he seemed like the same person he was before he left. But he couldn't talk about Mother Teresa with rowdy Robby, his partner in young men's adventures. Nor could he easily discuss the plight of the impoverished in South Asia with his cousin Anthony, the brother he never had, when they were busy teasing and taunting each other. He might have discussed it with Sasha, who saw the emotionally susceptible, artistic John with insights into human relations and life, but she wasn't that interested in politics.

John's friend from his summer working on the Wyoming cattle ranch was a generation older. John could sit there all night listening to Barlow's outrageous tales, but he could also have deeply felt philosophical conversations with the man. Barlow understood things that others might find grandiose, silly, or self-important.

"I've been reading a lot of biographies lately, and it occurs to me that most of the great men I read about were not really good men," John said to Barlow soon after he returned from India. "It would not be that difficult, given my circumstances, to become regarded as great, but I think a much more interesting challenge would be to be a good man."

Barlow would always remember those words as a key to his friend's life. Barlow had been around rock bands and movie stars enough to know what the blazing light of celebrity often did to goodness. He had traveled to enough places and seen enough things to realize how rare a quality true human goodness really was, and for a man like John how difficult it would be to attain. John had done nothing notable in his life, yet he had fame unlike that of any young man of his age. John had wealth, looks, education, and people ready to help him do whatever he wanted to do in life. And he had decided he wanted to be good. Barlow knew that lots of people wouldn't understand this goal and that even if his friend succeeded, others might not appreciate what he achieved.

John's mother had not brought up John as a self-indulgent rich boy. She pushed him toward a series of jobs that he performed with minimal diligence. Jackie set him up at the 42nd Street Development Corporation, where he showed up wearing jeans and cowboy boots. She also had Tempelsman introduce John to Communities in Schools, a program working with kids in the South Bronx. He excelled in that job. He did it full-time for a number of months, and nobody knew about it, not the media, not even many of his friends from Brown. "He rolled up his sleeves, and he was happy that most of the kids up there had no idea who he was and couldn't care less," recalls Bill Milliken, founder of the program.

As John sprinted around the city, he was vulnerable to everything from a coked-out psychotic on the subway to a celebrity-obsessed driver running into him on his bicycle. In May 1985, a seemingly inebriated man called the police in Herndon, Virginia, saying, as the FBI reported later in a memo, that "he and seven other individuals intended on kidnapping John Kennedy that evening." That turned out to be probably nothing more than a drunken boast. Ten years later, in July 1995, the FBI believed there was a creditable threat that professional criminals intended to kidnap John as he rode his bike and hold him for ransom. Threats like that would have driven many celebrities and wealthy heirs to travel by liveried car and at times with a bodyguard, but nothing deterred John from the life he intended to live.

John lived a life that he called free and more cautious souls might have called reckless, or heedless of dangers. The obvious psychological conclusion would be that in living with such cavalier bravado, he was running away from the memories of his father's and uncle's violent deaths, trying to throttle his inner fears. As irresistible as that thesis might be, he never talked about such fears, not even in a veiled way. When he did reflect on his father, it was not the nature of his death that obsessed him, but the fact of growing up without a father.

In his restless journey, John at times seemed to exemplify Jung's archetype of the puer aeternus—the divine ethereal youth, an ageless Peter Pan who swings down from his high mysterious perch to pluck people up for a thrilling journey only to set them down again. No woman is unique enough to keep his interest. No job is compelling enough to hold him. Nothing is ever quite good enough. He lives life in the moment, but always in the future there lies a finer woman, a richer journey, a livelier night. Somewhere ahead lies a time of great accomplishment, but that will just have to wait. And thus he flits endlessly onward, always just out of reach.

"The image of the mother—the image of the perfect woman who will give everything to a man and who is without any shortcomings—is sought in every woman," writes Marie-Louise Von Franz in *The Problem of the Puer Aeternus*. "He is looking for a mother goddess, so that each time he is fascinated by a woman he has later to discover that she is an ordinary human being." John sought out fascinating women but even when he seemed to find one and stayed with her for a few months, he always left her and flew off alone.

"There is always the fear of being caught in a situation from which it may be impossible to slip out again," writes the Jungian analyst. "Every just-so situation is hell. At the same time, there is a highly symbolic fascination for dangerous sports—particularly flying and mountaineering—so as to get as high as possible, the symbolism being to get away from reality, from the earth, from ordinary life. If this type of complex is very pronounced, many such men die young in airplane crashes and mountaineering accidents."

There were always people to pick up after John, and he had a rich man's carelessness. In the summer of 1987, when he was working as an intern at the Justice Department in Washington, a family friend lent him a beautiful apartment. He left the place a shambles, not even making a cursory attempt to clean it up. That same year, when he gave up his lease on his apartment on Eighty-sixth Street, the proprietor found the place in "obnoxious condition," with patched-up holes in the walls and damaged carpets.

John was careless with his lovers and careless with his friends. The women he saw got used to sitting around waiting for John to show up an hour late, sometimes more. Most of his acquaintances accepted this, and it taught him lessons he should not have learned. "You're such a Dick," Dan told John when he forgot a promise to call him, employing their mutual nickname. "I do it to everyone," John said, as if that excused him. "Just because you're a Dick to everyone doesn't mean you're not being a Dick to me," Dan countered, his logic indisputable.

John may have often been late and forgetful, but he was unfailingly loyal to his friends and his lovers. In August 1985, he entered a half-decade-long relationship with his Brown housemate Christina Haag. They got together when they co-starred in Brian Friel's *Winners* for six nights of invitation-only performances at the Irish Arts Center. When John took to the boards, he never squandered his time on the trivial. Friel's play was a tragedy in which he and his pregnant girlfriend drowned, a modern Northern Irish version of *Romeo and Juliet*. John was excellent, mastering the specific dialect and inhabiting his role. He had offers to take the play and his co-star with him to off-Broadway, but he backed away from proposals that were more about him than the play. It was the last time he ever acted.

Haag was a lovely woman, but John could never forget that he had first known her as his housemate at Brown, sitting across from him at breakfast and dinner. She was almost too comfortable, too nice, too much a friend and not enough of a lover—at least as John saw the world. She represented none of the dangerous excitement that John craved. He contemplated marrying Haag, and perhaps if he had met her under different circumstances, that would have happened. They went together for five years, but he knew that it was not quite right.

John was proud of being monogamous. Yet he was a uniquely handsome, celebrated single man in a city full of single women. Sex had everything he liked in an activity. It was intense, orgiastic, athletic, intimate, inventive, surprising, and didn't take much more than an hour. The straight and narrow road was hard to keep to when there were so many diverting side roads. John rarely talked about his sex life, but his trysts with Julie Baker, a sensuous model with the Wilhelmina Agency, were so intense that he told his friends about them in awesome detail.

Men are not the only sex proud of their conquests. Usually John was careful that he did not become merely another notch on somebody's belt. On a kayaking trip to Maine, one woman on the trip kept flirting with John. The group usually stayed at a bed-and-breakfast on Dead River, but the place was overbooked, so the owner let them camp outside in their sleeping bags and cook in the kitchen. While most of John's friends were preparing dinner, a couple staying at the B and B came hurrying downstairs to say that their room was locked. The owner charged up the stairs and knocked. Nothing happened. A few minutes later he returned to find the door open, the sheets disheveled, and John and his flirtatious new friend locked in the bathroom. As much as the owner liked bragging about his famous guest, he threw the group out in the rain.

Romance was the greatest game of all, and John added his own twists and difficulties. He started by falling for complicated, difficult women. The reality was, as he admitted to his friends, that he was more comfortable with strong women than with strong men. What he considered strong, however, his friends at the time thought was merely bossy, pushy, needlessly demanding, or endlessly neurotic.

John's affair with Haag dawdled on far longer than it probably should have. Then, from the late eighties on, John began seeing the Hollywood actress Daryl Hannah. By 1992 Hannah had become the most passionate as well as the most troubled affair of his bachelor years. The actress, best known for her role as the mermaid in *Splash*, was hardly a typical movie star. She did not have a regular publicist and abhorred personal publicity. She generally dressed with a hip Orphan Annie look, but on screen or in publicity stills she metamorphosed into a blond screen siren. Daryl came from an upper-class Chicago family that wintered in Palm Beach in a great house on Golf View Drive, where the social arbiters at the Everglades Club down the road considered the Hannahs several cuts above the arriviste Kennedys.

Daryl had the intense, unpredictable manner that John found fascinating. The actress had been in a decade-long relationship with Jackson Browne, a singer whose music provided anthems for much of John's generation. It may even have added to John's sense of challenge and conquest to know that when she returned to Los Angeles she was living with Browne, who was ten years older.

In September 1992, thirty-one-year-old Daryl was back at the Santa Monica home that she shared with Browne, moving out her belongings. That day her representative told the press, "She received serious injuries incurred during a domestic dispute with Browne for which she sought medical treatment." A picture of her with a black eye appeared in the tabloids.

John flew out to Los Angeles to comfort Hannah. In the next two years they had many joyous times together, skiing in Canada and Colorado, attending the Clinton inauguration, Rollerblading in New York, swimming and sailing in Hyannis Port. Daryl was a woman of bewildering juxtapositions, one minute needy and insecure, the next demanding or twisting an emotional knife into John's heart. On one occasion he had planned to go off on a major trip with Daryl; when she called the day before and said she

320 SONS OF CAMELOT

couldn't make it, he was on the verge of tears. On other occasions he could be cavalier in a way that was brutally hurtful to a sensitive woman. Whatever their problems, she would show up a few weeks later in New York, or he would dramatically fly out to Los Angeles, and everything would be fine.

"The reason it wasn't working was not that she was a movie star and he was the son of a president, but that they had a bad dynamic in the relationship from the beginning," says Barlow, who saw more of the couple than any of the other friends. "I said from early on that they weren't good for each other. They could never inhabit the same emotional space. They could never find a relationship of parity. They always were thinking that the other one was the pursuer."

I n July 1986, John called Dan and asked him to go off with him on a wilderness adventure. Dan invited his friend Jim Clark to come along. John was not going to settle for a genteel walk in the woods, even if Dan and Jim had never gone for a real hike in their lives. He wanted to make a trek to Asgaard Pass in the North Cascades. It was not technical climbing with crampons, but it was a tough hike.

John and Dan stopped at REI to pick up the requisite gear, but Jim arrived in Stan Smith tennis shoes, hauling his gear in a gym bag. As they entered the woods, Dan noticed a transformation in John. Gone was the watchfulness of his public personality. He took on a joshing persona that Dan took as John's true self. Dan cramped up on the first day's trek upward, but he was game. As Dan struggled on, he played the role of the inept comedic sidekick to perfection, allowing John to play Mark Trail, outdoorsman. They were not climbing Annapurna or trekking across Tibet, but as they saw it, they might have been. These were city-bred men in their midtwenties, and they were like boys playing soldier in the woods.

John displayed his woodsman skills making a fire, boiling water, and showing his comrades the right way to roll out the sleeping bags. "John was as competitive as I am, and we managed to turn our adventures into huge competitions without really saying anything about it, be it a better fire, tastier stew, or a prettier campsite," Dan recalled.

Out here there was no shadow of attention following John wherever he turned. Not that he wanted to sit endlessly contemplating his momentary good fortune. He preferred to lace up his boots and go. On the second morning they entered the glacier fields and headed straight up Asgaard Pass. The path was unmarked, and it was a brutal climb. At the top they sat down to congratulate themselves, doubly so when John pulled out the guidebook that Dan remembers as saying that the trail was "extremely difficult, if not downright suicidal," and they hadn't even taken the trail. John managed the considerable feat of losing the map, and when their indomitable leader missed the narrow pathway as they headed back, they were forced to trailblaze. They crawled across rock ledges and waterfalls and moved gingerly on logs on frigid streams.

As they struggled onward, John enlivened matters further with an interminable soliloquy about the murderous stalker who was said to be loose in the woods. The assassin's next choice for a victim surely would be John F. Kennedy Jr., after which he would surely finish off his two companions. Down they stumbled, moving over one obstacle after another; it was night now, and they were still not down. They linked arms to make it across one last raging stream and struggled onto a remote road, where two passersby in a pickup truck gave them a lift.

As they rode back toward Seattle, Dan realized that John had become different the moment he saw the macadam road signaling the end of their wilderness journey. When the driver recognized him, John had changed his psychological wardrobe and grown more distant and wary.

From then on, Dan was ready to head off whenever the bugle sounded, but Jim had had enough of John's adventures for one lifetime. A real outdoorsman would have laughed at John's casual ineptness in losing the map and blundering his way down. It was sheer luck that somebody did not break a leg. For some, John and Dan were just the kind of foolhardy city boys who should not be in the mountains, but Dan didn't see it that way. "Because of

my adventures with John, I know how it feels to travel for a sustained period
with each step being precious, potentially the difference between life and
death," Dan says. "Do not mistake any of this for a person living on the
edge. This is the portrait of a person whose soulful and spirited sides combine to encourage a full life."

This was Dan's first adventurous trip with his friend, and it set the pattern for later journeys, as well as for John's travels with other friends. There
remained a self-conscious element to John's risk-taking. Every time the two
friends went out they ended up in hair-raising situations that years afterward
they relished reliving together. There had been the time they stared a rattlesnake in the eyes in Arizona and then came close to walking straight off
into the Grand Canyon. And the trip to Glacier National Park when they had
sung "Sympathy for the Devil" in the middle of the night hoping to attract
bears. And who could forget that frigid night without a tent in Colorado
when Dan could have died from hypothermia?

Some of John's friends, though not Dan, believed that John was consciously upping the ante of risk to places where few would have taken it. To
them it was not casual ineptness that made him lose the trail map or head out
without the proper equipment, but a calculated attempt to make the journey
"real." And real meant that you could get yourself killed.

John went off with Robby on a trip to Ireland, where the two friends
linked up with Michael and his bride, Victoria Gifford. The two Kennedy
cousins went to the courtyard of their hotel in County Galway. When
Robby went outside, he spied the two young men climbing a clock tower.
Robby fancied himself an intrepid soul, but this was beyond anything even
he would have considered. "There was a sort of one-upmanship in this
arena, of being a little crazy," Robby reflects. "If you weren't a daredevil,
you weren't a man."

J ohn lived fully in New York too, but it was a different kind of life. At
times John and Robby were like the wild and crazy Festrunk Brothers
from *Saturday Night Live* who think that every night is party night. Robby

could tote his share of drinks and more, but he had a stomach as strong as zinc. John rarely drank, though when he did he often tied one on. He was always game for a toke or two of weed, but no matter how late they stayed up, he was up in the morning ready to roll. They had great youthful times, but great youthful times never last very long. Robby was in love and soon would get married, and Jackie was beginning to look askance at her son's partying ways. So in September 1986, John entered New York University Law School and got a place by himself in the West Nineties.

The legal profession is a natural choice for a young man contemplating one day entering politics. But a law degree is not easy to come by for a young man who has trouble reading and concentrating and spells poorly. (On the invitations to his twenty-fifth birthday party at the Nirvana Club, he handwrote that the event would be the "lessor of two evils.") But he had a sound intelligence that he had managed to hide from many of his acquaintances, and it helped pull him through the tedium of classes, mock trials, books, and endless minutiae. It was doubly tedious because he knew he could have been acting in a play off-Broadway, doing something a lot more exciting than reading *Houston East and West Texas Railway Company v. United States*.

In July 1988, twenty-seven-year-old John went to the Democratic convention in Atlanta to introduce his Uncle Ted. He was not the only young Kennedy there that week; indeed, it seemed then that the future of the Democratic Party would be brightened by a whole legion of Kennedy family members. Joe, a freshman congressman and heir to Ted's throne in Massachusetts, was there. Patrick had flown in from Rhode Island, where there was already talk about when he would move up from the state legislature to a seat in Congress. Teddy had not run for office yet, but it seemed only a matter of time before he would go for it. Kathleen had lost in Maryland in her first attempt at a seat in Congress; although she looked more like a dowdy, policy wonk than a public figure, she had her ambitions too.

John had shown no public interest in pursuing a political life, but he was the great Kennedy hope at the convention. As John stood on the podium at the Omni Center prepared to talk, the convention was awash with nostalgia.

He moved to speak, but wave after wave of applause played over him. "Please, please, thank you, thank you," he kept saying, "Please, please, thank you, thank you."

When John got together with his friends on Martha's Vineyard soon afterward, Billy was asked if he wanted dessert. "Please, please, thank you, thank you," Billy said. "Oh, shut up!" John yelled as everyone laughed. For months John could hardly get through a night without someone repeating the phrase.

Billy was there to tease him again when in September 1988 *People* magazine named John the "Sexiest Man of the Year." He was in a drugstore on the East Side with friends when they saw a copy. "Get your eyes off that man's extraordinarily defined thigh!" the magazine enthused in gushy *Cosmo*-like prose. "Get your eyes off that man's derriere!" John might have acted as if the title was a laughable, lamentable misfortune, but Robby sensed that his friend secretly loved it.

America had long had an obsession with celebrity, but the *People* cover was a watershed. John was not a movie or television star, as the previous subjects had been; he was a law student. Yet John became the most popular of any of the magazine's annual choices. From then on, the paparazzi dogged his steps as if he had a price on his head, which of course he did.

Robby was probably right that celebrity gave his friend more pleasure than pain, and he would have been disappointed if the photographers had packed up their bags and disappeared for good. Celebrity was just another game to John, one that was always available. It was fun outwitting the photographers, fun being chased down the street, fun snubbing the tabloid reporters, fun creating a buzz wherever you went.

Robby was forever catching his friend secretly glimpsing himself in any mirror they passed. They were traveling together in London once and were having an argument. As they passed a gilded mirror in the British Museum, John stared ahead. "Missed one," Robby said.

John was such a gorgeous figure that he might have looked at himself until he drowned in the image. He enjoyed the prerogatives of celebrity more than he let on, be it a good table at Elaine's or great seats for the Giants football game. When he was with other celebrities, he wanted to be around the A team and enjoyed Hollywood parties where almost every face in the room was famous.

John may have sometimes gazed too long and too lovingly into the mirror of celebrity, but sooner or later he would turn away, never fully mistaking that image for himself. If he spent hours working out to tone his body, he spent even more time working out at being a normal human being. "He never bit off the *E!, Entertainment Tonight* way of life," says Robby. "What he considered important was family, friends, and God; the rest was fluff."

Many celebrities cannot tolerate the idea of being alone. John craved his moments of solitude. He could have them while Rollerblading through Central Park, bicycling along Riverside Drive, or, later, flying a plane. He had been groomed for celebrity since he was a baby, and he had a rare control over the psychological realities of his life. For the most part he was alone when he wanted to be alone, and with others when he chose to be. If he didn't feel like being recognized, he affected a slouching, anonymous persona and wore an absurd hat. Someone might look a moment and think that the bizarre fellow sitting there reading the paper on the subway looked something like John F. Kennedy Jr., but of course that was impossible.

D uring the summers the anointed sexiest man in America pursued his legal education. One year he worked for a Los Angeles law firm. Another summer he went down to Washington to do an internship in the Civil Rights Division of Reagan's Justice Department under Assistant Attorney General William Bradford Reynolds. As his Uncle Ted saw it, John's new boss was the enemy, so much so that two years earlier, the senator had helped block Reynolds's elevation to the number-three position at the agency. Despite knowing that and listening to his friends wonder how he could possibly associate with such a person, John consciously chose to work with the conservative official. He found Reynolds to be a humane, thoughtful man. John saw that a person could care deeply about civil rights and still oppose such things as school busing or the kinds of new federal mandates that his Uncle Ted sought.

Reynolds had an impression of John as a young man only reluctantly studying law. "His mother hung the moon, and her wishes were going to be adhered to," said Reynolds. "I think that deep down law school was a career path he was doing on her account, not on his account. He wanted to act. He talked about it a lot. He kind of indicated he probably would never go there, but he was fascinated by it."

John graduated from law school in May 1989 and that fall went to work in Manhattan as an assistant district attorney. He had one small problem. He had just failed the New York State bar exam. His cousin Bobby Kennedy had failed the exam too, but the media had not noticed that, or the fact that of the 6,853 candidates who took the exam, 2,188 failed it right alongside John. John's problem—and it was a serious one—was that because of his ADD, he did not have the kind of focused, mechanical approach to learning that was crucial in memorizing all the data that came up in the exam. His ADD made it hard for him to crack the books for hours on end. Beyond that, if given a choice between going out and having a good time or sitting reading dreary tomes, he would make the choice he had always made.

John called Reynolds to discuss his prospects. Reynolds had the feeling that this young man hadn't wanted to become a lawyer and that now that he felt he had shamed himself and his family he was thinking of packing it in. "Look, you know you can do it," Reynolds told him. "And I know you can do it. And a lot of other people know you can do it. So what do you have to lose to give it one more shot?" "Yeah, damn it, I know I can do it," John said, sounding only half-convinced.

John went ahead and took the exam a second time the following May, and again he failed. Now he was becoming a minor joke—John, the dumb Kennedy flunking his way through life. "The Hunk Flunks," headlined the *New York Post*. Jay Leno had a go at him on *The Tonight Show*. "Well, I see that John Kennedy Jr. failed the bar for the second time," he told his audience. "That's the bad news. The good news is that it looks like the Democrats finally have a vice presidential candidate."

After John failed again, he called Reynolds for another pep talk, and the assistant attorney general sensed once again that John was seriously thinking of giving law up and moving on. The stakes were even higher now, for if he failed a third time, he would have to leave his position as an assistant DA. John crammed for the exam with a special tutor and then, to get practice, took and passed the Connecticut state bar exam. With this behind him, he was given the special privilege of taking the New York State test in a special room at the Jacob K. Javits Convention Center on July 24, 1990. Tabloid reporters who know nothing of John's ADD difficulties whispered that he was placed there so he could be fed answers. In this private setting free of distraction, he finally passed.

Dan teased his friend the assistant DA about the crummy working conditions at the district attorney's office in downtown Manhattan. "My God!" Dan exclaimed in his patently exaggerated manner. "I can't imagine working in a place where they have to have signs on the wall saying 'Please Do Not Spit.'" John's larger problem was that he did not have the rapacious instincts of a prosecutor, though in some ways he did not have to have them. "The perps kept confessing to him because they wanted a friend, and it just killed him," Robby recalls. "They figured this guy can help me out. They'd start out talking about how their mothers beat them, and then they'd admit the crimes. What's the challenge to that? He almost never got to the courtroom. It was embarrassing."

When he did get to the courtroom, he was not prosecuting serial killers and Colombian drug kings. "John asked me to come and sit in on one of the cases he was prosecuting," Amanpour says. "I was of two minds because I was curious, obviously, but also what if he wasn't any good, what was I going to say? Well, he was good, and if I remember correctly, he was prosecuting some Afghan street vendors who had set up shop without a proper license. I remember telling him he was too hard on these poor frightened foreigners trying to make an honest living."

D uring his years as an assistant DA, John decided to use seed money from the Kennedy Foundation to get involved with a social service project. The twenty-eight Kennedy cousins each had a chance at the money. First John had to come up with solid plans for the up to $50,000 grants and convince his skeptical relatives that his project deserved to get the award. John was competing against relatives who would give full-time attention to their proposals, but he was ambitious in his plans.

John talked to Jeffrey Sachs, a consultant who had been an adviser on health and human services to two Democratic governors. "I don't really know what to do," he admitted. "My interest is in the working poor, but the foundation is dedicated to mental retardation." Sachs suggested that John seek a project by talking to foundation officials, social service leaders, union people, and academics. Everyone had something different to offer, and John and Sachs heard enthusiastic proposals for everything from playgrounds to photography.

Then one of the experts told them that half the workforce in the mental retardation field left each year in part because their low-wage jobs led nowhere. These workers were so poorly paid that they often left for a menial job with only a little higher pay, or better hours or conditions. These caregivers had at best a high school education, and until they got academic credentials, they weren't going anywhere.

"John really began to get very excited about this issue," Sachs recalls. "He always used to say, 'It's so simple, but other people find it hard to get their minds around it. People need to have opportunities to go get educated to advance. It's such a simple concept that people look for more complicated problems and solutions.'"

John's concept was to provide classes so that these workers could get the credentials to build real careers in the field. It was a simple, elegant idea with the potential to change the lives of thousands of caregivers, who in turn would change the lives of tens of thousands of the developmentally disabled.

John could have gone about this in the classic American celebrity do-gooder way—organizing a fund-raiser, inviting his mother and a few movie stars, passing the proceeds on to the professionals, and then reading about his beneficence in articles placed by his publicity agent in prominent newspapers. Instead, John used his celebrity to win a hearing with important officials in New York and to push them to work together.

John and Sachs continued meeting with various experts, focusing on this one issue. They discovered that the owners of care facilities, union leaders, social service functionaries, and academics rarely sat down and talked things out. The business executives were suspicious of the labor leaders. The unions worried about a program that might educate workers right out of union solidarity into management. The scholars were concerned about compromising their academic integrity and also, if they were supposed to teach where the workers worked, about having to give classes all over the city.

It was not easy to get these people together, and it took all of John's political skills. "What we wanted to do was to rewrite the DNA within the City University of New York and get them to change their programs and get the state to change their programs and get the unions to change their programs," reflects Sachs. "John methodically met with all these people, and no matter how deeply you dig in this program, you're going to find his fingerprints."

After more than a year of discussions and negotiations, Reaching Up, Inc., was founded in 1989, headquartered at CUNY, and administered by Dr. William Ebenstein, a professor with an expertise in workforce development. Reaching Up not only provided college-level courses but also instituted the Kennedy Fellows, a scholarship program that gave caregivers the resources to commit to getting a higher degree. In the first ten years, ten thousand people enrolled in certificate programs across New York at fifteen public colleges. Over four hundred Kennedy Fellows received $2,000 grants that not only allowed them to work for a higher degree but often helped them evolve from competent caregivers to staff people with leadership potential.

During the first years John spent about an evening a week working on the program. "Whether you were rich or poor, college professor or unassuming, he treated people with respect," says Ebenstein. "And naturally people loved him. These caregivers are not treated with respect. They're just low-wage workers. They're like people with disabilities that way. They're not part of the power search. Nobody listens to them. He gave them a voice. He said, 'You're the most important person in the system.'"

Despite the media obsession with John, nothing substantive was ever written about this work. The few times the organization did try to get publicity, the stories seemed to focus on John's hair or personality. When John was coming to a meeting, the other participants were warned not to talk about it or some paparazzi might end up crashing the meeting. That never happened, because nobody ever broke that trust.

In the early 1990s, John came tearing into one meeting at CUNY on his in-line skates, carrying his tuna fish sandwich dinner in his backpack. When he first arrived, everyone was nervous and excited about meeting such a famous man, but after a while they realized that he was just a human being who cared about them. "He knew all of us, our names," reflects Seth Krakauer, then one of the Kennedy Fellows. "We had dinner and lunches together. He knew I was graduating at a certain time. He really took an interest. It wasn't bull. He didn't have to do it. Jesus, I don't know if I would have done it if I was born with twenty million dollars in my pocket."

CHAPTER THIRTY-ONE

Jungle Waste

hile John quietly went about his good work at Reaching Up, his cousin Bobby was becoming arguably the most visible environmental leader in America. He had long since reached beyond the Hudson River to focus on environmental issues throughout America and the world. Just as his father had envisioned himself as the leader of a worldwide movement, so in his way did his son.

Although Bobby continued his work with the Pace University legal clinic, he also had become an attorney for the Natural Resources Defense Council. The NRDC had been founded in 1970 as a public-interest law firm for the environment. It was a perfect setting in which Bobby could combine his passion for the natural world with his ability as a litigator. Bobby was pulled in half a dozen different ways, pursuing ongoing litigation in the United States, giving speeches across the country, and meeting family obligations. He was no longer shying away from publicity and was willing to use the capital that his name represented.

In July 1990, Bobby and two other NRDC staff members flew down to Quito for a four-day visit to the remote regions of the Oriente in eastern Ecuador. Judith Kimerling, a Yale-educated attorney, met the threesome in the Ecuadorian capital. Bobby considered Kimerling "one of the heroes of the environmental movement." As an assistant attorney general in New York State, Kimerling had confronted Occidental Petroleum for its notorious waste sites at Love Canal. After half a decade of fiercely fought litigation, she was burned out on that issue and had decided to look at how an American lawyer could help the environmental movement in Ecuador. She spoke no foreign languages, had never lived outside the United States, and brought no special expertise. Kimerling went to the NRDC, where a senior staff attorney said that she could use the organization's name while he looked for a small grant to fund her research on the impact of World Bank activities in Ecuador and to look into the state of the rain forests. That was the only vague promise she could get from any environmental group.

Kimerling had been in Ecuador barely a few weeks before she realized that the region was being ravaged by the oil companies, an issue of far more immediate concern than her initial interests. It had been little more than two decades since outsiders had first entered the Ecuadorian Amazon to tap the vast oil resources. In the first years, Texaco had dumped oil waste into open pits that often leaked into pristine streams and rivers. When the Texaco executives were confronted with their actions, they argued that it was standard operating procedure with everyone including Petroecuador, the national company.

Kimerling estimated that 10,000 gallons of oil leaked out of the pipelines each week, and every day the oil companies dumped another 4.3 million gallons of toxic waste into the streams and rivers. When Petroecuador realized what Kimerling was doing, the company sought to have her escorted out of the country, but she stayed and continued building her indictment. Part of the time Kimerling lived in an Amazonian Indian community, where she became so sick with shigellosis that she had to spend many days in bed.

Just as Kimerling's savings were running out, the NRDC came up with a $7,500 grant. That allowed her to pay some of her bills and to prepare a report that she delivered to NRDC headquarters in Washington. Kimerling had hoped that her shocking research would lead the NRDC to promote an effort to stop the devastating oil exploitation in Ecuador, but the environmental organization had other concerns. "I was basically told good-bye and good luck," Kimerling says. "They were not interested in continuing the work. I went back to Ecuador to close up my apartment. After I got there, I got a phone call saying they were going to do a rescue-the-rain-forest campaign to increase membership, and they were going to use Bobby Kennedy as their celebrity spokesperson. But before they did the campaign, they wanted to send him down to visit the rain forest. I was asked to take him to the rain forest and told to be sure to show him not only the oil contamination but also the pretty places."

As Bobby bounced up and down rude roads and traveled by canoe, he saw the realities of life in the oil-producing region of Ecuador. Wherever the oil companies built their roads, poor people from other regions in the country followed behind, seeking free land and opportunity. They burned and slashed the jungle for a plot of land. They built shacks in the little towns pieced together in part from whatever the oil companies had discarded. The towns filled up with a motley array of people: oil company engineers, roustabouts, and settlers from the highlands, as well as indigenous people— barefoot tribesmen with stretched earlobes, wearing T-shirts and hefting blowguns and shotguns. The oil companies washed the dusty roads down with oil. There was such a scent of oil in the air that it seemed that it surely must rain petroleum.

In the village of Shuara, Bobby and his associates were accompanied by two local priests. They came upon a group of shirtless Siona Indian workers using shovels and their hands to scoop up spilled oil and contaminated soil into open pits. "Two nearly naked men stood neck deep in oil in the center of the black pit," Bobby wrote. "Periodically they dove under the oil in an effort to attach hemp rope to a submerged tree stump. As they pulled the tree toward the shore their eyes gleamed white against the black ooze that coated their skin and hair. At the end of the day, the company hosed them down with gasoline to remove the crude oil."

In his rudimentary Spanish, Bobby asked the men how much they earned. They said they earned two dollars a day. "When they are sick, the company fires them," Father Jesus said. NRDC attorney S. Jacob Scherr stood back, videotaping the group. Bobby moved to shake an oily hand, but the worker jerked his hand back rather than shake the clean white hand of this foreigner. Bobby reached his hand forward and grabbed the man's hand.

Kimerling saw Bobby off and returned to the United States with a rough draft of her Ecuadorian research. There NRDC staffers helped research and edit the material into an important book, *Amazon Crude,* which the NRDC published that fall. The short book began with an eleven-page account of Bobby's four-day trip to the region. Nowhere was there a similar rendering of the extraordinary, often dangerous eighteen-month effort Kimerling made to obtain the material that formed most of the book. Bobby's account was vividly written, but compared to Kimerling's experience, it was little more than a tourist picture postcard home. Bobby and the NRDC had the name and the reputation to draw people to Kimerling's work, however, and she considered that a fair trade-off. The immediate question was whether Bobby had stayed long enough and learned enough to transform what was essentially Kimerling's research into action that was true to her experience and to the complicated social and political realities of Ecuador. Kimerling's answer was an emphatic yes, and she went to work for the NRDC in the United States on this issue "with great hopes that we would be able to do something."

In October 1990, the NRDC held a press conference to publicize Kimerling's research and to condemn further oil production in the Oriente. "We saw rivers that were on fire," Bobby and Scherr told the reporters. "Animals have disappeared. There are no fish left. Among the people, we found 80% malnutrition in an area where formerly there was none, all the result from oil exploration and drilling." In Ecuador, underlying mineral rights were owned by the government and Conoco, a subsidiary of DuPont, had won the right to begin exploitation of Block 16, a 49,000-acre tract that included most of the jungle home of the Huaorani people, an indigenous tribe of about 1,500. The NRDC called upon Conoco to back off and to leave the land and these people alone.

Soon after the press conference, Conoco executives met with the NRDC team and made a dramatic, unprecedented proposal. Instead of merely opposing oil development, as the environmental groups were wont to do, why didn't the NRDC consider working with Conoco to develop a model plan for the production of oil in Block 16?

Exploratory drilling had gone extremely well. Conoco estimated that there were 200 million barrels of heavy crude beneath the surface. The company wanted to go ahead full speed exploiting its discovery, but it was faced with a number of serious problems. For hundreds of years the indigenous Huaorani had wreaked bloody retribution on anyone so brash as to enter their homeland. Five evangelical missionaries who had arrived in 1955 were butchered and cast into the river. In 1987 a Catholic bishop was found dead with eighty-nine spear wounds in his body. Since then the Huaorani had not employed these traditional means to expel intruders, but they represented an immense problem to the American company.

Conoco could not do what oil companies had previously done—push the Indians out, pacify them with trinkets, and sometimes kill those who resisted. Conoco vowed to deal with the indigenous peoples in a humane way and to set a new standard for environmentally safe oil production. Of all the participants in the meeting with Conoco, no one was more skeptical than Kimerling. "They said the right things," she recalls, "but their exploratory drilling practices were no different than the other companies, leaving their wastes in holes in the ground."

For Conoco, the benefits of such an alliance would be enormous: if it could become the environmentally sound choice for oil development in Ecuador, it could readily expand its operations throughout South America. Of the $600 million it anticipated investing, the company said it was prepared to spend $30 million to $60 million on environmental concerns. The Conoco master plan included unprecedented rigorous controls over the drilling, the pipelines, and the disposal of waste. The company also promised to do whatever was necessary to minimize the uncontrolled colonization of the region by outsiders following in its wake. To make sure that Conoco kept its promises, the environmentalists would be there from day one, witnessing that all the safeguards were put in place and maintained.

This was a serious proposal, and after thinking it through, Bobby called the oil company lawyers. "Look," he said, "in addition to doing the independent environmental oversight and management plan, would you also be willing to give a portion of your profits to the Indian community for rain forest preservation and for development projects?" In terms of the money Conoco contemplated spending, this was perfectly doable, and the Conoco lawyers replied enthusiastically.

Kimerling considered this proposal a betrayal of promises the NRDC staff had made, and the very words that Bobby had spoken at the press conference. She saw herself as the surrogate for the unheard voices of the Oriente and would not still her protest. "I checked with a number of people in Ecuador, and they all said not to negotiate with Conoco," Kimerling says. "I simply didn't think it was right to go ahead while our partners in Ecuador opposed it."

Bobby seemed to Kimerling like just another arrogant gringo imposing his will on a distant, powerless people, but he had thought deeply and taken his own serious position on this issue. The NRDC attempted in a gingerly way to see if the indigenous peoples and the Ecuadorian environmentalists wanted to go ahead with this unprecedented agreement. The organization took its own soundings in Ecuador and found groups in favor of considering the Conoco proposal. Bobby's position was not unlike his father's hardhearted take on social issues, a position that was deeply aware of human corruptibility and the often convoluted journey of human progress. Kimerling might fancy that she and a coterie of environmentalists represented the Ecuadorians. But who were we, Bobby asked himself, to tell the Ecuadorians that they could not sell part of an oil resource "when half of that oil goes directly to our country, and while production is driven by Ecuador's giant debt, a portion of which is owned by U.S. banks." The Ecuadorians would continue to exploit their oil treasure, no matter how loudly the environmental purists in Quito and the United States shouted, and here was a unique opportunity to try to see that it was done right—and with a bounty of millions of dollars for the environmental movement.

Bobby was moving beyond the conventional wisdom of the environmental movement. He was risking his reputation in seeking a new, enlarged, and dangerous agenda. Whereas up until now activist organizations such as the NRDC had played the role of principled, passionate critics of oil exploration in remote pristine regions, Bobby and his colleagues would help set up righteous monitors of a company attempting to profit from the responsible exploitation of one of the world's crucial resources.

It was the most important issue with which Bobby had ever dealt. He brought to it both his virtues—intelligence, principle, daring, inventiveness—and his vices—nervous impetuousness, moral certainty, and aggressiveness that at times shaded into willful arrogance. He looked down at a political landscape that he did not realize was as fragile as the ecological structure of the Amazon.

Bobby and Scherr set up a meeting with Conoco to discuss the specifics of how the oil company's plan might work. "I said to Bobby, 'I'm not going to the meeting to negotiate an agreement,'" Kimerling says. "Bobby said, 'Don't tell me your problems. Your job is to get it done. Your job is to convince your friends [in Ecuador].' I said, 'You've got it backwards.'"

On February 5, 1991, Bobby and Scherr met with the Conoco executives. When Kimerling continued to refuse to attend unless the Ecuadorian activists agreed, she says she was fired. Bobby and Scherr said she was dismissed for "unprofessional performance" and called her "a disgruntled former employee with strong self interest in claiming the reason for her dismissal was philosophical." The two NRDC attorneys were worried enough about the ramifications of the firing that, according to Kimerling, in return for two months' severance pay, she was asked to sign an agreement to misrepresent the circumstances of her departure from the NRDC. Kimerling did not sign any agreement, and she became one of several public relations problems.

Bobby should have known enough about how power worked in America to have realized that it was almost impossible to keep the details of the NRDC's meeting with the Conoco executives secret. Within a few weeks someone leaked Conoco's confidential minutes to the media. Here was Bobby saying that while his organization opposed drilling in national parks, "NRDC believed that it is better to have Conoco do the development than anyone else." Here were the two activist attorneys strategizing with the Conoco officials about how to deal with other more "black and white" groups such as Rainforest Action Network (RAN). In exchange for this agreement, Bobby stated that he considered anything less than $25 million for the environmentalists to be insulting. Bobby was talking about money as a way to gauge Conoco's seriousness, but it sounded to other environmentalists like the NRDC was making an unseemly mercenary agreement.

The environmental groups were competitive over fund-raising and vied with one another in professing their commitment to the highest, uncompromising principles. To many of Bobby's colleagues in the movement, the NRDC meeting with Conoco looked like an unseemly betrayal of everything for which they stood. "If you didn't know otherwise, you'd certainly think Kennedy and Scherr worked for Conoco," James Ridgeway wrote in the *Village Voice*, a sentiment roundly seconded by environmental activists.

To a coterie of environmentalists in Ecuador, Bobby was the progenitor of a new imperialism, carrying a green flag with bulldozers following in its wake. When Bobby returned to Ecuador, he was greeted by a small picket line and condemned as a malevolent gringo know-it-all trying to impose his will on the Huaorani nation. To the Ecuadorian environmentalists, Bobby was one of the "ecological imperialists"; to Bobby, these Quito protesters were a "tiny, urban elite." He went on to meet with indigenous groups in the Oriente, who asked him to represent them in further discussions with Conoco. That was promising, but in the war of images the half-dozen activists who greeted him were portrayed in the media as an army. Bobby was bewildered by what he faced. "I have no idea what happened, or what the internal politics were," he said."

In the life of an environmental activist, as in the life of a politician or a scientist, there may be only one transformative idea in a lifetime. Bobby and the NRDC were attempting to infuse a whole new element of tough-minded, realistic politics into the environmental community, but it was his lack of political judgment that in the end helped to doom this venture. Barbara Bramble, director of international programs for the National Wildlife Federation, told the *Washington Post* that the NRDC's first mistake was to forge on before developing a crucial consensus among the environmental activists and Indians in Ecuador. "Mistake two was to look like, whether they intended it or not, that they were hiding things from people who were known to be naysayers or skeptical—because you can't get around them," she said. "I don't quarrel that that may have been the best solution, but they ruined it."

The leaked minutes created such a firestorm of controversy that in June 1991 the NRDC formally backed off its attempts to broker an agreement with Conoco for environmentally sound oil drilling. "NRDC is committed to following the lead of the indigenous people," Bobby announced. The proclamation was no more than a futile attempt at face saving. Although Bobby weathered the damage to his credibility in the environmental movement, he had lost an enormous opportunity. If Bobby and the NRDC had devoted the time and resources to developing a deep understanding of Ecuadorian realities, they might have been able to bring together the environmentalists and the indigenous peoples. If they had brokered an agreement monitored with fierce resolve and the threat of a boycott against Conoco gas stations in the United States, Bobby and the NRDC might have been able to provide a new model for the exploitation of natural resources in the poorer nations of the world.

With its retail facilities in America and high public profile, Conoco was far more vulnerable to pressure from environmental groups and bad media than most oil companies. The company had been serious about an agreement with the NRDC and watched with dismay as Bobby and his associates failed to deliver anyone—not the other American environmentalists, not the Quito activists, and not the Huaorani. In the end Conoco backed out of its plans in the Oriente, walking away after investing millions. Maxus, a Texas company, took its place. Although Maxus promised to follow through on the measures that Conoco had proposed, the company was almost a billion dollars in debt, and there was no group such as the NRDC overseeing its follow-through on its promises. Maxus formally negotiated with the Huaorani to be able to drill oil on their lands, an agreement that was more symbolic than legal. At the signing of the document, the daughter of the Ecuadorian president was overheard comparing what her country and the American company were doing to pacifying native peoples with "trinkets and beads."

In 1995, YPF, an Argentinean company, acquired Maxus and began extracting oil in the Oriente with little apparent concern for what outsiders thought. As for Bobby, he rarely again took such a seemingly compromising position on an environmental issue. Instead, he developed a reputation as the purest of the pure, an environmental Jeremiah raging eloquently against corporate despoilers.

CHAPTER THIRTY-TWO

A Man Apart

As his nation and much of his party moved to the right in the Reagan years, Ted held firm. "After the '80 campaign," he told his biographer Burton Hersh, "even before the inauguration, it was very, very clear to me that the focus and attention of the Reagan presidency was to undermine the basic construct of the human-services programs, which I had always thought were not to be sort of a handout but a hand up."

Ted saw himself not simply as the great protector of his brothers' legacies but of a half-century of social and other legislation that had largely defined the role of the federal government in American life. It was often a losing fight, but in some places he had helped patch holes in the protective underpinnings. He maneuvered through legislation allowing laid-off workers to keep their company health insurance for eighteen months. The employees would have to pay for it, but for the two million Americans who took advantage of the new law, it was an immensely important benefit.

Ted also had a very different conception of foreign policy from Reagan's. The new president famously called the Soviet Union "the evil empire"; Ted's anti-Communist bona fides were real, but he viewed Russia as a nuclear power with whom America would have to make reluctant accommodation. Reagan increased defense spending in part to force the Soviet Union to try to match American efforts and in so doing drive its economy into ruin; Ted took almost the opposite tack, fighting unsuccessfully for a freeze on new nuclear weapons by the two superpowers. The administration secretly backed the right-wing contras' war against the socialist Sandinista government; Ted was a powerful leader in the fight against what he called "President Reagan's secret war in Nicaragua." To the conservative Republican, any foe of communism was a friend, even if he was a dictator. Ted thought otherwise. He traveled to Chile, where he confronted the oppressive regime of Admiral Augusto Pinochet in its own lair, praising those who "have stood up for those bowed under the weight of tyranny and torture." He was a leader in overturning Reagan's veto of legislation authorizing economic sanctions against South Africa.

No one was so vilified by the far right. As vitriolic as many of those attacks were, they were in one sense not unwarranted. Ted was not only the best known but the most effective liberal legislator in America. What was so startling was the extent to which he not only worked with conservative Republican colleagues in an increasingly partisan Congress but enjoyed their company. Wyoming Senator Alan Simpson and Ted drafted immigration reform legislation. Ted and Senator Orrin Hatch of Utah became an even more important legislative team, passing bills on such diverse matters as providing funding for AIDS research, preventing private employers from using lie detectors, and deterring long-term dependency on welfare.

Ted also had a natural affinity with another conservative colleague, Senator Dan Quayle of Indiana. Although as vice president Quayle would become Mr. Malaprop, an endlessly derided figure, Ted found him an intelligent, sensible politician. The two men worked together to pass the Job Training Partnership Act, a piece of legislation that welded together the ideals of the left and the right. It authorized the training of several million workers, not by creating another federal program but through private organizations.

It was one thing to work with Republican colleagues, but quite another to foster goodwill from Jerry Falwell, the Baptist minister who ran the Moral Majority. One of Falwell's aides, Cal Thomas, wrote Kennedy half-jokingly inviting him to come to Falwell's Liberty Baptist Church in Lynchburg, Virginia. Ted not only accepted the invitation but asked to speak. When Thomas told the preacher, Thomas recalls that Falwell "turned white as a sheet and said, 'What!' He said some of his people thought Ted Kennedy was the Devil Incarnate."

Falwell envisioned an army of right-wing fundamentalists chasing the money changers out of the temple of American democracy and bringing *his* God into political discourse. That was in some ways exactly what native-born Americans had feared the Irish and other Catholics would do, letting their priests and their pope tell them not only how to live but how to vote.

Standing before an audience that included four thousand students at Liberty Baptist College in October 1983, Ted pleaded for understanding. He told the group that the very term "Moral Majority" "seems to imply that only one set of public policies is moral," but, he reminded them, at times liberals had acted as moral absolutists too. He warned about misusing religion by employing it too easily or broadly in the pursuit of secular goals. He said that all of us "must respect the independent judgments of conscience" and "the integrity of public debates." As he looked out on the youthful audience that had been taught that evolution is a lie, he said that "faith is no substitute for facts."

Ted was speaking as a politician living in a Washington in which the dialogue had become shrill and intemperate, the partisans of both parties full of ill will and rancor. Many of those in the audience thought of Ted as a polemicist shrieking his liberal message with a partisan bullhorn; in truth, he had at times added his shrill voice to the disharmony. But as his colleagues appreciated, he understood the parameters of civil debate. As passionately as he argued an issue, he almost never questioned the goodwill or honesty of his opponents.

"It may be harder to restrain our feelings when moral principles are at stake," he told the audience in conclusion, "for they go to the deepest wellsprings of our being. But the more our feelings diverge, the more deeply felt they are, the greater is our obligation to grant the sincerity and essential decency of our fellow citizens on the other side."

Ted took pleasure in dramatic political fights, especially when they occurred under the klieg lights of the media. In 1987 Ted led the battle against the confirmation of Judge Robert H. Bork to the Supreme Court. The former Yale professor was a man of forthright conservatism, often brilliant, usually combative, and inevitably convinced that he spoke with sound judicial logic.

Bork had fired Solicitor General Archibald Cox during Watergate. Ted considered that action so shameful that it rendered Bork unworthy of sitting on the nation's highest court. On the very day President Reagan nominated the judge, Ted gave a short speech on the floor of the Senate. "Robert Bork's America is a land in which women would be forced into back-alley abortions, blacks would sit at segregated lunch counters, rogue police could break down citizens' doors in midnight raids, schoolchildren could not be taught about evolution . . ." The litany went on and on. Doubtless these words had been penned by one of Ted's partisan aides, but he had known what he was reading. Many of the charges were unfair, exaggerated, and in some measure malicious, a perfect countervision of the malevolent caricature that the right had created of the senior senator from Massachusetts. This rhetoric was being expounded not in a fund-raising letter or a stump speech but in the United States Senate. It was precisely the kind of merciless stereotyping that Ted had condemned in his speech to Falwell's congregation, and it had the same pernicious consequences.

The Senate had unanimously confirmed Bork for his judgeship on the U.S. Court of Appeals in D.C., and at first it seemed certain that he would be approved again. But Ted raised the political stakes to an almost unprecedented level for a Supreme Court nomination, and in the end Bork was rejected. For the liberals, it was a great victory, since Bork might have been the swing vote on issues such as abortion and affirmative action. In politics memory is a dangerous thing, however, and for the conservatives, it was a defeat that would become a rallying cry.

I n the mid and late 1980s, Ted gave his conservative foes plenty of examples of bad personal conduct that they used to condemn his political activities, while those who admired his political achievements suggested that his private life should be kept private. In 1985 Ted and his drinking buddy Connecticut senator Christopher Dodd amused themselves at La Colline, a popular Capitol Hill restaurant, by breaking the glass picture frames holding their autographed photos, which the restaurateur had proudly placed on the wall. If it had been anyone else, there would have been a quick call to the D.C. police. Later that year the two politicians showed up at La Brasserie, another Washington haunt. While Dodd balanced a waitress on his knee, Ted rubbed himself against her. Two years later, during a luncheon at the restaurant, a waitress walked in on the senator mounted on a congressional lobbyist.

In January 1989, Ted was in a New York bar whose very name, American Trash, suggested that it was not the typical haunt of leaders. When one of the clientele insulted President Kennedy, Ted threw a drink in his face. Then Ted invited the man outside. At four in the morning on a New York street, the fifty-six-year-old senator defended JFK's honor by fighting the man.

Ted was a man without a wife or a consistent lover, haunted by ghosts of his brothers several times their real size. He constantly faced the psychological pain of his own past, of opportunity lost for him and for his brothers' legacies. Then there was the physical pain of his back, which came on as unexpectedly as did the emotional pain, all of it deadened with booze and sex.

Most of the time Ted was a disciplined, serious man, but on those occasions when he was supposedly amusing himself, he danced along on a precipice of disaster. Almost everyone around Ted enabled him, shushing up the problems and accusing those who mentioned them of being tabloid voyeurs. By the late 1980s he was indulging in excesses often enough that just as in the months leading up to Chappaquiddick, it again seemed just a matter of time before another tragedy or scandal would occur: a woman selling her story to the tabloids, a DUI arrest, another automobile accident.

In August 1990, just before Hurricane Bob struck the coast of Massachusetts, his Hyannis Port neighbor Larry Newman saw Ted taking his boat out into the roiling waters. "I just couldn't believe it," Newman recalls. "The only thing that I could think of is that the guy had a death wish. The guy wants to die."

Good Friday

O n the afternoon of Wednesday, March 27, 1991, Ted flew down from Washington to Palm Beach for a long Easter weekend. Even his memories of the good times in Florida were tinged with melancholy. It was here that his mother had suffered a stroke in 1984; she now lay bedridden in Hyannis Port, unable to return to the southern winters. Helga Wagner, the lover he might have married, lived here with her artist husband. So did another of his lovers, Dragana Lickle, the thirty-six-year-old ex-wife of Garrison Lickle, a socially prominent Palm Beach attorney. That romance had just ended in the midst of an embarrassing custody fight in which Dragana's ex-husband accused her of "dragging the children around in his [Ted's] presence" and "exposing them to the lifestyle of the jet set and to notorious characters."

In the winter Ted and his siblings each had the estate for three weeks. These were Jean's days, but she had invited her favorite brother along with young Patrick. Her husband had died of cancer the previous August. Now that Steve was gone, Ted was more alone than ever, and being with his sister reminded him of all that he had lost. Despite the fact that much of the world imagined that all the Kennedys hung out together, that was no longer true. Joe was spending the holiday in Palm Beach, but he saw no reason even to come by to say hello.

As frenetic as those last days in the Senate had been, on the very night he arrived for the Easter break Ted decided to go out on the town. At fifty-nine, he was still a handsome man, but he was so portly that his distinguished features seemed to be curtained by his jowls. He approached eating the way he did sex, devouring everything set before him, fast food, chocolates, and ice cream. He was such a compulsive eater that his staff tried to keep the more calorie-laden delicacies away from him. Each January he began a dramatic—and probably dangerous—liquid diet that drained the pounds. And then he started up again.

Ted was not about to head out to the nightlife of Palm Beach by himself. This midweek evening he took his son Patrick and his nephew Willie Smith with him. Twenty-three-year-old Patrick was a senior at Providence and in his second term in the Rhode Island House of Delegates. Patrick had not been comfortable with his classmates at Providence, and now that he was a professional politician, he had an excuse to spend almost no time among them. When a student reporter wanted to do a profile of Patrick for *The Cowl*, the student newspaper, he turned down the request, saying that his advisers told him it would not be a good idea. It was the kind of comment that rankled people and suggested that Patrick was not only self-important but too insecure to make his own decisions.

Patrick was in fact congenial and gracious to his seniors, which meant almost everyone he met. He was one of a small group of insurgents seeking to reform the Rhode Island House. His case was strong, but it would have been stronger if he had not spoken in public in a voice with only two settings, a dull monotone and a high-pitched screech, the latter the perfect vehicle for his self-righteousness.

Patrick's cousin Willie was an even more melancholy figure. His father had not paid much attention to his sons. His mother shared the Kennedy competitiveness, but in her there was a bitter core that she and hers had not received their due. Willie's older brother, thirty-three-year-old Stephen Smith Jr., had gone to Harvard and had a law degree from Columbia. On sheer credentials and family name, young Steve should have made a bold mark on the world, but he had fallen far short of his promise.

Willie was a young man who had grown up with privileges that had not so much enriched as isolated him. He had graduated from Duke and was finishing up medical school at Georgetown. He might have been handsome, but his features appeared almost off-center. He had small, messianic eyes and a weak chin—all in all, perhaps, a face too suggestive of his character.

Ted drove his car south from the Kennedy mansion to the town center, where at around eleven o'clock the three Kennedys walked into Au Bar. The upscale disco bar was the kind of place where patrons fibbed about their age, their residence, their wealth, their marital condition, and their reason for being there. When three men walked into Au Bar, they were as likely to be trolling for women as three men carrying fishing poles on the beach were to be going fishing.

The group did not stay very long, and Patrick and Willie drove to the house separately from Ted. Patrick said later that when he made his way back to the house, he found two women there with his father. Patrick called a taxi for one of them and apparently went to bed. Later the other woman was in such a questionable condition that caretaker Dennis Spear, Patrick learned, "drove the woman back in her car and found a means of transportation back to the house on his own."

Two evenings later, on Good Friday, Patrick, Willie, and several house-guests went out, but by midnight the two young men were in adjoining beds in their pajamas with the lights off. The door opened, and there stood Ted. He had been out on the veranda talking about the past with his sister. It was the kind of emotionally troubling conversation that he tried to avoid. He was so haunted by memories that he was unwilling or unable to go to bed. When Ted asked the two young men to go out with him, they knew immediately that no matter how tired they were, they should honor this request. They dressed, then drove with him the ten minutes down along the ocean to Au Bar.

Patrick and Willie roamed the dark room while Ted stayed at the bar. Willie immediately started a conversation with a woman. Her name was Patricia Bowman. She was a thirty-year-old single mother with a sick child at home, wearing a new dress for her evening out. By her admission, she did not talk the seductive palaver of singles bars but said things that would have driven many men away. She talked to Willie about some of the difficulties she'd had with her father growing up and about her baby's premature birth. The sadness in this woman coursed out of her.

Patrick hadn't had the good fortune of his cousin, and he decided to sit down at the table with Willie, who was carrying on a conversation with the woman. Patrick motioned for Ted to come join them. The senator was glad to do so, since his back was hurting and he couldn't comfortably stand up much longer. As soon as Ted sat down, he started introducing his son to another woman at the table. The woman, Anne Mercer, had just met Patrick. It rankled her that Ted fancied he could set up his son with her. "What do you think you are doing, coming here and thinking that you could introduce me?" Patrick recalls Mercer asking.

"You know Patrick is a representative," Ted replied. In some places, Ted might have expected deference for his son for being in the Rhode Island House of Delegates, but he could hardly expect it in a club where looks, youth, and suaveness were the highest currencies. As Ted was being royally berated, Patrick noticed that a woman who had rejected his invitation to dance was gesturing to him, and he happily left the table. A few minutes later Michelle Cassone agreed to come to the estate with Patrick for a nightcap. Cassone was a waitress in Palm Beach looking for better things. Willie had disappeared somewhere into the disco haze; he was making his way back to the estate with Bowman. Ted and Patrick also drove back up North Ocean Boulevard, followed by Cassone in her car.

Ted went to bed after donning his knee-length, striped nightshirt, but he soon got up and started walking around the house. There was a gothic weirdness about it all. He wandered barefoot through the dimly lit rooms while his sister slept in her room and her son was having sex with a woman he had just picked up, by his testimony first on the beach and then on the lawn almost beneath his mother's window.

Ted found his way to Patrick's room, where he and Cassone were sitting on separate beds talking. Ted stayed only a few minutes before finally retiring. Cassone and Patrick kissed a few times. Then after a few minutes on the beach he walked with her to the parking lot to say goodnight. While there, he noticed that Willie was also seeing Bowman off, although a few minutes later she mysteriously returned to the house to have further conversation with Willie.

"This girl is really strange," Patrick remembers Willie saying when he finally returned to the bedroom. "She is yelling, calling me another name. I wish we had gone to bed earlier." Willie left the room, and Patrick fell asleep, but he was awakened later when his cousin returned. "I can't believe it," Willie said. "She is saying she called the police." In most instances such a dramatic statement would have led to a lengthy conversation, but Patrick says that the two men went to sleep.

At around four-thirty in the morning, Mercer picked up the ringing phone in her West Palm Beach home and heard the plaintive voice of her friend Bowman. "I've been raped. Please pick me up. Bring another man with you." Thus began an episode that savaged two lives and dishonored many others.

Although Ted asserted that he heard nothing that weekend about the allegations of sexual misconduct, Patrick distinctly remembered his father being present Sunday evening when Bill Barry, a friend, former FBI agent, and sometime bodyguard, told them that "there's an allegation, a complaint by a woman who was visiting the other night, that there was a sexual assault or sexual battery." Barry said, "Well, I better talk to Willie. I think it's good that he get an attorney." Ted responded, as Patrick reconstructed the dialogue, "Well, you know, I guess, you know, I had better talk to Willie." At that point on Sunday, Willie had already flown out of Palm Beach before the police had had a chance to talk to him. The next day Ted and Patrick also left without talking to the police who had already begun their investigation that would lead to Willie's indictment on a felony charge of sexual battery, the Florida term for rape, as well as a misdemeanor charge of battery.

T ed would always regret that he had taken Willie and Patrick to Au Bar, but he had not picked up Bowman, brought her back to the house, and had sex with her on the lawn. And yet many of the newspaper stories and accusatory headlines ("Ted's Sexy Romp") almost implied that he was the one who had had sex with Bowman on the beach. There were sordid accusations about his activities that evening that, while not true, were devastating to his reputation.

The Kennedys protected their own, and the family formed a praetorian guard around Willie, but some of them knew that he was a troubled young man. His cousin Max's former girlfriend said in a sworn deposition that she met Willie in 1983. One evening she was staying over in a spare bedroom in the Smiths' New York apartment when she alleged that he attempted to rape her. She said in a deposition that he "tackled" her and "tried to kiss me and in the struggle that followed, they [his hands] were on my breast and trying to get up my dress." She said that she told Max about the incident, swearing under oath that "it was a violent attack" and she was "very scared and upset all night." She said in her deposition that Max's response was to say "that it wasn't a big deal . . . it was just the kind of thing that I should expect to have happen to me if I was attractive." She later broke up with Max, but after the events in Palm Beach he called to apologize. "Sounds like Willie has a really big problem," she recalls Max telling her. "You know, there's another girl that Willie attacked."

Initially Ted and Patrick attempted to put a *cordon sanitaire* between themselves and Willie. Ted's first statement was so judiciously neutral that it apparently infuriated his sister Jean, who blamed Ted for having taken her son to Au Bar. Ted then offered a second statement far more supportive of his nephew.

I n the fall Ted was back in the Senate taking part in the Judiciary Committee hearings over the nomination of appellate judge Clarence Thomas to the Supreme Court. The conservative jurist was one of the least distinguished candidates to be nominated to the Court in years, but he was African American and had as his champion Senator John C. Danforth of Missouri, an articulate, moderate Republican. Thomas seemed assured of nomination when Ricki Seidman, a staffer on the Labor Committee chaired by Ted, tracked down a rumor leading to Anita F. Hill, a University of Oklahoma law professor, who told of Thomas's importuning her with vile sexual asides when she was his subordinate at the Department of Education and at the Equal Employment Opportunity Commission. The Democrats on the committee thought that this was the weapon they could use to derail the nomination.

The stakes were high: the balance of the Supreme Court might depend in part on whether Thomas had talked in front of Hill about pubic hair on a Coke. Although Thomas's defenders will forever disagree, Hill was probably telling the truth and was savagely libeled by Thomas. That case should have been forcefully made, but no progressive senator had the eloquence and courage to expose Thomas's equivocations. No one on the committee was better prepared by his staff and more astute in the use of power than the senator from Massachusetts. But the same brush that the Democrats were using to blacken Thomas's reputation had already been used on Ted. He was hardly the one to lecture the jurist on sexual morals. Ted sat largely silent, as if doing public penance for the debacle in Palm Beach, his stillness serving as eloquent testimony to how private misconduct can affect public policy.

Ted finally took to the floor of the Senate to say things he had not said in the committee. "We do not need characterizations like 'shame' in this chamber from the Senator from Massachusetts," retorted Senator Arlen Specter of Pennsylvania, seeking to drive Ted back into the hole of humiliation from which he was attempting to emerge. "Anybody who believes that," added Hatch, a man Ted considered a friend, "I know a bridge up in Massachusetts that I'll be happy to sell them." Hatch said almost immediately afterward that he had meant to say "bridge in Brooklyn," and that may have been true. But the image of Chappaquiddick shadowed Ted, and Hatch was certainly not the only one to see it that day.

On October 15, 1991, Thomas was confirmed by a vote of 52 to 48. The new justice was a physically vital man who would probably be on the Court for several decades, almost always voting against the issues and causes that had been at the center of Ted's political life. And every one of those votes would be a reminder of what might have been if Ted had been living a different kind of life and had not taken Willie to Au Bar. It was unfair that he was being blamed for his nephew's conduct, but he had no moral chits left, and his conduct over the years was bad enough that his enemies believed only the worst of him.

In the days of the hearings Ted had become the least admired member of the Judiciary Committee; only 22 percent of respondents in a Gallup poll had a favorable view of him. A poll by the *Boston Herald* in October 1991 showed that 62 percent of Massachusetts voters thought it was time for someone else to serve them in the Senate. Ted had to do something dramatic to stem the decline. Late in October he gave a speech at the John F. Kennedy School of Government at Harvard University in which he apologized for his shortcomings. He stood in front of a full-length portrait of his brother. Most of the address was nothing but the familiar litany of causes and issues that mattered to him and his party. Then he headed onto the most uncomfortable of grounds, talking about his personal life. "I am painfully aware that the criticism directed at me in recent months involves far more than honest disagreements with my positions, or the usual criticism from the far right," he said, reading from a text. "It also involves the disappointment of friends and many others who rely on me to fight the good fight. To them I say: I recognize my own shortcomings—the faults in the conduct of my private life. I realize that I alone am responsible for them, and I am the one who must confront them. I believe that each of us as individuals must not only struggle to make a better world, but to make ourselves better, too."

Many of the journalists had expected Ted to arrive in sackcloth and thought he would detail his weaknesses and faults in such specificity and detail that the story would make the front pages of papers across America. Conservatives had hoped he would throw himself on the funeral pyre of American liberalism and announce that because of his manifold sins he was resigning from public life.

If Ted could not speak comfortably about personal matters in private, he could not disgorge himself in a public purging. Had he talked in specifics, mentioning Chappaquiddick, how he felt he had failed his brothers, or the foolish evening at Au Bar, he probably would have broken down in tears. As much as he loved politics and as much as he believed he belonged in the Senate, that was not something he was going to do.

A s the various Kennedys arrived at the Palm Beach County Courthouse for the December 1991 trial, they ran past cameras that circled the main entrance and poked out from the second and third floors of the garage next door. Scores of journalists had also arrived to cover the trial; 150 reporters were ensconced in a special media center on the third floor of the courthouse. The trial was broadcast live on Court TV and CNN and was in many respects a precursor to the O. J. Simpson trial.

John had been especially close to Willie when they were growing up, and he came down one day during jury selection, but neither his mother nor his sister joined him. All of the Kennedy sisters, except for Rosemary, were here. Ethel was frequently in attendance, and her sons Bobby and Michael showed up as well. Christopher Lawford flew in from Los Angeles, where he'd been working for years as an actor on a soap opera. Sarge sat in court on occasion, as did Tim.

In the seven months since the events at the estate, Bowman had been portrayed in the media as a Palm Beach heiress who had dressed for her night of sex in Victoria's Secret panties and bra, as if that was proof of her amorality. The woman who walked into the courtroom on the third day of the trial wore a conservative blue suit, a white, neck-hugging blouse, and almost no makeup. She was a plain woman with a dour complexion and a slight smile that was forever sliding into a grimace. Her dark frown lines had been there long before her night with Willie.

If the prosecutors had not gone ahead with the case, they would have been accused of knuckling under to the power and wealth of the Kennedys and their long history of hiding sexual misbehavior. In that sense, it was not just Willie but the Kennedy men who were on trial. When Willie's great-grandfather John "Honey Fitz" Fitzgerald was running for reelection as mayor of Boston in 1913, he backed out after he was warned that if he did not retire from the race his alleged affair with "Toddles," a cigarette girl, would be exposed. Willie's grandfather Joe brought women back to the house in Hyannis Port even when his wife was there. His uncle Jack had scores of affairs, including one with a suspected East German spy in the White House, and his uncle Ted had had scores of relationships as well.

As prosecutor Moira Lasch led Bowman through a recounting of her evening with Willie, she was a devastatingly convincing witness. "I was yelling 'No' and 'Stop,' and I tried to arch my back to get him off of me," she said as Willie stared at her, his eyes unblinking. "He pushed my dress up and he raped me, and I thought he was going to kill me. He said, 'No one is going to believe you,' and I got extremely angry: 'What do you mean no one is going to believe me? You raped me.'"

Willie's defense team was headed by Roy Black, a prominent defense attorney, assisted by at least half a dozen other lawyers and experts. When Black cross-examined Bowman, he began with a voice as smooth as a French liqueur. He attempted to ask why, when, and where she had taken off her panty hose. Every time Black probed too deeply or sneered too obviously, Bowman broke into tears, and her tears trumped any inconsistencies. No matter what he tried, Black was unable to shake Bowman's testimony. His badgering only reinforced the image of the physical assault that she had described with such shattering intensity.

There were three creditable women prepared to buttress Bowman's case by testifying that Willie had sexually accosted them, including Max's former girlfriend and two other single women with professional careers. In their pretrial depositions, they described a disturbed young man who had taken the Kennedy sense of entitlement and applied it to his sex life. When he wanted sex, he wanted it. Judge Mary Lupo refused to allow the witnesses to testify, just as she had barred those who would have talked about the alleged victim's sexual past. And thus the trial came down largely to who was more believable, Patricia Bowman or Willie Smith.

Ted was the first of the Kennedy men to testify. When he entered the courtroom, it was clear that a grand personage had arrived. He gave a slight wave to Willie, who sat with his hands clasped. It was a measure of how important Ted viewed this day, or of how anguished he was, that, as one reporter noted, he seemed to have "lost twenty-five pounds in the last month."

Ted wore a fine blue suit, the cuffs of his white shirt showing a perfect half-inch. As he took the oath, he had an aristocratic aura that said, either treat me well or do not treat me at all. His hair was silver and black. His face was florid. Although he had what looked to be an alcoholic's red nose, the discoloration was actually a decade-old scar from cancer surgery.

Despite the relentless way in which Lasch pursued other witnesses, she could not bring herself to confront Ted, to challenge his answers, or to prevent him from doing little but describe the events of the evening of Good Friday. Ted referred to thirty-one-year-old Willie and twenty-four-year-old Patrick as "boys," and when Lasch asked Ted to read testimony, he took out gold-rimmed glasses that he almost never wore in public. She asked him pointedly about why on Good Friday evening he had wanted to go out. "This was really the first time all of us had been together since the death of my brother-in-law," Ted said. "And we were visiting in the patio after dinner, and the conversation was a very emotional conversation, a very difficult one."

Black put a richly empathetic cast on Ted's conduct that Good Friday evening. "You said that that evening you had an intense conversation." Ted held back his tears, while a single tear rolled down Willie's cheek. "I think I described it, described it earlier," Ted said. Then he once again fought for composure. "I described it earlier," he said finally.

Ted did not rescue Willie that day, but he rescued something of the Kennedy family by making everyone in that courtroom sense how deeply he felt and how much he had lost.

Patrick testified later the same day. He had been worried that the scandal might hurt his political future. He had first sought to distance himself from Willie by pointedly telling the *Providence Journal-Bulletin* that the accuser "was a guest of one of my cousins," leading one Kennedy friend to assert that "Patrick hung Willie out to dry." As the weeks passed, Patrick started sounding more loyal to Willie and his tale of innocence.

To prepare for his day in court, Patrick paid $2,300 out of his campaign money to hire a professional speech coach, while attorney John Cul-

ver, a former senator and one of his father's friends, provided more substantive ministrations. However the duo prepared him, the very qualities
that made Patrick such a lame public speaker made him a good witness. He
spoke in plain, unvarnished terms about what he had seen that infamous
evening.

Ted and Patrick were both superb witnesses for the defense, but when
Willie testified, the trial probably hung in the balance. This was the first time
he had told his story publicly. He had prepared for this moment for many
months. Every inconsistency had been sewn up, every bump leveled, every
doubt battened over. He told how he and Bowman had walked on the beach
after she apparently took off her panty hose and shoes in her car. He spread
a towel on the beach, and they lay down. "She unbuttoned my pants, I took
her panties off with her help, and we embraced and I could feel her," he said.
"I put my hands on her, and she was, uh, excited, and I asked her if she had
any birth control."

He told how they had sex a second time on the lawn, when he mistakenly called her Cathie. "The minute I said it, I knew it was a mistake," he
said. "She got very, very upset, and she told me to get the hell off of her."

When Black finished leading Willie through his testimony, Lasch came
forward to attempt to tear apart this sordid saga of a thirty-year-old medical
student having unprotected sex beneath his mother's window with a woman
he had just met. Like Ted's sisters, the prosecutor had been educated by the
well-born nuns of the Sacred Heart. They helped impart to her a parochial
view of sex that made her unprepared to deal with these crudely intimate
details. "What are you, some kind of a sex machine?" she mockingly asked
of his claim to have had sex twice with the woman in forty-five minutes.
Again and again she tried to tear at his composure and budge him from his
story, but she was totally unsuccessful. Willie proved every bit as credible as
Bowman had been on the stand, and his testimony was consistent with most
of the physical evidence.

The jury took only seventy-seven minutes to come back with a verdict
of not guilty. Despite all the ugliness of Willie's admitted sexual conduct,
there was no compelling physical evidence or inconvertible testimony that

he had physically attacked Bowman. Whatever happened that night in Palm Beach, there had not been a strong legal case against Willie, and he probably would not have been indicted if he had borne another name.

Willie walked out of the courtroom as a free man who in winning acquittal admitted that he had dismissed Bowman as a person and employed her as a receptacle, an object of service to be utilized and discarded. Willie said a few words to the media, and then he, his mother, and a priest with a puffy red Irish face walked to the old paneled station wagon in which they had arrived each day. As Willie turned to get into the car, a voice shouted, "What about the three other women?" Willie pretended he did not hear, and the car drove away.

CHAPTER THIRTY-FOUR

Love, Loyalty, and Money

Some of Willie's relatives might mouth platitudes about justice prevailing in the end, but as Bobby Kennedy came out of the courtroom, he would have none of that. To him, the rape was of his family, not of Patricia Bowman. He told reporters that the charges were "unreal and unsubstantiated" and said that the prosecutor should have known this, but "I think she had a different agenda."

Bobby saw the world in black and white in the way that certain film directors do, as if color hides the starkness of the truth. To him, the environmental movement was a heroic struggle in which the people fought the overwhelmingly powerful and greedy corporations. No other leader possessed quite his passionate conviction, and no one was better at inspiring local groups. He spoke with poetic language reminiscent of the legendary naturalist John Muir and with the messianic conviction of Ralph Nader.

The greatest casualty of Bobby's years of troubles was his marriage to Emily Black, who had stood with him through it all. Bobby got a hurried Dominican divorce and immediately afterward, in April 1994, married Mary Richardson, an architect who had been part of the floating world around the Kennedys and with whom he would have four more children. The six offspring were brought up on an eleven-acre estate in Bedford, New York, large enough for falcons, peacocks, and other animals and the reasonable illusion of a rustic redoubt.

From the house at Pace University that served as the Environmental

Law Center, Bobby traveled across the country and throughout much of the world, promoting local causes. He flew to British Columbia to help lead a controversial effort to stop the logging on Indian lands. He traveled to Quebec and to Ottawa to add support to the successful efforts of the Cree Indians to stop a massive hydroelectric project that would have flooded 3,400 square kilometers of forest. The Riverkeeper concept expanded until by 2004 there were over one hundred organizations from Santa Monica to Puerto Rico and from Cape Fear to Cartagena.

While Bobby addressed issues across America and much of the world, he still had a compelling interest in policing his own neighborhood. Nothing mattered more to him than cleaning up the Hudson. This was one of the singular achievements of the modern environmental movement, and Bobby was one of the primary architects. When he went sailing on the Hudson now, he could look down and see clear, clean water. There was still a serious problem with PCB contamination, but fish were coming back and far more people were in their boats or on the banks enjoying the great river.

"I was on the river this weekend, and there's a cove north of Colton Point that last year had fifty boats in it," Bobby reflected in June 1999. "The year before there were maybe ten. And the year before that you'd see almost no one except occasionally somebody who came from Florida or somewhere. This weekend it had four hundred boats in it. So there's a growing constituency on the river, and there's more and more people using it. That's heartening."

In the early and mid-1990s, Bobby became a leading participant in a complicated issue over the future of the New York watershed. It risked dividing New Yorkers into two intransigent, militant factions. On one side stood the residents of the city and the politicians who served them; to them, the water they drank was precious and had to be protected from any and all who would pollute it.

The upstaters were for pure water too, but not at the cost of stifling their region. Many of the reservoirs that served New York City were the lakes where they fished. They sought to develop the desirable land around the reservoirs so that they could share in some of the wealth of their urban brethren. They brooked few limits to their freedom. If a plan could not be worked out to adequately protect the water supply, New York City would be forced to filter its once-pure water in new plants costing upward of $9 billion.

To promote an agreement safeguarding the watershed, Bobby and his associates orchestrated a multifaceted lobbying and public relations campaign similar to what corporations had done for decades to promote their interests. The effort began with a salvo of newspaper and television stories orchestrated by Bobby and others. They were behind the *New York Post* article reporting that the Croton reservoir had been shut down because of pollution that the EPA insisted was not sewage but "organic material," but they had nothing to do with David Letterman saying on *The Late Show* that evening that the *Post* story "scared the organic material out of me."

Bobby was able to transcend his political identification as a partisan Democrat and work with New York's Republican Mayor Rudolph Giuliani and Governor George Pataki. Bobby was not trusted by the don't-tread-on-me conservatives from upstate who worried that the environmentalists cared more about water than human beings. Bobby took their side in some matters, saying that Delaware County was right in insisting that the city come up with money to help develop the region. Bobby and his colleagues were tough, honest brokers as they moved between the two groups. Bobby had an especially good relationship with Michael Finnegan, the governor's chief counsel and a crucial player. At the news conference on November 2, 1995, announcing the groundbreaking agreement to protect the watershed, Bobby was the only nongovernmental person to speak.

Another person who might have taken a few accolades for the historic agreement was the *Post* reporter who had written the series of stories about New York's water supply that galvanized the paper's readers. The investigative reports detailed the discovery by the state Department of Environmental Protection of cryptosporidial cysts in the water supply. The DEP had not notified the public, and there had been notable increases in the number of hospital patients with cryptosporidiosis, a virulent diarrhea that sometimes leads to death. Strap-hanging New Yorkers reading the *Post* on their way home from work were startled at the idea that they risked being poisoned by a mysterious killer.

Most readers never realized that the Douglas Kennedy who had written the stories was Bobby's youngest brother. He was the forgotten brother in his family, overlooked, even neglected. "We're an Irish Catholic family," Doug says. "The eldest gets the land and the youngest gets to leave. I think there's something of that in our family."

Like several of his cousins, Doug had gone to Georgetown Preparatory School, where the Jesuits taught him that thought was the great defender of faith, not its opponent. From there he had attended Boston College for two years. The Jesuit school has one of the best philosophy departments in America, and one of its most distinguished teachers, Professor Peter Kreft, remembers the penetrating questions that Doug asked. Boston College had a moral seriousness lacking in many secular colleges, but after two years Doug transferred to Brown to test his Jesuit-tempered philosophy at a school in which the modern cult of moral relativism had taken firm hold.

Doug showed the same pugnacious, feisty character in the Brown classrooms as he did defending his turf on the football field at Hickory Hill. He came away deepened in his beliefs. "Everything was culturalism," reflects Doug. "In one of my first classes the professor was arguing that there was no difference between the text of a modern-day soap opera and the text of a Shakespeare play. He said that we have just been conditioned by society to think that Shakespeare is better. And everybody in the class was like, 'Oh, yeah.' And I'm arguing against the entire class. I'm saying that if you believed that and society has conditioned you to think what is right and wrong and good and bad, then you have no ability to say that the Nazis were intrinsically bad."

Unlike several of his older brothers, Doug neither endlessly bemoaned the fact that he was shackled with a famous name nor used the moniker as a free pass. He did not anesthetize himself from pain and tragedy, for he believed that in doing that he would be fleeing life itself. He struggled to make clear what he meant, and that struggle indicated how fiercely and deeply he had fought to understand this. "The only way to live your life with depth is to open yourself up to the experience of life, which is a lot of pain, grief, sadness, joy, whatever," he asserts. "That's the only way you really have love that you can give to another human being."

After graduating from Brown, Doug worked as a freelance writer for the *Herald* and the *Beacon Hill News* in Boston. Then he took a position as a reporter at the upstart *Nantucket Beacon*. When the summer was over and the tourists had left, Nantucket reverted to its insular, shuttered ways, which made the series on AIDS on Nantucket that Doug wrote for the weekly paper in the fall of 1991 even more sensational.

The first article published soon after Doug joined the weekly was headlined "AIDS: A Test of Acceptance." It was a test for the islanders themselves, who had largely ignored AIDS, treating it as if it were a disease of the sinners and the fallen. It was a test for a new newspaper confronting an issue that many people thought not only unspeakable but unmentionable. And it was a test for twenty-four-year-old Doug, who had the audacity to take on a matter that had been universally ignored.

Doug knew as a writer that the best way to tell the story convincingly was to personalize it. He did so by profiling a twenty-five-year-old Nantucket resident dying of AIDS. He interspersed data about nineteen other Nantucket residents suffering from the disease and the health community's limited ability to fight it. But it was his searing portrayal of the last weeks of Troy Maher's life that was best remembered. Maher had played football in high school for the Nantucket Whalers, the island's great pride. Doug was there at his bedside next to Maher's mother during her son's last days ("She has noticed a spot on his hip where the blood has clotted. She applies ointment to the sore and to other parts of her son's body").

No matter how homophobic a reader might be or how politically myopic, almost everyone who read Doug's words cared deeply about the young man lying unconscious at Nantucket Cottage Hospital. "Do you know what's the scariest thing about this disease?" Maher's mother asked Doug. "It's I know I won't be the last mother on Nantucket to sit by a hospital bed and watch her son die." The series helped to bring AIDS out of the closet on Nantucket to such a degree that the island's prominent citizens began raising money to combat the disease. The articles brought Douglas several journalism awards.

With the series on AIDS and the devastating pandemic on his mind, Doug attended a memorial service in New York City for a number of gay men who had died of the disease. One of the speakers, a handsome, articulate actor named Dan Cronin, had AIDS. Afterward, Doug joined a small group for dinner at Dinastia, a Chinese-Latin restaurant on West Seventy-second Street.

Cronin considered Reagan and Bush anathema to everything he valued. He had hardly sat down when he began castigating the conservative Republicans. Doug was the most argumentative of dinner companions, and debate was his favorite means of communication. He set out against Cronin even before the waiter brought drinks. Doug said that Cronin's baby-boomer generation was selfish and self-involved. Their idealism was nothing but mindless rhetoric.

Cronin had no idea who this intemperate young man might be, but he lectured him: "You're just a kid, too young to have any idea what people like Muhammad Ali and the Kennedys gave us." It had only been a few months since Dan learned he had full-blown AIDS, and that news had made him even angrier at people like this young man spouting inanities at a time when he believed the Reagan and Bush administrations were ignoring the tragic realities of AIDS in America. As the dinner concluded, Cronin thought he might have been a mite too tough on the kid. "You're okay," Cronin said.

Two days later Cronin received a phone call from a mutual friend who had been there for the dinner, saying that Doug was sending him a poster of his father, Robert F. Kennedy. After that the two men became the closest of friends. Cronin thought that he only had a few months left. He couldn't work, he went on welfare, and he had terrible relapses, but he always seemed to come back, and he was still alive a decade later. Every day he and Doug talked. They talked about politics and journalism, but almost never about AIDS.

D oug moved on to the *New York Post* in the fall of 1993. When Doug arrived in New York, he called his brother Bobby's close friend Chris Bartle and asked if he could put him up in his East Village apartment. Bartle, who would have done almost anything for Bobby, said that Doug could stay in his extra bedroom. "I didn't really expect him to take me up on it," Bartle says. "But literally at two in the morning, him and his friends arrive with furniture and his big dog and everything, and I wake up and the place is chock-full of his stuff and he's in the bedroom sleeping. Five years later he moved out when I sold the place."

Although Bartle enjoyed having Doug as his roommate, he found his squatting companion "hard on a dollar." Doug was not comfortable in the fancy restaurants once frequented by his parents, the places where his brother Bobby went to court wealthy, powerful benefactors to the environmental cause. Instead, wherever he was, Doug sought out the cheapest eating places. Unlike most of his siblings, he never applied for a start-up grant from the Kennedy Foundation for one of the philanthropic ventures that he often found so fraudulent.

Doug had moved further away from the traditional liberalism of his family than any of his brothers and cousins. He had spent his youth bemoaning the fact that he was born too late to have the Beatles and the Rolling Stones as heralds of his generation, to march against war and poverty, and to believe that getting stoned and making love was a political statement. Now he felt conned by the ideas of a generation that included his own older brothers and cousins and even by implication the ideas of his Uncle Ted and to some measure of his own father. Doug was outraged at the world these supposedly self-indulgent, profligate baby boomers had left his generation.

"You guys went on this spending spree that had no accounting for the fact that you were leaving that debt to future generations," Doug argues. "This is the most selfish act that one generation has committed upon another generation in the history of this country. Period. Your generation is self-centered and self-involved and has ideas that have no basis in the real world. You ruined the fucking environment. And you're running our society by passing these huge bills on to your children."

Doug proudly called himself a libertarian, celebrating liberty in all its manifestations. Although he revered his Uncle Ted, he would gladly have walked behind undoing much of what the senator had accomplished. To Doug, government was an instrument that if allowed to grow unchecked ended up shackling freedom.

Doug was one of the founders of Third Millennium, a group of Generation Xers who set themselves up as leaders of their peers. The organizational meeting was held at Hickory Hill in March 1993. Many of those present were conservatives, a species of humanity rarely seen in the McLean living room. Beyond their often intemperate attacks on their elders, the group publicized what would become some of the crucial political issues of the next years, including reducing the deficit, decreasing government entitlements, shrinking the size of government, dealing with the crisis in Social Security by limiting benefits, and emphasizing private acts of service.

Politics, however, was merely a diversion for Doug, who loved the gritty life of a street reporter. The *Post* was like a firehouse with bells going off every five minutes. As soon as one fire was out, another one was blazing down the street. Doug had found the perfect sport in which to compete. He was a cocky competitor who considered himself the best street reporter in New York.

In one story celebrated in the newsroom, if not by the Pulitzer Prize committee, he wrote about a deli owner who had been shot by a burglar. Doug found a ballistics expert who said that the deli owner might have been saved by the hero sandwich that deflected the bullet. On another occasion, while covering a murder, he left the police and other reporters and wandered down to the cellar, where he stumbled upon another body and a witness hiding there. "I took that guy back to the *Post* so nobody else would get him," he recalls fondly. "That was the life of a tabloid reporter. You get the source and steal him from the others and people are chasing you. It was great. It was fun stuff."

When Doug went incognito for two months to investigate Medicaid fraud, he had to do little more than wear his standard outfit of running shoes and scruffy slacks to make a reasonable impression of a down-and-outer. The healthy twenty-seven-year-old bought or rented Medicaid cards on the street and went to doctors and clinics across the city. In pursuit of his story, he took physicals and received injections and X rays. Often he only had to make a few vague complaints to be set up with prescriptions for expensive medicines, which he then resold to corrupt retailers. He collected a trove of back braces, asthma inhaler pumps, and enough medicine to stock an infirmary.

Doug's reportage was so influential that in May 1995 he testified before the human resources and intergovernmental relations subcommittee of the House Committee on Government Reform and Oversight. His mother sat in the audience, as she had done three decades before when her husband testified before Congress.

In the summer of 1996, Doug joined Fox News, at the beginning of its rise as the most important new television network in years. If Doug's last name had been Reagan, the assignment desk editor probably would have greeted him enthusiastically, and the camera and sound people would have vied to work with him. As it was, one of the editors took perverse delight in mentioning Chappaquiddick, shouting the word when Doug passed him in the newsroom. Nor did he get along with some of the crews. He was used to working alone and had none of the schmoozing skills so necessary for survival in television. It took him a good year or more to learn enough to become an effective on-air reporter.

Doug was not a natural for television. He did not have his brother Bobby's public persona or his cousin Joe's authoritative presence. Not only did he have nondescript looks, but his hangdog manner suggested that he was embarrassed that people were looking at him. Moreover, his voice was far too high to evoke the resonant tones of a TV anchor.

Doug was a Kennedy reporting for Fox News, and as outstanding as much of his work was, he was about as close to anonymous as an on-air reporter could be. Doug was as aggressively opinionated as Bill O'Reilly or any of the other personalities who came to dominate Fox, yet none of that seeped into his on-air manner. He got out of the way of his stories and never sought to make himself the center of his pieces. When there were two strong sides, he did not skew the tale but attempted to be fair-minded. In essence, this student of philosophy was saying that he believed in the judgment of the viewers; when presented with an honest rendering of the facts, they would come to an honest conclusion.

For a while he produced a series of "American Stories" about social and political dilemmas in small-town America, including: a debate over Bible classes in a West Virginia school district; a controversy over a Pikesville, Kentucky, ordinance requiring employees of escort services to wear name cards identifying themselves; an atheist activist complaining about the Hagerstown, Maryland, Suns baseball team offering discounts to those who brought their church bulletin to a Sunday game. In 2002 he produced a devastating investigative series on the dangers of the antidepressant Paxil.

As hard as Doug worked on his stories, he was not consumed with his career. In 1998 he married Molly Stark, a special education teacher he had met when he was working on Nantucket. They soon had two children.

Doug was such a good family member that he not only would pen a cover story about the five surviving sons of Robert and Ethel for *Esquire* in June 1998 but would allow Joe to take the byline. Although Doug was close to all of his brothers, he inevitably had the deepest kinship with Max and Chris, who were the closest to him in age.

M ax was an anomaly among the young Kennedy men. Although he had the catlike nervousness of his brothers, there was a spacey, ethereal quality to him. He met his wife, Vicki Strauss, when they were students at the University of Virginia Law School. Vicki's father was the chairman of the Philadelphia-based Pep Boys auto parts and accessories stores.

After graduation the couple moved to Vicki's hometown, where Max became an assistant district attorney. The Kennedys are considered America's first family of Democratic liberals, a curious honor since so few of them are pure liberals. Even as progressive as Max's father became in his last years, Bobby was as tough-minded as anyone on issues of crime. So was his son. Max served in the Pennsylvania city in the midst of a crime wave so humongous that the jails ran twenty-four hot bunks, three inmates sharing a bed eight hours at a time. Max rarely prosecuted someone he did not believe eminently worthy of being prosecuted. He saw that it was the poorest Americans who suffered the most from crime not their distant affluent neighbors in the suburbs; in filling the jails with those who richly deserved to be there, he believed he was serving the least privileged Americans. "Probably the best anticrime measure is to have a strong economy," Max says, and then in a conspiratorial whisper, "*but another one is to build big jails.*"

Max brought too much of the painful emotions home with him, and after three years he quit and moved to Los Angeles. His father was omnipresent in his life in a profound, untroubled way. When his two children were about the same three years of age that Max had been when Bobby had died, his sixth-born son decided to write a book about him. "I wanted to have something to give to my children," Max says, "and I thought there perhaps was a way to convey a lot about a person in their life in a small number of words."

Make Gentle the Life of the World: The Vision of Robert F. Kennedy is an exquisitely rendered book of Bobby's favorite quotes and notations from his journals. If it did not fully justify Max's receiving the title "author," the small volume was precisely what he set out to do, in a few words to convey the best of Robert F. Kennedy.

O f all the Kennedy grandsons, Chris was the only one with a passion for business. He was a brash, unruly child who had a business card when he was twelve that read "Christopher Kennedy, Entrepreneur." He and Brad Blank, a Hyannis Port neighbor two years his senior, started a business renting small sailboats to tourists. It was the younger boy's idea, and they incorporated themselves, got the proper licenses, hired a lifeguard to watch over their customers, and made back their initial investment of around

$10,000 by the end of the summer. "Chris being the executive that he is, we hired people to do the work for us," Blank recalls. "We didn't even go on the beach."

Chris attended Boston College, where he got up before dawn on the coldest of Massachusetts winter mornings to win his letter on the varsity ski team. He fell in love with an older student, Sheila Sinclair Berner, who agreed to stay in Boston until he graduated if Chris would then move to her hometown of Chicago, where she planned to go to law school. It would have been unthinkable for a Kennedy man of his father's generation to follow a woman's ambitions, but Chris happily ended up in Illinois.

Chris saw what Joe was doing with Citizens Energy in Boston, bringing cheap fuel oil to the poor. After graduating from Boston College, he had the idea of doing the same thing with food, cutting deals with suppliers, cutting out the middlemen, and making a difference in millions of lives. As excited as he was about the idea, he couldn't envision himself working with Joe at Citizens. Instead, to learn about the business, Chris had the considerable audacity to call Blaine Andrews, the CEO of the leading agricultural company Archer Daniels Midland, which hired him for an executive training position. Chris took the same pleasure in deal-making that his grandfather did. He moved around the Midwest a week at a time buying and selling corn, soybeans, bean meal, bean oil, and wheat.

"It was the greatest," Chris reflects. "I mean, I've always regretted in some ways that I didn't stay there." Chris had entered the business world at a time when fortunes were being made beyond anything even his own immensely wealthy grandfathers had envisioned. Chris probably could have started something on his own, whether a charitable Citizens Agriculture or a for-profit company. As much as business schools have attempted to teach entrepreneurship, it is not a series of skills or a quantifiable book of knowledge, but a mind-set and a passion. It is a hunger that cannot be satiated, a hunger that rarely comes to the children of wealth and privilege.

Instead of going ahead with a venture on his own, Chris followed what had become the pattern of the young men of his generation. In 1987 he entered the family business and went to work for the Merchandise Mart in

Chicago. Joe had purchased the immense building in 1946 for $13 million, and it was the Kennedys' primary business property and largest asset.

Four years after Chris began work there, *Forbes* opined in 1991 that "the Mart is probably a dwindling asset." That was the essential dilemma facing not only Chris but the entire family. Members of his family wanted to get hold directly of their inheritance. The only way to do that was to sell the Mart and divide the proceeds. First the value had to be enhanced. With its mix of wholesale showrooms, office rentals, and trade shows, the Mart was vulnerable to a volatile real estate market and dramatic changes in consumer tastes and business practices. Chris was adept at working with various clients, adjusting rents, and tinkering with merchandising concepts.

In 1997 the Merchandise Mart, the Apparel Center in Chicago, the Washington Design and Office Centers, and some undeveloped land were sold by the Kennedy family for $625 million, far beyond what the properties had appeared to be worth a decade before. The deal was structured to minimize the tax consequences and pay out the capital gains taxes over a long period. The Kennedy family had been held together by love, loyalty, and money, the third leg the unmentionable one. With the sale of the Mart and the distribution of much of the money, family members could go their own ways. Eventually the young Kennedys would not have to settle for a mere $70,000 to $100,000 a year from trust funds but would reap millions of dollars from the estate.

If ever there was a time for thirty-four-year-old Chris to leave and forge his own way in the business world, it was after the sale. Instead, as part of the deal, the Kennedys negotiated a management contract in which Chris would stay on as executive vice president, later becoming president of Merchandise Mart Properties.

CHAPTER THIRTY - FIVE

Team Play

he four Shriver sons were trying to carry on the deepest and truest part of their family inheritance while living what were truly their own lives. "They all four are extraordinary and so different and yet very similar," says their sister Maria. "All four are very dedicated to our immediate and extended family. They all four understand that and believe in it, and believe in doing whatever it takes to keep the family unified. They all work in terms of the family, whether it's Best Buddies or Special Olympics, and they all are very dedicated to our parents. They've been molded by both of them. They get along incredibly well. Whenever we get together, they hang out together. They are bonded in a way that I have rarely seen with brothers. They all have individual relationships, and as a whole they are together as a pack."

Their struggles for identity were not preordained to be successful. For Bobby Shriver the battle for a separate identity was an almost overwhelming endeavor that he dealt with in part by moving across the country. It took all that geographical distance from his mother and father before he could begin to have psychological distance. Tim had begun his adult life living far from his parents or an identity as a Shriver or a Kennedy, but he had returned, a dutiful son, but he knew that one day he would have to leave again to assert his own life somewhere else.

Mark, the third of the four sons, neither suffered through these complicated struggles over the meaning of family and identity nor even fully comprehended them. He had the happiest and least emotionally complicated of any of the four brothers' lives. When Tim had returned from his summer trips with tales of life on a South Dakota reservation or in a mountain village in Guatemala, he hardly mentioned the difficulties to his younger brother but played up the excitement of it all.

Mark had been enough younger that his parents were home much more than they had been for his older brothers. He was ten when his father ran for the Democratic presidential nomination in 1976, old enough for it to be fun walking with Sarge through the Faneuil Hall Marketplace in Boston as people called out his name and gave him fruit, but not so old that he felt the pain of his father's loss. As he saw it, he had been blessed with a noble heritage, a deep faith, a happy childhood, and a marvelously loving mother and father. "They do what they think is right and they don't blow their own horns," reflects Mark. "And these things work their way out. People give you credit eventually anyway. There's no point in going out and clamoring for it. And they don't."

The one aspect of Mark's personality that might have seemed incongruous was an athletic aggressiveness that stood out even in the Kennedy family. In those football games at Hyannis Port that often set the Shrivers against the Robert Kennedy sons, Mark was the Shriver kamikaze. In the middle of the field stood a rosebush that the other players made a point of avoiding. When Mark went out for a pass, he often wanted the ball thrown right at the bush. He didn't care if he risked a broken leg, as long as he held on to the ball. If the Shrivers lost, his brother Bobby recalls, Mark would sometimes be so upset that he would jump in a car and drive to Boston until he calmed down.

Mark had a bad knee, but that didn't prevent him from playing football his senior year at Georgetown Preparatory School. He would have gone out in previous years, but he was worried about blowing out his knee for good. As it was, he hurt his ankle badly enough early in the year to be moved from quarterback to defensive back.

"Mark was a super team player," says coach James G. Fegan, who could have been Pat O'Brien's understudy in *Knute Rockne: All American*. "He

wasn't big, he wasn't extremely fast, but he was always where he was supposed to be and doing the things that he was supposed to do. And there were
some kids playing behind him who had more God-given talents."

Fegan coached the sons of many prominent politicians, businesspeople,
and journalists, but he had never seen parents quite like the Shrivers. It was
not simply that they were there for all the games, but that they were a distinctive presence. At one game Fegan looked up in the stands and saw
Eunice taking notes on every play. It made no sense to the coach, and he was
fascinated enough to ask her about it afterward. "Well, you know," she confided, "I'm the offensive coordinator at Hyannis Port, and I'm just trying to
get some new plays."

After a victory against archrival St. Joseph's, the stands emptied out of
cheering, exhilarated parents and friends, all except for the Shrivers, who
walked through the aisles as if they were looking for something. Fegan
thought this was strange, until he realized they were policing the grounds
and picking up garbage.

Mark went to Georgetown Prep in part because it was a Catholic school.
In the previous Kennedy generation, the women had gone to religious
schools and the men to secular institutions, but that was changing. Mark may
not have contemplated becoming a priest, as Tim had, but he saw himself as
a Catholic man.

In the fall of 1983, Mark entered Holy Cross, a small Jesuit college in
Massachusetts. The Jesuits had always been the Church's greatest teachers,
and most of his education had been with them. "Clearly there was a frame-
work from which they operated, and obviously it's Christianity, it's Catholicism," Mark says. "And they were terrific. They were personable guys that
could teach you Melville and talk about existentialism over a beer, and
believe profoundly when the next day they got up to preach to you on what
Christ meant in their lives and what he should mean in your life. In the Jesuit
idea of manhood, there's not just constant action without contemplation, but
neither [is there] the extreme of just sitting around thinking about stuff all
day and [being] gripped by the burden of it."

Like his brother Tim, Mark was discovering that helping people was immensely fulfilling, a way to experience the broader reaches of the world. It was a crucial part of a full, happy life. In the summer of his freshman and senior years, he drove over to New Haven to work with Tim tutoring inner-city kids.

When Mark graduated in 1986, he worked as an intern in the office of Maryland Governor William Donald Schaeffer. The state was closing down many of the institutions that housed teenage offenders, and it lacked ways to help these youths once they got back in the community. Mark saw this as an opportunity to start a program helping teenagers make that transition so that they might avoid returning to prison as an adult.

Tim was the family expert on troubled teenagers. "You're going to have a hell of a tough time dealing with kids who have broken the law," said Tim, striking a cautionary note untypical for him. The reality was, as Tim also could have told his younger brother, that it was unlikely that a wealthy, untested young man could help with some of society's most intractable problems. Mark reflects that it took a combination of naivete and craziness to pursue his idea, but a more generous interpretation would be that he brought to it a grounded idealism and the kind of can-do/will-do optimism he had learned from his parents.

Mark did have advantages that a young man with another name would not have had, including an initial $44,000 grant from the Joseph P. Kennedy Foundation that along with other private and state monies gave him the $128,000 to fund the program initially. He had ample help from state officials to structure the program, but it was still a formidable effort to get enough funding from the state, foundations, and private industry to put together a pilot project in the Cherry Hill section of Baltimore. The only white men who usually walked these streets were cops, process servers, and bill collectors, and this was hardly an easy way to have one's résumé punched as a social do-gooder.

Mark decided to call the program Choice. "The kids had choices, but they thought they were hopeless and helpless and had to break the law to get by," Mark says. For the first two years Choice was housed at PS 180, where Mark had a broom closet office not unlike the one Tim had had at

Hillhouse High School in New Haven. Mark had a number of "team coordinators," recent college graduates willing to work in the Baltimore ghetto for $17,000 a year. These were not traditional social workers or parole officers making occasional visits to their clients. They were more like big brothers or big sisters obsessed with their siblings. They dropped in on their charges night and day, goaded them back to school, pushed them away from bad companions. In the first months Mark had his own tough choices to make, including the firing of all his initial coordinators. "One guy lifted me up by my throat in a public school, a humongous guy with the same build as Mike Tyson—scared the crap out of me," Mark says. "I fired him after that."

Mark soon realized that despite everything he and his team were doing, some of these young people were probably not going to make it. He had to do what Tim had done at Hillhouse—kick out the troublemakers. When Tim kicked out incorrigible students at Hillhouse, he was turning them out into the streets and uncertain futures. When Mark threw out some of the Choice teenagers, he asked that they be locked up. "That was difficult," he says. "But after a while it's clear that you couldn't save every kid, and some were going to be successful and some were not. I wouldn't call them unworthy or whatever, but that's just the way it is."

Mark's draconian solution was made harder because he was the only white person working in the program. On his twenty-third birthday in February 1987, his staff brought him a birthday cake with twenty-two black candles and one white candle. Mark lived in a Baltimore row house with two roommates, including for a time Martin O'Malley, who is now the city's mayor. Taking took only two weeks of vacation in his first three years, Mark built Choice into a life-transforming program. The team coordinators had only moral authority with their charges, but in many instances that proved to be a more powerful force than any mandate the law could give them. "The kids know no fear," reflected Leon Dickerson, one of the coordinators. "They need limits." The vistas of these young people were so circumscribed that some of them had never traveled two and a half miles down to the Baltimore waterfront. The program helped teach them that they had choices they had never recognized and that they could travel farther than they ever imagined.

A t a Holy Cross football game in the fall of 1990, Mark met Jeannie Ripps, who had graduated the year after him and lived in Boston. Jeannie was not only pretty but had an exuberant personality that immediately attracted Mark. In June 1992, the couple married, and while Jeannie continued her work at Merrill Lynch, Mark studied for a master's degree in public administration at Harvard.

Unlike his older brothers, Mark felt no need to get away from his parents. After a year the newlyweds moved not only to Maryland but to Montgomery County, within a few minutes of his childhood home. Mark commuted to Baltimore, where he continued working with Choice.

Mark decided that the best way to expand Choice was not just to travel to Annapolis to lobby lawmakers but to become one himself. In 1994 he decided to run for the Maryland House of Delegates in Montgomery County. Two of his cousins were also running for office that fall, Kathleen for lieutenant governor in Maryland and Patrick for Congress in Rhode Island. Kathleen had begun publicly calling herself Kathleen Kennedy Townsend, highlighting the primary reason she was on the winning Democratic ticket. Patrick, for his part, was almost shameless in exploiting the family name. As for the thirty-year-old candidate for Maryland legislator, he was Mark K. Shriver, the middle initial like a monogram or designer label, subtly insinuating the Kennedy connection.

In the spring of 1994, Mark began knocking on the doors of his potential constituents, in the evenings after work and all day Saturday and Sunday. That summer, after taking a leave of absence from Choice, he went from morning till night seven days a week. Before he was done, he announced that he had knocked on twenty thousand doors. He had knocked on more than that, but one of his friends said nobody would believe him.

Mark created an image of himself as a populist campaigner, though he raised more money than any previous first-time candidate for the House of

Delegates, over $138,000, while most candidates got along on about $15,000. He had seven hundred campaign workers. His parents knocked on doors too. "I'm not running for president, Mom," Mark told Eunice, unable to slow her efforts. As for Sarge, he would go into a home and still be there thirty minutes later, caught up in an intense conversation that he would write about in memos that Mark kept as his most cherished campaign souvenirs.

Mark handily won the election and went to Annapolis as the most scrutinized new member of the Maryland legislature. No matter how many doors he knocked on, or how large his contributions to society, there would always be some who saw him as little more than the spoiled son of a decadent family. Mark subtly ingratiated himself with his new colleagues. If anything, he was almost too gracious and deferential to his senior colleagues, and too willing to tip his hat to others.

Although Mark was unfailingly civil, at times he exhibited his impatience over trivial or inconsequential matters. He was full of nervous restlessness in committee hearings, sometimes popping out of the room or, if the subject was too tedious, not even showing up. That behavior irritated some of his colleagues, but for the most part he was liked as a politician who was paying his dues, happily becoming one of House Speaker Casper Taylor's trusted deputies, and never taking credit for accomplishments that were not his.

Mark thought of himself as a lobbyist for people who didn't have lobbyists. His proudest moments were when he heard about an issue from troubled constituents and then worked to create legislation to resolve their problems. He authored a bill that got delinquent fathers to pay back child support by revoking their driver's licenses. One couple with adopted children told him that they couldn't take a leave from their jobs the way birth parents could. That didn't seem fair to Mark, and he worked to pass legislation to end that distinction. After listening to other constituents, he became a leader in pressing for more money for early childhood education, including an innovative bill that provided funding for after-school programs. Many of his cousins would have considered a state legislature beneath them and their celebrated name, but Mark was glad to be of service.

CHAPTER THIRTY-SIX

Best Buddies

he Shrivers were as competitive as the other Kennedy families, yet when it came to those with mental retardation, they had a gentle, knowing empathy that was never patronizing. The Shriver children knew, even if their mother could never admit it, that this profound concern for the developmentally disabled came from Eunice's love for her retarded sister, Rosemary. Even before Rosemary was lobotomized in 1941, Eunice had been a second mother to her.

When Rosemary came to visit the Shrivers, as she often did, everyone in the family appreciated their time with her, but it was flighty, unpredictable Anthony who was closest to his aunt. Anthony spent hours with Rosemary, who, because of the botched operation, had the mental age of a five-year-old. Growing up, Anthony alone of all the Kennedy grandchildren flew out to visit his aunt, who lived in a small house on the grounds of St. Coletta, a facility for those with mental retardation in Jefferson, Wisconsin. As an adult, Anthony built special facilities for Rosemary in his home so that she would feel comfortable on her visits.

Anthony had the rakish good looks of Uncle Jack, his father's endless exuberance, and his mother's shrewd judgment on the uses of human beings. He had carefully coiffed hair, an easy smile, and the long face and piano key teeth of a Kennedy. Anyone who saw him at a party or in his endless put-downs of Mark, Tim, and Bobby would have found it hard to imagine that he would have time for an elderly, stolid woman who could barely utter comprehensible words.

"Rosemary was around a lot, and if you spend significant amounts of time with anyone who is profoundly or severely disabled, there's a certain sort of inner peace about them," says Anthony. "And it moves you in a really unique way. I find that you've really got to spend a lot of time, though. If you spend only two hours or three hours, you see them as someone who's handicapped and is nonverbal, or is not ambulatory, or has to wear a diaper. So with Rosemary it's one of the better experiences. We're best buddies, and I guess that sort of evolved as a young kid. Everything I do now goes back to what I did when I was a kid."

If Bobby had suffered the common fate of the oldest child in being the one his parents experimented on, Anthony suffered the fate of the youngest—being spoiled the way his four older siblings were not. The Shrivers wanted their children to have close friends, and they helped instill the value of such friends by inviting them along on trips or for extended stays at their homes. During Anthony's teenage years, his friend Jimmy Shay spent most of the summer at the Shriver house in Hyannis Port. Their days began with breakfast and an hour of reading, followed by sailing lessons, lunch with Eunice, swimming and other athletic lessons, and a late-afternoon softball game, the sports activities overseen by a professional coach. There was a ten o'clock curfew until the boys were sixteen, a tough hour to have to run home from an evening with your friends.

"On occasion we would blow off curfew," Shay recalls. "We would be running around town, and when we got home there would be notes on a yellow legal pad with Mrs. Shriver's familiar handwriting, 'Boys, 10:05. You're still not in. Come and see me when you get in.' And then further up the stairs. '10:25. Still not in. Be sure to wake me.' '11:15. Haven't been able to sleep. Where are you?' We know we better go in and wake her up. And she would come down the hall and say, 'I want you to say your prayers.' And we'd kneel down beside the bed and say our prayers. And then it was lights out, no talking. And if there was talking, she would sit outside the door. And she would usually have her rosary beads with her, and she would just say, 'It's getting late. You have a big day ahead of you, no talking.' You know, and it was just discipline. It was discipline all the way."

Anthony reminded even his mother of her beloved Jack, and her youngest son had chits that his siblings did not have. "Anthony could do no wrong," says Shay. "So if something went wrong, you were blamed and Anthony was exonerated. If you're with him, you're in trouble. All she's going to do is talk to you, but you've never been more fearful of someone talking to you in your life, because she's going to give you a dressing-down like you have never had in your life."

Anthony matriculated at Catholic Georgetown University. For a time he worked as a banquet waiter under a pseudonym, but for the most part he saw the Kennedy/Shriver connection as an immensely useful tool to advance him in life. "If you had the choice of either being it or not being it, I don't know many people in the world that would rather not be it than be it," says Anthony. "So I mean, I find it to be excellent."

Anthony was a mediocre student more interested in a pickup game of basketball or working out than studying. If there was one quality that set him apart from his peers, it was not his Kennedy blood but his friendship with his Aunt Rosemary. He gave much, and in return he received an empathy that was the most useful tool in his life. He had been part of the Big Brother program at Georgetown, mentoring underprivileged youth. While a sophomore in 1987, he decided to start an organization in which college students could have their own experiences with a developmentally disabled friend like Rosemary. He called the program Best Buddies and started chapters at Georgetown and at Catholic University. He was an astute student of human nature, and he succeeded by turning Best Buddies into "a cool, hip thing to do on campus."

In the spring of 1989, the newly graduated Anthony decided to turn Best Buddies into a national organization with programs in colleges across America. Unlike his brothers, twenty-three-year-old Anthony was shameless in the way he played the Kennedy card, flashing it wherever it would bring in money. Anthony was an immensely shrewd promoter who developed a money-raising strategy that he rarely varied. For each charity event, he needed a wealthy benefactor willing to pay royally for having his name linked with the Kennedys and Shrivers. He needed a number of celebrities, television stars, football players, track stars, artists, and movie stars; B list or not, they added to the mix. He needed publicity in the newspapers and hundreds of people who would be delighted to pay a few hundred dollars for an evening.

Anthony's Uncle Jack had famously said in his inaugural address: "Ask not what your country can do for you; ask what you can do for your country." Anthony had deep insight into the motivations of the wealthy. He saw that a more accurate admonition would be: "Ask what you can do for your country that will benefit you as well."

"There's nothing for free in life," reflects Anthony. "So if you're just expecting someone to write you big checks and they get nothing in return, it's not going to happen. You know, people don't just swarm into Best Buddies events because they care about those with mental retardation. To a lot of people it has some attraction because it's a Kennedy thing and everybody's attractive and there are money people around and they can network and go to fancy homes and see fancy stuff."

When Anthony started developing Best Buddies, he worked out of a cluttered office at Special Olympics, just down the hall from his parents' offices. He could not envision himself, however, living in the shadow of Sarge and Eunice. So early in 1992 he moved himself and Best Buddies to Miami.

South Florida was a perfect place for a brash, bold, sports-loving young man. "The people I know here are my people," he says. "I can get tickets to little things like the Heat games. I get those because of me, not because the guy is a friend of my dad's. And I come here and I introduce my parents to people and I take them around and I show them stuff. We all get our fulfillment in different ways, and a lot of people wouldn't need to have that kind of reassurance or build their self-confidence that way, but I do."

Shortly after he arrived, Anthony met Alina Mojica at a cocktail party given for him by a friend. For a man of Anthony's background, Alina was exotic, a Cuban American and former ballet dancer, with a baby at home from her first marriage. The men of his father's generation would not have married a divorced woman, but the younger Kennedys lived in a different world. The couple married in June 1993. Anthony adopted the baby and was an affection-ate, hands-on father to his stepson, as well as to the two children who followed.

As a fund-raiser, Anthony had mastered the technique of first shaking a hand and then opening his palm. He was extraordinarily good at it, but as much as he believed in Best Buddies, he sometimes tired of always hitting wealthy folks up for money. When he went fund-raising, he saw a world of luxury beyond that of his parents or most of his cousins. He fancied himself with his own private jet and the ability to write a check to buy the Miami Heat. He spent about half of his time on various business ventures, hoping to make his own fortune.

Most CEOs of a charity would have considered it unthinkable to approach their donors to become part of their personal business deals. But Anthony figured that in this world of mixed motives that was the best way to succeed, and nobody at the Best Buddies events did a better job of network-ing than the chairman. Anthony did not draw a salary from Best Buddies, and he utilized major contributors in his business deals. He sometimes hit pay dirt but unfortunately, a series of ventures did not work out, including a company to sell prescriptions on the Internet and a construction business. Investors lost money, and Anthony sought more chips to go back to the high-stakes table again.

"I'm not sort of going after the little things where I'm going to make fifty thousand or one hundred thousand dollars," Anthony says. "I'm trying to do things where you're going to make ten million dollars or you're going to make fifty million dollars, you're going to make one hundred million dollars. Bigger deals take more time and they're more challenging, and the chance of failure is much more significant. So if you're shooting for the stars and you're trying to make it big, you're not going to get it every time. But I'll get it. Eventually I'll get it."

"If my mother gave me five million dollars in venture money to start something off and it was successful, I would feel very different than if I went

out and raised five million dollars," says Anthony. "So it's harder, and sometimes I wish I just could call my mother and say, 'Can you give me three
million dollars just to get this thing moving?' And it'd be easier and make
my life happier, and I could hire a bunch of people. I could get the thing
going. I know it could be successful, because I know I'm capable and I know
I'm a smart guy. I know I will get something big, and it will hit and it'll be
worth a ton of money. I know that will happen. So it's a matter of just time."

Anthony was more the natural politician than his brothers. He twice
seriously considered running for mayor of Miami Beach. He was gregarious,
with a memory for names and an ability to ingratiate himself with almost
anyone. He had an in-your-face candor that was a tonic to those tired of the
bland proclamations of many politicians. And he had a Cuban American
wife in a city dominated by Latinos.

Yet there was a perpetually adolescent quality to Anthony, and a
passive-aggressiveness that popped out of him at the most inopportune
moments. In early 1993 he had shown up uninvited at a Palm Beach party at
Donald Trump's Mar-a-Largo estate. He drove his Jeep Cherokee across the
exquisitely kept grounds, scattering guests and damaging the lawn.

"I was having a good time," Anthony recalls. "I guess I'm probably not
the first guy that's driven around the lawn there. I drove through a couple of
bushes. Not the end of the world, I don't think. Trump didn't like it. He
called the police. I talked to him, though, on the phone after it. I think he likes
to create excitement, and the more excitement and stuff that happens at Mar-
a-Largo, the better off he is. He was pretty gracious about it afterwards."

Anthony did not drive a car through the crowds at the Best Buddies Ball,
but he insisted year after year on making fun of Tim, mocking him for the
dress shirt he always wore emblazoned with bug designs. The put-downs
might have been amusing at the Shriver dinner table, but most of the guests
knew very little about the Shriver sons, and Anthony's antics took away
from his opportunity to explain and expand Best Buddies.

Anthony always had to show that he bowed to no man, though it was
also one of the themes of his life that everyone had to pander to other people. "I have a job where I have to kiss a lot of people's asses," Anthony says.

"Everybody kisses ass. Arnold Schwarzenegger [married to Maria] always claims how he never kisses anybody's ass and he's a big shot and he's got money and he's a superstar, yada yada yada. We went to this Special Olympics thing on the lawn of the White House when George Bush Sr. was president. I was watching Arnold walk around, kissing George Bush's ass like I've never seen him kiss anyone's ass in my life. And I'll never forget it. I turned to my brother Bobby and I said, 'You know what? There you go. Everybody kisses someone's ass. Has to suck up for something.' "

U nlike his youngest brother, Tim did not want to become dependent on his family name to do good in the world. After the 1995 Special Olympics, he stayed on in New Haven, closing up the accounts while working on his Ph.D. thesis. He loved living in New Haven and contemplated running for mayor.

In Washington, Special Olympics International had lost its CEO, and the organization was in the midst of a rough transition. With a budget close to $200 million and about a million participants in 150 countries, Special Olympics had become one of the largest international social organizations in the world. Tim's father may have been the chairman of the board, but it was his mother who was the all-powerful force. This single-minded woman held on so tightly to the organization she had founded that it could not become an institution larger than her. Eunice was unable and unwilling to see that the expanding organization needed professional management. Anyone who tried to take the necessary steps would have had to sidestep her authority and in doing so was bound to be struck down.

Myer "Mike" Feldman, chairman of the executive committee of Special Olympics, headed the search for a new CEO. "I tried to get somebody other than a Shriver, and we hired a consulting firm to find someone," he recalls. "They gave us fifty names, and it didn't seem that any of them really fit. That's why I said finally, there is only one person, Tim Shriver." The selection might have appeared to outsiders as blatant favoritism, but to the former Kennedy administration deputy counsel, it made irresistible sense.

"I asked him first of all to come down even just part-time," Feldman recalls. "We needed someone, and three times he rejected the whole idea." As a young teacher in New Haven, Tim had forged his own identity while remaining true to the best of his family's values. Now he was being sucked back into his parents' world. "No matter what happens, it's always Mother's," reflects Anthony. "I've said that to Tim. He knows going in that this is Mother's thing. It's Mother's thing for the rest of history."

Tim understood that perfectly well. "I came because I feel I could make more of a difference than anyone else during this period of time," he says. "I feel like I know what I want to do here. I feel like I know what I can do here. I feel like it's an important contribution. But I'm here in part as a member of a family, not as an individual. And there are positives in that and negatives."

When Tim took over as CEO of Special Olympics in the fall of 1996, there was an undertone of resentment among many on the staff. While they worked hard for modest salaries, they worried that the organization was being turned into a lucrative sinecure for the Shriver sons. Former school-teacher Tim was being paid $200,000 a year, two-and-a-half times his previous salary. Even some of those who supported the move looked on it cynically as a way to maintain the Kennedy/Shriver identity so useful in fund-raising. The staff's disenchantment increased when Tim sent out peppy memos that some of them found embarrassing.

Tim did not have the flamboyant charisma of his big brother Bobby or of young Anthony, both of whom had inherited their father's engaging public manner. He would always have something of the high school social studies teacher in him. If people were at times less than impressed with him on first meeting, it was in part because it was rarely his intention to impress; he preferred to put forth his well-defined agenda or merely to listen. He had learned in New Haven to work with all kinds of people. By listening and learning, he slowly assuaged the doubts of the staff in Washington and learned the magnitude of the problems.

Every four years thousands of Special Olympians arrived from around the world, but in most of these countries there were neither year-round activities nor broad awareness of the ideals of the organization. People looked to the Special Olympics office in Washington not simply for inspiration but for leadership, and it was unprepared to give it. "We couldn't respond well when these countries had challenges," Tim says. "We just didn't have the capability. We'd get faxes from places, they'd go into black holes. We understood that Special Olympics was life-transforming And yet we didn't have the capabilities to really take that message to scale."

The new president tried some organizational reforms, but these efforts were ineffective and just the kind of half-failures that would have aroused the ire of Eunice if anyone else had been in the CEO's office. Tim had freedom no one else would have had. Beyond that, he had the self-confidence to recognize his own shortcomings, to understand his own limitations, and to not be afraid of them.

To Tim, it was clear that Special Olympics needed to bring in some high-level consultants. Tim made a cold call to Bruce A. Pasternack at his office in San Francisco where he was a partner at the Booz Allen consulting firm. Tim had just read and admired the book that Pasternack had co-written, *The Centerless Corporation: A New Model for Transforming Your Organization for Growth and Prosperity*. He wanted to see if the consultant would help. Pasternack, whose two daughters taught ice skating to Special Olympians, did not have to be convinced of the value of the organization. And though Pasternack had served in the Ford administration, he had voted for the McGovern-Shriver ticket in 1972 and admired the Shrivers. The two men talked long enough to know that they had much in common and decided to continue the discussion in person.

Tim had been brought up in a political family with a prescient understanding that life is a matter of mixed motives. To get a person to do what you want him to do, you must find something that he believes will have equal benefit to him. In this instance, Booz Allen agreed to commit resources and time worth probably $2 million in a reorganization whose goal was to double the number of athletes by 2005. In exchange, the company wanted make a videotape of parts of its efforts to play at company meetings and at colleges and help brand Booz Allen as a consulting company with an altruistic limb.

With Booz Allen's help, Special Olympics decentralized, forming four regional offices across the world. Pasternack joined the most diverse board in America. The members included Eunice, Sarge, and several others of their generation who remained emotionally committed to the traditional themes of Special Olympics; Special Olympians who were not tokens but active participants; academicians with expertise in the developmentally disabled; foundation people, corporate leaders, professional athletes, journalists, and politicians too. With so many people and so many perspectives, it could have been a cacophony of voices and interests, but Tim was able to orchestrate it. Special Olympics now had not only a great vision of itself as a truly international organization but a well-marked road to get there, with the prospect of a million Special Olympians in China alone and a different attitude toward mental retardation throughout the world.

Tim bought a house near his parents and Mark in suburban Maryland. He worked hard, but he was far from being a workaholic—he enjoyed weekends with his wife and five children. His $200,000 salary would have looked good to most Americans, especially considering that he had a trust fund that brought in at least another $50,000. But he also had five children in private Catholic schools, and his income was not a lot compared to what the CEOs of many other major charities received. Yet every time the board tried to raise his salary, he refused to accept it.

Tim lived far better than most Americans, but he had always had the most modest lifestyle of all his siblings. Maria flew around the country in a private plane, Bobby lived high, Mark walked away with $1 million from his short stint in the telecommunications industry, and Anthony had his own cook. Tim worried about getting a good deal on a four-year-old car.

Adrenaline Addicts Anonymous

obby Shriver had the frenetic, in-your-face demeanor that made him a natural New Yorker. He enjoyed living in the city and earning embarrassingly large amounts of money as an investment banker, but in 1991 he decided to start anew in Los Angeles. He felt he was moving west not in the proverbial quest for new opportunities and freedom but to save himself, to move far enough away that he might finally reach beyond the shadow of the family. He had had many advantages growing up a Shriver and Kennedy, and his struggle to establish his own sense of self may have sounded like a psychological cliché, but that did not make it any less powerful an emotion. In his mind it took the broad expanses of the continent to give him enough room to have an identity apart from his family.

Bobby arrived in southern California with enviable advantages. Los Angeles was home to many of his old newspaper friends as well as to Maria, his journalist sister, and Arnold Schwarzenegger, her movie star husband, and from the day he arrived he was part of elite Hollywood circles. Bobby's grandfather had understood that for a man to become wealthy in America he needed to gain proximity to those who held it. Bobby had contacts not only thanks to his family connections but far beyond that.

Back in the summer of 1980, while at Yale Law School, Bobby had clerked at the prestigious Washington law firm of Williams and Connolly. He had not only worked with family friend Edward Bennett Williams but with Larry Lucchino, president of the Orioles and a partner at the presti-

gious Washington law firm. Like both Shriver men, Lucchino was a Yale
Law School graduate. He had also dated Bobby's sister Maria. In 1988
Lucchino called Bobby to ask his help in buying the team from the dying
Williams. The partnership would include Lucchino, Bobby, and an outside
figure who would provide the money. Lucchino alerted Bobby well before
Williams expired from cancer. A number of other suitors were approaching
the owner and Lucchino hoped to be firmly ensconced as the new president
and CEO.

Bobby presented the investment to several wealthy individuals, including Ronald Perlman and Peter Peterson, both of whom had turned the idea
down. But industrialist Eli Jacobs, yet another Yale Law School graduate,
had expressed enthusiastic interest. Bobby worked with Jacobs hiring
accountants and rigorously investigating the team's finances before making
an offer. The final $70 million deal included a 4 percent slice of the team for
Bobby, who gave one of the points to his father. Washington cognoscenti
figured that Sarge had been the rainmaker, making a few phone calls for
which he was richly rewarded. But the truth, as Lucchino told the *Washington Post*, was that Bobby "pulled the eventual pieces together." When the
team was sold, Bobby ended up with about $5 million, far more of his own
earned money than any of the other young Kennedy men.

In Los Angeles, Bobby set up offices to make Special Olympic records
and was paid an annual salary of $125,000. In the next decade he produced
four more Christmas albums, raising an additional $40 million. He corralled
many of the top musical artists in America to give their time and creativity to
Special Olympics. Performers are inundated by requests from charities, and
it was a singular measure of Bobby's persistence and shrewd understanding
of human nature that he succeeded so well for so long.

Bobby's work was little appreciated or understood, however, by many
back at headquarters in Washington. Brash Bobby barreled through the
office with his L.A. attitudes, flaunting his superiority. As the development
staff saw it, all Bobby seemed to have to do was make a few calls to his
celebrity friends to produce the records while they spent day after day dialing for dollars. Bobby gave no indication of how hard he worked to get

those records produced, and it was understandable why so many at head-quarters considered him an obnoxious dilettante.

Bobby worked with Bob Bookman, a fellow Yale Law School graduate and CAA agent, to create an idea for a spy movie. Bobby loved James Bond pictures. When he settled on what he thought was a workable idea, he came over to his sister's house to show his brother-in-law a French comedy, *La Totale*, about a secret agent living a double life. Arnold liked the idea and hopped up with high-tech flourishes, the film became the 1994 hit *True Lies*, with Bobby credited as executive producer.

Bobby did not develop a serious career as a producer in part because he didn't work that hard at it. He spent most of his time on the Special Olympics records. In 1997 he bought a house in Santa Monica not five minutes from Maria and Arnold's home. He had a reputation as a charming bon vivant who gave some of the most eclectic parties in Hollywood. Arnold and Maria would almost always be there. Warren Beatty and Annette Bening. The late Herb Ritts, the famed photographer. Bobby's friend from the *Times Herald* days, Wanda McDaniel, and her husband Albert S. Ruddy, producer of *The Godfather*. Russell Simmons, the hip-hop impresario. Beyond that, there were rock 'n' rollers, movie stars, writers, journalists, social activists. During these evenings, Bobby ran from group to group throwing in a provocative quip or confrontational phrase that would heat the moment up a bit. The evenings were dinner at the Shrivers bottled up and shipped west.

Bobby had a string of girlfriends, but none of them lasted, and afterward he would usually make the woman a friend. He was the center of an intense, caring group of friends. He was a great friend, but he was not an easy friend. He might call you at two in the morning, or his number would pop up on your cell phone in the midst of an important meeting. He showed up so late sometimes that his friends would stomp out of the restaurant where they'd been waiting.

Bobby tried to advance his friends in the world, as did his sister Maria. In 1988 Jackie called to say that her sister Lee Radziwill, who was working for Giorgio Armani, was looking for people to open the first designer outpost on the West Coast. Shriver friend Wanda McDaniel was hired, and thanks in part to her inspired public relations, Armani pioneered the fusion of fashion and entertainment, and the clothes became practically a uniform for sophisticated stars. "All I know is that the monumental relations with this key group of people in my life started sharing a telephone with Bobby when we were working at the *Herald-Examiner*," McDaniel says. "Maria is one of my closest friends too, and I don't even remember if it was Bobby or Maria who called about the position. All I do know is that all things in my life trace back to Bobby."

Sometimes more than a new career was at stake. Bobby met Russell Simmons when he was working on the first Special Olympics record. He went to the young president of Def-Jam Records, and talked with him about putting Run-D.M.C. on the record. At that time many Americans considered rap little more than the crude call of the ghetto, not mainstream music that deserved to be represented on a Christmas album by leading artists. Bobby realized that hip-hop was breaking out, becoming the most important development in popular music in several decades.

Simmons had not become the most important African American in the music industry since Motown's Berry Gordy without being a quick judge of people. Simmons moved to L.A. in 1992, and he and Bobby began spending a lot of time together. "Bobby took me to yoga, and yoga changed my life," Simmons says. "He's a dear friend, one of the best men at my wedding, but the real thing is that he helped me find a path not only in yoga but in life. He reminds you of real priorities, service to others. I used to help other people, but he taught me about service. He knows that's what pays the bills, the heart happiness thing, he knows that's what makes you happy. He focuses on the real shit. People don't know the spiritual thing that we all have, but he knows it. Bobby said to me, 'You want to talk about your 30,000-square-foot house and your new Rolls-Royce. But, Russell, what about the starving children? You're not here for the 30,000-square-foot house. You're here for the starving children.' "

Russell saw Bobby almost daily at their yoga class in Santa Monica. Like most of Bobby's closest friends, he was not about to speak well of him without also giving him the shiv. "There's a funny thing about his big funny Catholic ass," Simmons said, sitting with Bobby at a Starbucks a few minutes before class. "The fact that he goes to yoga and reads the scriptures, he's not a good Catholic. That's a good thing to put in the book. He's got this happy look, tolerance, love child. But he's not adhering to any of this doctrine."

W hen Bobby was on one of his manic jags, those who did not know him were often convinced that he was on drugs. In November 1999, some of those who saw him promoting the new record *A Very Special Christmas Live* on *The Today Show* were convinced he was high.

"We should mention that you've just come from Howard Stern," Katie Couric said as she introduced Bobby, not telling the audience that he had rushed into the NBC studio seconds before airtime.

"I was a little wound up because I walked into the green room and there was a naked woman. Then I came on the air, the only thing the guy asked me about was, like, 'How much money did you inherit?' And they put down six pieces of paper, saying, 'Now, here's a list of all the girls you've dated.' And I go, 'Oh, my God! It's going to be a long morning.'"

Couric was well prepared for the interview, but it was hardly a natural transition from a naked woman to *A Very Special Christmas Live*. Couric wrenched the subject back to what it was supposed to be. Bobby ranted on, the words piling up one over another.

"Okay, stop talking," Couric said finally. "We're not on Howard Stern. Hold on."

"We're not on Howard Stern?" Bobby said in mock relief. "Thank God!"

Bobby pitched the public to go out and buy the CD as if he were selling Vegematics on an infomercial. "All right, Bobby Shriver," Katie said finally, trying to ring the curtain down.

"Thank you for having me, Katie. You look pretty in black."

"Limit the caffeine," Couric said.

"I haven't had any caffeine. That's the funny part."

395995

H is adrenaline *is* his drug," reflects Mary Ann Dolan, one of Bobby's closest friends. "He doesn't need a drug. I've had so many people come to me so often and say, 'You've got to do something. Bobby's really, really in trouble now.' And I say, 'You know what? I swear to God, the man will never need a drug as long as he lives.' That said, there ought to be an AA for him, Adrenaline Addicts Anonymous."

Bobby did not anesthetize himself with liquor, drugs, or self-pity. He was a man of deep spirituality, but it was born of skepticism and intellectual struggle. He read deeply, and he discussed questions of God with the utmost seriousness. In Los Angeles he had found his only addiction, yoga. If he did not have his daily hour-and-a-half session, he was back to his nearly out-of-control self.

When Bobby was working on a project, he could be brutally blunt and impossibly demanding, truly his mother's son in the endless impositions he made on those around him. When Dolan worked with him on one of the Special Olympics television productions, she came close several times to walking away from Bobby.

"Why are you doing this?" Warren Beatty asked Dolan. He was a close friend of both of them, yet he didn't understand why people tolerated Bobby's excesses. "Why do you do this? And why do you care about him?"

"Because he is who he is, and he's an important creature on this earth. And he will do unbelievably important things, which people may or may not understand. And I love him and I see what he is evolving into."

"Bobby will never be who he is until his mother dies," Beatty said.

Bobby had already begun to make that journey while his mother was alive. He did so by learning what so few learned: how to tell Eunice no. Anyone who was a longtime friend of Eunice set up certain boundaries, but for a long time Bobby couldn't do that to his mother. When she wanted something from him, it rarely had to do with a personal matter but instead was about some transcendent cause so important that surely he would set aside whatever he was doing and help her. He learned finally that he didn't have to take his mother's call when he was busy, and that life went on and the love between mother and son was not diminished.

For all Bobby's shrillness, he was also a quiet soul. He was quiet in that he did so many things that nobody ever knew about: advising his family about financial and political matters; managing money for Special Olympics, raising millions with the Christmas records. He was quiet too in that few people understood his spirituality.

Bobby did not have his mother's certain faith. One of the books that helped him and that he passed out to friends was Stephen Mitchell's *The Gospel According to Jesus*. Mitchell prunes away all the words and sayings falsely attributed to Christ and attempts to show just what Jesus said and how he lived. This is a Christ who is the son of God but also the son of man. This is a Christ who as a child is considered a bastard and suffers the ostracism visited on illegitimate children.

"He would have always been feeling badly about it, like, 'Who is my father? What is that about? How come people don't like me?'" Bobby reflects. "When he goes to the Jordan River, the voice from Heaven says, 'You are my son.' The primary emotional thing in his life is that I'm a bastard. And suddenly, 'I'm your father. You are my son.' And really, it's the real thing. Everybody has a thing that they are kind of ashamed of, humiliated about, whatever it is. And the force of God lifts from you your shame."

CHAPTER THIRTY-EIGHT

Michael's Way

ith his large teeth, lean frame, and tousled hair, Michael looked more like his father than any of his brothers. That may have been part of the reason his mother seemed to favor her fourth son over the others, but it was more than that. Michael was a spirited mix of all the qualities a son of Robert and Ethel Kennedy was supposed to have. He was as dogged and strong as Joe, as intelligent and witty as Bobby, and as sensitive as his late brother David. Nothing seemed hard for him, from the brutally difficult obstacle course at Hickory Hill that stymied even Redskins football players to his studies, first at Harvard and then at the University of Virginia Law School.

In this family where athleticism and physical courage were the mark of a man, Michael had no equal. On the ski slopes, there was rarely another skier like him on the double diamond runs. "He was the best skier I've ever seen," says Lorenzo di Bonaventura, his freshman roommate at Harvard and one of his closest friends. "He had this spirit of willingness. He was always ready to go. We'd be standing there at the top of some incredibly difficult run, and I'd be willing to go, but first I'd think about it. Michael would just go for it. He carried that attitude in everything in his life. He had an unquenchable thirst to experience every sensation to the max." In the two decades during which the two friends went whitewater rafting on some of the most difficult rivers in the world, di Bonaventura recalls only one time when Michael agreed to walk around dangerous rapids, and that was only after a lengthy discussion.

Michael had met his wife, Victoria "Vicki" Gifford, when he was a teenager at St. Paul's School in Concord, Massachusetts. She was a beautiful young woman who, when she married Michael in 1981, had never dated anyone else. Her father was the football star and sports commentator Frank Gifford, her stepmother the television personality Kathie Lee Gifford. Michael and Vicki had three children, Michael LeMoyne, Kyle Francis, and Rory Gifford, and what to all appearances was a deep, true marriage that seemed to refute all the sordid stories of the Kennedys and their compulsive womanizing.

When Michael graduated from law school in 1984, he joined his brother at Citizens Energy. The organization was growing rapidly, spinning off a number of profit-making subsidiaries that were supposed to pour a stream of money into the nonprofit companies for social beneficence. Soon after he arrived, Michael set up a meeting with John Rosenthal, who was working in his family business, Meredith Management, a major Boston real estate firm. The two men talked about one of the profit-making offshoots of Citizens Energy that had a program for low-income apartments. Meredith Management owned a large apartment portfolio, including a four-hundred-unit apartment complex that would be the biggest project that Citizens sought to retrofit with energy-efficient equipment. The two men struck up an immediate rapport and began seeing a great deal of each other. "We completely hit it off, in part because of this for-profit, nonprofit balance that he tried to maintain, and I was doing the same thing," Rosenthal recalls. "We became best friends."

Michael was impressed by Rosenthal's life story, in part because it was the sort of autobiography that the young Kennedy was supposed to be living himself. Rosenthal had dropped out of Syracuse University to become a leading antinuclear power and weapons protester. He had spent three months in jail for civil disobedience, an experience that to Michael was a way of proving your political manliness.

Michael also enjoyed Rosenthal because he was an excellent athlete, ready to go rafting, kayaking, skiing, or sailing with Michael at the hint of an invitation. As Rosenthal first grew close to Michael, he realized that he hardly existed to those around the Kennedys. "I was enamored by the Kennedys," says Rosenthal. "Who wouldn't be growing up in Massachusetts? But then I saw the downside, the price that you pay. I started going on those rafting trips with all of his close friends and not-so-close friends, with people who knew him better or were related to the family, and there were various other family members. There was a real sense of family and adventure camping and rafting. I recognized that I was new, but I felt like I had accomplished some things in my life even by then, but no one seemed interested. No one, except Michael and Rory, appeared the least bit interested, because I didn't have celebrity status. And all these people were friends of Michael or friends of Rory or friends of Bobby, and it seemed like their lives revolved around the celebrity of the Kennedys. And it was just odd. I had never felt so invisible in my life."

Rosenthal was in some ways fortunate, since Michael and his siblings often invited one new acquaintance whom they could belittle. The idea was to make one person an outcast so as to foster solidarity in the rest of the group. There was a subtle undertone of fear in this device, since all but the most secure worried that it might be his turn next time, the last stop on the way out.

When Joe left for Congress, Michael and Wilbur James divided up the Citizens empire. Michael became president and CEO of Citizens Energy, the nonprofit charitable end of the organization; James was president and CEO of Citizens Corporation, the umbrella company for the profit-making concerns. It was a natural division of the men's talents and interests, though Michael straddled both sides, earning part of his salary from work for the companies run by James.

Michael and Rosenthal saw even more of each other. Once Rosenthal settled back into Boston life, he began devoting about half his time to social beneficence, especially to an organization he founded that same year called Friends of Boston's Homeless. The charity bought a dilapidated crack house in downtown Boston for a dollar, then renovated it for about $200,000 into ten apartments for working homeless men.

Rosenthal knew that he was hardly the only one susceptible to Michael's aura, and he asked his friend to join the board of Friends of Boston's Homeless. Rosenthal pushed Michael to the front of the stage, acting as if Michael were his partner in running the charity. Michael had what Rosenthal called "political savvy and a true commitment to those less fortunate," but Rosenthal did the heavy lifting and Michael largely showed up for public events. That was the way it almost always worked with the Kennedys. With all the access, publicity, and excitement among contributors that Michael engendered, Rosenthal considered it a better-than-even trade.

It became obvious as the years went by that Rosenthal was a better exemplar of what the young Kennedy men were supposed to be than Michael himself. In a 2003 portrait titled "Mr. Nice Guy," *Boston* magazine described John as "a walking oxymoron: 'a businessperson with ethics, big time,' [who] attests to the rewards in this town of doing good."

Michael accepted that he was little more than the drawing card at a fund-raising party. He did not seem to grasp that that could have been the beginning, not the end, of his contributions. He appeared so brave on the ski slopes, yet he remained deeply fearful of moving the borders of his own family. He could have done his own unique work in a charity, left Boston, or stayed there and done something other than working for the company his big brother had started.

"John, look what you've done, look what you've created with the Friends of Boston's Homeless," he told Rosenthal. "You have really done it. And you're a no-name person. You just did it. No matter what I do, everyone will always say it's because of my name, not my individual accomplishment. You must feel so good!"

I n 1989 Joe's old friend and the president of Citizens Corporation, Wilbur James, told *Boston Business* that "access is the key to our ability to survive these days. Frankly, this company would not be here today if it weren't for Michael Kennedy." The nonprofit company that had begun bringing oil to some of the Massachusetts poor had in 1988 delivered no fuel oil at all. Citizens Energy had moved on to a worldwide series of deals in which the Kennedy name was the crucial advantage.

Michael had been visiting Angola since the early 1980s, and it was the most important foreign country in the Citizens Energy plans. The southern African country had been in a civil war since 1975 that in the next quarter-century would take as many as one and a half million lives. The conflict was one of the last tragic spillages of the cold war. The Marxist-influenced Popular Movement for the Liberation of Angola (MPLA), with its contingents of Cuban troops, controlled the capital of Luanda and most of the country. In the south the rebel forces of the National Union for the Total Independence of Angola (UNITA), supported by the Americans and South Africans, tried to push northward against a determined foe. The former Portuguese colony may have appeared rich only in blood and poverty, but the foreign businessman who frequented the capital knew that the country was also wealthy in oil and diamonds.

Michael and Citizens cast their lot with the MPLA. It was a shrewd move, for the leftist regime had few friends in Ronald Reagan's administration, and Michael became an unofficial emissary. When Michael's plane set down in Luanda, he was greeted with as much pomp and deference as the leader of an important nation. He often brought members of Congress with him, on junkets paid for by Citizens. They were the elected officials, not Michael, but the Angolans paid the politicians little attention as long as Michael was there.

In December 1988, Michael and Wilbur James brought along four members of Congress and Thomas Gagen, a *Boston Globe* reporter. "Wilbur seemed smarter than Michael," Gagen recalls. "He was a big business person interested in a way that Michael was not. They were both trying to reach out far beyond what they had done, and Wilbur was reaching the farthest." Wilbur was a lean, dour man with a gigantic mustache that made him look like an extra in a Pancho Villa film.

Citizens was trying to win the rights to exploit a major 1,833-square-mile oil concession. The profits from such a venture could easily have paid for fuel oil for all the Massachusetts poor and turn Citizens into a major

American foundation, if that was the intention. Citizens had teamed up with Conoco, which already had the rights to a concession of offshore oil in Angola; the American oil company was willing to form a partnership because Michael had such a close relationship with the MPLA leadership.

"Citizens deserves to get the Angolan contract," Gagen wrote in the *Boston Globe*. "Kennedy and James really have the best interests of the Angola people at heart and a reliable source of oil would assure poor people in Massachusetts of a long-term supply of inexpensive fuel. No matter what happens, Kennedy promises that Citizens will survive. The question is whether Citizens will thrive, and that decision is in the hands of the Angolan government."

Citizens did not win the concession, but Michael continued to push his Angolan connection, especially at the beginning of the new Clinton administration, when his connections were even more valuable. The organization continued ingratiating itself with the leftist leaders; the largest gift in Citizens' history was $1.5 million to build a college not in Massachusetts but in Angola.

James had a clear vision for Citizens that the Kennedy brothers did not. The Peace Corps veteran talked about turning Citizens into "a kind of modern, business-driven Peace Corps of the 90s," but what he was primarily concerned with was turning it into a modern business. Michael surely knew what was going on. Joe remained close to the operation through not only his brother but his dear friend James and several other of his closest friends at Citizens.

Joe says that he had no idea what was happening to the company he founded. "I made a very hard but very firm decision when I chose to run for Congress to sever everything," he says. "The only thing I learned about was what salaries were being made when I read it in the paper. I never discussed it." As unlikely as that may seem, Joe had tunnel vision about his life and pursuits, and he usually cared little about the past.

Michael was paying himself a generous salary, working up to $662,000 in his best year not including all of his lucrative stock options and other benefits. He had it all—the public perception of him as a noble, selfless do-gooder and a great deal of money in the bank. Joe, for his part, was living in a dumpy apartment on Capitol Hill and had nothing of the upscale lifestyle of his little brother. Citizens Energy had been Joe's idea, and it upset him that his brother and closest friend were making enormous amounts of money exploiting a company that he had started.

Citizens began selling off profit-making subsidiaries to companies that then hired the Citizens executives to run the operation. The first of these deals took place in January 1987, just as the new leadership team of Michael and James was taking over. Citizens sold Citizens Heat and Power to EUA Cogenax for about $1 million. The head of the Citizens subsidiary, Joseph Fitzpatrick, became the president of a new company that in the next decade grew 20 percent a year, reaching revenues of about $100 million by 1995.

This was the archetypal spin-off for numerous Citizens companies. Wilbur and Michael believed that they were better at starting up companies than running major corporations. Citizens should take its profits and move on. The Citizens executives who left to run the new companies had an incentive to get in on the ground floor with stock options in companies poised to take off dramatically. Although there was nothing illegal about any of the deals, it was striking that after the sales, one company after another became enormously successful.

The Medco subsidiary had taken off dramatically and with the rich deal that Joe had offered John Doran to run the company, the executive was becoming by any measure a wealthy man. The executive had not joined the organization as a social reformer, but to make a good living, and he was perfectly justified in reaping the benefits of his efforts.

The problem was that this wasn't supposed to be what Citizens was all about, and there was discussion about what to do. A lot of it focused on the idea "that you can't allow this to take place for anyone else in the enterprise," recalled Larry Kellerman, then the president of Citizens Power & Light, another for-profit subsidiary. "This creates an environment in which the outside directors as well as the public think that Citizens only exists to create a platform for people to get wealthy."

Kellerman and others at Citizens saw a moral dichotomy between Michael, whom they believed was true to his brother's ideals, and James, who they saw as out to make himself rich. "Sometimes Wilbur claimed, 'Well, gee, you know, the Kennedys cheat me out of this and the Kennedys cheat me out of that.' " Kellerman said. "And one of his sayings was, 'Well, when I first joined this organization, Joe Kennedy and I cut a deal. Joe would get all the credit and I would get the money.' "

James may have rationalized being so much in Michael's shadow, but that did not make him happy. "Wilbur's got a major league ego which is one of the reasons why he's successful," says James Hilliard, a longtime associate of Citizens. "He's very smart, and I'd hitch my star to his wagon any day. There were constant disagreements between Wilbur and Michael. Wilbur was making the money that allowed Michael to give it out for charity, and Wilbur was getting no credit at all. Everyone referred to Citizens as the Kennedy company. Wilbur would say, 'Goddamn it, why do they call it the Kennedy company.' "

Michael was all for doing well himself, while James took care of the details involved in running the profit-making companies. In 1988 and 1991, Michael and James received the majority of the shares of the Medco stock options awarded to Citizens employees and executives.

When Medco was renegotiating its deal with Citizens, the board was told that the medical company was insisting that Michael, James, and Doran be tied with golden handcuffs, deals so lucrative that they could not afford to walk away. The board members were impressed and there is no evidence that one of the principals might have suggested the deal.

Most of the options—over 60 percent—went to Michael, James, and Doran. The options made the three men a fortune. The *Wall Street Journal* estimates that depending on when they and a few other Citizens employees exercised the options, the total would have been between $19 million, if exercised at the worst point, and $75 million, if they sold at the highest point and held on to all their unexpired options until 1998.

Their good fortune must be contrasted with the amount that Citizens gave to its various charitable activities, ostensibly the raison d'être for the organization. The first year, when Joe and a largely ragtag group of volunteers were running Citizens, it gave out about $4.13 million. Despite its massive growth under Michael and James, Citizens never once reached that total again. It was giving something just over $3 million a year by the late 1990s. Until 1997, when other scandals focused a harsh new light on Michael, every single major newspaper story, magazine article, and television profile lauded Citizens Energy as a nonprofit, with nary a hint of the apparently massive shift in the company's purposes.

S oon after Michael joined Citizens he began drinking heavily, and he was drinking like a man who did not know he had a problem. His thirtieth birthday celebration in February 1988 was a night out with the boys that included several executives from Citizens and one woman who worked there. The group was having a raucous time at a Boston restaurant when Michael and the woman went off together to a restroom. Michael returned first, waving his finger in front of the other men's noses, making them smell a whiff of sex. When the woman came back to the table, she accepted the smirks as a small price to pay for her time with Michael; she would become one of his mistresses as she continued working at Citizens. Michael's associates and friends laughed admiringly at his conquests, let him borrow their apartments for his assignations, and listened to his sexual boasts.

Michael had not one addiction but three: alcohol, sex, and danger. He treated his addictions as interchangeable, drinking until he was high, getting it on with a new sexual partner, taking extraordinary chances on the ski slopes, in his car, or out on a boat. Each addiction gave him a great rush followed by a crashing letdown that he got out of by heading into one of his addictions again, as often as not using his Kennedy name to indulge himself or get himself out of trouble. He wanted others to sense what he was doing and either to applaud or to tolerate whatever conduct he felt like exposing them to.

"I watched Michael slowly, as he became more powerful, how that power became intoxicating and he needed to drug and drink in order to sort of deal with that intoxication of power and opportunity and risk," reflects one close friend. "He just kept pushing the envelope because the envelope kept expanding."

Michael's friends and associates tolerated behavior that if he had not been a Kennedy they would have found mindlessly dangerous, hopelessly self-indulgent, or simply repulsive. Michael's wife, Vicki, saw things in a different way—or more accurately, she did not see things at all. The term "co-dependent" is in some ways an unfortunate one, for it suggests a kind of moral equivalence between the addict's actions and those of the person with whom he lives. That said, Vicki had all the classic signs of a co-dependent person. Her father had his own troubled personal history, having left his second wife and Vicki's mother for Kathie Lee. On her daily television show, Kathie Lee was a cloying Pollyanna bubbling about her perfect husband living with his perfect wife and their perfect children in a perfect house.

Like many children of divorce, Vicki wanted little more than to have a good loving husband and children, and she thought she had them. She was fiercely protective of all of them. Michael was drinking heavily, snorting coke, and keeping strange hours. It took a willful, studied turning away not to see that something was wrong. She had always been good, but now there was a blind goodness. She revered the ideal of family, and she poured even more of her energies into making a strong home in Cohasset for Michael and the children.

Across the bay sat the nineteenth-century mansion of Paul and June Verrochi. The Verrochis were a bit flashy and nouveau for a town that prided itself on the conspicuous inconspicuousness of old New Englanders. But the Verrochis had one quality that even the most traditional of the town's families still admired: they were one of the wealthiest families in Cohasset. Paul had gone from cleaning buildings himself to building his roll-up company, AMR, into the largest ambulance company in America. He was a large, expansive man who won his way into the esteem of his neighbors by massive philanthropy. When Michael came calling, he gave $50,000 to one of Michael's causes, handgun control, and joined the board of Citizens. The two families began socializing. June Verrochi, a former schoolteacher, was a fortyish blond who had already had too many face tucks and wore clothing that was too young for her. She was an alcoholic, who was having difficulties mothering her daughter, Natalie,* and two younger sons.

In Natalie Vicki saw a young teenager who needed to get away from a troubled home; she invited the girl to come and live in her house and do some occasional baby-sitting. The young teenager was only five years older than Michael's oldest son, Michael Le Moyne. She became another member of the family, with Vicki as her surrogate mother. One of the side benefits for the Verrochis was that the arrangement not only separated their daughter from June's problems but gave them an even closer relationship with the Kennedys.

There are friends of Michael's who assert that Natalie was fourteen going on twenty-one, that she dressed in the provocative manner of contemporary teenagers that would have been unthinkable for a previous generation of high school students. That may have been true, and it surely was not unthinkable that a teenager might have had a fantasy about a handsome man twenty years her senior in whose home she happened to find herself. A few years back, Michael probably would not have touched Natalie, but he was consumed with sex now. There was a terrible loneliness in him. His sexual conquests were often greased by liquor. He might go from an adulterous tryst to whitewater rafting just as others might go from steak au poivre to crème brûlée.

Natalie told others that she first had sex with Michael three or four times when she was fourteen and fifteen, after he had begun secretly fondling her. He took her and his children with him on family camping trips to Maine, driving north on narrow highways at ninety miles per hour, sometimes while smoking dope. Michael's brothers and sisters were at times on these trips and had the opportunity to observe that there was something peculiar about his relationship with Natalie.

On one of these outings, Michael sat talking and drinking around the

---

*Natalie is a pseudonym for Paul and June Verrochi's daughter.

campfire in the evening. "Daddy, I want to go to bed," Michael Jr. told his father. Michael nodded to his son but kept right on doing what he was doing, more concerned about his relationship with Natalie than about his son. A while later little Michael Jr. toppled off the bench fast asleep, waking up when he hit the pots and pans on the ground around him.

N o one in his family would have understood Michael's relationship with Natalie, not his grandfather Joe, who even when he was in love with Gloria Swanson had never been so obsessed as to risk destroying his life. His Uncle Jack had seen a lovely young woman at a campaign speech in 1958, a highborn Radcliffe student. He would sweep down on campus in his car and pluck her away for an occasional tryst, but he never picked up his dates at middle schools.

The one man who would have understood Michael was not even a figure of politics but of literature, Humbert Humbert in Vladimir Nabokov's classic novel *Lolita*. Michael and Humbert were the same age, both lived in New England, and each found his beloved nymphet in his own house. Humbert had been interested in young girls all his life. For Michael, this was a more recent taste, though he surely would have agreed with Humbert that "mid-twentieth century ideas concerning child-parent relationship have been considerably tainted by the scholastic rigmarole and standardized symbols of the psychoanalytic racket."

Michael's obsession with Natalie did not blinder his lustful eyes. Single or married, secretary or executive, he remained on the prowl. Yet surely Michael would have seconded Humbert's belief that "there is no other bliss on earth comparable to that of fondling a nymphet. It is *hors concours,* that bliss, it belongs to another class, another plane of sensitivity. Despite our tiffs, despite her nastiness, despite all the fuss and faces she made, and the vulgarity, and the danger, and the horrible helplessness of it all, I still dwelled deep in my elected paradise—a paradise whose skies were the color of hell-flames—but still a paradise."

CHAPTER THIRTY-NINE

A Child of the Universe

hen John ventured out, there was often the whiff of betrayal in the air. One evening John volunteered to help stuff envelopes for Naked Angels, a theater group, upstairs at Tatou, a Manhattan restaurant. R. Couri Hay, the publicist for the restaurant, secretly alerted the paparazzi, and in the midst of the evening a troupe of photographers charged up the stairs. Hay played John's protector, hiding him in a closet. "So naturally the next day I couldn't resist a little joke, which I think was on 'Page Six' [in the *New York Post*] that John Kennedy was caught hiding in the closet," Hay remembers fondly.

John tried to get away from that world as often as he could. One of his favorite things was to go off kayaking. Sometimes it would just be an hour-long jaunt around New York Harbor; other times he would go off on adventures in distant regions in the world. The first of these major foreign kayaking trips took place in the summer of 1991 in the Åland Archipelago in the Baltic Sea between Finland and Sweden. "Most of us had had the packaged 'wilderness experience' in which we were anesthetized by the convenience of guides and cooks, and towed through the outdoors without ever washing a dish or getting lost," he wrote in the *New York Times* afterward. "This trip was to be different—four desk jockeys in search of manageable danger."

These were amateurs then, good fellows heading off into the wilderness. John understood perfectly that if he and his friends were to run such risks, they should do so in the proper way. "We accused him of being kind of anal about the particular preparation," says Jim Alden, a friend from Brown who went on several of the trips. "He would have a different color food bag for each meal, to ensure that we would leave no refuse behind. We had to take all the food out of the packaging and put it into reusable containers to ensure that there would be no impact to the environment. We always used clean-burning fuels for our stoves, everything would be just sort of meticulous. And the safety equipment, that was my department, and we always had satellite beacons and radios and all the other things that we needed." On this trip one of the "desk jockeys" capsized his kayak and could well have been lost, but John helped to haul him out of the icy water.

One of the paradoxes of John's life was that when he was doing something clearly dangerous—flying a plane, kayaking in the Arctic, climbing a sheer cliff—he was meticulously careful. But when something was only mildly risky, John upped the ante by approaching it in a cavalierly heedless manner. Whether it was because he bored easily during more mundane jaunts or was simply fatalistic, it was a different John within sight of civilization.

John became a familiar figure at the Downtown Boathouse in lower Manhattan, especially after he purchased a stark SoHo loft for $700,000 in 1994 only a few blocks from the dock. John would jog down the street, grab his kayak, set it in the water, and paddle away, leaving paparazzi, phones, girlfriends, obligations, and onlookers behind on the dock.

While leaders of the sport were at first excited that John was becoming one of them, they increasingly became uncomfortable. A picture appeared in the newspaper of John out in his kayak in cold weather without a life preserver or a wet suit. That could have been a fluke, but it was not. On one occasion a seasoned kayaker went up to John and told him that he was going out improperly prepared. "He really didn't practice what's called safe kayaking," asserts Ralph Diaz, a kayaking authority and author. "No life jacket. No concern about cold-water gear. No extra stuff. No pumps on the boat. No paddle support. It all adds up to a person kind of oblivious to his surroundings."

Robby may have been a novice kayaker, but he wasn't about to be bested by his friend. One stormy spring evening he paddled with John across to

Brooklyn to grab some pizza at Patsy Grimaldi's on Fulton Street right under the Brooklyn Bridge. The current was up, and Robby wasn't too adept with the paddles. He was swept downriver before he clawed his way back. After eating their fill of brick-oven pizza, they headed back to Manhattan. The trip should have been only mildly exciting, but John led them across the path of the Staten Island Ferry. "He's about six feet away from the ferry," recalls Robby, who insists that they were wearing proper gear. "It honks and starts taking off. He's already in front of it. So he hauls ass. I wasn't there, but I was worried about the wash. Well, after that I think that someone told the harbor master, because the other ferries were honking and going around in circles."

When John's old friend Dan visited from Seattle, the two men took kayaks out on a January morning with snow in the air. Dan was a man of natural enthusiasm, a dream of a friend to have on a chilly day like this when few others were on the wind-driven waters. They had just made it around the Statue of Liberty when John paddled furiously away. Dan was valiantly trying to keep up when John yelled out: "Watch out for the ferry!"

When Dan looked and saw that the Staten Island Ferry hadn't even left its moorings, he figured that John was fooling with him. Then suddenly, to Dan's eyes, it appeared that the ferry bolted out into the water like a speedboat. "What the fuck, Dick!" Dan exclaimed as he saw the Staten Island Ferry bearing down on him, using the nickname that at that moment seemed especially appropriate. "I almost died," Dan recalls. "John was laughing hysterically. The ferry actually had to stop. I thought I could beat it because it was moving slowly, but it started taking off."

Those who took the major expeditions with John, be it kayaking in the Arctic or vagabonding around Vietnam, were willing to head out again with their friend. But Dan had become one of the few friends still willing to go off with John on these seemingly less risky journeys. One Fourth of July, John and a friend kayaked into the middle of New York Harbor to watch the fireworks. The harbor police ordered them back to port, but John insisted that they stay there, without lights, in the black immensity of the harbor. After

the fireworks they were practically run down by the scores of boats returning to port.

"He tested everyone," reflects one friend. "He put us all in mortal danger. He had that weird thing going. I don't think he was afraid of death. I
think he didn't think he was going to live very long."

It was not so much danger that John was seeking but life at its purest and
most intense. As he grew older and the world was so much with him, that
purity and intensity became even more important to him, and he was willing
to do whatever he had to do to achieve it. John continued regularly smoking
marijuana. In New York City he was not going to find himself frisked and
busted for having a few joints in his pocket, but he was John F. Kennedy Jr.,
and his activities could easily have become tabloid fodder, which would have
been devastating to his public image. The drug was important enough to
John that on one of his skiing trips to Canada he carried some across the
border, an act that his companion found far more dangerous than any of
their heli-skiing.

John sought the edge also in his sense of humor. When he was with one
of his closest friends, a black woman with a wit so sharp it could disembowel
you, she ridiculed the picayune size of John's penis in comparison with that
of any African American male, and he tore at her as a member of a lackadaisical, lackluster race. Neither happened to be true, but an outsider would
have thought he was coming upon the beginning of a race war.

John was awesomely good at mimicking others, his accents almost perfect
renderings of his victim's voice. "He loved role-playing," remembers Alden,
thinking especially of John on their kayak trips. "He had one very funny joke
that he would tell where he would play the role of the Queen of England. He
would sort of roll out this dialogue in a perfect sort of cockney dialect."

That was good, but John's rendering of Arnold Schwarzenegger's Austrian accent was brilliant. John took exquisite pleasure in imitating his cousin
Maria's husband. Arnold was the other mega-celebrity in the extended
Kennedy family. John saw Arnold most often at Sun Valley, where the actor
had a big house and John went to ski. On one occasion John brought his dog

Friday with him. In the middle of a party Friday defecated on a white rug on the floor of Arnold's prized personal bathroom. Maria called John the next morning to render Arnold's outrage. "Gee, why call me? Rob Reiner did it," John said, mentioning one of the other guests, a leading Hollywood liberal. "He wanted to show how much he dislikes Arnold's politics."

Another year John was at Sun Valley over the New Year's holiday with a group of friends. Dan had never been on ice skates, and at a skating party Arnold tried to help by giving Dan a gentle push. Arnold's idea of gentle was enough to send Dan sprawling. He lay on the ice in considerable pain, his spill memorialized in an enormous black bruise on his buttocks.

Four evenings later the fifteen guests at Arnold and Maria's New Year's Eve party included Tom Hanks and the ubiquitous Reiner. "Hey, Maria, have you seen the bruise on Pinky's butt?" her brother Tim asked, using Dan's other nickname. "No," Maria said. With ample prompting, Dan displayed his black-and-blue bottom to the other guests.

John could tell that Hanks was unhappy that he had been upstaged. The movie star ran into the kitchen. He reappeared in the living room, pulled down his pants, and displayed his buttocks on which he had placed two steaks commandeered in the kitchen. When that didn't get enough of a laugh, he went back into the kitchen to grab a fork to plunge into the steaks. From then on, Hanks joined the repertoire of the characters John made endless fun of, imitating his voice and his presumptions.

John continued his rancorously funny put-downs of Anthony Radziwill even after his cousin was diagnosed with cancer. If anything, John was tougher on Anthony, and Anthony tougher on him. John had gone through Billy's cancer, but this was worse, one operation after another. Anthony would bounce back and resume his work as a producer for ABC, then sooner or later cancer would come visiting again. And John would be back in a hospital room, heaping endless verbal abuse on the sickly form in the bed. Anthony had a sweet feline quality that sometimes turned to truculence. John and Anthony were like an old married couple in which each partner knew the other's weaknesses and mercilessly probed them.

Anthony's mother, Lee, who called her memoir *Happy Times*, was not the ideal parent to deal with a sick son. Thus, it had fallen to Jackie to take care of Anthony at her Fifth Avenue apartment, and while he was recuperating, she brought him to her home on Martha's Vineyard.

E very time one of the guys married, they had a bachelor party, a ritual that was really about commitment and love and life going on. For Dan's, they had gotten together in Reno. That alone was a weird setting for the party. Then somebody had the idea of driving out to the Mustang Ranch, the most famous legal brothel in America. The men had no intention of frequenting prostitutes, but the idea of simply going there was irresistible, a hysterical caricature of a bachelor party. Nobody flinched at the idea, not even Tim, the old priest himself. Tim was driving, and John was at his side, when they realized that the valet parker at their hotel had given them the wrong car. "This is great!" John exclaimed, looking at Tim. "Willie Smith's on trial for rape, and we're going to get busted in a stolen car on the way to the Mustang Ranch."

John was thirty-three, and life was moving on and his friends with it. John was perfectly aware of the natural trajectory of life. The man who at twenty is the life of the party becomes at thirty an amusing diversion, and by forty is silly old Harry, the perennial bachelor over there in the corner with a drink in his hand.

Nothing was going right, not his tedious work at the DA's office, not his tortured romance with Daryl. Nothing. When in July 1993 John left his position in the district attorney's office, the good news was that he had won every case he tried; the bad news was that in three and a half years he had tried only six cases. And one of his former colleagues told the *London Sunday Times* that these were ones that "any lawyer's three-year-old could win."

John was talking about the possibility of starting a magazine, but it was mainly talk, and he seemed disconnected, more nervous and flighty than ever. No one had deeper insight into John's emotional life than Sasha, and no one felt sorrier for him. "I remember thinking, 'God, you gotta move on,' " recalls Sasha. " 'You're all over the place.' He was like two people. There was the John who noticed his surroundings and the John who couldn't stop turning the wheel of his own mind alone."

When Sasha gave birth to a son, John asked to be the boy's godfather, an offer that she joyously accepted. He was walking one day with Sasha and his godson Phineas. "You guys are the best parents," John said, referring to Sasha and her sculptor husband, Philip Howie. Sasha looked down at the little boy and realized that he was just three years old, the age John had been when President Kennedy was assassinated. "Oh, God, just imagine if Philip was just gone today from his life, imagine what that would be like," she said in her breathless, girlish way. "That's what happened to you." John stopped and turned toward his friend. "And you know, you never get over it."

John still loved Daryl, and even if their romance had become weekly tabloid fodder, it was the best thing in his life. Of course she was difficult, but of all the women he had known, she was the most adventurous, skiing routes that few men would dare and traveling with him on trips to such places as Vietnam and the Philippines. They always managed to outwit their pursuers and be off on their own sojourns.

Then early in 1994 John learned that his mother was dying of cancer, and nothing mattered the way it had. Jackie had wanted a son who was both a true man and a gentleman, and in his mother's last months he was both. He was strong and he was stoical. He took Jackie for walks in the park, pushing through the gauntlet of photographers and reporters. He didn't dare even to say out loud that there was a time to live and a time to die and it was his mother's time to die, and she should go gently into that good night.

Jackie had felt that Daryl was not right for John, but in the last weeks of his mother's life, Daryl was strength and resolve. One day about two weeks before his mother's death, John went for a run in Central Park. That was his Valium, what he did to sweat out his anxiety. He took Daryl's dog with him, and when he was returning to her West Side apartment, a car struck the dog down as it followed John across the street. This was the careless, thoughtless part of John that Daryl abhorred, and in some ways after that it was never quite the same between them.

When his mother was too sick to walk in her beloved Central Park, the family and friends convened in the Fifth Avenue apartment. John was full of restless energy. He and a few friends standing vigil with him hurried out of the building to walk in the park. "He was immediately swarmed and almost crushed by the clamoring media camped outside, shouting inane questions at him, yelling in his face," recalls Christiane Amanpour. "I was so enraged that I forgot for a second that I was a journalist myself, or maybe I was ashamed. But I just screamed at them, 'It's his mother, for chrissake.' It made no difference of course, and John gave no outward sign of distress. He just put his head down and concentrated on getting out of there."

On May 19, 1994, John came down from his mother's apartment to tell the reporters below that his mother had died. In some ways this was as much a defining moment in his public life as was his salute to his father thirty-one years before. "She was surrounded by her friends and her family and her books and the people and the things that she loved," he said. "And she did it in her own way and on her own terms, and—and we all feel lucky for that. And now she's in God's hands."

John went back upstairs. Later in the evening, when most of the guests had left, he and a few friends walked out on the balcony. As they stood there in the spring night, cheers wafted upward from the two or three thousand people who still stood below. John found this not an intrusion on private mourning but an expression of how much his mother was loved. "He was blown away that they were still there, and how much his mother meant to America," remembers Billy, who was standing next to him.

During those first days of mourning, John went Rollerblading. To some that was an unacceptable activity during the dignified formalities of mourning. But only in physical action could John seek some release. In the aftermath of his mother's death, John needed time and space to sort out the complex emotions he was experiencing. He was deeply distraught at his mother's death, but as alone as he felt, he had freedom he had never had before. All his life his mother had measured him against her own standard. Now he would be judged only by the yardstick that he chose, and only in the times and places that he considered important.

Robby saw the change in his friend: "I mean, he loved his mom. She was a hell of a beacon. Her passing he dealt with from a perspective of, How can I honor that gift? What gift? The gift of love and care and just plain old great mothering. Now he was alone. He was cast out alone. He was America's kid, but he was alone, and it caused him to do further in-depth introspection, which in the end made him better."

During the summer of 1994, it seemed as if every last bachelor John knew, from Anthony to Billy, was being herded toward the altar. John was supposed to bring Daryl with him to Billy's wedding, but at the last minute he said that he was coming alone. Billy sensed that John figured that if he brought Daryl, she would see the wedding as a dress rehearsal for her own. By not bringing her, he was signaling that it was over between them.

John was the only man by himself in a sea of couples. At the party beforehand, John gave everyone a T-shirt with the words "Big Bad Bill" emblazoned on the front, along with the infamous shot of hotshot Billy protecting his buddy from the paparazzi assault on his eighteenth birthday. On the back were the words "Sweet William Now," with a picture of Billy and his love, the red-haired colleen Kathleen Maguire from the Irish enclave of South Boston. John had written new lyrics to the Ry Cooder song "Big Bad Bill (Is Sweet William Now)," and wearing their T-shirts, the guests sang the song.

At the wedding, Tim read from the scriptures with an earnest intensity that few priests could equal. Then John recited the inspirational poem "Desiderata." It was as if he were saying the words to himself and to Timmy and to all the friends, as well as to Billy and Kathleen:

> . . . You are a child of the universe, no less than the trees and the stars; you have a right to be here. And whether or not it is clear to you, no doubt the universe is unfolding as it should. Therefore, be at peace with God, whatever you conceive Him to be. And whatever your labors and aspirations in the noisy confusion of life, keep peace in your soul. With all its sham, drudgery, and broken dreams, it is still a beautiful world. Be cheerful. Strive to be happy.

John's personal and professional lives were both deeply unsettled. He was not a man to cheat on a constant lover, but unbeknownst to Daryl, John was seeing another woman. Her name was Carolyn Bessette. Like Daryl, she was a tall, sprightly, beautiful woman. He was seen with her in public as early as November 1993, when they attended the New York marathon. John was still immersed in his relationship with Daryl, while Carolyn's lover was model-actor Michael Bergin. On at least one occasion John had a date with Carolyn when Daryl called to say she was flying in from Chicago. John had a friend explain to Carolyn that he would not be able to keep the date.

Once Carolyn became John's regular date, she took a profound proprietary interest in him, beyond what any of his previous lovers had shown. Carolyn was an expensive woman, not only in the clothes she bought and the money she spent, but emotionally. She had a withering intensity, and whether she was the most interesting woman he had ever dated or merely the most difficult, she mesmerized John. When she turned her gaze on you, it was like being in an interrogation room lit by a bare lightbulb. She had a gift for honesty, a staggering candor with her friends. To strangers, she might have appeared a natural denizen of the fashion world, where an air kiss is a sign of affection and beauty and style are the hardest currencies. But it was her understanding and awareness of the world around her that made her so compelling to John. "She, of all his girlfriends, was the most honest and the most capable of delivering that honesty," says Robby. "She was so sensitive—sensitive to the pain and joys of life, and that made her vulnerable to ups and downs. She could parcel a human down within thirty seconds to a three-by-five card. And be 99 percent right."

Carolyn had a thoroughbred model's bearing, perfect posture, and a sleek, easy walk. Some would write about her later as a Cinderella—her upbringing was so different from the world of the ultra-chic New York woman she had become. She had not been brought up like John, but then nobody was. She came from privileged surroundings in Greenwich, Connecticut, where her mother divorced her orthopedic surgeon husband when Carolyn was in elementary school. She had attended St. Mary's Girls School and, like John, was the most nominal of Catholics. She went to Boston University, where she planned to become an elementary school teacher.

Instead of teaching, Carolyn took a position in sales at Calvin Klein's retail store in Chestnut Hill, a Boston suburb. She was only in her early twenties, but she had both a natural sense of style and a demeanor that appealed to the wealthy women of the Boston suburbs. She was so successful that she was invited to move to New York. There she worked as a private shopper for Calvin Klein's celebrity clients. Every designer in New York sought to have famous women, from movie stars to socialites to television personalities, wearing their clothes and enticed them with wholesale prices. Carolyn had a natural imperiousness about her that intimidated the more insecure of her clients while ingratiating her with others. She soon became an integral part of New York's fashion world, all before she was twenty-five years old.

"She just blew you away with this self-confidence," reflects Paul Wilmot, the former Calvin Klein public relations director who hired her in New York. "And she was ebullient. She exuded this aura, this positive light that shone from within. It was like 'I know I'm gorgeous, I know I'm smart, but I'm interested in you.' "

Carolyn worked her way up to become head of public relations for the main collection, another formidable achievement for such a young woman. If she had in large measure invented the Carolyn Bessette with whom John was falling in love, that was fashion's dream, and it was a splendid invention. They were both immensely intense, emotional beings, and they circled each other warily, sensing what a deep involvement would mean. It was not until they had been dating for more than a year and a half that Carolyn moved into John's SoHo loft, in July 1995.

One of the first things Carolyn did with John was to get him to dress well. That was only the beginning of the makeover that she sought to do on her John. She felt that several of John's friends weren't good for him, and she pushed them away. John was no slouch at understanding the dubious motives of much of humankind, but he had put up with these men for the fun he had being around them. Now John decided that he had better things to do, and he no longer spent much time around them.

"Carolyn, when she wanted to be Miss Modern, Miss Elegant, she was very good at that," says one friend. "But when my little girl wanted to get down and dirty, oh, she could get down and dirty. And there were times when she would use that down-and-dirtiness to intimidate people a little bit, because she was very protective of John."

John lived for intense, passionate experiences, and in that respect Carolyn was a natural match. John enjoyed his times with A-list celebrities, but he could be himself when he was out having a good time with Carolyn and a few friends, and she was almost as much into it as he was. There was no such thing as a typical evening with John. One night the couple, along with a woman friend of John's, drove out to Great Adventure in New Jersey. AOL Time Warner had taken over the giant amusement park. Most men in their midthirties did not find excitement in thrill rides, but John ran from Skull Mountain to the Scream Machine, from Nitro Taz to the Twister. In the middle of the evening John realized his wallet and keys were missing.

Carolyn had gone through this with John endless times, whether it was keys, plane tickets, or credit cards. This was hardly a major crisis; at worst they would have to call for a locksmith, but Carolyn became extremely upset. While John rummaged around on foot, she commandeered an electric car and zoomed around the grounds as if that were the way to find the lost items. John managed to find the wallet and keys under the Scream Machine, where they had dropped during his ride. He took this as some kind of achievement, conveniently forgetting how he had messed up in the first place.

On the ride back to Manhattan John decided to stop for a drink. The New Jersey hangout that they chanced into was a karaoke bar that looked as if it belonged in *The Sopranos*. John waltzed into the room, with a handsome black woman with a stare that said don't even think it on one arm and on the other arm a model-gorgeous blonde with legs so long they seemed to have no beginning or end. John had his wool cap pulled down over his forehead, and nobody recognized him for anything but one lucky guy. He sat facing the room, something he rarely did, and ordered a pitcher of beer. At first Carolyn was nervous and ill at ease, but the three of them proceeded to get loopy listening to some excruciatingly bad singing and watching some characters who rarely showed up in Manhattan.

When John took off his cap, people in the place looked over at his table as if he were an apparition. "Are you? . . ." one of the waitresses asked. "Yes, I am," John said. That brought pitchers of free beer and louder bad music, but no intrusive calls for autographs. In the early morning hours the trio made their way drunkenly back to New York City, just another night on the city.

CHAPTER FORTY

Games Kennedys Play

hen Joe returned to his Boston home for weekends, matters had become so bad in his marriage to Sheila that even their aged dog Loco had become a center of controversy. Both Joe and Sheila loved the dog, but Joe thought that it was time for the blind, half-crippled dog to be put down. Sheila, who had refused to have more children, treated Loco like a dying relative who had to be seen through to the end. Joe got tired of the endless concern over a dog, and Sheila considered his anger just another mark of his endless selfishness. On one occasion, Joe exploded over his wife's endless ministrations to a pet. "Joe, nobody's asking you to compete with Loco dying," she told her husband. "I'm taking the dog to the vet."

Sheila noticed in her husband "an inability to behave as an ordinary family member. All week he would be surrounded with people who believed he could walk on water. Sharing household chores was beyond him. It wasn't as if he could sit down and discuss these problems. I did not feature in his weekday diary."

Sheila did not express her discontent, but let it fester. "I had simply become afraid of him," she said. He abused her verbally no more than he did others around him. He was not a man of physical violence, though Sheila feared what demons rested within. As Sheila saw it, his anger was triggered by the smallest of matters—opening the refrigerator and finding only two of his favorite soft drinks instead of the three that were supposed to be there, or discovering that his wife had bought English breakfast tea instead of Irish breakfast tea.

Sheila was a stoical Episcopalian, a member of what she called the "Frozen People." Her anguish came out not in confrontation but in illness. In the summer of 1988, she was stricken with bacterial pneumonia that came close to killing her. "My attitude up until that point had been 'I will make this marriage work if it kills me,'" she reflects. "And then suddenly it dawned on me that it was going to do just that."

I n March 1989, Sheila left the house with their two sons and moved across the Charles River to Cambridge. Joe announced a separation and at the same time decided not to run for governor in 1990. A year and a half later the couple filed for a no-fault divorce, which was granted in January 1991. Soon after the separation Joe began living with Beth Kelly, a scheduler and secretary in his office who devoted her life solely to Joe's career and personal needs.

In his first years in Congress Ted had tried to tutor Joe, but he could not easily impart his natural legislative skills to his nephew. Joe had to learn for himself, and in his inimitable fashion he did so by butting his head against the door of Congress until he realized that the door could be pushed open with the help of a few colleagues. He would never be like his Uncle Ted, a partisan who was nonetheless able to have fast friends on both sides of the aisle. Joe could, however, work with like-minded colleagues and even reach out to those well to the right of him. As a Kennedy, he had a natural constituency in the poor and the deprived. Although he took that as given, he also had sympathy for the lower-middle-class families that his father would not have cared that much about. He thought that the system was tilted in favor of the rich and powerful, and he sought to address that imbalance.

For the young generation of Kennedys, the election of Bill Clinton in 1992 was in several respects the most important election of their lives. It was not only that Clinton revered JFK and patterned himself after the former president, but that he created an atmosphere in which their aspirations could prosper, not least of which was the Democratic majority in Congress. Joe was not a politician who thrived eating the humble pie that is the minority party's daily fare, being gaveled down by a Republican chairman, and routinely being defeated on everything from committee votes to major legislation.

For Joe, the brightest reward of the new era was the chair of the banking subcommittee on consumer credit and fees. In the subcommittee hearings he was merciless with those who came ill prepared and endlessly sympathetic toward everyday citizens who testified about their various plights; he was also ever aware of the exigencies of the media, whose lights he needed to shine on the matters that concerned him. In the first months of the Clinton administration Joe was becoming an immensely popular figure in Massachusetts, outshadowing even his uncle. A March 1993 *Boston Globe* poll showed that 53 percent of the Commonwealth's citizens thought favorably of the young congressman, while only 25 percent viewed him unfavorably. As for his Uncle Ted, 41 percent viewed him positively, but almost an equal number, 40 percent, saw him negatively.

As Joe rushed along the corridors, he might have a cigar in one hand, his drugstore reading glasses in the other, a splotch of his lunch on his tie, and scruffy shoes on his feet. He was the first downscale Kennedy. People in his district loved it that he was one of *them* but was also a Kennedy. The politician of his generation he resembled the most was Governor George W. Bush of Texas, an aristocrat who had reinvented himself as a middle-class Texan with a schoolmarm wife. Like Bush, Joe wasn't merely playing the role of the average Joe. He was authentic in his role, and the congressman's empathy toward his constituents was real.

I n April 1993, during Easter week, Sheila went to the door and received a certified letter from the Archdiocese of Boston asking her to respond in writing to her ex-husband's contention that "his marriage to you is null under the provisions of ecclesiastical law on the grounds of: Lack of Due Discretion of Judgment." Reading only far enough to understand that Joe was seeking an annulment, she ran to the bathroom and threw up.

Joe was in the Caribbean on a vacation with Beth. When he returned, he told his former wife that he had planned to tell her first but that the letter arrived sooner than he thought it would. He was angry at Sheila's lack of understanding that he simply wanted to get on with life and marry Beth in a church wedding. "How can you be opposed?" Sheila recalls Joe saying. "What right do you have to be opposed? How can you prevent me from going on with my life?"

Sheila may have been tiring in her moral certitudes, but her rectitude and high principles and doggedness were not outfits that she wore only to church. Sheila decided that if she did not fight this, she would be the same "doormat" that she had been in her old life. Joe could not wait for the outcome but in October of that same year married Beth in a civil ceremony while he continued to press for an annulment.

The Kennedys and the Catholic Church in Massachusetts had grown up together. Once poor and despised, they had sought to become the equals and then the superiors of the Protestants who dominated the Commonwealth. They had largely achieved what they had sought, and Sheila faced what appeared to be an impossible challenge.

T ed watched Joe's divorce with immense attention, for its effect on both the future of the Kennedy family and on his own immediate political career. He was delighted that the matter had been quietly negotiated with only minimal damage to the Kennedy status in Massachusetts, and especially to Ted's reelection prospects in 1994.

Ted announced that Michael would run his reelection campaign. This was the family apprenticeship, and it said much about how deeply Ted cared about his family's political future that election after election he passed out this important job to one young relative after another. In most of these races, Ted had only nominal opposition, and the campaign manager hardly mattered. This year, however, Ted was facing the most difficult election challenge of his career.

Ted had served his constituents in the most heavily Democratic state in the Union long and well. By that criterion, his reelection should have been almost a given. It was not Ted's personal life per se that made him vulnerable. It was that his opponent, businessman Mitt Romney, appeared to underscore every one of Ted's liabilities. Handsome and fit, Romney looked a good decade younger than his forty-seven years. Rotund, sixty-two-year-old Ted appeared at times as if he had just stumbled unaware out of the pages of *The Last Hurrah*, Edwin O'Connor's brilliant novel of Boston Irish American politicians. By all appearances, the Mormon Romney had a loving marriage, five photogenic children, and a caring, untroubled family. Ted's legislative record was found in the *Congressional Record*, but over the years his nocturnal escapades had been chronicled in the *National Enquirer*. Ted's proponents argued that that was all behind him, but if he had won election the first time largely because of the Kennedy image, thirty-two years later it looked as though he could lose it for the same reason.

Ted had one immense plus, both politically and personally. He had married thirty-eight-year-old Victoria Reggie in July 1992 in the living room of his home in McLean. She was a tall, dark-haired sensuous divorcée twenty-two years his junior who had been raising her two children in Washington while practicing law. Her father, Edmund Reggie, was a Louisiana bank president, judge, and Democratic Party leader who had known the Kennedys for years. Reggie was a full participant in the more dubious aspects of Louisiana politics. He was a wealthy man, but his six children and a nephew received thirty-four years' worth of tuition waivers at Louisiana public universities, estimated by the *New Orleans Times-Picayune* to be worth $644,028 in contemporary dollars. There was nothing illegal about it, though the scholarships were supposed to fund poor students who otherwise would not have been able to attend the university. There was something illegal about Reggie's practices in his bank, so much so that in 1992 he was convicted of bank fraud. The following year he pleaded no contest to another charge of bank fraud for misapplying money from a savings and loan company that he founded in 1959; the judge ordered him to be detained in his home for 120 days and to pay a $30,000 fine.

None of this tainted Vicki personally, and Ted had actively pursued her.

During Easter week of 1992, on a vacation on St. Croix, Ted set his engagement ring in a bed of coral, and when the couple went scuba diving, he directed Vicki to look there. "I saw this big fish, a grouper, going down there, sort of headed toward her," Ted recalled. "And I said, 'My God, if that fish eats that ring I'll be certifiable,' not knowing that I'm putting a shiny sparkling thing down there that could attract a fish."

Although Vicki found it "the most romantic thing," she went into the marriage with her head as much as her heart. Ted was not only good with her children but found them a double blessing. He attended the operas that Vicki adored and trusted her more than he had Joan.

The newlyweds gave a series of interviews soon after their wedding, as if he had scaled off his past and the problems that had obscured his political achievements. She was a strong-willed wife, eternally watchful of her new husband, but she did not attempt to change him in ways he would not change. He was still a big drinker, still capable of pratfalls and boorish moments, but comfortable with Vicki and their new life.

T ed watched Michael's managing of the campaign with increasing discomfort. It was a mark of how cavalierly Michael took his task that he brought in his cousin, Michael Skakel, to chauffeur him around and provide continual diversion. Skakel was useful to Michael, as he arranged Michael's various assignations around his campaign duties.

Michael knew that authorities in Connecticut had recently reopened their investigation into the bludgeoning death of the Skakels' next-door Greenwich neighbor, Martha Moxley, in October 1975. Skakel was a primary suspect, and it took a certain audacity to have him sit with the family at the Democratic state convention in July and appear at other public functions.

Ted was beside himself at the casual ineptness of the campaign. The lackluster effort was apparent just from walking into headquarters. There was no excitement, no firm hand, at times nothing but milling, disconnected campaign workers. Ted was a generation older than almost all of his aides. Although they treated him with deference, there was a sense among them that the senator was so uncertain of syntax and so flabby of form that the best way to get him reelected was to keep him as far from a direct contest with Romney as possible. The strategy was for him to hide behind TV commercials and position papers. They could not take the chance of giving interviews to the Boston television stations or having major, high-visibility debates.

The aides discussed this behind Ted's back like coaches debating strategy before the big game with a quarterback prone to throwing interceptions. Ted had insecurities rare in a politician of his seniority and accomplishments, and his aides' doubts played into them. He was an endless questioner when he was upset. Michael and others began hearing that voice on the phone, second-guessing them, but second-guessing himself as well. He knew the campaign was going wrong, and he wasn't sure he knew how to fix it.

Vicki was used to bare-fisted Louisiana politics, and she convinced her husband to hire the Investigative Group, Inc., a controversial Washington firm that set out to gather whatever negative information it could on the supposedly prissy clean Romney.

T ed wasn't about to embarrass Michael by firing him, but by early October he had gotten so nervous that he knew he had to do something. He called his old administrative aide David Burke, who had gone on to a major career on Wall Street. "Will you ride in the car with me?" Ted asked. In the old days, that had been the administrative aide's natural place, sitting there in the seat, feeding information to Ted who always sat in the front passenger seat. Burke was no longer in the thick of politics, but he was a man of the soundest judgment. Ted didn't want or need Burke to take charge of any one aspect of the campaign, but just to be there.

That first morning in the car, Burke took care of a few simple things. Ted had been sitting there working a couple of phones, peppering campaign staff and others with questions. Burke arranged it so there would be one phone that he kept hold of most of the time and one person in the office answering it. That way Ted was no longer on the phone as the chauffeur drove him through the campaign day, a minor change that had a major impact on the candidate.

Other former Washington aides and staff members arrived too, a devastating commentary on the way Michael had been running things. In 1952 Robert had taken over for Jack's previous campaign manager, but this was the first time that outsiders were rescuing the Kennedys from themselves. Another of the crucial new advisers was political consultant and speechwriter Robert Shrum. He was a liberal Democrat of overwhelmingly partisan passions. Shrum and his associates in his campaign consulting firm learned from the research of the Investigative Group that two years before a company chaired by Romney had fired a number of union workers when it took over a paper plant in Marion, Indiana. Romney had been on leave from his company for six months when this took place, and he had nothing to do with the decision. But that was a subtlety that didn't seem to concern Ted's campaign. The TV spots of the fired workers proved devastating.

Romney made a stumbling reply to these accusations that had nothing to do with him. If he had responded in partisan kind, he would have aired campaign spots alluding to Ted's troubled personal life and scandals, the great unspoken issue of the campaign. Although the Republican ads were rightfully tough on political issues, Romney's campaign barely grazed such matters.

Ted told Burke that he had decided against debating Romney, and Burke told him he had no choice. "To stop all that negative noise, the first thing we're going to have to do is debate," Burke said. There was an overwhelming sense that if Ted lost the first high-visibility debate, he would lose the election, and that if he won, his victory would surely come. Ted was rising in the polls already, and he could have avoided a direct confrontation, but he decided that Burke was right: he must indeed debate Romney.

Staffers fed Ted tough questions for hours at his home in McLean. Everyone knew that the health issue would come up at some point. Nobody in the Senate knew more about health or had advanced more government programs than the senior senator from Massachusetts had. On that one issue alone, his advisers felt they could create the most devastating of contrasts.

For Jack's first presidential debate, one of his associates had pushed for small podiums that would display JFK's elegant spare frame in contrast to Nixon's stolid body. In this instance, Ted's aides insisted on massive, three-foot-wide podiums to hide the senator's girth. "You could stay overnight in this podium," Ted quipped.

On the afternoon of the debate Ted had the chauffeur drive him to the Kennedy Library in Dorchester. Carrying a stack of papers and briefing books, he walked around the glass-sided building and sat outside next to Jack's sailboat, the *Victura*. When evening drew near and he walked back to the campaign van, he did not tell Burke whether he had been reading the briefing books or just sitting there communing with his brother and the past. Burke figured that Ted knew better than anyone the magnitude of this evening and that he wanted to reflect on many things there beside his brother's boat near the water that he loved.

"I remember when I went out there," Ted said. "I think it was just a thoughtful time. You're sort of reviewing your own thoughts and organizing. You want to collect your thoughts. You're thinking about broader kinds of themes about your life and what it's meant and what your values are and all the rest. You need the time to give some thoughts and some values in order to sift through. And that's what you do from time to time."

On the drive to the debate at Boston's Faneuil Hall, where Jack had made his last campaign speech in 1960, no one in the van said a word. With each mile along the crowded highway, the tension grew more palpable. When the car finally reached the narrow, snarling streets near Quincy Market, the sidewalks were full of supporters. The legions of preppy suburbanites bearing the Romney banners were overwhelmed by hundreds of rowdy, sign-waving stalwarts from the carpenters' union, looking like they had arrived out of some early-twentieth-century torch-lit rally.

"Senator, before we get there, can I ask you something?" Burke asked in his most solicitous voice.

"What!" Ted snorted, seemingly irritated at this intrusion.

"How come Steve Breyer is on the Supreme Court," Burke asked, men-

Anthony Shriver and Best Buddy Gen Burke. (BEST BUDDIES)

Patrick Kennedy wins election to Congress from Rhode Island in 1994. (NEW YORK NEWS SERVICE INC.)

Douglas Kennedy, a Fox News correspondent, with his wife, the former Molly Stark. (NEW YORK NEWS SERVICE INC.)

Christopher Kennedy, an executive at Chicago's Merchandise Mart, with his wife, the former Sheila Berner. (NEW YORK NEWS SERVICE INC.)

Robert Kennedy Jr. with his son John Conor on his shoulders.
(NEW YORK NEWS SERVICE INC.)

John Rosenthal (second from left) *and Michael Kennedy, cofounders of Stop Handgun Violence, with Massachusetts politicians at the signing of a comprehensive gun-control law.* (JOHN ROSENTHAL)

Timothy Shriver, *president and CEO of Special Olympics International, with Chinese authorities and Special Olympians.* (SPECIAL OLYMPICS)

*O*ne of Robert Shriver's closest friends was the late photographer Herb Ritts, who took
this photo. (HERB RITTS)

The teenage John F. Kennedy with his thick head of hair.
(NEW YORK NEWS SERVICE INC.)

John playing football with a group including Robert Kennedy Jr.
(NEW YORK NEWS SERVICE INC.)

The young John F. Kennedy in Boston with his sister, Caroline.
(NEW YORK NEWS SERVICE INC.)

John Kennedy Jr. kissing his college girlfriend Sally Munro. (NEW YORK NEWS SERVICE INC.)

John Kennedy Jr. and his mother, Jacqueline Kennedy Onassis, walking through the new museum in his father's honor at the John F. Kennedy Presidential Library. (JOHN F. KENNEDY PRESIDENTIAL LIBRARY)

John Kennedy (right) *off camping with his friend Dan Samson.* (DAN SAMSON)

John loved nothing more than to be out in the wilderness. (DAN SAMSON)

John Kennedy Jr. Rollerblading through his beloved New York streets.

*J*ohn Kennedy and Carolyn Bessette off on a boat named after his father's PT boat.
(New York News Service Inc.)

William S. Noonan and Kathleen Maguire on their wedding day, with Timothy Shriver and John F. Kennedy Jr. (WILLIAM NOONAN)

Good friends off in John's plane: Robert Littell, Dan Samson, Jody Samson, John Kennedy Jr., Carolyn Bessette Kennedy, Franny Littell, and two Littell children. (DAN SAMSON)

John F. Kennedy and his beloved friend Sasha Chermayeff on her wedding day.
(J. LANDEY)

John Kennedy Jr. and his friend Gary Ginsberg with former president Gerald Ford.
(DAVID HUME KENNERLY)

John Kennedy Jr. and his dear friend Christiane Amanpour on his wedding eve. (MARTA SGUBIN)

John Kennedy Jr. and Carolyn Bessette Kennedy attending a formal event at the John F. Kennedy Presidential Library. (JOHN F. KENNEDY PRESIDENTIAL LIBRARY)

John Kennedy Jr. and Carolyn Bessette Kennedy with their dog, Friday.
(JOHN HENRY BARLOW)

tioning another old Kennedy associate, "and I'm stuck in the back of this
f——ing van with you?"

Ted turned and let out a gigantic laugh just as the door was pulled open
and he was engulfed in well-wishers. In the debate that followed he was
overly made up, and his perspiration was positively Nixonian. On occasion
he shouted as if unaware of the microphones. In some of his responses he
described the minutiae of bills in numbing detail.

The overwhelming sense, however, was that this was a man of substance
who knew about the issues that faced his constituents and cared about them.
When he got the chance to talk about health, he made Romney look hopelessly unprepared. And when he came to talking about his family, he had the
best, most memorable, most seemingly irrefutable line of the evening. "The
Kennedys are not in public service to make money. We have paid too high
a price."

For all his myriad faults, Ted had a remarkable way of performing well
when he had to. All victories are sweet, but his overwhelming reelection two
weeks later by seventeen percentage points was especially so.

Ted had one other campaign whose results that evening he awaited
with immense anticipation: Patrick's race for the House of Representatives from Rhode Island. Twenty-seven-year-old Patrick had decided he
was not going to squander his youth in the minor leagues of American politics when he could use his family to help him become the youngest member
of Congress.

Despite Patrick's three terms in the state legislature and hours of speech
coaching, he had developed little as a public presence. Early in the campaign,
when Governor Bruce G. Sundlun heard Patrick speak at an NAACP dinner
in Newport, he felt sympathetic toward such a sad public figure. "Patrick had
his speech all written out on a piece of notebook paper with blue lines on it,"
the former governor recalls. "And he held the paper with both hands in front
of him, and you could practically hear the paper crack as his hands shook.

And it was really a rather pathetic performance. But I don't know, I admired his courage, because you see he was under a lot of tension and distress, as it was not something he had done much of or did well. He was lousy. He was so nervous giving it, it was painful. And I made my mind up, for whatever reason, I'm going to endorse this kid for Congress. And I did."

Patrick's Republican opponent for the open seat, Kevin Vigilante, trumped Patrick in his accomplishments in almost every regard. Vigilante was an emergency room physician known for his work treating AIDS patients. He came from a middle-class background similar to that of most of the voters. He earned about $100,000 a year as a doctor, while Patrick received roughly the same amount from his trust funds. Vigilante had worked summer jobs, including a stint in a steel mill, while, other than his legislative position, Patrick had not worked a day in his life.

Patrick's campaign suggested, probably unfairly, that Vigilante was exaggerating his work in the mills. "Young Kennedy questions how close Dr. Vigilante really got to the coke ovens at the steel plant that employed him one summer," wrote Froma Harrop in the *Providence Journal-Bulletin*. "This is not a wise strategy. Had Dr. Vigilante spent the summer buttering popcorn at the Avon Cinema, his sweat record would have still dwarfed that of his Democratic rival."

In September vandals broke into Patrick's campaign office and spray-painted the word "Die" on a campaign poster and covered the walls with graffiti. Some campaigns would have dismissed this as a stupid adolescent prank, but Patrick had good reason to view the matter more personally. He had already received two death threats during the campaign, one in a letter and a second in an anonymous phone call. The matter was serious enough that in the last week of the campaign Patrick had special police protection.

Vigilante was overwhelmed by the money that Patrick dispersed, the weight of his celebrity, and the tough professionalism of his campaign. Patrick's press secretary said that to win they would have to turn Vigilante "into Attila the Hun rather than Mother Teresa." Vigilante attempted to reply in kind, but the doctor was more comfortable with the blood in an emergency room than that drawn in political warfare, and Patrick won, 86,904 to 73,527. He was not the only young Kennedy to win: his cousin Kathleen scored her own victory in the race for lieutenant governor in Maryland.

Patrick's victory had gone against the tide that had swept a Republican majority into Congress. While many of his new Democratic colleagues were bemoaning the loss of departed friends, Patrick arrived in the capital with an entourage worthy of a conqueror. Almost a hundred Rhode Islanders rode down to Washington with Patrick in two special train cars. They were there for a special reception in the Senate Caucus Room, where Patrick's Uncle Jack had announced his presidential candidacy in 1960. Not every new legislator's arrival was marked by the presence of the vice president of the United States. Al Gore was there, along with Patrick's father serving as master of ceremonies, his stepmother, his aunts Pat and Ethel, and his cousins Michael and Joe. "No father in this country is prouder," Ted said, and by the trill in his voice, none would have doubted him.

When Patrick spoke, the *Providence Journal-Bulletin* reported that the freshman congressman "led with a joke at the expense of his cousin Joe, well-known in his freshman term for breaking some china and irritating his seniors as he bulled into the House." Joe had to sit there and smile as his younger cousin made his little dig, just another indication that political life was becoming a trial for the congressman from Massachusetts. His loud, intemperate voice would often be stilled in a House controlled by Republicans, and when he did speak, he would not be the only Kennedy contending for attention.

Joe's Uncle Ted had at crucial times helped Joe in his political advance, but now there was another Representative Kennedy who took precedence. As a new member, Joe had been unable to win the important committee assignments he sought, but even before taking his seat, with the help of his father's friends, Patrick had won a seat on the powerful Armed Services Committee, renamed the National Security Committee. Patrick showed an astute awareness of the nature of power in the House when he considered his committee assignments "the most critically important political decisions I made." Rhode Island was a small state with major military contractors and defense installations, and the Armed Services Committee was in some ways the most important committee on which its congressmen could serve. Patrick would vote liberal on many domestic issues, but he was a conservative hawk when it came to defense appropriations containing monies for Rhode Island.

When Patrick was sworn in on the floor of the House, he had plenty of time to look up in the distant reaches of the House gallery where a woman sat by herself. As he saw his mother so far away, he reflected how "when she was twenty-five years old, she was married to the youngest senator. . . . Now all this time has passed. My dad's remarried; I've been elected. And here she is. It must be so weird."

Although his mother felt immense pride, other sentiments surged through her as she watched Patrick and his cousin Joe on the House floor. She thought about what it had been like for poor, sickly Patrick to attempt to stand up to "all that rough and tumble with his cousins, especially the Robert Kennedys, with all those dogs and horses." She had taken on a full measure of the Kennedy family competitiveness and thought that her sons stood above the rest. It was her other son, Teddy, who had always seemed the more likely candidate to be standing there. In 1991 her eldest son had checked himself into the Institute of Living in Hartford, Connecticut, to deal with his alcoholism. In 1993 thirty-two-year-old Teddy married thirty-four-year-old Dr. Katherine Gershman, a psychiatrist who was nonplussed by the Kennedy name. Now Teddy was studying law at the University of Connecticut.

Joan remembered how, on one of her birthdays when her youngest child was no more than fourteen, he had been in bed with the flu. He seemed always to be sick with one thing or the other, and that he was upstairs by himself was not unusual. Then he showed up in the living room with a suit and bow tie. He took a seat at the piano. And there he played Louis Armstrong's "Satin Doll" for his mother.

F or the two Representative Kennedys, the House was a very different place. Joe lost his chairmanship of the banking subcommittee on consumer credit and fees, and the subcommittee was disbanded. Speaker of the House Newt Gingrich and his minions did everything they could to show that Joe's brand of populist activism was not wanted except for salting the earth beneath his feet.

At precisely the time when Joe had the seniority and the savvy to have an important impact on policies that affected most Americans, he was pushed to a back bench to watch the Republican ascendancy. He was not content, however, to sit there observing what he considered the dismantling of government and was thinking of running for governor in the heavily Democratic Commonwealth, where he thought he could truly have an impact.

"There are two ways to do this, the right way or the way I did it," Joe told Patrick. This was good counsel, but it was hardly needed. Unlike his older cousin, Patrick was adept at ingratiating himself with his elders. It was unthinkable that he would strut through the Capitol irritating his colleagues as he elbowed his way past them.

When Joe arrived in Congress, he had been wary of the media. But no one in the Kennedy family had ever been quite as suspicious around journalists as Patrick. He had seen his parents become tabloid entertainment. He was not about to let that happen to him. Patrick took the unusual step of stationing his press secretary, Larry Berman, in Providence, where the aide would spend almost all his time dealing with local Rhode Island media. He traveled back to his district on weekends and was a familiar sight at senior citizens' gathering and American Legion halls. He was still a frightfully bad speaker, and wherever he went he was kept protected and sheltered by his aides. Although his ambitious staff was blamed for hiding him away, they were merely following Patrick's orders.

Patrick did not want to be profiled in the national media and turned down interviews with everyone from Peter Jennings to David Letterman, from *People* to *Vanity Fair*. Patrick's reticence convinced his colleagues that he was not a media show-off like Joe but just one of the boys. "*Vanity Fair* calls up wanting to do a feature on Patrick," says Senator Jack Reed, then the other Rhode Island representative. "Well, I've been a congressman, and no one is calling me, you know. And Patrick turned it down."

As much as he had fled into adulthood in part to avoid endless hazing, Patrick found himself the victim of it in the House. Representative Neil Abercrombie of Hawaii describes the atmosphere in the House as being "a lot like high school or college, where friends rag one another. You work each other over. There is no mercy in here. And people see that. They test you right away. They come at you, they'll work you over something fierce."

There was no better way to prove your macho bona fides than by putting down this slight, hangdog Kennedy in a loud voice heard by as many of your colleagues as possible. "We'd do the *Wayne's World* thing—'We're not worthy, we're not worthy,'" says Abercrombie. "And then when he would laugh about that, we'd turn around and say, 'You punk kid. You think you really have something to tell?' Or, 'Oh, here comes Kennedy. Oh, everybody stand back, it's Kennedy. Oh, Great One, what do you have to tell us?' Or, 'I see they wouldn't let you into Massachusetts. You had to try to outflank them by coming from Rhode Island? Who are you kidding?' All the time he's got a big grin. And he's disarming that way. Politicians read body language, and right away you could see that Patrick had no pretensions. Joe was a little more distant from people at first, but Patrick was more a House person right away."

CHAPTER FORTY-ONE

Humbert Humbert

During the week when Rose was dying in January 1995, Michael spent his days at Citizens Energy or off on various other political and social activities. His 104-year-old grandmother had been the last link to the strong beginnings of the family. The Kennedy matriarch had attempted to give her nine children rigorously disciplined lives. She had believed overwhelmingly in the destiny of her family. She understood, as had her husband, that just as great fortunes can be dissipated, so can the personal blessings and opportunities that come with that wealth. It troubled her immensely that so many of her grandchildren lived lives of what she considered moral lassitude and ease.

Michael was not only cheating on his wife but had developed all the sly deceptions of a veteran alcoholic, hiding his bottles in the shower or on bookshelves and pouring his drinks with a casual air that suggested he didn't care whether he had a drink or not. Nothing that had happened to Michael recently—not his dispirited efforts running his Uncle Ted's campaign, not the changes at Citizens—altered his belief that he could live precisely as he wanted to live. He could have long weekend camping trips with his three kids while using the time equally for trysts with Natalie. He could have a heroic image as the selfless CEO of a nonprofit company serving the needy while he took home a salary and benefits that he would have had a hard time matching in the corporate world. He could have a loving wife and an ever-changing harem of women for casual diversions, plus the delicious Natalie at home.

On the last night of his grandmother's life, Michael was lying in bed next to his wife. Once Vicki fell asleep, Michael quietly got out of bed. When Vicki woke and found herself alone, she knew where she had to look. Natalie was like another daughter to her, and it was all so unthinkable. But she walked up those stairs and into Natalie's room and found Michael in bed with her. Although Vicki later denied the story, both Michael and Natalie told others that was what happened.

Michael proceeded to admit what he had to admit. He suggested that though he had been found in Natalie's bed, he had stumbled into it in a stupor unaware of what he was doing. He said that he was an alcoholic. That was true, but alcoholism was also a brilliant excuse; it was as if a third party had walked up into that bedroom.

To most wives, this deception would have been as transparent as glass, but Vicki no more wanted to know the truth than Michael wanted to tell it. The truth was dammed up, and if it ever burst, it would send a flood of sewage, ruining innumerable lives. Vicki appeared to believe her husband and sent him off to be healed of his alcoholism after agreeing to stay with him at least until he had worked that out.

O n the very day Rose died and one era ended, Michael Skakel drove thirty-six-year-old Michael Kennedy to Father Martin's Ashley, a Catholic rehabilitation facility in Havre de Grace, Maryland. The upscale institution was proud of its fine cuisine in a dining room looking out on Chesapeake Bay and featured a holistic approach that sought to treat "body, mind, and soul," relying in part on the twelve-step philosophy of Alcoholics Anonymous.

The story of Michael's treatment became public knowledge both in the Boston papers and in the national media. Although there was much understanding for Michael's difficult trek back to sobriety, while he was in rehabilitation his father-in-law Frank Gifford gave a speech to a conservative group in which he condemned the "freaks and rehabbers who populate and adorn our nation's media."

By mid-March, Michael was back at his office at Citizens, and back at home with Vicki. He returned with what Rosenthal called "religion around recovery." Michael was like a born-again Christian most comfortable among professed sinners like himself who could lead the fallen down to the river to be baptized. Skakel, who was now working at Citizens, was a deacon in the church of recovery. His one great accomplishment was his recovery and the succor he gave to others who had fallen. Skakel spent much time with Michael and tried in his way to keep his cousin on the path of righteous sobriety.

Alcoholics Anonymous is an immense spiritual force among the young Kennedys. Bobby considers it a key to his spiritual salvation. Joan counted up more than a dozen young Kennedys who had come to her with their alcohol and drug problems, seeking help mainly through AA. For the most part, the Kennedys did not attend meetings with other alcoholics outside the family. They feared their stories might appear in the tabloids, but there remained a precious, insular quality to their sincere belief in AA. At AA meetings, young and old, rich and poor sat together, finding commonality and trust in their common affliction. When the young Kennedys had their meetings at Hyannis Port, the friends and relatives who did not take part felt like outsiders to an exclusive club.

When Michael took off the robes of his new faith, it was the same man beneath. Michael was a wizard with words—he appeared to be telling his dark truths but kept the important secrets to himself. Natalie was not so immediately available to him any longer, but this man of endless, uncontrollable lust started secretly seeing her again. He also saw other women. Without the balm of liquor, his thirst for sex and adventure took even greater hold. He was invited to a friend's wedding and beforehand used the bridal suite for an assignation with one of the guests. He flew to New York for a business meeting and had a rendezvous in an apartment with a woman before flying back home. He was eclectic in his pursuit, everything from highborn ladies to sexually acrobatic professionals with Russian accents provided by one of Bobby's old friends.

At Citizens, Michael continued playing the role of the most charitable of men, while the dismembering continued. One of those watching with apprehension was Larry Kellerman, president of Citizens Power & Light. Kellerman saw great possibilities in lowering electric costs for consumers and helping the poor pay their power bills. At first his fledgling unit was largely ignored among the big money dealers at Citizens, but with the deregulation of electricity, the subsidiary was right there in the midst of the action. Kellerman felt Joe's old friend and Citizens Corporation CEO Wilbur James standing behind him, watching his moves.

There were rumors and meetings dealing with a possible sale of the subsidiary, much of the discussion kept even from top executives. Kellerman recalls James telling him, "If Michael asks you anything about the business, how the business is going, what deals you are working on, don't tell him. Direct him to talk to me." The Citizens Power & Light president had the sense that James was trying to keep Michael from knowing the value of the company and the extent of its activities. "Now, don't worry," James told him. "I'm going to make us a lot of money with this deal, and you stay out of it and I'll deal with it."

While Michael was running Ted's senatorial campaign, Citizens Power & Light merged with Lehman Brothers Holdings Inc. James took over as CEO of the joint company while Kellerman became president. Joe may have been in Washington, but he had some responsibility to watch over the company he had started and to see that the struggle over the direction of the company would go toward maintaining Citizens initial ideals.

James had a masterly mind for the minutiae of business, especially when it came to cutting a deal that could make him his second fortune of the decade. He had his own right to be angry at Michael for his dangerous misconduct, and the two men were no longer getting along.

In March 1995 the organization sold off another profit-making subsidiary, Citizens Conservation, to EUA Cogenex Corp., the company that had previously purchased Citizens Heat and Power. The Citizens subsidiary was a privately owned company, and no one was under any obligation to disclose the details of the sale. The eighteen Citizens employees who had worked lowering heating costs in public projects were the next day working for EUA Cogenex.

Michael was more and more the public man, as if the spotlight were the best place to hide. In 1995 Michael's friend John Rosenthal had founded an important handgun-control organization, Stop Handgun Violence, but it was Michael who was often out front, as if it were all his idea. He drew contributors to fund-raisers and attacked House Speaker Thomas M. Finneran for "dirty politics" for trying to prevent the passage of a bill controlling handguns. This garnered him important media exposure.

Michael was thinking of running for Congress in 1996 for the seat of the retiring Garry Studds. The district would have been exchanging Studds, who was censured in 1983 by his colleagues for having sex with a seventeen-year-old male page, for Michael, whose predilection was for a teenager of the opposite sex.

When Michael decided that he was not interested in running, his younger brother Max announced that he had "taken soundings" about entering the race. Max was a former assistant district attorney in Philadelphia and a recent arrival in the Boston area. To many it appeared that the Kennedys considered the seat theirs for the having. Max says he was not serious about this, but his political musings had not made a good impression on many Massachusetts voters.

While Michael was contemplating running for Congress and promoting gun control, Wilbur was leaving taking much of Citizens with him. "Wilbur left because Wilbur and Michael were at each other's throats in the office," says James Hilliard, a longtime associate of Citizens. "With Wilbur's ego and ambition he wanted to do things his way. Michael was far more socially conscious and they kept butting heads. They had major arguments over how to run the profits and they were literally kicking the shit out of each other in board meetings. Michael said finally, 'I can't work with this guy.' And so they took a look at the companies, which of the profits would stay and which would go. Wilbur cherry-picked, and based on financial information we had at the time, we felt we didn't want to keep these companies."

Before he left Citizens for good, James negotiated the purchase of 80 percent of three of the remaining profit-making subsidiaries, including the oil trading and oil exploration units. To finance it, the new company received a $4 million loan from Citizens at far below market rates. He also received $100,000 in severance pay and half a million dollars in another deferred bonus.

Then James began lobbying the board to buy partial interest in Citizens Lehman Power LP, the electricity trading company in which Citizens Power & Light and Lehman Brothers Holding Inc. were partners. "Wilbur was actively working the rest of the board members, saying this was how it was going to be beneficial for the entire company," Kellerman recalls. "I don't believe that the board members were conscious of how much it was that they were prospectively giving up."

In January 1996, for half a million dollars, Citizens sold James a 16 percent interest in Citizens Lehman Power LP. Citizens lent him all but $100,000 of the purchase price. Michael and the board at Citizens officially valued the entire company at slightly over $3 million, meaning, by their estimates, that James had probably overpaid for his share. James brought a number of employees into the deal, which gave them every incentive to be behind the sale. While he owned 16 percent, he told the *Boston Globe* that another 20 percent was "held by 25 employees who were given shares or options as deferred compensation." That meant that 72 percent of the Citizens half of the company was owned by James and twenty-five other employees.

The sale came to haunt Citizens.

D uring its first year, the new company achieved operating profits of $14.5 million. James's share of that paid off his purchase price in less than three months. In the following year Energy Group PLC of Britain bought the company for up to $120 million based on stock value paid over five years. James's take would have been as much as $19.2 million. Kellerman was so distraught at what he saw happening that he left the company, though he reflects that his share would have been $3 million if he had held on three more years to become fully vested. Most of the other employees stayed, and James was not the only new millionaire at Citizens.

For his decade of work at Citizens, James was walking out with a fortune probably larger than that of almost any of the Kennedys. The Kennedys had always had loyal retainers, lawyers and executives who sought to advance themselves through association with the family. They often succeeded, but in almost all instances the Kennedys used others more than others used them.

When asked about the sale, Michael noted that neither he nor his brother was benefiting and said that it "will allow us in the future to fund more non-profit activities that will help low-income people."

M ichael was no longer getting high on booze, coke, or marijuana but was on antidepressant pills. The medication should have settled him down emotionally, but when he took his children out water skiing, he went from intense concentration on their efforts to total disregard. Little Michael was standing putting on his skis when Michael gunned the boat. The boy fell off the small vessel, and his father laughed as if to say "Gotcha!" There was something new and disturbing in the father who guided this boat, and Vicki was reluctant to allow her husband to take their children out on the water.

In June 1996, Michael and Vicki attended the graduation of eighteen-year-old Natalie from Thayer Academy, where she was an honor student. They came as friends to celebrate Natalie's achievement, but the affair continued. Natalie was a talkative teenager. She told her friends about her relationship with this handsome, charismatic Kennedy, and they told their friends, and the matter became part of the underground of gossip, ready to erupt somewhere. She suggested that Michael was ready to leave Vicki to marry her, an indication of the crazed unreality of their affair.

B obby says that in October 1996 at an RFK charity golf tournament on Cape Cod, Joe asked him to speak to Michael and convince him to end his affair. If the relationship ever became public, Joe would suffer more collateral damage than anyone else in the family. In the Commonwealth he was respected as an attack dog of progressive politics and as the founder of the benighted Citizens Energy. With the popular Republican governor William Weld probably retiring, most observers felt that Joe had a lock not only on the gubernatorial nomination but also on the election, and Joe and his advisers were preparing for the race.

Of all the brother, Bobby had the deepest connection to Michael's problems. There are so many different emotions that can drive a person to addiction. It is not one problem but an enormous, immensely varied family of problems. What Bobby and Michael shared with most of their siblings was a manic intensity, exacerbated by all the mammoth expectations of their lives and the endless temptations. They learned to control the intensity with temporary palliatives, be it liquor, sex, or drugs. A cynic would add AA to the list, a quick pop of spirituality. Politics was another attraction strong enough to still the beast within. "He [Michael] came to me because of my own history of recovery from addiction, and my experience in helping other family members who were battling those kinds of issues," Bobby says.

Bobby had a self-righteous quality that sometimes irked those around him, but his sincerity and empathy were unquestionable. "I was not the only one who confronted Michael at that time," Bobby says. Family members had tried, but "there is a limit on anybody's ability to control another human being's behavior." If Ted had truly been the unquestioned patriarch of the family, he might have done something. But the senator had no authority to stop this matter, and no one looked at him as the crucial guide. Unfortunately, there was no one in the family—no Joseph P. Kennedy Sr., no Steve Smith—with the resolute strength and determination to force Michael to stop, not only for his own sake but also for the sake of the family and its future. Michael didn't have to listen to Bobby to know that the road he was traveling was rutted with the tracks of several other Kennedys who had followed a pathway to destruction or disgrace, but he still headed onward.

Natalie was now a freshman at Boston University, with a life that did not include Michael. As she drew away from him, he was, if anything, even more obsessed with her. In the fall of 1996, he prowled around looking for her, on campus, in Cohasset, outside her parents' elegant home in Boston's Back Bay. It was a comedown for a man who had been the master of sex to be creeping around Boston University looking for his eighteen-year-old paramour.

N atalie's mother, June, had no idea about her daughter's affair. She started receiving anonymous phone calls in September 1996. June chalked them up to pettiness and jealousy. What was unthinkable was unthinkable. Then her friends started telling her of the rumors, their voices gentle and protective, and she began to wonder.

This was to have been a special season for the Verrochis. With Natalie off to Boston University, her parents planned to spend a lot of time in their new million-dollar condominium at the beginning of Commonwealth Avenue, overlooking the Public Garden. At the Crohn's and Colitis Foundation of America dinner in November, the Verrochis were to be honored for their fund-raising. The event would signal a large measure of social acceptance for June and her husband.

One of June's closest friends was Lorraine Paolella, whose husband had sold his New Haven ambulance company to Verrochi's growing company. June told Lorraine about the rumors and made sure she came up to Boston for a cocktail party in October for the major donors to the big November dinner. Michael and Vicki were among the guests, and Lorraine watched them all evening. The couple arrived and left separately, but what bothered Lorraine was the look that Michael gave Natalie during the cocktail party. It was a look he had been giving her for at least three and a half years, but Lorraine was the first one to see it for what it was and to act on it. "I watched the way Kennedy watched your daughter," she told her friend. "Something isn't right."

The following night the Verrochis took their eighteen-year-old daughter to dinner to ask her if she was having an affair with Michael. In any circumstances, this would have been a horrendously difficult moment, but for Natalie's parents, it was even worse. Paul was a leading Democrat and an admirer of the Kennedys who had treated Michael almost as the reincarnation of his sainted father. He had contributed to Michael's causes and sat on the Citizens board because he thought so much of him. June had sent her daughter into the Kennedy home because she thought that Natalie was safe in the sheltering arms of a noble family. As for Natalie, she knew her mother's emotional vulnerabilities, her drinking, her multitude of problems, and what the truth might do to her. For several years Natalie had received fine training in the craft of deception from Michael. She looked at her parents and suggested it was nothing but a cruel rumor.

In the morning Natalie called her mother and said that she was coming to see her to tell her something. Natalie told her mother the truth because she had begun to realize that her relationship with Michael could not go on. Natalie was becoming afraid of her obsessed lover. He was shadowing her, haunting her, stalking her. She saw the sickness of the man and realized this was no Hollywood romance.

In the early morning of November 7, 1996, the day after Natalie told her mother about her affair with Michael, June became so distraught that she climbed up onto the roof of the Commonwealth Avenue townhouse. She stood in a white nightgown in her bare feet in the cold rain, looking to one of the rescuers "like an angel standing on the gutter." Firefighters raised a 110-foot ladder to the roof and brought her down where Natalie stood next to Skakel.

Skakel was there because, as Natalie said later, he had taken her "under his wing." Skakel directed the college freshman to a therapist in Cambridge, who helped wean her away from Michael. Skakel was a good and generous friend to the troubled Natalie. But in doing so, he was either betraying his friendship with Michael or asserting that he was his moral superior. That Skakel dared to side with Natalie was also a mark of the increasing vulnerability and weakness of the Kennedys.

It was an emotional revenge for all the rebukes that Skakel felt his family had suffered at the hands of the Kennedys and all the demeaning duties that had been his sad lot. In his own mind he contrasted the inspiring story of his grandfather, George Skakel, who had "founded the Great Lakes Carbon Corporation and revolutionized the world's aluminum industry," with what he called "the whiskey-running gangsterism of Joseph P. Kennedy" and his "calumny and slander" against a man who was far superior.

"Michael and I fought bitterly," Skakel wrote in a book proposal for an unpublished book. "He claimed that I'd threatened him physically. He even tried to claim that I was the one who had been stalking [Natalie] Verrochi in an attempt to smear and blackmail him. He was a desperate addict caught in a trap of his own devising. He had used up all his options. He was dangerous."

That same November 1996 Joe received a phone call from John J. Casey, a Boston executive who was one of his major supporters and a close friend of Paul Verrochi. Casey was calling to put together a meeting between the aggrieved father and the congressman to try to end the affair. Joe had the impression that Casey was making the call without having talked first to Verrochi. Joe said later that "no one ever approached me saying they were an emissary from the family or with any message from them." Joe hung up on what was probably the last chance before the matter became public.

CHAPTER FORTY-TWO

A Tattered Banner

icki had three lovely children and a husband whom she truly loved, but it was devastating beyond imagination that Michael had lied to her all over again and was still seeing Natalie. This time he told his wife that he was a sex addict. This was true, though it was merely another manifestation of the spiritual malaise that so affected him. He had successfully pushed the problem away again. It was as if he had picked up a bizarre malady and only needed to find the proper medicine and all would be well. In this case, the proper medicine was a stint in a rehabilitation center focusing on sexual addiction.

In December, Michael headed off to Scottsdale, Arizona. This program employed the twelve steps practiced by those in AA. Step eight, and one of the most difficult of all, is not only to "make a list of all the persons" one has harmed but also to become "willing to make amends to them."

AA has been so successful in part because these steps, as simple as they may seem, are psychologically profound. Thus, step nine is to make "direct amends to such people wherever possible, *except* when to do so would injure them or others." Sometimes by confessing one's sins, the sinner renders new damage. A person seeking personal salvation must know when to speak and when not to speak.

Michael carefully followed step eight, but not so carefully step nine. He wrote down the names of all the women he had slept with, a lengthy accounting that reportedly included one of his sisters-in-law, with whom he had allegedly been intimate when she was single. The list also contained a number of family friends. Michael allegedly went to Vicki and showed her the list. It was as if betrayal was everywhere. Vicki had stayed with Michael through all his difficulties, but soon after reading the list she told Michael that she wanted a divorce. Vicki's lawyer had Michael in the most unenviable position: he was an admitted adulterer who for the sake of his family name was willing to do almost anything to keep the whole sordid business quiet.

I n April 1997, a few days after Michael and Vicki announced a separation after sixteen years of marriage, the *Boston Globe* broke the story of his affair with the teenage baby-sitter. Although the paper did not mention Natalie's name, it named her parents, allowing anyone interested to figure out who Natalie was. That was something the newspaper would have been unlikely to do without having received June and Paul Verrochi's permission, as *Boston Globe* editor Matthew Storin "strongly hinted" to the *Washington Post*. The Verrochis had apparently decided to punish the Kennedys in the forum that would hurt them the worst.

The Kennedys, once they could not deny that a relationship had taken place, suggested that the affair had begun after Natalie's sixteenth birthday, when in Massachusetts Michael would no longer have been guilty of statutory rape. The *Boston Globe*, for its part, said that Michael "began a relationship with the girl when she was 14," citing anonymous "friends" as its sources. Boston's leading paper had a long tradition of supporting the Kennedys; it almost certainly would not have published the story unless the editors believed that the girl had been underage and that this was not merely a titillating sex scandal but a crime story. The prosecutors who began an investigation did so for the same reason: they were seeking to determine whether Michael could be indicted for statutory rape.

T he story in the *Boston Globe* was a blood call to the tabloids, and they and their more elevated colleagues descended on Cohasset. They knocked on neighbors' doors, purchased yearbooks from recent Thayer graduates, and interviewed self-professed "Kennedy friends," but as always in family scandals, intimates defined themselves primarily by their silence. A reporter for the *Star* chased Vicki and Rory, her sister-in-law, down Jerusalem Road in Cohasset. "Tab Team Terrorizes Victoria Kennedy," headlined the local paper, the *Patriot Ledger*.

Michael was pursued by the tabloids too, but to him it was an adventure straight out of a movie, switching cars, hiding out at friends' homes and offices, speeding around to meetings and rendezvous with lady friends. On one occasion he hurried into Rosenthal's offices in suburban Boston. "I gotta use your car, John," Michael said. "I gotta be incognito here."

"Well, did anyone follow you?"

"No. Are you kidding? I beat them."

Just as Michael was bragging about how he had outraced *Hard Copy* and the *Boston Herald*, the phone rang from Rosenthal's father. "I was just watching the news," he said. "Michael just walked into your office."

For all the attention given Michael, there was one diversion. Michael was having a difficult time with his outraged father-in-law, who had puffed himself up into a posture of gigantic moral superiority. In the midst of Michael's travails, the *Globe* published a story and pictures of sixty-six-year-old Gifford having sex with an airline stewardess in the Regency Hotel in New York City. The tabloid had paid the woman to set up the sports commentator where he could be secretly videotaped. The spectacle did no honor to anyone, not the tabloid, not the woman, and not Gifford. What so amused Michael was that during his videotaped tryst Gifford confided to the woman how upset he was with his adulterous son-in-law who had slept with a fourteen-year-old baby-sitter.

Michael looked with amusement at his father-in-law's utter insincerity, but he did not look hard at his own life. Most people rarely change their fundamental conduct unless they are forced by the most dramatic and unpalatable of circumstances, and sometimes even that does not change them. There was so much that Michael could have done during 1997 to restore the balance of his life. He could have returned to Citizens with the firm intention of turning the organization into a far worthier philanthropy. He could have fallen to his knees over the shame of it all, the hurt he had caused, and the life he had lost.

When Michael went to visit Rosenthal in the hospital where he was grievously ill, he noticed that the window looked out on a girls' private school. Looking down on the playing field below, Michael made a crack about Rosenthal's good fortune. It was just a joke, and that's how Rosenthal took it, but it was a peculiar joke for Michael to make.

In June, Skakel was called into the district attorney's office to be interviewed. Michael's cousin told prosecutors all that Natalie had confided in him, detailing a sexual relationship that had indeed begun when she was underage.

Natalie and her parents were faced with the common dilemma of victims of child molestation or rape. If she came forward, the prosecutors would almost certainly indict Michael. Natalie would have to testify in a public trial in which the main question would be her age when she started having sexual relations with the defendant. A number of people have said that Michael and Natalie were intimate well before her sixteenth birthday. Some of them were likely to testify. Michael's attorneys would have to argue that their client had the high moral standards and shrewd good sense to wait until Natalie's sixteenth birthday before he plucked her innocence. The defense attorneys would have to impugn Natalie's honor by suggesting that she was lying about the year the affair began. Even if Michael escaped a guilty verdict, the public exposure would be so devastating that it would surely end the political prospects not only of Michael but of all the brothers.

Natalie did not come forward. She did not want to be dragged through muck, nor did she want her mother to suffer more than she was already suffering. Without Natalie's testimony, the prosecutors had no choice but to walk away from the case.

❧

I n 1999, the Verrochis divorced. Soon afterward, Paul married Kelley Bowen, executive director of the Crohn's and Colitis Foundation, which had honored June and Paul Verrochi as "Man and Woman of the Year." In August 2001, fifty-one-year-old June would die of what the family said was "a long illness."

It became indisputable what Natalie's testimony would have been when in 2002 twenty-four-year-old Natalie appeared in court in Skakel's behalf when he was tried for the murder of Martha Moxley. Natalie testified that she had met Skakel when she was fifteen and already "in the middle of a personal and family mess, and it involved Michael Kennedy."

I n the summer of 1997, a reporter from the *Boston Globe* came to interview Michael about the money he made at Citizens. This was a story that probably would not have been published without the previous scandal. In the midst of the interview, one of Michael's dogs jumped up into his lap. Michael began carrying on an imaginary conversation with the animal, saying that the article would charge that he had made "millions of dollars." The reporter tried to get Michael to answer his questions, but Michael kept talking to the dog. "It's going to make me look sooooo bad," he told the animal. The reporter "was struck by his apparent failure to grasp how serious the matter was," but the truth was probably the opposite. Michael understood precisely how serious the matter was, and he was attempting to lead the reporter away.

That year, when Michael knew that he was under immense scrutiny, he and the other two top paid employees, John Doran and treasurer Joseph A. Moran III, received a combined $1,603,321 in salary, benefits, and deferred compensation. One law firm, Connor & Hilliard, received $175,770 and two consulting firms over $90,000. That same year Citizens Energy paid out $2,513,523 in programs and contributions. If those figures became public, it would be clear that the noble philanthropic ideals were largely a sham. The outside directors who approved the corporate financial statements were themselves receiving $333 an hour for attending board meetings.

The story that appeared in the *Globe* in June 1997 pointed out that in 1995, when Michael earned $313,000 in salary, not including stock options, he was already earning more "than the heads of Harvard University and Habitat for Humanity, the American Red Cross and the Salvation Army, the United Way of Massachusetts Bay and the Smithsonian Institution in Washington." Joe's former head of staff and close friend, Chuck McDermott, expressed Joe's discomfort, presumably at Joe's request. "I have been witness on more than one occasion to Joe expressing disappointment and disagreement with the kind of compensation figures" Michael has received, said McDermott. The article was a solid exploration of the financial realities, but it did not deal with stock options in the same detail as the salaries, and Michael could feel a sense of relief.

Michael might be able to make a game out of the media's relentless pursuit, but Joe could hardly outrun them or shut himself off in his house. He had his duties as a legislator. And if he was to run for governor, he was going to have to lift off the burden of the scandal before it crushed him. In April 1997, when Governor William Weld announced he would not be running for reelection, 17 percent of Massachusetts voters told a *Boston Herald* pollster "they're less likely to vote for Kennedy for governor because of his brother's alleged affair."

Joe had to deal not only with Michael's scandal but with his former wife's new book detailing "a woman's struggle to stop the Catholic Church from annulling her marriage." Although only a few pages in the book detailed their marital problems, but those were inevitably at the top of the stories. *Shattered Faith* was not a personal vendetta but a serious book that aired the issue as had not been done before. Sheila took the reader along with her through the hearing of the Church's private court, which in October 1996 granted Joe's annulment, a decision that his ex-wife appealed to a higher Church court in Rome.

In the pages of the book, Joe's hypocrisy and the seeming hypocrisy of the Church blended one into the other. In one year alone, 1990, the archdiocese spent $255,604 on these matters, while granting 90 percent of the annulments. And here was Joe, who elsewhere described himself as a "cafeteria Catholic." He didn't like the Church's stance on abortion, divorce, or women as priests, but he took a big helping of annulment because he thought it would be good for him.

"Sheila Rauch Kennedy writes of being intimidated, even browbeaten, by her husband," wrote *Boston Globe* columnist Eileen McNamara. "How many staff members who have rotated through the Congressman's office might make the same claim? For the congressman to have persisted in this annulment charade even after his first wife's strenuous objections, even after his marriage to Beth Kelly in a civil ceremony, is to invite the voters to question his good sense."

At the Democratic Party issues convention in June 1997, in front of the two thousand delegates, Joe wrenchingly tried to free himself from the weight of the scandal. His grandfather and namesake had never publicly apologized for anything in his life, but Ted and now Joe lined up at the public confessional. Joe's was a Clintonian apology, letting hang out what he had no choice but to let hang out.

"I had a marriage that didn't work out," he said, though it had worked out for a decade, and he had two sons by Sheila. His new wife, Beth, was seated behind him on the podium, and she presumably found the sentiments edifying. Her husband was telling the world that she was his first chance at a successful marriage. If things worked out, she would be Joe's first wife in the eyes of the Church as well. She was devoted to Joe, and without children or a separate career to divert her, she gave him her undivided attention. "I can't tell you, and I can't put into words, how sorry I am about that," Joe said. "I said things that I wish I'd never said, and I did things I wish I had never done. I've told you, I've told Sheila, I've told anyone who cared, how sorry I am.

"On the matter of my brother, I am sorry, so very sorry, for what has happened to the Verrochi family," he went on. "I extend to them the deepest apology that I can summon. I love my brother very much. I will always love my brother, and I will stand with my brother."

That was not Joe's apology for *his* actions but for his brother's. What he did not address was his culpability in ignoring the problems for so long. What he could not say then or probably ever was that he had grown alienated from his beloved brother, enraged at what Michael and his other dearest friends were doing with Citizens. For the most part he had little contact with his brother except on mandatory family occasions. As for what he said about his marriage that day and other days, what was important was what he did not say. He never told his side of the failed marriage. Nor did he allow his aides surreptitiously to peddle *his* story to the media. His discretion was immensely honorable, doubly so for a man of such a volatile temper, and his reticence doubtlessly helped his sons to weather the divorce.

In July, Joe went off to the beach at Hyannis Port with his sons to shoot off some fireworks he had purchased. It was just the kind of thing that sixteen-year-old boys love to do, even if it was against Massachusetts law. Joe shot off the first of the fireworks and gave the next one to Matt, who in setting it off severely burned his arm. Joe drove his son to the emergency room at Cape Cod Hospital. He told reporters, "Matt's injury wasn't serious and the doctor says he will be fine." It was serious enough that the doctors considered skin grafts.

CHAPTER FORTY-THREE

"Ladies and Gentlemen, Meet *George*"

In early 1993, at the beginning of the Clinton administration, John sensed a country politically energized in a way it had not been since his father was president. He did not suggest that this was because Bill Clinton had modeled himself upon the JFK he had met in the Rose Garden as a teenager. He believed "this reawakening is due less to changes of heart than to changes in how the elected communicate with the electorate." The signs were everywhere that politics and popular culture were coming together. Politicians went where the people went, and the people were not hanging out at political rallies but staying home watching the tube. Clinton played the sax on Arsenio Hall's show. George Bush wooed young voters on MTV. And when Ross Perot announced for the presidency, he did so on *Larry King Live*.

John saw just how obsessed America had become with celebrity. The *People* magazine that had anointed John the sexiest man of 1988 was the most successful magazine of his lifetime. *People* legitimized the national fixation with celebrity, celebrating the triumphs of the stars and chronicling their woes, while elevating others with its attention. Other major magazines, including the newsweeklies, began regularly putting entertainment stars on their covers. For the most part, publicists made sure that magazines portrayed their clients in glowingly inoffensive profiles that promoted their latest film or project. TV programs such as *Entertainment Tonight* portrayed Hollywood with a lens that rarely showed blemishes. Journalism had become part of celebrity culture. Media news personalities made their way largely through the forceful assertion of their own opinions, while serious reporters turned up on television to speak in sound bites, the common idiom of the age.

John and his friend and partner, public relations executive Michael Berman, were looking for a business project. While John was working in the DA's office, they began talking seriously about publishing a new kind of political magazine that would "deal with politicians and political figures as pop icons and politics as an aspect of popular culture." John attended a two-day seminar on "Starting Your Own Magazine," but the two partners had talked and planned for a good two years before proceeding. They brought in David Kuhn, a former *Vanity Fair* editor then working in the film industry, as a consultant. John had known Kuhn since his days at Andover and trusted his judgment.

"I always thought he [John] had a good journalistic gut because he was naturally and understandably wary of people and their motives—though he tended not to show it—and had a great bullshit detector," says Kuhn. "He had the ability to see the story behind the story of people and events that others did not. I think that was in part because he was very bright, and a good reader of people, and also because of his need to constantly figure out who wanted what from him and for what reason.

"I told him that I thought he had what it took to be a very good editor. He was concerned about his lack of experience, but I encouraged him not to worry as much about his official training and to trust that he had a good story sense, knew a lot about the world he was going to cover, and would be able to get access others might never get—all things that a good magazine editor needs."

John had a political background unique in America and awesome celebrity. Although Berman had little interest in politics, he brought media savvy and an often cautionary note to John's exuberant aspirations.

"I grew up feeling that people in public life were heroes," John said, "that their lives were difficult, they were often misunderstood by people on the outside, and that this illusion of power and high living was really misinformed." John was taking his idealized vision of his own family and projecting it out onto the world. His father was a brilliant model of this meld of politics and popular culture, though what was called in the sixties "the Kennedy style" was both celebrated as worthy of emulation and condemned as trivial. "I didn't understand why movie actors were the only ones who could sell magazines and why people in entertainment were the only heroes of popular culture," John mused. "I thought if I could parachute behind enemy lines, in a way, and join the journalistic profession, that I could begin to let my perspective about politics seep in and maybe influence the presentation of politics."

John wanted to publish a forthrightly bipartisan magazine, with no slant to right or left. He intended to focus on the intimate political lives of the nation's leaders while leaving their personal lives where he thought they belonged, behind their closed bedroom doors. He believed that most elected officials worked hard and tried to serve their country well, and he wanted to write positively about them.

Most new magazines fail, and John was taking a considerable risk. He could have minimized the risk by publishing a forthrightly "serious" magazine that would have served as an enviable platform for the political ambitions that he secretly harbored. Instead, he was reaching out to an audience that would never pick up *The New Republic* or *Washington Monthly*. In doing so, he knew that his new magazine was likely to be snubbed by what he considered inbred, insular Washington, where journalists thought they were talking about the world when they were only talking shop.

The title, *George*, boldly announced that this was not your Uncle Harry's political magazine. John would eschew the dreary design of political magazines, from *The National Review* to *The Nation*, and instead seek a look more like *Elle* or *Vanity Fair*, one of those "really inviting, accessible, exuberant youthful magazines." At a time in American journalism when if you wanted to succeed the best place to stand was at an ironical distance, *George* would have a sun-flecked optimism about politics and politicians.

John and Berman took their plans to a number of potential investors. It was an unlikely concept for a mass magazine, and if John's name hadn't been affixed to it, they would not have even gotten in many doors. When they showed their proposal to David Pecker, the CEO of Hachette Filipacchi, a successful second-tier magazine publisher, they found their man. Pecker was a shrewd, cynical executive whose eyes focused on the bottom line. The French parent company had the parsimonious spirit of a petit bourgeois shopkeeper. Hachette was not about to pony up $20 million to become half-owner of the magazine because of an abiding interest in American democracy. John had begun with somewhat more modest aspirations for his magazine, but with Hachette's infusion of capital, he was quickly in the big-stakes game of magazine publishing. His name on this new magazine was buzz enough that the chicest advertisers lined up to be in the first issue, 175 pages in all, the largest ever for a magazine start-up.

Editor-in-chief John would call big advertisers and ask famous authors to write for the magazine, but he needed a team of talented editors to handle the grunt work of putting out a magazine. John and Berman were amateurs, about both politics and magazine journalism, and they clearly needed a staff of professionals knowledgeable in both arenas. Instead of signing up a hands-on editor who knew the Washington world, they hired thirty-eight-year-old Eric Etheridge, the executive editor of the *New York Observer*, a studiously irreverent, upscale weekly that observed the capital from a wary distance.

Elizabeth "Biz" Mitchell came from *Spin*. One of the other senior editors, thirty-year-old Richard Blow, had at least lived in Washington as the last editor of *Regardie's*, a local business magazine. John's friend Gary Ginsberg came on board too as another senior editor and the magazine's attorney at a third the salary he was earning as a lawyer. He was graduating from a stint as an attorney in the Clinton White House and Justice Department and had the best political instincts of any of them, but he had no expertise in journalism. The other staff, almost none of them over thirty, knew practically nothing about politics. John considered ignorance a virtue for a magazine that would move beyond the narrow, clichéd world of Washington journalism.

There had never been a magazine editor with John's celebrity. He needed to know that his intimate words and actions were not going to find their way into the *New York Post*'s "Page Six" or be sold to the tabloids for a few pieces of silver. The staff had to sign a confidentiality agreement promising that they would not write or talk about John. Of course, if you hired the right kind of people, you did not need such an oath. If you hired the wrong kind, it did not matter what you asked them to sign.

John had never in his life worked as hard or as long as he did preparing the first issue of *George*. He put in sixty straight days, often fifteen hours long. He pulsed with nervous energy, and afterward he usually either headed out to Central Park to throw a ball around or worked out at a health club. No matter how ravenously he ate, he was losing weight. He rarely saw a doctor, but he realized that this was serious enough that he went to a specialist for a series of tests.

John would have loved to put off flying to Montgomery, Alabama, to interview former Governor George C. Wallace until he got the medical results. But John's interview was going to be a central feature of the first issue, and it could not be left out. John's father had confronted the diminutive Wallace in June 1963 when the governor had vowed to "stand in the schoolhouse door" to prevent blacks from attending the University of Alabama. The racist Wallace had touched a disturbing chord in American life while articulating the deep, unspoken fears of many white Americans who felt that a liberal elite was controlling their lives. In 1972 a would-be assassin had left Wallace paralyzed in a wheelchair for the rest of his life. In recent years the politician had disavowed his segregationist past and won reelection with the help of substantial numbers of black voters willing to forgive his past.

On his first morning in Montgomery, John went for a run with Gary, who had accompanied him to Alabama. John usually outran his friend, but this time Gary left John far behind. When Gary looked back, he saw John slumped against a tree. That wasn't like his friend, and Gary worried that John might be seriously sick.

John was so weak that he would have been wise to reserve his strength for the Wallace interview. But before leaving New York, he had arranged to spend four full days learning everything he could about Alabama politics and culture, and he could hardly back out of that commitment. He was hoping to play the journalist, but he found himself treated like a celebrity. When he went to see Governor Fob Jones at the state capitol, half the politician's staff appeared to be in the room. Everywhere he went, from the mayor's office to the Dexter Avenue Baptist Church where Martin Luther King had been preacher, there were crowds and photographers.

The next day John and Gary went to Wallace's home. John was amply armed for this meeting with his father's nemesis. He had spent hours with Nicholas Katzenbach, a Kennedy Justice Department aide who had dealt directly with Wallace. John had interviewed scholars, read history books, and set down a long series of questions on three-by-five cards.

When John and Gary arrived, there were video cameras ready to record the interview and a staff aligned behind Wallace. The former governor came alive in the glare of lights and public drama, but he was a sick old man shaking with Parkinson's disease and hardly able to rise out of his torpor. This would likely be one of the last public moments in the man's life. The former governor was largely deaf and could hardly speak, and his eyes were wet.

The two fledgling journalists had hardly sat down before they realized that it would be impossible to get a decent interview out of this dying old man. They knew that they wouldn't even have been here if the interviewer had been anybody but John; the facilitator had wanted to bring the bright light of John's celebrity to Montgomery.

This was John's first experience as a working reporter, interviewing a Wallace who was so sick but still instinctively shrewd that probably no interviewer could have discovered what was the meaning of those tears. John failed at another level, for he could not bring himself to ask the tough, legitimate questions that he had written on his cards. Each time during the three sessions that there was a probing, difficult query, he passed the card over to Gary, who shouted it into the governor's ear or wrote it on a slate. Wallace's mumbling, stumbling answers were not a politician's professional dissembling but a measure of just how incoherent he had become. John might have pressed harder, though, and in a question twice asked found an answer in a silent response, but he was simply incapable of performing part of the essential process of journalism.

When John left the interview that day, he and Gary went off to play tennis. John was neither upset over the interview nor terribly interested in the game on the court. "You can't tell anybody, Gary," John said, as he opened a small box to disclose an exquisite diamond engagement ring. "You gonna do this?" Gary asked. "Yeah, I'm gonna do it." John had kept his intention of marrying Carolyn secret. It was a mark of how deeply he cared that after the failure of this important interview he was thinking not about the first issue of *George* but about the woman he wanted to be his wife.

John and Gary went back twice more to interview Wallace. It was an exhausting, frustrating four days. On the last evening John wanted nothing better than to be left alone. Anyone else would have chewed out the Alabama contact. John not only refrained from reaming the man out but instead agreed to have dinner with him and two friends at a roadside restaurant outside Montgomery.

When John and Gary arrived, there stood about a hundred excited Alabamans who had been invited to have dinner with the famous visitor. John made a hurried excuse and rushed off into the men's room with Gary. John did not want to live like a politician, with a frozen smile on his face and an eternally outstretched hand, but this trip had turned into something that had nothing to do with journalism. He had put up with it until now, but he was weak and weary.

"I gotta get out of here," he said hunched together with Gary in the men's room. "I gotta get out of here." Gary knew that if John walked out now, he would look like a snotty New York upstart. "You can't do it," Gary concluded. "You've got to go back out there." And so John walked back into the restaurant and played the Kennedy card all night long. Everyone loved him.

As soon as John got back to New York, he presented Carolyn with the engagement ring and asked her to marry him. Carolyn said that she needed time to think about it. John was a man who became fickle when romance got too easy. Carolyn was making a statement about her own worth and what a marriage to her would be like. Beyond that, she had an inkling of what it would be like to be married not just to John the man but to a mega-celebrity whose fame would, if anything, grow even larger. It was not a decision to be made on a romantic whim, and it took her a month to give her definitive yes.

While John waited and worked on the final details of the first issue of *George,* his doctor told him that he had Graves' disease, an incurable but treatable hyperthyroid condition that usually afflicts middle-aged women. The malady sapped his energy and made him even more nervous than he was normally. The disease was successfully controlled by drugs that John continued to take for the rest of his life.

The Wallace tapes were transcribed, and as John feared, it was a hopeless, dreary mishmash. John secretly called in Bruce Kluger, the longtime editor of the justly celebrated *Playboy* interviews. Kluger looked at the pages and saw that he could make nothing out of it. He recalled that when Wallace lost his first run for governor as a relative moderate, he had notoriously vowed "not to be outniggered again." "Did you ask him to define what the word 'nigger' is?" asked Kluger. John said he had not asked the question.

Gary and Blow flew back to Alabama and talked another time to a Wallace who had little more to say. Then Kluger took the pile of transcripts and did a masterful job creating a workable interview, but there was nothing in it of either the chilling menace of Wallace in his prime or the incoherent, pathetic man John had met. The scene in Montgomery had been something out of a political *Sunset Boulevard,* the paralyzed Wallace in his small house lying in a room dark except for a big television, watched over by two former convicts to whom he had granted clemency. None of that made it into John's introductory essay. It was a measure of how reluctant John was to confront dark realities that in his preface he did not use the "n" word but wrote that Wallace had vowed "not to be 'outsegged' again."

O n September 6, 1995, an impeccably dressed John F. Kennedy Jr. strode onstage at Federal Hall in lower Manhattan to applause and flashbulbs. He looked out on an overflow audience of journalists. "Ladies and gentlemen, meet *George*," he said, opening a lighted display of the cover with supermodel Cindy Crawford posing as George Washington with a bare midriff.

"Our first choice was Alan Greenspan in a Speedo," John said. For all his intentions to join the fraternity of journalists, he was as primed and scripted for this moment as his father had been when he had announced for the presidency. John had been prepared for this media conference by Michael Sheehan and Paul Begala, both of whom had done the same for President Clinton. They had staged a mock press conference in which Gary, playing a hostile journalist, asked John: "Mr. Kennedy, you failed the bar two times. Are you dumb or just lazy?" John laughed—and kept laughing for years afterward when his friends delightedly asked the infamous question again.

John read the scripted laugh lines as if he were hosting *Saturday Night Live*. "I don't think I've seen as many of you gathered in one place since they announced the results of my first bar exam," he said. Then he moved out ahead of the questions that were bound to probe everything from his romantic life and whether he wore boxers or briefs to his political aspirations. He answered them before the reporters could even pose their queries. "Yes. No. We're only good friends. None of your business. Honest, she's my cousin from Rhode Island. I've worn both. Maybe someday, but in New Jersey."

Up until this day John had never courted the media. Even now he tried to eschew personal publicity by talking about *George* and not himself. The one moment that appeared to touch him deeply was when a reporter asked him about his mother. "She would have been mildly amused," he said, "but she'd be glad to see me standing up here."

John's stellar performance matched the startling sales of *George*'s first issue. The half-million copies sold out, as did another hundred thousand that were rushed into print. That was an overwhelming success, but it would be hard to hold on to an audience in part drawn by the novelty, and nearly impossible to increase the circulation.

The first issue was an eclectic mixture that included an essay by novelist Caleb Carr, a piece on Julia Roberts's misbegotten attempt to do good in Haiti, and Madonna on what she would do as president ("The entire armed forces would come out of the closet"). Crawford made three appearances in the premier issue—as the cover girl, in a full-page Revlon ad, and in a bitchy, witty conversation with designer Isaac Mizrahi on politicians and style. The Cindy Crawford cover announced that John was not the least bit embarrassed to sell his magazine by posing half-draped actresses, sexy models, and handsome actors, usually with some ersatz political connection.

John intended *George* to be the *Sports Illustrated* of politics, but his critics said that it was more like *People*. Clearly, the initial issue did not come close to John's conception for the magazine. If he backed off those aspirations or got only partway there, *George* risked becoming just another glossy new face on the newsstand, adding to the celebrity-addled society by treating politicians and journalists with the same fawning adulation that movie stars already received.

Even if everything had been going swimmingly, John would have had a full agenda overseeing the publication of a bimonthly and soon to be monthly magazine. As it was, he had enormous problems, and they began at the top with both his partner and his top editor. Berman had strong ideas about the direction *George* should take: he wanted John to interview the leading political figures in America every month. John, however, was not one to be pushed on stage day after day to speak lines written for him. Berman was John's partner, but no one in an enterprise with John was going to be his equal. Beyond that, others at the magazine felt that Berman could not move beyond his friendship with John and see *George* as a business partnership.

John began to assert himself over the creative and marketing decisions, and Berman grew more upset. John was the least confrontational of men in his professional life, and he moved his office away from Berman's. Within a few months staff members walking in the hall heard the two friends screaming at each other behind their closed office doors. Finally, in mid-1997, Berman left the magazine for good.

John did not have a strong relationship with Etheridge either. The evening before the editor arrived, John had worked with Gary until two in the morning finalizing a list of story ideas that they had been working on for a year. Etheridge had the sophisticated patina of a successful New York editor while they were amateurs, intimidated by the idea that a real editor was showing up. Two days later, when Etheridge had not even commented on their list, they approached him. "I haven't focused much on it," the editor said somewhat dismissively. The next week Etheridge told them he was not excited about any of their ideas, but he had a winner—the decline of the Mexican aristocracy. The article might make it in *The Economist*, but it was not quite what John had in mind for *George*.

For John, it was a minor epiphany. Maybe the experts weren't so expert. Beyond that, Etheridge had made a fateful mistake by so cavalierly dismissing John and Gary's ideas. John was a man of instinctive, immediate judgments. Rightly or wrongly, he felt he had taken Etheridge's full measure. When the article about the Mexican elites arrived, John killed the piece. That may have made sense, but with both Etheridge and Berman, John showed that he found it difficult to deal directly with men in the power structure, preferring to wait them out or to subvert them in other ways.

John had articulately set out just what *George* was supposed to be, but Etheridge had a more upscale vision of the magazine. John was not capable of candidly discussing his differences with Etheridge, and it probably was only a matter of time before he would decide to take over editorship of the magazine in fact as well as in name. In February 1996, after only three issues, Etheridge left *George*. John promoted Mitchell to a new position, executive editor. Thirty-year-old Mitchell was talented, but she was unlikely to challenge his authority.

J ohn was reluctant to put his name on the cover every month with the *George* interview, even though he knew that his name sold thousands of magazines. He was conscious of the way Hachette was using him, touting his name together with theirs, shuttling him out to meet advertisers who would be pushed to buy ads in a group of magazines of which *George* would only be a part.

Many of the best pieces went against the celebrity concept of the magazine and took tough looks at American politics and journalists. The magazine published an article about Ruth Shalit, a twenty-five-year-old writer for

The New Republic who had plagiarized her way to what some called fame. *George* signed up another hot young writer from *The New Republic*, Stephen Glass, to do a profile of Clinton's friend Vernon Jordan. Glass proved to have been not only a plagiarizer but a fabricator. John later apologized in an editor's note, but he did not point out that Glass was yet another example of where this obsession with celebrity had led.

In April 1996, John interviewed Iain Calder, the longtime editor of the *National Enquirer*. John considered himself the tabloid's occasional victim, but he understood, as many of his colleagues did not, how the distinctions between the tabloids and other parts of popular media were becoming blurred.

Calder told John that he edited the *National Enquirer* for a "mythical Mrs. Smith in Kansas City [who] wants entertainment because her life, to some extent is boring. She likes looking at celebrities." There was this celebrity world out there where loves were deeper and life was richer. If one could not get there, at least one could enjoy it vicariously. In that world, John may have been a great icon, but he was also little better than a piece of meat to be fed to the hungry until they grew tired of the taste.

CHAPTER FORTY-FOUR

A Father and Son

very Thursday between 12:30 and 1:30, Ted and Patrick met for
lunch in the senator's hideaway office in the Capitol. The paneled
room was full of memorabilia from the Kennedys' rich and
poignant history. That these hours were so important to Patrick was a measure of the frenetic, restless father he had had growing up. Over the years
Patrick had gone back and forth, believing that being the son of such a powerful and controversial figure was both a curse and a blessing. He was in his
late twenties, and he still had not resolved many of the psychological dilemmas of childhood.

"For me it's opened up, personally, a whole dynamic of our relationship
that heretofore I haven't had the chance to have with him," Patrick said of
these lunches. Both men were uncomfortable with intimate emotional talk,
but their shared profession made it easier for both of them. "So as a son for
whom his father is the most important person in his life in a lot of respects,
this is a big change for me," Patrick said. Most twenty-nine-year-old men do
not consider their father the most important person in their lives, but that
was the reality of Patrick's life.

In his chosen profession, there was only one way that Patrick could rise
above his father's achievements, and that was to be president of the United
States. In 1988, when he was a twenty-one-year-old state representative, he
had admitted with aw-shucks nonchalance that he would like to run for pres-

ident one day. Many youthful politicians secretly harbor that goal, but few would admit to it, and those who did were often full of steely ambition.

Patrick's own father had slowly, deferentially made his way in the Senate, but from the beginning Ted was a strong, determined politician. Patrick's uncle Bobby had had a high-pitched, weak speaking voice, but by the sheer strength of his message and will had become a good and at times mesmerizing speaker. Patrick had grown more comfortable with his role as a congressman, but he remained inarticulate and prone to bouts of shrill moralizing that he attempted to pawn off as eloquence.

One of those who watched his career with acute interest was state treasurer Nancy Mayer, who, while winning justifiable acclaim for straightening up Rhode Island's pension system, was touted as a likely Republican opponent to Patrick in 1998. At an event in the fall of 1996 the two of them ended up sitting next to each other on a stage in a mall. If on a brilliant fall day she was unhappy to be seated inside for such a banal event, she imagined what it must have been like for him. "Patrick, what are you doing here?" she recalls asking him. "You're too young for this. You should be outside having fun with your friends." Patrick's face turned red with embarrassment. His uncles and father had at least seen two roads before them, one marked personal happiness and one political achievement, and they had chosen the latter road. But to Patrick the road to personal happiness had always been blocked from sight.

Patrick was a profoundly partisan Democrat but had few issues that he made indisputably his own. One of his deepest concerns for the most profound of personal reasons was gun control. In March 1996, he stood up in an impassioned, if forlorn attempt, to stop passage of a Republican-sponsored bill to allow citizens to own certain assault weapons. "My God! All I have to say to you is, play with the devil, die with the devil!" Patrick said, letting out what Mary McGrory in the *Washington Post* described as a "primal scream." "There are families out there! You'll never know—Mr.—Mr. Chairman, you'll never know what it's like because you don't have someone in your family killed. It's not the person who's killed, it's the whole family that's affected. It's not about crime, it's about the families and victims of crime. That's what we're advocating in proposing this ban, and that's why we should keep this ban in place."

Gerry Solomon, a combative Republican from upstate New York, rose up to tussle with his overwrought colleague. "Mr Speaker, before the gentleman leaves the floor—you know, I just want to say to him, I have a great respect for he and his family but when he stands up and—and questions the integrity of those of us that have this bill on the floor, the gentleman ought to be a little more careful. And let me tell you why!"

"Tell me why!" Patrick interjected.

"My wife lives alone five days a week in a rural area in upstate New York! She has a right to defend herself when I'm not there, son."

"And you know the facts about it," Patrick yelled. "You've got guns in your home . . ."

"And don't you ever forget it! Don't you ever forget it!"

Solomon invited his twenty-eight-year-old colleague to go outside and settle the matter with their fists. It was an invitation that Patrick's father might well have accepted, but Patrick turned down the offer, and the contretemps passed into the lore of the House.

What set Patrick apart from his colleagues was largely his name, the audiences that name brought with it, and the money that they contributed. In the 1996 campaign Patrick traveled across the country speaking for his party's candidates, raising large sums of money, and ingratiating himself with local leaders.

Many of his constituents in Rhode Island felt slighted by their itinerant congressman. "He doesn't really seem like a congressman from Rhode Island anymore if, in fact, he ever did," wrote Bob Kerr in an acerbic column in the *Providence Journal-Bulletin* in October 1998. "He is simply using the state as a return address, a place to come back to and pick up his mail after taking his name on the road. . . . He doesn't even stick around for his own campaign. Kennedy is so much not a congressman from Rhode Island that he won't even debate his opponents in this year's congressional race."

I n his first term and a half in Congress, Patrick proved a competent, hardworking member and a team player. When Senator John Chafee decided to retire in 2000, it seemed inevitable that Patrick would run for the seat and in the heavily Democratic state win handily. At the age of thirty, he was better prepared to run for the Senate in Rhode Island than his father had been to run in Massachusetts at a similar age. No Kennedy had ever turned away from such an enviable opportunity.

Patrick turned down the race, backing away against the advice and strong admonitions of his father and against everything that his Kennedy blood and heritage told him he must do. "Part of me worried about getting in front of myself, sticking my neck out for something that might not fit me now," he said. "It can be scary, getting too far out there. I don't want to live with that anxiety of not feeling like I'm living in my own skin."

It was one of those unknowable conundrums: was Patrick displaying a profile in courage or cowardice? Some believed he had shown courage in facing up to the realities of his own life and needs and choosing, despite four generations of Kennedy ambitions and his father's wishes, what was essentially his own life. Others thought him cowardly, such a feckless, feeble person that he could not risk assuming a position that might expose him for what he was.

Patrick was not without strong ambitions, even if he did choose not to run for the Senate. In November 1998, thirty-one-year-old Patrick became the chair of the Democratic Congressional Campaign Committee, the central figure in raising money and backing candidates for the 2000 congressional elections. Democratic members might elevate a celebrated young member to a prestigious committee for which they considered him largely unworthy, but they were not about to chose the DCCC chair for any other reason but the member's ability to raise money and do the job. By any measure it was a high honor, making him fifth in the House Democratic leadership, a formidable achievement for a third-term member. Patrick oversaw a staff of 126, arbitrated squabbles, and apportioned limited resources.

D uring the administrations of Bill Clinton and George W. Bush, politicians who once had proudly called themselves "liberals" fled from that term as if it were a badge of shame or a mark of embarrassing naivete. Ted was a shrewd enough politician not to emblazon everything he did with the newly despised word, but he continued to walk down the same streets he had always walked and to stand for the same issues. When Ted won, he often greeted his defeated opponent with a guffawing laugh that shook his whole corpulent body. If not for his superbly tailored suits, he would have looked like a nineteenth-century Irish American ward boss, with whom he shared many of the same skills.

Ted was an aged warrior with a young man's energy. He knew that a battle was not a war and that what looked like the end of a war was often only the beginning of a new battle. If he could not win in a frontal attack, he would attempt to outflank you, or he would sneak into your camp at night with some sort of shrewd gambit, perhaps attaching his agenda to a Republican-sponsored bill.

Ted had never balanced a checkbook and rarely looked at a price tag, and he had brought that disregard for the bottom line into his politics. When Clinton vowed to reform welfare, Ted saw that as a euphemism for cutting away the safety net beneath millions of disadvantaged Americans. He did not grasp that the price of welfare may have been too high, both for the taxpayers who funded it and for the recipients demeaned by a life of handouts. In 1995 the Senate voted on a bill to limit welfare to five years in a lifetime and to put half the recipients to work by the year 2000. Teddy was frenzied in his objections, trying to hold back his colleagues by arguing that "the Senate is on the brink of committing legislative child abuse." Despite such dire prophecies, the Senate voted 87 to 12 in favor of the measure.

Welfare reform proved nothing like the disaster Ted envisioned. In a bountiful economic time, the streets were not filled with the desperately poor. Christopher Jencks, a progressive academic, argued in *The American Prospect* that those who claimed that the reform "would cause a lot of suffering no longer have much credibility with middle-of-the-road legislators, who see welfare reform as an extraordinary success."

Ted had seen what long-term care meant in his parents' long invalid lives. When his son Teddy had his leg amputated, Ted learned "what parental leave is really all about." Those were both reasons why he was such a proponent of health care. Ted was for a patients' bill of rights to allow Americans great latitude in suing their HMOs. He eloquently attacked the medical establishment. Despite what Ted argued, those Republicans who proposed a more limited initiative were not merely the hirelings of the health industry. They had legitimate points to make about excessive, exorbitantly expensive litigation driving up health costs. A dispassionate person might conclude that both positions had their virtues, but dispassion was not a quality that Ted cultivated.

As much as Ted rarely moved beyond the received wisdom of traditional liberalism, he was always seeking new means to achieve his goals. In 1996 he and Senator Orrin Hatch cosponsored a bill raising cigarette taxes. The money would be used both to reduce the deficit and to give children health insurance. Ted had wanted a far bigger tax, but it was only the Utah senator's restraint that allowed the partners to create a bill that won other Republican cosponsors. Trent Lott, the Senate majority leader, considered even this lesser bill an assault on his struggle to push through a balanced budget amendment with the help of the administration. When the majority leader told President Clinton that the Kennedy-Hatch bill was a deal breaker, the president turned off his support and the measure failed. Ted came back again but failed in his attempt to raise the tobacco tax by forty-three cents a pack and had to settle for the twenty cents that Hatch had favored.

Ted continued to push until the Senate passed a bill that would, in his words, "give every child access to affordable health insurance." It was a great moment, but only a moment, and once he had taken the appropriate bows and thanked the appropriate players, he was on to a new issue. He focused entirely on playing the next down. "It's really 3 yards and a cloud of dust with him," says Patrick, an appropriate metaphor for his father's in-the-trenches politics.

There are senators who make their greatest contributions when their party is in the majority, and others who make their impact felt when they stand in the minority. Ted was almost unique in that he was a player equally adept at both roles. In the years between 1995 and 1998, the Republicans in the House, led by Speaker Gingrich, vowed to pass a "Contract with America," which would cement the conservative agenda into a broad series of laws. In early 1995 the speaker maneuvered the bills through the House in a hundred days, but when the legislation reached the Senate, Ted was waiting. "Nothing gets through the Senate, because Kennedy, basically, organizes the forces of resistance, and piece by piece, he brings the Democrats back, because he remembered what they stood for," recalls Nick Littlefield, his administrative assistant during most of the nineties.

In early 1996, when Senator Bob Dole was Clinton's putative opponent in the fall election, Ted set out to make the Senate majority leader look ineffective, mean-spirited, and vindictive. On issues such as the minimum wage and medical savings accounts, Ted helped maneuver his colleagues so that the Kansas senator appear a bumbling neophyte. Dole signaled his dismay by deciding in mid-May that he would leave his Senate seat. Ted had neither publicly vilified Dole nor indulged in vicious debate on the Senate floor. His actions were observed most closely by the man who benefited the most.

"A lot of people have great skills in Congress and aren't great politicians in an electoral sense," Clinton told Elsa Walsh of *The New Yorker*. "Then there are a lot of people who are great politicians who don't understand the moves, you know. He's as good at what he does as Michael Jordan is at playing basketball. I mean, he can always see the opening. He's got lateral vision and it's uncanny what he can do."

No matter what he was doing, Ted was always scanning the horizon for new issues and new possibilities. His hand touched everything from a peace plan in Ireland to his inevitable call for a higher minimum wage. "I'm in awe at his ideas and determination," says Melody Miller, his deputy press secretary. "My God, let's try this, let's try that. You don't want to ride in the car with him, because he'll say, look into this, or, we ought to call so-and-so. He has so many ideas, and that makes him happy. He thinks big."

Beyond his political astuteness, Ted had a hard-won emotional wisdom that drew people to him. When there was death or grievous sickness to deal with, he was often called. When he looked into your eyes and said that he knew what it meant to lose a loved one, and that life did go on, you knew that he was telling the truth.

CHAPTER FORTY-FIVE

John's Best Shot

I n February 1996, John and Carolyn went for a walk with their dog, Friday, in Washington Square Park. They were engaged now, living together, and as they strolled along they began to argue. Most people learn that their partner has certain subjects that, no matter how angry you get, you do not go there. Carolyn, however, knew just where to touch John to hurt him most. She raged at him and pushed him, and he pushed her back. "You've got my ring," he yelled at her. "You're not getting my dog." Then he collapsed in tears. Carolyn consoled him gently. He got up, and they walked away, followed by Friday.

John got so angry with Carolyn that he could hardly speak, but he would not only forgive but forget. Carolyn, for her part, collected every slight, every argument, every neglect, and stored it away, keeping it ready to bring forward at the next appropriate moment. And the longer they were together, the larger her emotional database became.

The scene in the park would have been painful enough if no one had seen or heard them, but wherever they went they had to confront the possibility that they were being watched. Their argument had been secretly videotaped for a tabloid show and publication in the *National Enquirer*. Millions of Americans took their voyeuristic pleasure in John's pain, while the *Los Angeles Times* suggested that it might have been little more than a public relations stunt. "This man who supposedly relishes his privacy has dated a string of actresses certain to juice the paparazzi," the paper said though he had done no such thing. "One has to wonder: knowing he's regularly tagged by the tabloids, why didn't he take his squabble indoors?"

When Carolyn quit her job, her overwhelming preoccupation became the man she had agreed to marry. During her years in New York, Carolyn had bobbed along in the floating world of celebrity, wealth, and high fashion, whereas John had a number of deep friendships. "God, I can't believe John's friends are so normal," she said to one of them. Much of Carolyn's old life and acquaintances were like clothes that now seemed overdone and vulgar.

John flew into Chicago with his friend Gary Ginsberg to attend the Democratic National Convention in August 1996. "You went to a convention where you essentially had twenty thousand Democratic Party groupies all in one little arena," Gary Ginsberg recalls, still in awe at the memory. "And it was like Mick Jagger walking through a Rolling Stones trade fair or Mr. Spock at a Star Trek convention." John did only one interview that week, with Tom Brokaw on NBC. As he talked in the sky booth visible to the delegates, down below the crowd whistled and cheered.

"John, I mean I've been around you a lot in public settings," Gary said as they drove away from the arena, "and I've never never seen anything like that. Did that freak you out?"

"It would have freaked me out more if there wasn't that reaction," John said. Maybe he would never play Hamlet, but he was nonetheless an actor who walked the boards on the greatest stage of them all. And he didn't like playing to a half-empty theater any more than Olivier did.

John's dilemma was how to take the media fascination with him and turn it into interest in his magazine. After a year the circulation was barely holding, advertising was down by two-thirds, and most of the talk in the media about *George* was negative.

In September, John put Drew Barrymore in a low-cut, sequin-studded gown on the cover with the line "Happy Birthday, Mr. President." The issue

starring Barrymore playing Marilyn Monroe on the evening that she sang a
sensuous birthday greeting to President Kennedy sold 200,000 newsstand
copies, twice what the previous month had done. Nonetheless, some of the
editors were stunned that John seemed to endorse the idea that his father had
been Monroe's lover.

John wasn't so much exploiting his father's love life as having fun, jok-
ing around. As he fielded criticism over the cover, he was working to pull
over a magnificent scam on the media. John and Carolyn were getting married, and John had decided that he was going to have a private wedding in
which only friends and family would stand witness. The press had covered
every wedding that had taken place in the Kennedy family since his Uncle
Bobby married in 1950. The idea that on the most important day of his life
John could veer off that media runway with his bride, family, and friends
was as sweetly irresistible as it was improbable. Not only would everyone he
invited have to keep quiet but also the scores of others involved in planning
the event. Carolyn gleefully entered into the spirit of things, tiptoeing quietly as she prepared her wedding gown.

John arranged three months before the September 21, 1996, date to
reserve all the rooms at the small Greyfield Inn on remote Cumberland
Island off the coast of Georgia. There were accommodations for only forty
guests, and John and Carolyn had the difficult chore of winnowing the invitation list down.

Christiane had to be told weeks early so she could fly back from covering the Bosnian war for CNN in Sarajevo. She kept the secret, as did everyone else. The media were thrown off even further when Carolyn flew off to
Paris instead of joining John at the Democratic convention; that had led to
speculation that the romance was over. As John, Carolyn, and the forty
guests were flying south, the tabloid *Star* was being printed with its headline:
"JFK Jr. Dumps Galpal Carolyn."

The guests were ferried over from the mainland to the seventeen-and-a-
half-mile-long island. It was not simply a wedding but an exquisite party
with eclectic, joyful participants representing the range of John's friends.
For years in New York he and a friend had planned their birthday parties

together. John had considered every detail, from the decor to the guest lists to the music. He had given up on those parties, however, not because he had outgrown them but because outsiders and media had found out about them and the events were plagued by crashers and photographers. He was back to planning a celebration again, though, and this would be the ultimate party.

The inn hardly had enough room for all the guests. They shared rooms, waited for the bathrooms to free up, and bumped up against each other in hallways. The rehearsal dinner Friday evening took place on the inn's porch. John was poignantly aware that his mother was not there beside him, and he turned to Sasha, who was seated next to him. "That's where my mother would have sat," he said.

Later in the evening there was a great bonfire on the sands. The gazebo was full of an inexhaustible supply of liquor. The guests laughed and partied until early morning. One of the revelers got so drunk that he fell asleep in the pasture, waking up Saturday morning among the horses. One of the most high-spirited of the guests was Carolyn's friend Narciso Rodriguez, who had designed her wedding gown; the publicity would help catapult him into the upper ranks of American designers.

During the day on Saturday John and most of the other men did the sorts of things he had done all his life: playing Frisbee on the beach, running in the surf, carousing around. Then, in the late afternoon, everyone drove to the tiny First African Baptist Church. The wedding was running late, in part because as Carolyn was putting on her veil her eyeliner ran and her makeup had to be redone. It was late as well because in John's typical fashion, he had misplaced some of his clothing. "He couldn't find his tie," Billy recalls fondly. "He was going nuts about it. He thought I hid it on him."

Nobody worried about the time. What wedding had any of them ever gone to in which they were transported in dusty Hummers along a rutted dirt road? And what church courtyard was like this, a pig farm with rutting hogs and their special odor wafting along on the late afternoon air? And what manner of church was this rickety old building, a slave church that had heard the prayers and songs of deliverance?

Carolyn was still not ready, and while the guests waited, Bobby

Kennedy chased after an armadillo. John was growing especially close to
Bobby, going off with him on trips and talking to him with intensity and
depth in ways he rarely did any longer with his two other close cousins, Tim
and Willie. John regaled the group with his own euphoric take on the proceedings. "They told me they caught these two *National Enquirer* reporters
in the brush," John said, laughing at the just revenge. "See, they were all cut
up, sweaty, and desperate, with this map they couldn't make out, begging the
security to get them out of there."

When Carolyn finally arrived, it was nearly dusk. Carolyn wore a floor-
length dress of pearl-white crepe and bare-toed sandals. Even those who
had long lauded her beauty were stunned at her presence. The church had no
electricity, and John and Carolyn stood before the priest in the cramped
building with their faces lit by candlelight. John looked as if he were the
defining model of a handsome groom, but models don't display the open
emotionality of John that evening. He had waited to marry longer than his
friends, and he was marrying the woman he wanted to marry, in the way he
wanted to marry her—among the friends he wanted to be there as witnesses
and participants in his happiness.

Anthony was the best man, and though he had had a bout with cancer,
he appeared fine now. Of his fifteen Kennedy cousins, John invited only
Tim Shriver, Bobby Kennedy, and Willie Smith. Most of his other male
friends were there, including Robby and Billy and their wives and Barlow.

When the ceremony was over, everyone returned to the inn. John wore
his sentimentality lightly, except when his Uncle Ted got up and read the
poem that the Irish ambassador to the United States had read to his father
and mother soon after his birth:

> We wish to the new child,
> A heart that can be beguiled,
> By a flower,
> That the wind lifts,
> As it passes.
> If the storms break for him,

May the trees shake for him,
Their blossoms down.
In the night that he is troubled,
May a friend wake for him,
So that his time be doubled,
And at the end of all loving and love
May the Man above,
Give him a crown.

After dinner the partying began anew. John had flown in a DJ, and he told the man to play a heavy diet of funk with an emphasis on the early Jackson Five. It was party night at Brown all over again. Everyone was out there on the floor. John didn't have steps, he had moves. He and Carolyn were boogieing away when the all too familiar sounds of the Village People singing "YMCA" sounded through the Greyfield Inn. John stopped in his tracks. He stood with his friends watching the center of the dance floor where Ted, florid and tilting to port, danced with Vicki in what he considered abandon. It was so bad it was perfect.

"Bet you the next one is the Macarena," said one cynical friend. And sure enough, it was. After that not so golden oldie, John told the DJ to forget Ted's other requests and return to the original list.

John and Carolyn danced until the morning hours. "Well, I did it, I did it right," he told Robby. "This is my best shot."

When the newlyweds arrived home to their Tribeca loft after their sixteen-day honeymoon in Turkey and Greece, the paparazzi stood waiting for them. John was full of overwhelming civility and a sense that he could reason with almost everyone. "This is a big change for anyone, and for a private citizen even more so," said John as his wife grasped his hand. "I ask that you give Carolyn all the privacy and room you can."

When John headed off to work, Carolyn stayed home. The magazines vied to see who could be first to put her on their cover. The designers and upscale boutiques dreamed of her wearing their clothes. Restaurateurs were ready to offer her their best table. If she wanted a job, she could have stepped back into an elevated position in the fashion industry on her own terms.

A few days after their return Caroline hosted a party for the newlyweds at her Park Avenue apartment. John arrived attired in a business suit on his bicycle, smoothly rolling into the lobby. Carolyn showed up a few minutes later in a taxi, as stylishly dressed as her husband in a long black dress and pearls. "I can't see," she exclaimed as she ran the inevitable gauntlet of photographers. "Leave the Kennedys alone," a passerby shouted. Upstairs in the apartment Carolyn faced a formidable, threatening presence in Caroline, her new sister-in-law. Caroline was restrained and disciplined in all aspects of her life, hardly one to be overly sympathetic to a pouting, self-pitying Carolyn.

When John and Carolyn attended their first high-profile event a month and a half later, a benefit dinner at the Whitney Museum, Carolyn was no more comfortable with the attention than she had been when she first returned from her honeymoon. She ran through the explosion of flashbulbs and the intrusive shouts and whistles, not even giving the society reporters any details on her designer gown. The paparazzi booed, a sound that John was unused to hearing. Almost all celebrities, including John's mother, understood that there was a game to be played. They could complain, they could hide their faces much of the time, but if they wanted a measure of peace, they had to give up a part of their image. Carolyn was not playing, and she risked having the media portray her as rude, not cool, as neurotic, not private, and as gauche, not original.

Carolyn abhorred venturing out of the SoHo loft to be confronted by paparazzi. As she looked out the window of the expansive loft, she saw that the photographers had moved back across the street, but she knew that as soon as she stepped out they would prey upon her. There was no back door Carolyn could escape through, no doorman to help her rush to a cab, and once she left the apartment, she felt that she was on a stage from which she could not retreat.

When she had gone two weeks without leaving the apartment, John asked Sasha to talk with his bride. Carolyn was in tears. "I go out there and I'm talking to somebody, having a really incredible personal conversation, and one of the photographers would be walking backwards zooming in on my face," she said. Sasha got her relaxed enough to go out to dinner that evening.

For all of his adult life John had been resigned to the endless exploitation of his life and image. "When summer comes and the news is slow, it's always time for the tabloids to do their John stories, making up one thing or the other," John reflected to his friends. He could shrug that off philosophically, but Carolyn was a different matter. He was anguished at the way she flinched under the attention, recoiling back into privacy. On one occasion on the Cape he became so enraged at a photographer that he rushed at the man and broke his camera. "Fuck with me, but leave my wife alone!" he yelled practically in tears. The old John would never have acted in that manner.

John was so upset at the paparazzi that he talked to a lawyer and the district attorney's office about getting them out of there, but that was legally impossible. John had thought that Carolyn had exquisite training to be his wife, and he couldn't understand why she couldn't get used to the attention. He knew that Carolyn would have to change, for the world would not change around her.

John's friends sympathized with Carolyn, but they could not understand why she could not get on with it and was forever playing the drama queen. "John was patient, and he wanted her to develop because he loved her," reflects Sasha. "There were great things in their love, regardless of what anyone says. She was strong, and she had some wonderful qualities, but the question was whether she was going to get beyond this obsession with the paparazzi. Get over it. Move on. Smile for them. Frown to them. Give them the finger. Be the world's biggest bitch, Leona Helmsley with blond hair and a ponytail. Or be Lady Di. Who cares? Fill up your soul. This is not a dress rehearsal. Wake up. Wake up, and you better do it soon."

Dan Samson got along particularly well with Carolyn, but even he could not understand why she did not just live her life. He visited often from

Seattle, where he was building a regional ice cream brand. Once, when he was staying at the loft, he suggested that Carolyn accompany him on a cab ride to an antiques store on the Upper East Side. "Not a cab, they'll follow us!" Carolyn insisted.

"Come on, don't be ridiculous," Dan said in a tone that was more a challenge than a put-down. Dan was sure he was right when they grabbed a cab and whizzed uptown with nary a problem. Carolyn insisted that they were being followed, and Dan told her she was crazy. They had hardly gotten out of the cab when two photographers started setting up to film them. So they jumped back in and headed to John's office. Now Carolyn took over, ordering the driver to turn at unexpected moments and to zoom through yellow lights. At one traffic light the two photographers rushed up to either side of the cab, their cameras pointing into the backseat. The next day there was a picture and a story in one of the tabloids about Carolyn's latest run-in with the media.

"I loved her so much," says Dan. "She was the favorite of all my friends' wives. To the world she seemed like a quiet, demure person, but she was the opposite of that. She was a boisterous, in-your-face person. If you didn't know her well and you saw her yelling at John and John yelling back, you'd think they were about to break up. That's the way they were. They thrived on that. As hard as it was for her with all the publicity, she would joke about not being able to deal with the attention. Outside of John and the close circle of friends, she didn't want to give anything to anyone. When she went out, she always wanted to wear black. If she wore a new designer dress, she knew she'd be photographed and written about. When she went shopping for clothes, it was more for her friends than for herself."

When John came home at night from *George*, there were times when he was confronted by an enraged wife prowling the confines of the loft. Carolyn was on occasion doing cocaine. To her fashion world friends, it was not a problem; it was just part of the scene. She did not try to hide her drug use but did it in front of their friends. To most of them, it also was no big deal, just another of Carolyn's quirks. After her death there would be lurid rumors that she was a coke addict, an allegation contradicted by every one of John's close friends. "When was the last time a really good-looking, immensely healthy-looking drug addict passed your way," laughs Sasha.

John and Carolyn had many great times together, especially when they were alone or with a few friends in their loft or on Martha's Vineyard. Yet Carolyn's obsession with holding tight and not giving out anything of herself to the media who pursued her began to take its toll. As the months went by Carolyn seemed to grow even more uncomfortable with her life. John so much wanted children. Carolyn said no; she would not bring a child into this ugly world. Her therapist prescribed antidepressants that she took along with cocaine. She became obsessed with psychology, seeing it as the key not only to liberating herself but to freeing her husband from maladies of which he was blessedly unaware. She talked about ADD and depression, all kinds of categories and problems. John had a different way of seeing life and found much of what she said demeaning to the human experience.

Some of John's close friends worried that Carolyn was such a drag on John's high spirits that he too would be pulled down into darkness. Others thought that Carolyn was just going through a phase. "Some have made so much of her depression and subsequent therapies," Dan says. "True, she sought counseling, but it was not something that either she or John treated as a shameful necessity but rather as just part of a life in New York City. She'd say, 'Today I have to go to the hair dresser, the florist, my shrink, and then buy some groceries.' This is not to minimize what these sessions might have provided her, but given what she had to deal with, I hardly think that seeking counseling and therapy made her ripe for *The Cuckoo's Nest*."

As apprehensive as Carolyn was about leaving the apartment, she increasingly ventured outside to go places and see people about whom John knew little. Sometimes he came home and Carolyn was not there. He waited until she arrived at two or three in the morning. He didn't know what to do. He loved his wife. He was trying to be a good husband, and he was sorely tested.

"I felt sorry that the honeymoon period that they should have been entitled to somehow didn't exist," says Sasha. "I feel sorry for the drug culture of America that puts everyone on antidepressants without telling them that their libido is going to be shot. Puts them on the pill without telling them that their entire sense of smell is going to change and they may be attracted to people differently. I feel sorry for this whole culture of therapy that says we should go to shrinks who say that the world is all about me me me. And John was at such a different place, waiting to see if this beautiful young woman with all this potential would catch up. He was in the throes of a crisis, stressed out. He wanted someone where he could feel grounded. He didn't want someone coming at him and telling him endlessly about their problems, which really and truly are almost adolescent. He needed someone who was mature."

Carolyn needed something to do, a job, a charity, a cause, an avocation, something other than sitting in the apartment waiting for John to come home or heading out to her secret haunts.

"You know, girl, you need to go out and get a job," one of John's close friends told her. "You need to go out and work."

"But I've got to take care of John," Carolyn replied almost smugly.

"Excuse me, was John walking into walls before you met him?"

There was a problem when a friend felt compelled to say such things, doubly a problem because John couldn't say it himself. "Yeah, yeah, tell her, but don't let her know I know about this," he told his friend.

CHAPTER FORTY-SIX

Poster Boys for
Bad Behavior

O n the editor's page of the September 1997 issue of *George*, John posed seemingly nude, sitting knees up with an apple over his head. He was supposed to be Adam while on the cover a nude breadstick-thin Kate Moss played Eve, sitting placidly as a deer nuzzled her hand. What was so startlingly revealing was not John's photo but the words that accompanied it. "I've learned a lot about temptation recently," he wrote. "But that doesn't make me desire it any less. If anything, to be reminded of the possible perils of succumbing to what's forbidden only makes it more alluring."

When it came to the sins of the flesh, the Kennedy men were doers, not talkers, and this was a bizarre essay, veiled in its specifics but unlike anything any man in the family had ever written or possibly even thought. John made it clear that he was perfectly aware of the price to be paid for indiscretions, of giving in to what one author he had read called the "essence of our true selves—one that's impulsive and rude and ruled by passion and instinct."

"We can all gather, like urchins at a hanging, to watch those poor souls who took a chance on fantasy and came up empty-handed," John wrote in a pseudo-Nietzschean spiel. "Give in to yourselves, like Mike Tyson and chomp off your tormentor's ear, and become an outcast; conform utterly and endure a dispiriting and suffocating life." John then turned to his two cousins, Joe and Michael, who were in the midst of their own controversies.

"Two members of my family chased an idealized alternative to their life," John wrote. "One [Joe] left behind an embittered former wife." Joe's "idealized alternative" was an annulment to wipe out the slate of his marital past. Michael, "in what seemed to be a hedge against his own mortality, fell in love with youth and surrendered his judgment in the process. Both became literally poster boys for bad behavior." Michael had fallen not in love but in lust. He had fallen not for youth but for a youth, precisely, an underage young woman. In doing so, he had committed what the state called statutory rape. John thought that the "interesting thing was the ferocious condemnation that met their excursions beyond the bounds of acceptable behavior."

John soon found himself in a controversy over what he had not written. "John F. Kennedy, Jr., took the unusual step of slamming his own family in the pages of *George*," wrote the *Washington Post*, turning the phrase "poster boys for bad behavior" into John's single-minded condemnation of his cousins. He in fact wasn't so much criticizing his cousins as musing philosophically about the dangers of giving in to desire. The story of a public spat between the Kennedys was irresistible, and most papers followed the same story line.

Although Bobby Kennedy read the essay with philosophical distance and saw it not as an attack on his brothers but as intellectual musings, Joe did not have quite that distance. He was outraged. The truth was that of all his cousins, John liked Joe the least, considering him vulgar and overwrought, and it was wildly ungenerous to link Joe's divorce and annullment with Michael's affair with an underage baby-sitter. If Joe was a poster boy for bad behavior, so were millions of other divorced men.

When the press confronted Joe, he had his scripted answer ready. "Ask not what you can do for your cousin, but what you can do for his magazine," Joe said at a press conference. It was a deliciously quotable line, though foolishly confrontational and gave credence to the media's take on the essay. John said nothing in rebuttal, and the phrase "poster boys for bad behavior" became the most memorable line John would ever write.

The controversy created publicity for *George* that it had rarely received since the first issue. People talked about the magazine when John put Drew

Barrymore dressed as Marilyn on the cover or commented on his cousins' troubles in his editor's note, but they weren't talking about the content. Hachette was not about to put the kind of money behind *George* that would bring in major writers doing important pieces, and the magazine had neither the cachet to pay them largely in prestige nor the ability to create its own stable of stars. That was just one of the ways in which John was having difficulties dealing with Pecker and the other Hachette executives. By any measure, *George* was in trouble.

oe and his brothers had bitter feelings toward John, who had violated the family code of always showing a united front to the world. For so long he had been an amusing diversion, silly John squandering his life. Now that he was editing *George* and being seriously talked about as a political candidate, several of them reeked with jealousy, even anger. John hardly noticed the changing attitudes. He was probably the only member of the family totally immune to jealousy, a vice that gives no pleasure.

In the summer of 1997, when Ethel's family got together at the compound for Joe's annual barbecue, John was decidedly not in attendance. Everyone was dressed in cowboy outfits, and Michael danced to country music with his mother and led the others around the floor as if he had been flown in from Nashville for the occasion. They had their own family AA meetings, and by that criterion it appeared that Michael was confronting his problems. "It's a very different family at Thanksgiving or Christmas [or other holidays], when most people are sober," reflects Kathleen. "There's no question about it. And not only sober, but also spending an hour a day in self-examination. So that's a big change."

In the fall Michael entered a bicycle race in Cohasset at the last minute. The thirty-nine-year-old had never been in a bicycle race in his life, but he won. In September, Michael flew down to Ecuador with a group that included his old Harvard roommate Lorenzo di Bonaventura. The Hollywood executive hadn't seen Michael for several years, and it was like old times. "Michael is figuring how to get mountain bikes across these volcanoes in the middle of nowhere," recalls di Bonaventura, "and one of the friends falls off, and he's down like thirty feet, and Michael's looking down saying, 'How are you doing, doing okay, fine, see you later. Here we go.' He was just having a ball and dragging you with him."

When Michael returned to the United States, he wanted to get back with his wife and kids and resume the life he had known. His divorce would be final at the end of February 1998. Before that Michael thought he might be able to work things out with Vicki and life would go on as before as if nothing had changed.

In late December 1997, Michael flew to Aspen to be with his mother, brothers, and sisters. They were a family of rituals, frequenting the same few places each winter, skiing the same runs, and seeing many of the same people. There were three other mountains in Aspen, but for them the only mountain to ski was Ajax. You could ski out of your condominium to the gondola, zoom down the expert runs, and have some white-bean chili and apple strudel for lunch at Bonnie's at midmountain.

Another part of the ritual was the football game on skis that the Kennedys played on the last run of the day. They had been doing this each afternoon at Aspen for twenty years or more, within clear view of the indulgent ski patrol. They would arrive at the top of the mountain at close to three-thirty waiting with a number of other skiers for the ski patrol to order everyone down. The Kennedys did not listen to the ski patrol's admonitions to leave but considered it their right always to be the last people off the mountain. Then once all the other skiers had departed, they headed down knowing that there was just one sweep left, and if something happened, they would be on their own on the mountain as dusk descended.

They divided up into two teams, gave their ski poles to someone else, and then skied down the mountain passing a small plastic football as they went, using various markers along the course as goalposts. The game was a ski version of the football game the Kennedys played at Hyannis Port, in which the goalposts constantly changed and you could throw a pass anytime you wanted, even far beyond the line of scrimmage.

On New Year's Eve there were not enough Kennedys to make up two teams. So they got various friends and hangers-on to play with them. A few years back Max had hurt his ankle and been taken down by the ski patrol. But for the most part these daily scrimmages down the slope had proved a harmless, if edgy, bit of business.

Half past three is midafternoon in most places. On a December day on Ajax, however, shadows fall on the mountain, and soon dusk begins to descend. The last day of the year was cold, cloudy, and icy. By the time the gondola stopped and the ski patrol started ordering people down the mountain, there was a gray unpleasantness to the afternoon. Until a few years before Ethel would have skied with the players, but today she had a cup of cocoa in the Sundeck Restaurant and took the last gondola back down.

While Michael waited for the other skiers to set off, he skied around with his video camera filming his children and the rest of the group. Michael was the captain of one team, his youngest sister, Rory, captain of the other. The day before the game had ended in a tie. Kennedys did not believe in ties, and they vowed they would compete to the bitter end; one of the participants, longtime gossip columnist R. Couri Hay, recalls them vowing "death to the loser." Then they set off, a swarm of thirty-six Kennedys and friends, including several onlookers and two friends carrying the ski poles. Michael handed the camera off to someone else. For the children, it was immensely exciting, playing as equals with the adults. Michael's estranged wife and their three children—Michael Jr., fifteen; Kyle, thirteen; and Rory, ten—were staying with Vicki's father and stepmother down the road in Vail, near enough so that the three children were spending New Year's Eve with their dad.

The players grasped for the little ball and then heaved it down the slope to a teammate. Michael was a superb skier, but some of those who tried to keep up with him were rampaging out of control. One of the players smashed into the sign that was serving as a temporary goalpost. "Okay, new rule," shouted Michael, as Hay remembers. "We're going to make a twelve-foot radius around all the goalposts before somebody gets killed."

About halfway down the mountain a member of the group ran over the skis of Ted Widen, who was not part of the game. "If you come near me or

any of my friends, I'll kill you," Widen shouted out, as he watched incredu-
lously at "how many people were so close together and how dangerous it
was, especially at that time of day."

Michael raced ahead a hundred yards or so, getting behind everyone on
Rory's team. Blake Fleetwood reached his arm back to throw a pass to
Michael. Fleetwood was one of those people who had hung around the
Kennedys most of his life. He was always ready to go off with them, whether
it was rafting in South America or playing football on skis.

Fleetwood threw the ball so far ahead that it would have been a tough
catch on a real football field; it was almost impossible for a skier to grasp.
Michael reached out and snatched the ball out of the air, an incredible catch
that no one else among them would have even attempted. As he grasped the
ball, his ski tip caught the bottom of a tree, his body was propelled forward,
and his head smashed against a tree.

Rory was one of the first to reach him. His face showed no color. She
attempted mouth-to-mouth resuscitation, and when she pulled back from
her brother, her lips were bloody. She wiped her face clean in the snow and
went to pray with Michael's three children. They knelt in supplication to
God, repeating the words of the Lord's Prayer. They and the other children
were led fifty or so yards away, while everyone waited for the ski patrol to
carry Michael down Ajax. Michael had already expired from a severed spine
and massive head injuries, but it was part of the myth of the mountain that
nobody died up there. Not until the ski patrol brought him down to a waiting ambulance and he was taken to Aspen Hospital was he declared dead.

M ichael's body was flown to Hyannis Port in the actor Kevin Costner's
jet. For a day Michael lay in a closed coffin in the house that he had
come to every summer. The TV cameras and journalists stood in the street
behind barricades and shouted out questions, but the brothers were oblivious
to everything but the fact of Michael's death. John walked over from next
door with his dog by his side. Some of the other mourners took that as a sign
of disrespect. Joe walked by himself on the beach and asked only to be left

alone, and Bobby kept breaking into tears. For Max, Chris, and Doug, the grief was written on their still youthful faces.

The funeral took place at Our Lady of Victory Church in Centerville, where Caroline had been married twelve years before. The oak pews of the old church were full of family friends and politicians. This funeral was not for the nation, and not even primarily for intimate friends and other relatives, but for Ethel and her nine surviving children.

Joe's eulogy was an attempt to find meaning not only in Michael's life but also in their own. He began by talking about a Nike television commercial in which Larry Bird and Michael Jordan play a pickup basketball game. The ball bounces off the top of a building, a highway, and a scoreboard before whooshing into the basket. "People think that's a new concept," Joe said. "My brother Michael wrote the earlier version many years ago."

Joe then told how one day when Michael was at Harvard the brothers and their friends played an especially memorable football game. The group was sitting in the kitchen when Michael defined the field of play as the first floor of the house, the front lawn, and the street. The goalpost was the neighbor's front lawn on the other side of the street. When play started, Michael ran down the hall, jumped over the couch, and spurted out the front door. Joe cocked his arm, threw the ball down the hall, through a closed window to the lawn, where Michael grabbed the ball, spiked with shards of glass, and streaked across the street for a touchdown.

Everyone in the church laughed. Seated in those front pews along with Michael's widow were his daughter and two sons. Joe's teenage twin boys were there too, and other young Kennedys. Joe was implying that this was the way they should behave. "He was fearless—on the slopes, on water skis, wherever he could test himself at the edge," Joe said. "This was one of the glories of his life, and it should not be diminished by his loss."

"He was not made for comfort or ease," Joe said. "He was the athlete dying young in A. E. Housman's verse: 'Like the wind through the woods. . . . Through him the gale of life blew high.'" Despite what Joe said, Michael was a troubled middle-aged man, not a "smart lad to slip betimes away from fields where glory does not stay."

Joe almost broke down in tears as he continued with a litany of achievement, taking almost everything his brother had touched and turning it into singular success. Michael's tenure as Ted's 1994 campaign manager had been so questionable that a whole team of outside advisers had been required. Yet Joe asserted that, "as Senator Kennedy said, that grand victory belonged in a special way to Michael, who quietly took the criticism but never took the credit." Michael had helped to dismantle Citizens' idealistic purpose. Yet Joe called his brother a selfless force of good who was there wherever there were needy, providing "free heating oil" and "life-saving medicines." In Angola the rebel UNITA troops had not fully disarmed, in accord with the 1994 peace agreement, controlled the diamond region, and fostered sporadic violence. Yet, as Joe saw it, Michael "helped bring peace to an entire country—and because of him, thousands of children will live on and will not lose hands and arms and legs." His beneficence was endless. "Your energy and daring eroded mighty walls of indifference, poverty, and suffering."

In the best of moments Michael was an unfettered spirit who brought exuberant, joyful times to his family and friends. Joe could have memorialized him for what he was, not for what he might have been. But in this terrible time there was a need in Joe to pretend that Michael was things he was not.

When Bobby gave his own eulogy, he came no closer to the truths of his brother's life. "He gave his life to the poor," Bobby said of his brother, who had lived in a million-dollar house in Cohasset. He said that "Citizens Energy had had a spectacularly successful year under his leadership." Michael had had a spectacularly successful year earning $391,254, a far better year than the poor whom Citizens Energy supposedly served. Bobby said the "personal issues with which he struggled were not about malice or greed. They were about humanity and passion." As the mourners sat in that cold church, Bobby invoked Jesus in dismissing his brother's sins. "His transgressions were the kind that Christ taught us are the first and easiest to forgive."

CHAPTER FORTY-SEVEN

Clinton and the Kennedys

I n March 1998, Joe announced that he was retiring from Congress and would devote himself full-time to Citizens Energy. He was the first Kennedy to give up political office. Almost any other year, Joe's decision would have been a major news story. It happened in the early months of the Clinton sex scandal, however, and the story of the first Kennedy to retire from political office passed with only cursory notice.

The Clinton scandal played into the lives of the young Kennedys in ways both explicable and inexplicable. In the popular imagination, the words "Clinton" and "sex" and "Kennedys" and "sex" went together like "Shaq O'Neill" and "basketball." Michael's disgrace and death and Joe's annulment further lowered the esteem in which the family had been held. John had nothing to do with his cousins' behavior, but he too was part of the tittering asides about the endlessly naughty, tragic Kennedys stumbling along their cursed path.

John was distressed at the stories about Clinton's relationship with Monica Lewinsky, a twenty-three-year-old intern. John felt sympathetic to a president who not only shared his centrist grasp on politics but seemed the kind of witty, irreverent person with whom John enjoyed spending time.

John was strangely quiet in the editorial meeting when the staff excitedly discussed the allegations against Clinton. He knew that *George* would have to write about the charges, and he did not object as the editors planned an April issue focused almost exclusively on the seamy story. This was a perfect mesh of politics and popular culture, though not at the elevated level that John had envisioned. In the following months articles about the scandal in the *New York Times* occasionally read like the *National Enquirer* and the phrase "oral sex" came trippingly off the tongues of television reporters.

John was probably on the cusp of success with *George* if he could make the magazine's coverage of the scandal indispensable. Instead, he pushed his magazine away from the unseemly business. John, who refused to print the "n" word in his magazine even though Governor George Wallace had spoken it, could hardly publish a story detailing the tilt of Clinton's penis or a learned essay on just what constituted "sex."

However many times John discussed the scandal, the most important name was the name not spoken—his father's. A few years before, Christiane had asked John if he had seen Oliver Stone's *JFK*, a dark fantasy about John Kennedy being murdered through a right-wing conspiracy within the government. "That's my father, Kissy," John said. "I don't want to see it." There were many things about his father he did not want to see. Seymour Hersh had just come out with an exposé on the Kennedy administration, *The Dark Side of Camelot*. Hersh described a New Camelot of whores, bribers, and mobsters prancing where once patriots had walked. John could not even look at the book's excerpt in *Time* without throwing it aside.

As much as John could never publicly admit it, there was a natural connection between Clinton's sex scandal and his father's career. No one talked more to John about politics than Gary, but even his friend was subtly deferential, understanding that there were certain things that he could only suggest. "John, you've got to separate out your own family from this story," Gary told him. "You can't do it, but you've got to."

"Clinton is a human being," John said, never mentioning his father. "He's no different from you or me. He has needs. He has wants. And he acted on them. Just like you would. Just like I would. And who's too holy to have engaged in the kind of behavior he did?"

That was John's essential point, not only about Clinton but also implicitly about his father. In this summer of American discontent, he had finally begun to come to terms with his father. In the past three years at *George*,

John had received a matchless education in American politics, and it always came back to his father. Hersh and the revisionists had created a grubby countermyth that John believed was not true. As he saw it, his father, like all men, was flawed, but a flawed man could also be great.

As each day brought more revelations, John had dinner with Christiane and her fiancé, James "Jamie" Rubin, at the back of a Greenwich Village restaurant. John had met Rubin, Secretary of State Madeleine Albright's spokesperson, a year earlier. He was the first man about whom Christiane had been serious, and he had wanted to see if Rubin was worthy material. In the following months John tested Rubin by having Gary put items about him in his *George* political gossip column. Gary went along calling Rubin "pushy" and having "got into some hot water" over his love of publicity. At first Rubin thought that Gary had done this on his own, but he soon concluded that John was behind it.

This was a sport that most women and many men would have considered crude. But to John and his friends it was just part of the game. If Rubin had complained, John would probably have dismissed him as a suitable mate for his friend, but he did not, and John and Carolyn attended their wedding in Italy that August.

Inevitably, though, the conversation that evening turned not to the impending wedding but to events in Washington. Rubin was articulate and highly opinionated. He raged at Kenneth Starr, the independent counsel, for snooping into private lives.

From her perspective as a journalist, Christiane was equally appalled. Many of her colleagues had gleefully joined in the chase, knocking down bedroom doors, trumpeting their pathetic revelations and calling it news. John trumped them both in his outrage. Ranting at what politics was becoming, he was certain that the combination of Washington journalists, Republican investigators, and the FBI had brought the country enormous shame. The three talked louder and louder, topping each other in the fury of their rhetoric.

John got so riled up that he began discussing what he could do. Speak out. Write a column. Make his presence felt. As Rubin listened, he became less a cheerleader for John's activism and more the political operative. "It would be great for the Democrats," he said. "You can get an audience talking about it on *This Week with David Brinkley* and do a very compelling condemnation, but then they're going to do the same thing to you and your father and you just have to be ready for this."

John became silent. Rubin was not wrong, but in making the comparison with JFK, he was saying something that none of John's friends ever would have verbalized. Nor had members of John's own family dared to speak openly about his father's infidelities. His Uncle Ted had felt it his duty to lie to his sister Eunice, pretending that the tales of adultery were nothing but tabloid gossip.

To admit to such indiscretions in *George* or on Sunday morning television would probably elicit a fury of rebuke from his aunts. His own mother had been a primary architect of the myths of Camelot, so in confronting the facts of his father's life, he would be confronting his mother as well. He and his cousins had been brought up with a wary protectiveness about their family history. They learned that a flippant word or a revelation casually spoken might besmirch the family. If John was to go forward, he first had to look back, but he could not bring himself to do so. At least one of his friends believed that John also feared that if he stepped forward, the tabloids might start probing into his own marriage.

In September 1998, John and Gary rode out in a private car to the U.S. Open. The Starr report on Clinton's conduct was about to come out with its sexually explicit supporting materials. The smart money had it that Clinton would soon be an ex-president writing his memoirs to pay his legal bills. The worse things got for Clinton, the more vociferous John became in defending him. "You are totally missing what's happening," he told Gary, who was convinced that Clinton would be impeached. "Clinton is going to get off. He's going to emerge from this more popular than ever. He's become much more identifiable to many more people than he ever was before."

"You're supposed to be a smart guy," Gary said sarcastically. "You're totally missing this. You just don't understand the American people."

"I understand the American people," John said, turning toward his friend. "You'll see."

John had far deeper insights than many of the media commentators, and a far more accurate prediction of the outcome. Yet nothing of this appeared in *George*, a comment not merely on his editorial judgment but on his own intellectual courage. He had prophetically seen that politics and popular culture were coming together, but he had envisioned that happening at a positive level, not in the sordid spectacle of Clintonian Washington. John was unwilling to cast his magazine's vision down in the muck where the truth bubbled forth. He could have written an essay that was equally about Clinton and his father. He could have picked up the phone and gotten to some of the major players, as almost no one else could. He probably could have reached Monica Lewinsky, who had wanted to intern for *George*. He could have pushed *George* into the fray, but he did nothing and stood back. He ran a few scattered pieces on the scandal, but none of them mattered in the great debate that was going on across America.

In December, when impeachment loomed, *George* ran a fashion spread. There was a model playing Hillary meeting the press, and another model playing Monica in a chair with a cigar in her hands, her stockings exposed up to her thighs, while an ersatz Clinton leered over her with a lighter in his hand.

P atrick faced much the same psychological dilemma as John. All of his life he had avoided anything that touched on his father's sexual misconduct. Such matters were embarrassing to him as the loving son of a beloved father, as the loyal son of a mother who had suffered mightily from his father's misconduct, and as an ambitious politician tainted for conduct that was simply not his.

"We were afraid that people were going to say, when you were a little boy, wasn't your father involved with other women?" reflects Patrick's press secretary Berman. "We didn't want to get into that." Patrick tried to stay away from the issue, but in October 1998 he arrived back from a trip to Puerto Rico with First Lady Hillary Clinton as a full-fledged adherent of the Clinton cause. His father knew when to temper his self-righteousness, but when Patrick had an important point to make, he often made it in an emotional screed that pleased partisans, rarely moved the undecided, and irritated adversaries. "I'm at the point where we've had enough," Patrick told reporters in the Capitol. Clinton's behavior "is better left to God and family than the Congress and the media to judge."

Patrick said that Hillary had told him about "a series of lies that [independent counsel Kenneth Starr] perpetrated on numerous occasions." The charge would have had more resonance if Patrick had been able or willing to detail at least one of the lies. Two months later Patrick attacked Rhode Island senator John Chafee for "masquerading as a Republican moderate" while putting forth an "outrageous" call for witnesses at President Clinton's trial. Patrick was not wise in attacking an elder statesman admired by citizens of both political persuasions, but he could not resist making his digs.

In another time Joe would have been a fervent defender of Clinton, a bulldog of determination and will. But in those crucial months of late 1998, Joe was a lame-duck congressman who could not expect the same attention as those who were continuing in the House of Representatives. Beyond that, the brother of an alleged statutory rapist could hardly defend Clinton dalliances as trivial matters.

Ted took a very different approach to the Clinton scandal than did his son or his nephew. Probably no one in public life understood better what Clinton was feeling. Ted had not had the details of his sexual life discussed in a congressional hearing, but the tabloids had amply chronicled his sexual adventures. Ted had not faced impeachment, but there were millions who believed that he should have been driven from office.

During Clinton's months of personal and political crisis, Ted often talked to the president. "His advice is always simple," Clinton told Ted's biographer Adam Clymer in the midst of the crisis. "It's just sort of get up and go to work, just keep going, and remember why you wanted the job in the first place. He's a very tough guy and he understands that if somebody accuses you of something that's true, maybe you're your own worst enemy, and you have to hope that when people add up the score, there will be more pluses than minuses."

In the Senate the debate over whether to call witnesses in the impeachment trial threatened to divide on party lines. Ted was one of those calling for a compromise that would defer the whole issue so that the debate could move on. It was almost unthinkable that the fifty-five Republicans could find twelve of their Democratic colleagues to join them to convict the president by the mandated two-thirds majority. But there could have been such poisonous ill will as to paralyze government for months to come. Ted and a number of his colleagues acted in a manner that allowed Clinton to finish out his diminished presidency without the shadow of scandal above his every action.

In May 1999, after the impeachment drama had ended, John attended the annual White House Correspondents Association dinner. The event was a celebration of the mix of popular culture and politics that was the idea behind *George*, and John was in his element, but the bruising, collegial atmosphere offended Carolyn. Television networks, papers, and magazines tried to trump each other in the political clout and celebrity of the guests at their table.

The previous year *George* had been the only important media table with no invited celebrities or politicians. This spring John invited Larry Flynt, the publisher of *Hustler*. Much of the Washington press corps considered the quadriplegic pornographer no more than a sewer rat. Brian McGrory in the *Boston Globe* called John "something of a whore" whose "antics took a more disgusting turn" when he invited Flynt. William F. Buckley was one of those asking, "What does John Kennedy intend, by inviting Larry Flynt as his guest? If there is another answer to that question than: To shock everybody and draw attention to himself—come up with it."

Another answer would have pleased almost no one at the dinner. When Clinton's removal from office appeared likely, Flynt had infamously offered a $1 million reward to those who came forward with sexual tales involving Republicans. He hired Dan Moldea, a veteran investigative reporter, who turned up information on Speaker of the House–elect Bob Livingston that forced him to resign. Among Clinton's loudest critics Moldea discovered closeted gays raging against homosexuality, adulterers of all stripes, and one member who threatened suicide if Flynt outed him. In the end the Republicans probably backed off in part because of the feast of outings that appeared imminent.

For Clinton's supporters, it was unthinkable that a man who had published Jackie Kennedy's nude photos and whose own daughter accused him of molestation had helped to save the president. For conservative Republicans, the idea was equally unthinkable that their party had turned away from its principled duty to impeach a disgraced Clinton in some degree because of what the late *Washington Post* columnist Michael Kelly called "a loathsome campaign of sexual blackmail." Washington journalists were no more ready to admit that a pornographer and his hireling had pulled off this feat.

John may have been willing to have Larry Flynt sitting at the *George* table, but he did not consider the story of the pornographer's impact on the impeachment worthy of display in the pages of his magazine.

CHAPTER FORTY-EIGHT

A Life of Choices

 ll his life John had wanted to fly. He had started taking lessons his senior year at Brown, and he would fly forty-six hours with an instructor over the next six years, leading to a one-hour solo flight. Jackie, however, was so opposed to his flying that he dared not pursue it. Only after her death did he begin to contemplate getting a license.

The only magazine that John devoured was *Flying*. It may have been in those pages that he got the idea of flying powered parachutes. You did not even need a license to fly this craft, which was little more than a motorized cart protected by a bare metal cage with a parachute attached. John purchased the Cadillac of the lot, a one-seater Buckeye for $13,500. The owner of the company, Lloyd Howard, had several professional athletes as customers, but nobody of the stature of John F. Kennedy Jr. Howard traveled with his famous customer to a small airport in Albany, New York, in August 1996 for the four-hour training.

John flew the craft over fifty hours in the next year, primarily on Martha's Vineyard. There are those on the island who vividly recall hearing what sounded like a lawn mower and then seeing John flying past at house level, looking like something out of *Mary Poppins*.

The following year, when Robby showed up for the annual Memorial Day get-together, John buckled Robby into the craft on the beach. John's other friends had all begged off, but Robby was game for going up in the flimsy contraption. John knew that first-time pilots were supposed to take a short training course, but that did not stop him from egging his friend on to take the mini-craft up.

"So I get up in it, scared to fucking death, and I'm over the ocean, I'm staying about four hundred feet in the air," Robby fondly recalls. "I decide I want to come down right away. You know, real badly. So I'm lining up the beach below me, and a gust of wind comes up, blows me out of my landing zone. I gunned it, and the dune knocked the front wheel off the Buckeye, but I kept going. The chute stayed up, and I landed about a hundred yards past that. Greatest feeling I've had in the last five years, other than my son being born. Why? I had balls and I checked them. It was my feeling of grandeur. And I impressed my friend. That's why. We ran at each other just laughing and collapsed guffawing."

Robby was John's partner in their games of manhood, but he was losing his edge and he knew it. Part of it was that Robby was softening a little, while John was working out as rigorously as ever. Robby had a wife and two children, and he did not get the same rush out of their games anymore. But John did. "It's like drugs," reflects Robby. "If you smoke pot every day, and you want to get the same high as you did on day one, you've got to smoke more and better. The guy's got a huge adrenaline pipe. He's gotta fill it up. He's gotta risk himself in order to have fun. If you're not a daredevil, you're not a man."

D uring the summer of 1997 John flew to Oshkosh, Wisconsin, to attend the weeklong Experimental Aircraft Association show with J. "Mac" McClellan, the editor of *Flying*. John had been delighted to meet the editor of his favorite magazine, also published by Hachette. John had a spectacular week, flying with the legendary test pilot Chuck Yeager in a P-51, and going up in a B-17 as well. By the time he headed home he was obsessed with the idea of renewing his determination to learn to fly. "I'm not coming back here until I can fly myself," he said as he left.

His friends considered John an amiable klutz. To them, he was the least likely of pilots. John was perfectly aware that he suffered from ADD and knew that flying tested the limits of his abilities to concentrate and follow through on a series of explicit instructions. John was busy running a magazine, and if he insisted on flying, the obvious thing to do was to go to a flight school in New York on the weekend among middle-aged doctors and stockbrokers looking for a new hobby. It was a measure of how seriously John took learning to fly that McClellan arranged for John to be accepted at the Flight Safety Academy in Vero Beach, Florida. His fellow students for the most part would be flying for carriers such as Swissair and Air China or piloting corporate jets. The school made a special exemption for John, and he began coming down on weekends in December 1997.

The flight training was a release from all the pressures of New York. He stayed in the student dormitory and worked diligently. Occasionally Carolyn came down to Florida too. "We laughed many times about the news reports that his wife was afraid to fly with him, because, he always said, she was his first passenger," recalled McClellan.

John bought a twenty-year-old Cessna 182 after discussions with McClellan, who told him it was "one of the best airplanes for a new pilot in that it offers reasonable speed and capability with ease of flying qualities." Whereas the summer before he had flown just aboveground in his Buckeye, now he was soaring high above the world in his own plane. He still went off with his friends to kayak or play football, but he found flying the most exquisite way to get away.

Despite all the responsibilities of editing a monthly magazine, John insisted on having an adventurous life. When he flew to Vietnam to interview General Vo Nguyen Giap, the strategist behind the defeat of the French and the Americans, he spent ten days in the country early in 1999. "We had traveled ten hours north from Hanoi through the mountains to the Chinese border, and from there paddled down the whitewater stream," recalls Kyle Horst, an aid official who traveled with John. "Every hour or so we would come upon a crude, low bridge which marked human settlement, and a flock of Montagnard children would run to river's edge. The braver ones jumped in and swam out to climb up on our boats; one of them stayed with us for hours. Drifting past immense karst towers down a ribbon of water the color of beryl, John late that afternoon called it 'one of the most incredible days of my life.' " John's exclusive interview with Giap on the last day was almost an afterthought.

G *eorge* had begun with a "let's make a magazine" insouciance. By early 1999 it had become all business. It had been John's idea to elevate Mitchell to executive editor rather than bring in a top professional who might have challenged his sovereignty. The magazine's sales and advertising continued to flounder. John eventually concluded that Mitchell had neither the vision nor the managerial skills to handle the top slot, and she often kept herself shut off behind closed doors, blanketed in a cloud of cigarette smoke. And that was just the beginning of the problems. "It was a morass," reflects Elizabeth Kaye, a contributing editor. "You could not get anybody on the phone to discuss a story."

J ohn knew that he needed a new editor, but he cared deeply for Mitchell. It was one of those dilemmas in which business and friendship were so mixed together that one couldn't be ended without probably ending the other. He cried when he told one close adviser what he was about to do. While professing his loyalty to Mitchell, he secretly went around looking for a top New York editor to replace her—the kind of corporate duplicity that he abhorred.

Early in January 1999, Keith Kelly, the media reporter for the *New York Post*, learned from one of his regular sources that John was looking for a new executive editor. Kelly tried to contact John, who was unreachable on a plane. For all his gritty street reporter persona and the negative items he had written about the magazine, Kelly says that he was not without a sentimental soft spot for John. He knew the piece would hurt *George*, and instead of emphasizing the editorial problems, he wrote that John was "searching for a hands-on editor to try and bring it to the next level."

The *Post* story forced John to act prematurely. When he returned to the office, he had a private conversation with Mitchell, and she walked out of the office for the last time. John called Richard Blow into his office and told him that he had the job. "He [John] had been negotiating with Hachette over a television spin-off of *George*," Blow says John told him in their meeting. "Hachette had given away some rights that John had insisted upon keeping. When John found out, he fought to reclaim them. . . . The *Post* story, John said, was David Pecker's revenge." The Hachette CEO had supposedly left his fingerprints all over the story by talking about bringing the magazine to a higher level, though that had been the reporter's own spin on the story.

What John did not tell Blow was that Kelly's story had been correct. He had been looking for a new editor, and Blow had never been anything but a backup candidate. The primary reason he was elevating Blow was that the *Post* story obligated him to get rid of Mitchell immediately and fill the slot from within. John later told an editor friend that "he had mixed feelings about Blow, who he later found out had behind his back had some conversations with Hachette brass that may have resulted in the leak." Blow had not been the source of the leak, but it was no easy matter working with a man whose loyalty John may have believed was suspect.

John was not only the editor-in-chief but also half-owner of *George*. As such, he should have kept an eye focused on the business side of the magazine. He had paid almost no attention to the various publishers, the chief business executive, whom Hachette had assigned to *George*. The initial contract with Hachette would be up at the end of the year, and Pecker was not happy. When John had unveiled the magazine, the Hachette CEO had stood next to him with a celluloid smile, proud just to be on the same stage with John. Now Pecker was playing the bloodless banker and John was the farmer coming in behind on his mortgage.

Advertising was dangerously down, and the magazine was bleeding red ink. Pecker was leaving to head American Media, the publisher of supermarket tabloids. At least Pecker had shared the exhilarating early months with John, and his departure in April was another disquieting sign.

Most editors-in-chief of a major magazine would have shrugged off the departures as the natural ebb and flow of life in New York publishing. John, however, needed a higher comfort level, and he found it dispiriting news that in these difficult negotiations he would have to deal with a new, as yet unnamed CEO. He didn't like dealing with strangers. He needed someone like Gary at his side who cut through all the stroking and half-truths, but Gary had left to work for News Corporation. He missed Mitchell too, a bright, irrepressible soul whose company he often relished. John considered Blow a competent editor who had brought much to the magazine, but he feared his executive editor as a wannabe trying to get close to him. Rose-Marie Terenzio, his personal assistant, was still there. She was his bulwark, a woman of the fiercest loyalty and devotion, but beyond her there were rooms full of strangers and the merely deferential.

In the past John had turned down most interview requests largely because he was not sure he had enough to say. In March he did the cover interview for *Brill's Content*, a new media magazine. Stephen Brill was proud of his probing questions. Even two years before, John would have been unable to stand up to such thrusts, but he gave as good as he got. It was by far the best, most substantive interview he had done since starting *George*.

"The best training for any journalist is to have someone write a profile on you, and have it published in your hometown newspaper," John told Brill. "Then you understand everything about journalism: You understand how to be sensitive to taking quotes out of context; you understand how to be sensitive to doing captions, unfair headlines—the whole nine yards."

John suggested that he and his magazine had a better sense of how the American people felt than the media elite and the Washington political establishment. "The current political imbroglio has indicated that real people across America do not have the same investment in partisan politics as people in Washington do," he said to his interviewer. "You find people want solutions; people want government to work; people want to feel good about politics. They vote the person rather than the issue a lot of the time. And I think that happens increasingly. That is part of the reason why I think the Washington community maybe has not embraced *George* to the extent that readers across America have."

✒

Gary, who had advised John to do the *Brill's Content* interview, was now executive vice president for corporate affairs for News Corporation. Unlike his Uncle Ted, who grew petulant when a trusted associate left him, John always wanted the best for his friends. As much as he needed Gary around him at *George*, when his friend got the job offer, John had pushed him to take the high-powered position and test himself in an elevated arena. John warned Gary as he was leaving, however, that he was just too nice and would have to toughen up if he wanted to succeed.

Gary was still the man John went to when he had hard decisions to make. They rarely sat down for lengthy discussions but talked between sets of racquetball or after working out. When the issue of *Brill's Content* came out, John was out in Central Park playing Frisbee with Gary. "It was a fucking great interview, huh, Gary?" John said, more ironist than braggart. As the two friends walked back along Fifth Avenue, a bus passed with an advertisement for the magazine with John on the cover. "My God, look at the juice I got," he teased.

John was happy with the interview but consumed with the questionable future of *George*. "How's it gonna look if we fold it up?" John asked. It was a reasonable question from a reasonable man.

"You can spin it both ways," Gary said. You could almost see the headline in the *Post*: "The Hunk Flunks Again." There would be countless journalistic autopsies, everyone from *Time* to the *Star* dissecting the failure in pleasured detail. *George* was largely a joke among the publishing fraternity in New York, and it was not only unloved but unread in Washington. Yet with no journalistic background at all, John had built the most popular political magazine in American history. It had not been avidly read by Beltway boys and New York mediacrats, but the audience had been out there in what John considered the real America. No, with people like Gary spinning the tale and the immense goodwill people felt toward John, he might walk away

from this looking just fine. Beyond that—and this was something that journalists might not understand—he did not think that journalism was the be-
all and end-all of life. No way was he going to spend many more years doing
this anyway.

John had an even better way to spin the end of *George*, and this was the
most superb exit of all. What if he ran for the United States Senate in 2000
for the seat that Daniel Patrick Moynihan was vacating? John was serious
about this prospect. He saw his life in five-year blocks. Half a decade was
over. Time to move on. Since he was a little boy, people had endlessly
hounded him about politics, and he had always said that he was not ready.
The years at *George* had given him solid practical training for a political life.
Robby was always talking about "skill sets." John had picked up a score of
them. When he had interviewed Madeleine Albright, the secretary of state
had been impressed that he was asking the questions not of a journalist but
of an insider, questions that a senator on the Foreign Relations Committee
or a White House staffer on the National Security Council might have asked.
He had become a strong public speaker who was comfortable on the podium.

Gary decided that he would get an outside take on John running by asking Roger Ailes, the Republican strategist who was building Fox News into a
major force in the business. Ailes wasn't advising candidates any longer, but
the two friends figured that if John was serious, there was no more realistic
judge of his chances. Over lunch Ailes picked his way through the various
candidates and their pluses and minuses and concluded definitively that John
was "a viable candidate." That may not have sounded like a ringing
endorsement, but given Ailes's party affiliation and his cautious judgments,
it was a decided plus.

Gary reported back to John, who was receiving other positive reports
about his prospects. Other advisers were having discussions with New York
power brokers. There were secret promises of support from labor officials
and other New York politicos. John may still not have had the passion to go
for it, but there was an irresistible quality to the idea, to the risk and excitement of it, and to the chance to do something real and good.

Beyond his own uncertainty, there were two other potential glitches in the plan, and they both involved women. One was Carolyn and the question of how she would deal with political life. If she could hardly leave their Tribeca loft, it was hard to imagine her campaigning with him and handling the duties of a senator's wife. The other problem was First Lady Hillary Clinton, who was making an inordinate number of "nonpolitical" trips to such watering spots as Binghamton and Buffalo.

John watched with growing dismay as Hillary subtly insinuated herself into what he considered his state. By dipping her toe in and out of the political waters, she shrewdly held off others who were thinking of diving into the senatorial contest. An early poll showed that most New Yorkers thought John would make a better senator, but he was not going to get into a political fight with Hillary. There was still the governor's race in 2002, and John marked that on his calendar as the next possible race.

John's cousin Bobby Kennedy was also looking with interest at the 2000 senatorial race. In the last half of 1998, before Hillary had expressed even a tentative interest, Bobby was one of the top potential Democratic candidates. A November poll among registered Democrats by the Marist Institute for Public Opinion showed that Carl McCall was supported by 23.1 percent followed by Bobby at 21.4 percent and his brother-in-law Andrew Cuomo at 15.7 percent. It was impressive that an environmental leader who had never run for office should be almost as popular as the well-known African American state comptroller. That same month, forty-four-year-old Bobby announced that he was not running, pointedly saying that this decision "was for this time around" and adding that "if I wanted to run for something, there's always something to run for."

Bobby was so wired that his legs sometimes shook when he sat. "I don't really want to lose control of my life," he said. "I would do politics if I had to, if it was the only way that I could feel that I wasn't squandering this wealth in terms of access and influence that I've been given. But politics imposes a huge cost on the individual and the family. To run in New York State for a statewide seat, I have to raise twenty million dollars, which means you end up hanging out with a lot of people you don't particularly like or want to hang out with. You put on a back burner a lot of the friendships that are important to you. Your time no longer belongs to you. You end up doing a lot of stuff that you'd probably rather not do. It's just losing control over my life. So you don't have anything left. I like the fact that I can take my kids skiing or whitewater rafting."

Bobby remained a dogged environmental leader. In the early months of 1999 New York mayor Rudy Giuliani felt Bobby's sting when he backed off commitments to protect wetlands that Bobby had helped negotiate in 1997. The mayor appeared to be attempting to curry favor with upstate voters in his probable campaign for the Senate. Bobby attacked on several fronts, including an op-ed column in the *New York Daily News* in which he publicly denounced the mayor. "The gravest threat to city drinking water," he maintained, "is a mayor with statewide political ambitions." He added, "I've become increasingly disturbed that in every aspect of watershed protection, the city is in retreat." He also ghosted letters signed by Democrat politicians Mark Green and Fernando Ferrer and authored a report charging that the engineering staff of the city's Department of Environmental Protection had become "an agent of destruction in the New York City watershed."

J ohn observed Bobby's work with admiration, but he had no intention of devoting himself full-time to philanthropic pursuits. Since the early nineties, John had sat on the board of the Robin Hood Foundation, a board that early on included Paul Tudor Jones II, a brilliant young money manager, and Jann Wenner, the *Rolling Stone* publisher. The organization soon became the charity of choice for the nouveau glitterati of Wall Street, and the board the place for the most exalted of networking. As the decade went by, some of these millionaires became billionaires. In comparison, John had just gotten the last check from his mother's estate, and as wealthy as he was, it sobered him.

John wanted the cachet of being their equal, and *George* wasn't going to get him there. In March 1999, John called Dan in Seattle and congratulated his friend on having just sold his ice cream business. "You know, we've always talked about doing something in business together," John said. "Things might be changing here." Dan was a successful businessman, but he was hardly the kind of partner who was going to impress Wall Street, unless they were planning to take over Ben and Jerry's. But Dan was a true friend. He was smart, shrewd, and loyal, and those qualities trumped everything else. It said something about John too that he was not talking about giving Dan a 20 percent cut of a venture with the name of John F. Kennedy Jr. over the door, but a fifty-fifty partnership.

"I've got the idea," John said. "I just want you to mind the store." John still had his obligations to *George* as well as his political ambitions, and it was unlikely that this would be his full-time pursuit. But he had faith enough in Dan to believe that his friend could make it work.

"Yeah, but what's the idea?" Dan asked excitedly.

"Well, I have this idea that involves crossing different media properties," John said, explaining his concept for their boutique company. "It would involve content writing for the Internet and putting different properties of the media together." It was hardly more than a notion, and Dan could not explain the idea further then or ever. But in the midst of the dot com frenzy, using John's name and throwing the word "synergy" in there somewhere, would probably raise some real money. It said far more about John's personality than the strength of this idea that Dan was prepared to move with his wife and daughter to New York. And it said something about how uncertain John was about his future that he was considering such disparate possibilities as running for political office, revitalizing *George,* and starting a business in an area in which he was a novice.

Despite all of his uncertainties, John continued to take the steps of a man who one day intended to run for office. For years the director of the Institute of Politics (IOP) at Harvard's Kennedy School of Government had been a retired politician who saw the position largely as a sinecure. John wanted to get a young academic in there who would brown-bag it with the students, increase the institute's public posture, and provide him a strong link with Harvard. His choice was Professor Douglas Brinkley, the director of the Eisenhower Center at the University of New Orleans. John discussed with Brinkley "the possibility of creating an academic policy journal," a useful way for John to develop a more serious image. John had a shrewd awareness of how to make things happen, and in the spring of 1999 he set up meetings for Brinkley with his Uncle Ted and others, so when the current director, Senator Alan Simpson, retired, John's candidate would be wired.

John continued to engage in important social activities involving *George* with Carolyn on his arm. In mid-May, when Carolyn rushed into the dinner at the Alexander Hamilton U.S. Customs House for the Newman's Own/*George* Award for Corporate Philanthropy, she had the frightened look on her face of Bambi with the woods on fire, but once she got inside and sat down, she was the wickedly witty Carolyn who was anybody's idea of a great dinner companion. As their friends knew, despite the tabloid rumors, the couple still delighted in each other's company. Barlow was in town, and John figured that the former Dead lyricist would hit it off with Sean "Puff Daddy" Combs, the hip-hop impresario. On the other side of Barlow, he placed former Senator Alfonse D'Amato. John had just signed on the defeated Republican New York senator as the most unlikely of columnists, and this evening John assumed that Barlow and D'Amato would do everything but attack each other with bare fists; he placed Gary where he could watch the battle. No matter how bad the food or how boring the speeches, he would have an outrageously entertaining evening. The unpredictable Barlow took an intense dislike to Puff Daddy and spent much of the evening talking to D'Amato.

After the dinner John's close friends sat together at the bar, where Carolyn did a perfect imitation of Puff Daddy at the dinner. The friends who attended evenings like this one did not believe that John and Carolyn's marriage had become a facade.

John's magazine might be failing, but he loved to be in this kind of mix of politics and popular culture, be it attending an awards dinner, editing *George*, or on occasion talking about it in public. In May he rode his bicycle uptown to the NBC studios in Rockefeller Center for an interview with Katie Couric for *The Today Show*. Though John rarely did television, he was as natural for the medium as his father. Unlike JFK, who as likely as not led with his wit, John projected a natural warmth and humor that gently pushed away intrusive questions. Couric could not coax him into admitting that he one day wanted to run for office, but it was clear that the five years at *George* and the Clinton scandal had not diminished his belief in the nobility of politics as the highest calling in a democratic society. That sentiment shone brightly through when Couric asked him what he considered his father's greatest legacy and what he was personally proudest of.

"Well, I'm enormously proud of a lot of things," John said. "They do a vote of popular presidents and among eighteen- to twenty-five-year-olds he is the fourth most popular president. Now, these are people who were not alive when he was president and know nothing about him except what they see in the paper or read—you know, see on TV. And the fact that he still manages to inspire people, to make people think that public service is a good thing, that people are still interested in his life, and how he got into politics, is something that I'm very proud of. That he has a resonance that has endured beyond his years. People come up to me and say, 'Your father got me into politics,' or, 'He's a great man.' That happens to me several times a day, and that's a really special thing."

J ohn had the memory of his father to inspire him, but the Shriver sons had the living presence of their parents. Sarge and Eunice were in some ways even more formidable in their seventies than they had been four decades before, and as much a goad and inspiration to their children.

Eunice was alone in her upstairs bedroom of the Shriver great house in Potomac in June 1999 when it caught on fire, and she was driven out onto the grounds. Mark lived nearby, and though it was almost noon, he happened to be home. He hurried over to the family estate where eighty firefighters and twenty-five pieces of equipment were fighting a major electrical blaze that did $600,000 worth of damage. Mark ran up to his mother, who was stand-

ing back from the fire with a cell phone to her ear. "Mummy, how are you?"
Mark asked pleadingly. "Are you okay?"

"Be quiet, Mark," Eunice replied, irritated, turning to her son. "Can't
you see I'm talking to Hollywood about my movie?"

Whatever his mother did, she was consumed by it. This latest venture
was a two-hour TV movie on the life of Mary. Once Eunice had the idea, she
inundated her son Bobby for months with transcripts of conversations with
Marian scholars and pushed him to move her movie idea ahead. Bobby had
been having dinner in a restaurant on San Vicente when he spied David
Israel, his close friend from the old *Los Angeles Herald-Examiner* days. Israel
was with his wife, Linda DeKoven, who also happened to be head of movies
at NBC. Standing on the sidewalk outside the restaurant, Bobby pitched the
most unlikely of TV movies in a secular age. DeKoven spent most of her
days saying no, but she found herself saying yes, yes, yes. As for Bobby, who
had moved to L.A. to get away from his family, he found himself working
on a daily, sometimes hourly basis with his mother as the co-executive producer of *Mary: Mother of Jesus.*

The film brought Eunice and Bobby together in a way that once would
have seemed impossible. They were the most unusual of production
teams—Bobby, who envisioned Jesus as an outcast shunned by society, and
his mother, a neophyte who ran around like Sam Goldwyn reincarnate. "I
didn't know how involved they would be," says DeKoven. "They were
involved in every detail of this production." In the late spring Bobby and his
mother flew to Hungary for the filming. "I think she went through a lot of
very difficult times with her son," Eunice said. "Mary is a great symbol of
leadership and strength."

While Bobby worked with his mother on the movie, Tim was busy with
final preparations for the Tenth Special Olympics World Summer Games in
North Carolina. The 7,000 athletes from 150 countries made it the largest
sporting event in the world in 1999. At the opening ceremonies, which were
shown on national television, seventy-seven-year-old Eunice and eighty-
three-year-old Sarge, the chairman of Special Olympics, stood on the

podium. Tim stood back from the limelight, aware that this might be the last year that his mother would stand there and look out at what her idea had become, at how it had grown throughout the world and lived on in the lives of hundreds of thousands of Special Olympians and their families and in the lives of the dedicated minions, including her sons and daughter. Tim was proud of his achievements, but in a few days he would be forty years old, well into middle age, and he hungered for a sense of individual achievement.

CHAPTER FORTY-NINE

Night Flight

On Memorial Day weekend, John and his friends got together at the house on Martha's Vineyard. They had been coming together for these weekends at the end of May since their college days. "It was such a love thing," Robby says. "There was a buzz in the air, a real buzz, and it was like, goddamn it, life is wonderful."

This year, as usual, Robby was there with his wife, Franny, and their two children. Sasha and her husband, Phil, and their two children were also there, and a woman friend from John's Brown days. They had all become Carolyn's friends too, and she looked forward to these long weekends almost as much as John did.

During the day John led the men through a regimen of activities that included everything from tennis to water skiing, touch football to biking, kayaking to swimming. In the evening everyone sat down to dinner at a dining table above which hung an exquisite pair of mobiles that Sasha and Phil had put together with John from white whalebones that he had brought back from the Arctic.

When they talked over dinner or afterward, it was often the kind of discussion that most people had left in their college dorm room. They talked about what a good life was. They talked about fate and duty. They talked about God. They were all outspoken in their way. They had to be to be around John.

There was plenty of time for private conversation, and as they often did,

John and Sasha went off by themselves. John was preoccupied with a subject that he could not bring up even around this group of his closest friends. John knew that this was the last time for a long while that he was going to have the kind of fun on Martha's Vineyard that he relished; at the end of the weekend, Anthony and Carole, his wife, would be coming up, and they would stay until his cousin died. John's mother had invited Anthony to the Vineyard half a decade before to recuperate, and it was only right and proper that John had invited his cousin there to end his days. But this was John's playground, and part of him was uncomfortable with the fact that for the rest of the summer he would be dealing with death and dying, even on the most beautiful and blissful of days.

Anthony's cancer had come back, and he only had a few weeks left. In the previous five years Anthony had had dozens of operations. Now he was suffering through painful, debilitating chemotherapy, but there was no hope. "You don't understand, but terminally ill people are really difficult," John said, referring to both his mother and Anthony. John had been there after Anthony's last surgery when the surgeon told him that there was nothing more he could do. Anthony looked at John and said, "Well, okay, we'll do another operation and go in and take that off." John wanted to say that it was all over, it was time to die, but he could not say it. "Sometimes people have this almost disturbing will to live," John told Sasha. "They have it when it's all over. Anthony simply can't face that he's not going to be around anymore."

Almost everyone who has gone through the lingering death of a loved one has probably thought such thoughts. But few would speak them, and as much as Sasha felt that John's words were justified, she sensed a core of youthful selfishness, as if questions of mortality were only for others. John had not spoken those thoughts to his cousin. Anthony had been the brother John never had, and he had visited Anthony regularly in the hospital. As much time as John spent with Anthony, it was nothing compared to Carolyn. She was sometimes at the hospital around the clock, staying overnight or returning on weekends, when she so easily could have gotten away. For those many hours, there was nothing self-absorbed or self-conscious about her.

During those months Carolyn kept saying that Anthony's death was going to break John the way nothing ever had. There was an impenetrable core of John to which no tragedy, no sadness had ever reached, and as his wife saw it, John had to be broken to be whole.

L ate Sunday afternoon the group sat around in the living room watching the New York Knicks take on the Indianapolis Pacers in a thrilling first game of the Eastern Conference finals. As the game reached its last minutes, dusk was fast approaching and John headed out the door, saying nothing. When Robby heard the sound of the Buckeye engine revving up, he wondered what was going on, since John was supposed to be taking the powered parachute on its trailer down to the beach where there was plenty of room to take off.

Robby turned toward the window just in time to see John flying forward in the little craft just above the ground, then grazing a small tree and crashing earthward. The four children playing outside rushed toward John. His ankle had been broken, his hand hurt, and he was in pain. "There's no problem," he told the children, not wanting to frighten them. "No problem. Just ran out of gas."

Later that night John arrived back from the hospital in a cast. He faced surgery in the morning, and no one felt like going to bed. Sasha lay down next to John on the living room floor, where he rested with his leg propped up next to his crutches. Carolyn looked on from the couch as the two of them talked with quiet intensity, holding each other's hand.

Compared with the maladies and accidents that had afflicted his family in the past, John's broken ankle was nothing. But it made him aware of Anthony's illness in a new way. Anthony's imminent death was not breaking John, but it was bending him, and he saw and felt things he had not seen and felt before. "The reason I broke my ankle is because I'm gonna help out more now," John half-whispered. "I'm gonna be around, and I'm gonna sit up with my leg next to the bed, and we're gonna be these two fucking frustrated guys complaining and bitching about the world. That'll be how we'll pass the last weeks of Anthony's life."

"That's right," Sasha said intensely. "This terrible thing happening to Anthony sucks, and your summer is going to be rough, but your summer was going to be rough anyway."

John had myriad troubles—troubles with Carolyn, troubles with George, troubles with his brother-in-law, Ed Schlossberg, troubles with his foot. With Anthony dying and things so much in turmoil with Carolyn, he was buffeted by waves of uncertainty. He had always had a buoyant, resilient optimism. He did not try to talk his way through problems. He outran them. He was full of nervous energy that he regulated with a regimen of daily exercise, but he couldn't work out while hobbling around on crutches. And the crutches advertised his failure with the Buckeye, just another reason why he was in a foul mood. He told one friend who was doing particularly well that their lives were going in different directions, and then he changed the subject.

John did not talk to his close friends about his relationship with Carolyn. He apparently talked about his difficulties with one acquaintance who would later tell John's stories to the tabloids, which published them in a wildly exaggerated way. It may have seemed unlikely that John would confide to someone outside the inner circle of friends, but after his mother's death he had related his anxieties not to his intimates but to two acquaintances, Richard Wiese and John Rosenthal. On both occasions John was momentarily exposing the most troubled part of himself. He did not want to betray his wife in any way, and talking about her to his closest friends was a form of infidelity. But what he supposedly said to his acquaintance was only part of the truth. The reality was that he had both a troubled marriage and a blessed marriage. He had always lived with emotional ups and downs, but this relationship had become exhausting and without respite. The gossips whispered that the couple weren't even having sex any longer; that was the most unknowing cut of all, since no matter how their fights began, they ended in only one way.

On Thursday evening, July 8, John showed up for dinner at David

Kuhn's apartment on crutches. John had known Kuhn since his Andover days, when Kuhn had been a freshman at Harvard. Kuhn had been a top editor at *The New Yorker* for six years and was someone with whom John felt comfortable discussing the ups and downs of *George*. When John hobbled into the lobby of the eighteen-story building, Sasha was already there at the elevator, along with her houseguest, Jenny Christian, John's first love from Andover. John saw Jenny only infrequently, and he was especially happy to see her.

Except for Kuhn's partner, Kevin Thompson, the other four guests went back to John's teenage years, and it was bound to be an evening in which the past was more present than the future. New York in July did not have very many exquisite evenings like this one, especially sitting outside on a roof terrace in Greenwich Village, eating a fine dinner and talking.

Sasha and John always managed to have their secret conversations, even in the midst of others. "Look at those yellow pants," Sasha whispered, admiring John's lemon yellow trousers. "They're so chic." John drew closer to Sasha and whispered barely audibly, "They were my father's."

John had a personal matter on his mind, and later in the evening the two lifelong friends found themselves seated together alone on the terrace. "I've been thinking of my eulogy to Anthony, and it won't be easy," John said. "Well, what are you going to do?" Sasha asked.

John talked a little about the contradictions in Anthony's personality, a subject that any of their friends could have endlessly elaborated. Anthony was generous to a fault and mindlessly selfish; grumpy and exuberant; witty and crude. The whole array of contradictions had presented themselves in years of bickering, cantankerous friendship. John knew that within a few weeks he would be standing as a witness to the meaning of Anthony's life, and his words had to be right and true. There would be other eulogies, but his would be the one people would remember. John's obsession with the eulogy was in part probably his way of distancing himself from the pain of Anthony's death. He would have to tear Anthony down the way he had done when his cousin was alive. Then he would have to celebrate his life in ways he never had, capturing the essence of a man he loved as a brother.

The conversation turned inevitably to the question of whether Hillary Clinton would decide to run for the Senate from New York. John said he believed that she would run, and that if she ran, she would win. "Don't underestimate the degree to which it's basically about name recognition," he said. "And she's got that in spades."

John expressed his irritation at Hillary for promising to do an interview for *George* and then backing off to do it for the first issue of *Talk*. That had upset him immensely, but he had tried to shrug it off. John understood people almost too well, and in understanding came forgiveness, even when it was not always deserved.

In the midst of the spirited conversation, John got up and walked into the kitchen, where he could talk to Kuhn alone. John was obsessed with his magazine, not only with what he might do for it now but with what he had done wrong. "My biggest regret," he told Kuhn, "is that I did not assert more of my power with Hachette." John said that he knew he was the celebrity face of *George*. That was part of the bargain, but he had been too malleable for too long. As he looked back on it, he realized how badly he had been used. He cringed when he thought of the long, boring dinners with car company executives that Pecker arranged. John had sat through them, but then found the car company ads showing up in other Hachette publications, not in *George*.

"How'd you let that happen?" Kuhn asked.

"Pecker is smart and cunning," John replied. "But you know what I've come to realize? I have a problem with male authority figures."

"Come on, you've got to be kidding," Kuhn said.

"Well, I didn't have a father growing up, and you know, I do think that kind of makes sense."

T he following weekend Christiane and Jamie flew up to spend the weekend on Martha's Vineyard. John flew his own plane up, accompanied by a co-pilot. That was an important precaution, since John was navigating on crutches while wearing a "Cam-Walker," a molded sandal that strapped halfway up his leg. At the estate John saw another vivid reminder of his accident. The caretaker had left the broken tree just where it was. The man had decided that it would stand there, a reminder to John to try living more cautiously.

Christiane had risen to become one of the most famous reporters in the world, and she was the friend who had the best grasp on how celebrity often distorted life. "What always stunned me was how incredibly normal he was and what moderate appetites he had despite being offered everything and anything there was," she reflects. "He had everything literally thrown at him: celebrity, massive fame, incredible tragedy, money, opportunity, women, friends. You name it, John could have it. But he was never spoiled. How amazing is that? He was never overwhelmed, brought down, or brought to the heights of arrogance by all of this. He never overindulged. He took a little of this and a little of that, and he knew how to control himself. How rare is that? He was proud to be a Kennedy, without being slavishly devoted to his family's every foible. He was kind and considerate in ways that were not showy, just touching and meaningful. He was boisterous and fun and larger than life and shockingly handsome, and somehow he managed to stay on an even keel."

John didn't like to frequent the restaurants and clubs on the island very often, since the experience almost always turned unpleasant, with one photographer or another sticking his camera at Carolyn. Saturday evening, though, the two couples headed out in "the Goat," John's beloved 1969 Pontiac GTO convertible, to go for drinks at a couple of island bars, the Wharf and the Lamppost. For once people left them alone.

"Kissy, are you pregnant?" John asked, as they sat drinking margaritas.

"No, why?" Christiane asked

"Because you're not smoking," John said.

"We're trying."

John may have been even more intuitive than Christiane realized, for unknown to her, she already was pregnant.

"Oh, don't get pregnant, Kissy," Carolyn said jokingly to Christiane, "or I'll have to get pregnant."

Anthony was too weak to join the group. He could hardly stand, but John wanted him to enjoy the ocean one more time. Sunday afternoon John and Jamie lifted Anthony up and half-carried him to the car and drove to the beach. There John and Jamie helped him out and brought him to the edge of the water. John stood on one side of him and Jamie on the other, and holding Anthony up, they walked into the water. The waves buffeted them, and they grasped Anthony so he did not fall. And all the time that John was physically holding Anthony up, he was verbally putting him down. And Anthony was responding in kind, trying to top John in his withering insults.

When they got back to the house, Carolyn asked if the other guests would like to go water-skiing. Jamie thought that was a great idea, and so did John. "You don't want to go water-skiing," Anthony joked. "By the time John gets everything organized, it's going to be late, there are going to be bugs, and it's going to be a whole lot of trouble for a little spin." By the time John got everything organized, it was late, there were bugs, and it was a whole lot of trouble for a little spin.

On Monday John flew from Martha's Vineyard to Toronto for an important business meeting. He had a flight instructor along on the five-hundred-mile international flight. Even if his left foot had not been in the Cam-Walker, it would have made sense to have a co-pilot. John was flying his new plane, a Piper Saratoga that he had purchased two months before, and was still getting accustomed to it. Though the six-seater was technically a more sophisticated machine than his Cessna 182, McClellan, who had advised him on the purchase, considered it "the reverse of high performance, more like the station wagon or minivan of the air."

John had talked to all kinds of media companies in New York about coming in and taking over Hachette's half-ownership of *George*. This trip to Canada was an indication of just how limited his options had become and how little time he had left. The magazine was in precipitous decline: it was barely holding on to the guaranteed rate base of 400,000 readers and had fallen 30.3 percent in ad revenues in the first half of the year. The ad-thin recent issues looked more like a handout than a thriving magazine. There were 214 ad pages in the entire first six months of the year, only 39 pages more than in the first issue.

John had made some strange journeys in his life, but this was surely one of the most bizarre, flying to Ontario to attempt to sell half of an American political magazine to a Canadian auto parts manufacturer. Stranger still was that Keith Stein, vice president of Magna International, and Belinda Stronach, daughter of the Magna founder, said they loved *George* and sounded serious about buying in. After spending a few hours in Toronto, John and his co-pilot flew to New York City.

I think this is the real thing," John told Gary Tuesday morning when he got into his office and called his friend. John thought he might have found a new partner in Canada. "What do you know about them?" It suggested how naive John was about the business aspects of publishing that he had gone all the way to the Canadian capital without even checking out the bona fides of a potential partner. Sitting in his office at News Corporation, Gary ran some stats on the company on Bloomberg, then called John to tell him that Magna was a strong company. John became even more excited.

John still hoped that he could work things out with Hachette. After his return from Toronto he went for a crucial meeting with Jack Kliger, Hachette's new CEO. John carried bottled water, his version of worry beads, which he set on the executive's desk. This was a Hollywood-style meeting where the length of the meeting was a statement of the importance of the person, but all that mattered was the last five minutes. As Kliger and John exchanged their endless meaningless pleasantries, Kliger reached forward and started drinking out of John's bottle. John realized that Kliger was nervous and that nothing but bad news would come out of this. John learned that day that it was probably all over with Hachette and that if he wanted to keep *George* going, he would have to come up with a new partner.

Dan would be flying in at the end of the week to discuss their business plans. John called and left a message on his friend's answering machine:

"There's a chance that things [at *George*] might still go on, but I'm resigned to the fact that this is probably the end of something I've done. I'm looking forward to the next phase of my life."

On Wednesday afternoon Blow heard John yelling into the telephone. He could not follow most of the conversation, but then he heard the shrill, strident phrase: "Well, Goddamn it, Carolyn, you're the reason I was up at three o'clock last night!" That evening John did not go home but took a $250-a-night room at the Stanhope Hotel, near his mother's former apartment on Fifth Avenue. His loft did not have a separate office or bedroom where he could go off and closet himself. He knew from his nearly three years of marriage that he could not simply walk in and make up with Carolyn. That always took time, and he had a more immediate problem on his hands. The clock was ticking on finding a way for *George* to survive. He had to prepare materials for his potential new Canadian partners and figure out how to deal with the media reaction when on Monday he announced that he was seeking a new partner. It tore him apart when he and Carolyn had these fights, but for a few days he simply had to focus on the future of *George*.

At around midnight John called Dan, not mentioning that he was at the Stanhope. Dan was planning to fly into Boston on Friday; John would fly up to meet him so that the two friends could spend much of the weekend discussing their prospective business partnership. John spent most summer weekends at his home on Martha's Vineyard, but this time he had decided to go to his house in Hyannis Port. John said there had been a change in plans. He would still be making it to Hyannis Port, but he would be flying to Nantucket Friday evening to attend a dinner at the Club Car to celebrate the fifth wedding anniversary of his friend Billy. This would be a long evening after a long day, but he wanted to be there to drink a toast to Billy and Kathleen.

After midnight Julie Baker visited John. Like others old lovers, she had remained a friend. She was a temptation to him, as he had admitted a few months before to one of his friends; she was a voluptuous, passionate woman who may still have been in love with him. Some of the most intense, erotic times of his life had been spent right there at the Stanhope with her in the early nineties. Usually when he was emotionally troubled, sex was not what he thought about, not what he wanted, not his idea of a way out, not a way to forget. He may simply have wanted to talk to Julie.

Baker left an hour or so later. She arrived back at the hotel in the morning around 8:00 A.M. and had an intense talk with John over breakfast. John was operating in a world of uncertainty. The staff at *George* was rattled, nervous about their jobs, gossiping in the hallways, perusing the media columns in the papers for hints of the future. John knew that he had to keep them as informed as he could without telling them the details of what he was trying to do.

Later that day John got together with the magazine staff and told them that one way or another *George* would continue and they would hold on to their jobs. That was John's wish, but it was by no means certain. He apologized that he had not focused on the work, blaming a personal matter that he said would soon be resolved. The best news for him that day was that the leg device came off, though he was still limping around on crutches. He chanced into McClellan, the editor of *Flying*, and told him how much he was looking forward to heading down to Florida again to complete his instrument training.

That evening John went to Yankee Stadium with Gary in a town car to watch the Yankees take on the Atlanta Braves. Gary had been invited by Rupert Murdoch's sons, James and Lachlan, and Gary decided to bring John along. Baseball was never John's sport, whether to play or to watch. He needed action, and he spent most of his time talking to Gary and the Murdochs.

Gary was a good and trusted friend. Yet he saw only veiled hints of the personal difficulties his friend was going through. They started talking about Julie Baker, though John did not let on that he had just seen her twice at the Stanhope. The former model was making a living selling clothing, and Gary started ragging John about how his friend had convinced him to buy some of her custom-made shirts and none of them fit. John responded by saying that he had purchased four suits from her and none of them fit either. That was enough for John to call Baker on his cell phone. "You better get in another line of business, Julie," John said.

Riding back to Manhattan after the game, Gary looked at his friend silhouetted in the shadows, and he sensed that something was wrong. "So what's bugging you, John?" Gary asked. "What's going on?"

"God, it's tough," John said, looking out into the night.

"This is like God's testing you to see how much stuff you can take in your life," Gary said, knowing how John felt about Anthony's dying. Gary knew John wouldn't want to talk about that, and he turned the conversation to *George.*

"You know, I have a real problem with male authority figures, guys like Jack Kliger," John said, mentioning the new Hachette CEO he would be meeting with for the second time in a week in the morning.

"John!" Gary exclaimed. "What about that big lecture you gave me a couple of years ago about me having to become confrontational? You're the one who's got the problem."

"It's for obvious reasons," John replied. "Women, you know, I have no problem. But what about Murdoch? Why is Rupert so successful?"

John was not making chitchat but seeking insight into how the Australian born entrepreneur had built a worldwide media empire. John admired Murdoch, whom many thought of as a buccaneer. That was precisely the quality that he liked. The man was not a mere manager. John was always trying to learn from people, and this evening he was trying to figure out how Murdoch had done it. Instead of going to his loft in lower Manhattan, John had the driver let him off at the Stanhope.

A t most of the publishing firms in New York, a Friday in July meant at most a half-day of work and then a sprint out of the city. John was getting away too, but not until later in the afternoon. On Saturday, Rory, the youngest of his cousins, was getting married, and though he had not formally RSVPed, he could hardly spend the weekend in Hyannis Port without attending her wedding. More important, Dan had flown in alone from Seattle on the redeye Thursday night and was already sitting in John's Hyannis Port house.

Soon after John arrived in his office, he went in to see Kliger. John could tell that the executive had no emotional stake in whether *George* succeeded or not. John was going to put together his plans over the weekend and present them to Kliger. If it made sense to the CEO, Hachette would renew its partnership. If it did not, the publisher would probably go ahead and sell its share to somebody else. John knew that he had some serious Canadian investors who might be interested. He had also heard from Gary that Lachlan Murdoch had inquired about *George*'s subscription numbers.

John had lunch with Blow at Trionfo, an Italian restaurant in the office building. John was glad to have the device off his foot, but he was still hurting and propped his leg up during the meal. He was full of obligatory professional optimism, pumping up his executive editor, who in turn would pump up the rest of the staff. Blow was not someone in whom John confided, and he gave Blow no hint that he was thinking of a future that might not include *George*.

At around two o'clock Friday, John called Dan. "I'm running a little bit late, you know," he said.

"But what about that Billy thing?" Dan asked

"I think we will have to bag it."

John went on to tell about the newest change in plans. "Carolyn's sister Lauren decided to come. She's going to Martha's Vineyard. So instead of coming direct, we're going to drop her off there." That meant essentially that John had given up his friend's anniversary dinner so he could do a favor for his wife's older sister, an investment banker in New York. Carolyn was especially close to her sister, and given the fact that he had just had a rousing fight with his wife, it was probably not a bad peace offering.

That meant that John would be later than originally planned. "You know, Shriver called and said he'd heard a rumor that you were coming to Hyannis Port," John continued. "So he goes, 'Should I come?' I told him no, so I don't think he's going to come."

John was close to Tim, but the weekend was already shaping up as so complicated, with so many obligations, that he couldn't envision making room for yet another diversion. He had to work with Dan on a business proposal and prepare plans to show Kliger Monday that might determine *George*'s future.

"Hurry! Hurry!" Dan said, knowing how late John always was.

"I will! I will!" John replied, with the conviction of a man taking an oath. It was almost slapstick, but that was the way the two friends often talked.

John was always complicating plans, always making things almost too interesting. With everything that was going on in the office, it was not until around 5:00 P.M. when he called Billy to tell him he would not be there that night. That was typical of John, and Billy was as used to it as any of his friends. He had learned years ago to shrug it off, and add the anecdote to his endless repertoire of John stories.

At 6:30 P.M., shortly before he left his office, John logged on to Weather Service International on his computer to check out the weather. There was lots of haze out there, but the visibility was at least five miles, and there were no storms on the horizon. John got route briefings both to Martha's Vineyard and directly to Hyannis, suggesting that he may not have been certain he was going to drop off Lauren Bessette.

John drove out twenty-five miles to his plane at the Essex County Airport in Caldwell, New Jersey, in his convertible with thirty-four-year-old Lauren. It seemed like everyone in the city had decided to get out of town. The traffic was so slow that John's presence in the convertible was a welcome diversion to many of the other drivers. In contrast, Carolyn was being chauffeured in a Lincoln Town Car with tinted windows whose privacy no one could violate.

When John flew off by himself, it was the most exhilarating, freeing of experiences. The flight this evening was not likely to be one of those times. He had planned to be there early enough so it would still be light, but dusk was fast approaching. Carolyn arrived soon afterward, and John prepared to depart. He hadn't seen his wife for three days, and the last time he had been alone with her, they had argued. There was much to be resolved there, and it was not clear when they would have the private time they needed. Despite the marriage counseling and Carolyn's hours with a psychiatrist, they still confronted their problems with hysterical blowouts that left both of them emotionally ravaged. Somehow they had to learn how to keep the good times that they still had with each other, while getting rid of these scenes that ripped them apart.

John would have had a full weekend making up with his wife, but that was just the beginning of it. He had a passenger to deliver, prolonging this day. He had Dan anxious to see him, and Billy disappointed that he had missed the party. He had Rory's wedding. He had something else to do at Hyannis Port. His house had just been redone, and he would be seeing the work for the first time. John had wanted to do a good deed by giving the job to a local musician. John figured that the man would not cheat him, because the prestige of doing the house would bring him work all over the Cape.

John checked the Piper Saratoga out carefully before taking off. He was flying without a co-pilot this evening, and that was another added anxiety. Of the approximately 310 hours that he had flown in his life, almost three-quarters of the time (238 hours) a certified flight instructor sat at his side. John had flown roughly 36 hours in his new plane, but only 3 hours of that had been without a professional pilot next to him—and less than an hour of that was at night, as most of this flight would be. Most pilots with his ratings and hours had far more hours flying solo. Although John had been trying to fly safely by usually flying with a flight instructor, in doing so he had increased the risks of flying solo this evening.

Since hurting his foot, John had taken a professional pilot with him on every flight. When one of his instructors expressed his willingness to go along this evening, John told him he "wanted to do it alone." When this flight instructor had flown along to Martha's Vineyard for the Fourth of July weekend, John's Cam-Walker device had been so troublesome that the pilot had taxied the airplane and helped him with the landing. This evening John's foot was still troubling him enough that he was on a crutch. Even without considering John's ankle, the flight instructor said afterward that "he would not have felt comfortable with [John] conducting night flight operations on a route similar to the one flown on, and in weather conditions similar to those that existed." The pilot, who had flown thirty-nine hours with John, believed that he "had the ability to fly the airplane without a visible horizon but may have had difficulty performing additional tasks under such conditions."

In April, John had flown down to Florida on several weekends to take instrument training. He had gotten halfway through the course of twenty-five lessons, but he had to repeat one lesson four times before succeeding. Since he had not completed the training, if the weather closed in, he might find himself unable to put the Piper Saratoga on the autopilot that can almost fly itself to the destination.

John was not one to take seriously the lesson of that broken tree on Martha's Vineyard. Yet he was not setting off this Friday evening the way he had done Memorial Day weekend in his powered parachute, with a cavalier disregard for precautions. He had done everything that a pilot was supposed to do before taking off—learning the latest weather forecasts, checking and cross-checking the instrument panel.

It was after eight-thirty by the time John was ready to take off. Although there was summer haze, there were no negative weather reports or ominous cloudbanks to give him second thoughts. In his own way, he was a careful pilot, but he did not have a seasoned pilot's instinctive caution. He had not filed a flight plan. It was not mandatory, but it was a prudent measure that he could have taken. There were those like his sister, Caroline, who would not fly with him in part because of his reputation as an amiable scatterbrain.

Just before John took off, he called Hyannis Port and told Provi Parades, who had worked for his mother in the White House, that he would be even later arriving there than he had thought. As he flew through hazy evening skies, John did not know that he had just had another one of his legendary close calls. This time he avoided not the Staten Island Ferry but an American Airlines passenger jet. In the haze he had flown so close to the big plane that the pilot had to change his course.

When there was no haze, this was a beautiful flight; with the lights of the towns and villages below, the journey was straight out of Saint-Exupéry's

Night Flight. As he passed the outer reaches of Rhode Island, the haze thickened; he could see little of the land below. The prudent move would have been to turn northward, hugging the coast and probably landing at Hyannis. Instead, John flew out across the ocean toward Martha's Vineyard. There was nothing down below but water and this trackless haze. Many times before in his life he had flown through clouds and darkness, and always before the darkness had passed and the clouds had opened up.

Beguiled and Broken Hearts

o way had John died. No way had he taken his wife and sister-in-law with him. It did not matter how many hours had gone by with the search planes and the ships sweeping the sea, finding nothing. He would turn up, and his friends and family would chide good old John. He would come bouncing into the Hyannis Port house soon, asking what all the fuss was about and secretly liking that he had caused such a stir.

President Clinton ordered a search-and-rescue operation unprecedented for a private citizen. All day Saturday scraps of Styrofoam insulation, a cosmetic bag with Carolyn's prescription bottle, Lauren's luggage, and other debris drifted onto the Gayhead beach on Martha's Vineyard, not far from the house that John had loved so much. On Tuesday they found the wreckage of the plane seven and a half miles off the coast of Martha's Vineyard in 116 feet of water.

The National Traffic Safety Board report concluded that the probable cause of the accident was "the pilot's failure to maintain control of the airplane during a descent over water at night, which was a result of spatial disorientation. Factors in the accident were haze, and the dark night." John apparently could see nothing, not the ocean below, not the sky above, not land ahead, nothing. In that situation a pilot frequently cannot tell up from down, and the instruments read the opposite of everything his instincts are telling him. John had taken enough training to be able to fly the plane on autopilot, but he apparently was unable to apply what he had learned or to handle several tasks at once. About seven miles from the island, he began his descent, and started climbing again, making a left turn. Then he began to descend yet again, not a glide downward, but a nosedive—faster and faster and faster, burrowing into the ocean like a missile.

John had learned to fly in a serious, responsible way. Yet he was not a natural pilot, and no amount of training could make him so. He was full of brazen confidence that he could work his way out of any difficulty. If he had taken a professional pilot with him, the trip would have ended on the tarmac, not in the sea.

John's body, along with those of his wife and sister-in-law, lay in the cockpit of the plane. Some thought that the bodies should stay there, that no one should even attempt to bring them up. Ted told Caroline that he wanted the bodies brought up so that his nephew could be buried at Arlington National Cemetery next to his mother and father and near his Uncle Bobby. There the eternal flame would burn for him and for a family's future as much as for its past. For Ted, that memorial would honor not only John's life but the never-ending journey of the Kennedys.

The discussions encapsulated the age and its obsession with celebrity. If they buried him at Arlington, would it not become the Kennedy Graceland, a grotesque, Disneyfied memorial? But if they left the bodies encapsulated in the plane, was it not possible that the same breed of ghoulish predators who had attempted to invade his wedding would dive down to photograph his remains? As Caroline considered the alternative, she decided she wanted her brother's body brought up to be cremated and to have the ashes strewn across the sea.

Caroline would not take her children every July 16 to Arlington National Cemetery. There would be no cultish shrine where seekers would leave flowers and notes. It was a profound decision, for until now each honored Kennedy's death had been used to inspire the living while raising even higher the mythic importance of the family in the American mind. In the years since his death Caroline has reinforced her decision by vetoing all attempts at memorializing her brother in a major way, leaving him and his memory to herself and to those who had truly known him.

For a week the cable news networks devoted themselves largely to this one story. The papers were full of it. Wherever Americans assembled, in factories, offices, churches, and schools, they talked of John and his life. John was no president, no senator, and no war hero. He was a private citizen whose father was an assassinated president. John's greatest professional accomplishment was the editorship of *George*. It was doubtful that years from now historians would pore over its pages learning what politics and culture had been like in the nineties. Yet in the reaction of Americans to his death, it was as if a great and noble man had died. For most Americans, there was a terrible symmetry in a story that had begun with the death of President Kennedy and ended with the death of his son thirty-six years later. There was a sense of finality now, of promise eternally unfulfilled, of the last great hope of the family being gone.

In trying to understand, the television commentators drew on everything from Greek myth to Shakespeare, but none of it fit. In Greek myth, when Icarus flew too high, the wax on his wings melted, and he fell to his death, into the sea. But he carried no passengers with him. Nor did Shakespeare's Romeo mistakenly poison Juliet's sister. For the most part, the commentators rarely mentioned the deaths of Carolyn and especially Lauren. To do so was to bring up implicitly John's responsibility, and it was a terrible burden on his name. John was an icon, and the story of the three of them did not make for classic myth or the catharsis of classic tragedy.

John's friends did not call up Greek mythology, dust off metaphors about a Kennedy curse, lament the death of Camelot, or talk about America's version of Shakespeare's royal tragedies. His friends did not mourn a celebrity. They did not mourn America's child. His friends mourned a friend.

In the days following John's death Robby lay in a fetal position on the living room sofa, as his children tiptoed by. Dan slept in a different bed each night and wandered the grounds in Hyannis Port during the day. Billy remembered that in Ireland when someone died too young they cut a tree down and left the stump as a memorial to what had been lost.

O n Wednesday morning, July 21, Ted flew from Hyannis Port with his two sons, Teddy and Patrick, by Coast Guard helicopter to the Coast Guard station at Menemsha. From there they traveled on a forty-one-foot Coast Guard cutter out to the USS *Grasp,* a salvage vessel, to be there when the three bodies were brought up out of the ocean. Yachts full of news media and helicopters carrying camera crews were kept over a mile away, so they could see nothing of the mission on which the three Kennedys had come.

It had been five days since the accident, and only part of John's body was discovered and brought up. To ensure that there were no pictures, the bodies were bagged before they were brought up to the surface. Then it fell upon Ted, Teddy, and Patrick to identify the bodies, and to do so in the presence of military officials they did not know. Carolyn and Lauren had died no less than had John, but much less attention was paid to them, and that was yet another unfairness.

Ted did not want to prolong the pain and asked that the autopsies be done as quickly as possible. He also asked that no autopsy photos be taken since they could end up in the tabloids. The medical examinations were performed within a few hours.

Later that night the three bodies were taken to be cremated. The two brothers Doug and Max went to the crematory to be there until the bodies were nothing but ashes. It was the way of the Kennedys to stay with the body until the burial, and though some had stood watch at night over the body of Bobby in St. Patrick's Cathedral, no one had ever foreseen a night like this. Through the long hours they prayed and consoled each other and cried unashamedly.

In the morning the USS *Briscoe* sailed seaward, stopping about five miles off the coast of Martha's Vineyard. The boats of the media were kept far back so that even the most determined photographer with the most powerful telephoto lens would have no images of the service. On the deck were seventeen members of the Kennedy and Bessette families prepared to commit their loved ones to the sea. Among them in a wheelchair was Anthony, whose eulogy John had prepared.

When it was time, an officer in dress whites walked down a ladder to a platform just above the sea and gave a brass urn to Caroline, who cast John's ashes onto the water. He had dipped the paddle of his kayak in waters from the Arctic to the Baltic. From now on, when those who loved him looked at the sea, standing by the waters of lower Manhattan that John knew so well, sailing on the Cape, or kayaking in Seattle, they would often think of him.

O n the following day the family and friends of John and Carolyn attended a memorial service at St. Thomas More Church on the Upper East Side. This was the small church that John had gone to as a boy. Clinton headed the list of dignitaries, and though there were a few other politicians and celebrities among the 350 mourners, for the most part the pews were full of John's family and friends as well as Carolyn's.

The ceremony was not televised. Out in the eighty-five-degree heat thousands stood behind barricades, and all across America millions watched on television, learning what they could from the commentators. John was mourned as much for what he might have been as for what he was, and for what he was, not for what he had done.

Wyclef Jean, of the hip-hop group the Fugees, sang the reggae song "Many Rivers to Cross," and for those sitting there the rivers to cross were deeper and wider than they had been before. Anthony, so close to death himself, recited the Twenty-third Psalm that eight-year-old John had read at his grandfather's funeral thirty years before. "Our revels now are ended," Caroline said, speaking lines from Shakespeare's *Tempest* as her farewell.

> And, like this insubstantial pageant faded,
> Leave not a rack behind. We are such stuff
> As dreams are made on, and our little life
> Is rounded with a sleep.

Ted was the last to speak. Those who had been at John's wedding heard Ted speak the words of the same poem that he had read on that occasion, and the words had a poignancy almost impossible to bear. And then Ted talked of John's life.

"From the first day of his life, John seemed to belong not only to our family, but to the American family," Ted began. "The whole world knew his name before he did. The most famous little boy in America sat beneath his father's desk in the Oval Office. He was an actor posing for photographs when he could hardly talk. He was a boy who grew into a man with a zest for life and a love of adventure. He was a pied piper who brought us all along.

"He had a legacy, and he learned to treasure it," Ted said. "He was part of a legend, and he learned to live with it. Above all, Jackie gave him a place to be himself, to grow up, to laugh and cry, to dream and strive on his own. . . . He accepted who he was, but he cared more about what he could and should become. He saw things that could be lost in the glare of the spotlight. And he could laugh at the absurdity of too much pomp and circumstance.

"He was lost on that troubled night, but we will always wake for him, so that his time, which was not doubled, but cut in half, will live forever in our memory, and in our beguiled and broken hearts," Ted concluded. "We dared to think, in that other Irish phrase, that this John Kennedy would live to comb gray hair, with his beloved Carolyn by his side. But like his father, he had every gift but length of years. We who have loved him from the day he was born, and watched the remarkable man he became, now bid him farewell. God bless you, John and Carolyn. We love you and we always will."

CHAPTER FIFTY-ONE

Life Lessons

A t the end of August much of the Kennedy family returned to the Cape to celebrate Tim's fortieth birthday. It had been a summer of funerals, first John's, and then Anthony Radziwill's, who died a month after his cousin. What should have been a celebratory occasion had an undertone of melancholy to it. Tim's old principal at Hillhouse High, Red Verderame, became so sick momentarily that he had to be taken away in an ambulance. The event was not only a birthday but a christening for Tim's newest child, one-year-old Caroline Potter Shriver, on the beach at Hyannis Port, next to the ocean water where both John's and Anthony's ashes had been cast. Though no one talked of it, there was a poignant sadness to it all, an emotion rarely prevalent at Kennedy family gatherings. They had been brought up to believe that life lay in the future, but for many of them the horizon no longer glowed with promise.

The Kennedy cousin who probably had the greatest impact on the world in the years after John's death was Bobby Shriver. From 1999 on, he worked with Bono, the lead singer of the Irish rock group U2, in arousing the American conscience and convincing the American government to retire much of the burdensome debt that had made the term "developing nation" a misnomer in most of Africa. Bobby had met the rock 'n' roller through the Special Olympics Christmas records. They made an extraordinarily effective duo, with the articulate, charismatic Bono out front while Bobby carefully created the context in which Bono worked his public act. They sought to do what the Shrivers had done for mental retardation—to take a decidedly unsexy problem and turn it into a popular cause. To do so, they had to create a mass constituency behind debt forgiveness and convince conservative power brokers to support the idea.

Bono was on the cusp of forty in a young man's business, the lead singer in a rock group that some people thought had seen its glory days and was only a few years away from the land of golden oldies. Even if he had been at the top of his career, the kinds of people Bono had to deal with now cared nothing about the aristocracy of rock 'n' roll.

Bobby arranged a series of meetings for Bono with key players in the power elite, starting with World Bank president James D. Wolfensohn, with whom Bobby had worked for seven years as an investment banker. Only after fully grounding themselves in the issue did Bono and Bobby arrive finally in Washington. They did not follow the well-worn pathway of activist celebrities, the meet-and-greets and photo ops with the likes of Bobby's Uncle Ted and other sympathetic Democrats. Instead, they spent their time talking to conservative Republicans such as Senators Orrin Hatch and Jesse Helms. Bono came out of the latter meeting saying that Helms had cried over the plight of Africa's children. In the House of Representatives their mentor and guide was Representative John Kasich, the Republican chair of the House Budget Committee. Kasich feared that the Kennedy connection might be hurtful; he asked Bobby not to come into the meeting while Bono pitched Speaker of the House J. Dennis Hastert and Majority Leader Richard K. Armey. Bobby waited outside while Bono went into the offices, and then they moved on to other meetings.

Bobby lobbied the media as much as he did political figures. His fingerprints were on articles in the news magazines and stories on the evening news. His strategy was not just to make a call but to call again and again. He pressed. He connived. He begged. At times Bobby was so over the top that he irritated the person subject to his entreaties, but he pushed onward.

After Bono and Bobby lobbied the White House, Clinton announced during a speech to a World Bank/International Monetary Fund meeting that he had committed America to the debt relief program. Congress passed legislation appropriating $435 million to the World Bank for debt relief financing, almost ten times what was originally agreed upon in committee. That amount would leverage to about $30 billon in write-offs of outstanding debt. The European countries followed suit by being more generous in their contributions to debt relief as well.

On November 6, 2000, Clinton held a meeting in the Cabinet Room at the White House to sign the historic bill. Almost all of those who had played instrumental roles were there, with the exception of Bobby. His mother lay in critical condition at Johns Hopkins Hospital in Baltimore, and for the first time in many months debt relief was not on his mind. "I am sorry that Bobby Shriver, who also played a key role in this effort, cannot be with us today because of his mother's illness," Clinton said. "And I ask for your prayers for him and his family, and especially for his remarkable mother, Eunice, who has fought for so many good humanitarian causes in her long and rich life."

Eunice recuperated, and Bobby returned to his efforts in a new Republican administration. He and Bono immediately began cultivating young staff workers who eight months later led them to power players in the new Bush administration, wangling meetings with national security adviser Condoleezza Rice, Treasury Secretary Paul O'Neill, political adviser Karl Rove, and others to make their case. After over a year of largely behind-the-scenes efforts, in March 2002 President George W. Bush announced $5 billion of new developmental aid over the next three years. The president insisted that Bono be on the stage beside him. "I carry this commitment in my soul," Bush said. In terms of the magnitude of the problem and the percentage of their gross national product that Americans gave compared to other wealthy nations, the amount was piddling. Yet for a Republican administration wary of foreign aid and rightfully critical of the squandering of so much assistance, it was a large sum, and authorizing it an act that among many of Bush's supporters was unlikely to win any plaudits.

The issue had become not simply debt relief but the horrendous burden of AIDS, which, if left unchecked, would be worse than the Black Death of the fourteenth century. The debt relief that these nations had received was being soaked up by this one horrendous problem. There were 24.5 million people in sub-Saharan Africa living with HIV/AIDS, 5,000 of them dying each day. The disease had already left 12.1 million orphans in the region. AIDS was not a cause but a crisis of unprecedented magnitude. By that standard, the Bush administration's response had been little more than a few coins thrown in a beggar's box.

There was little political pressure on Bush to do more, and the only way to change that was to create a public groundswell with the moral fervor of the civil rights movement. Bobby and Bono formed an organization, DATA, funded in part by Bill Gates and George Soros, to push the issue. Moving beyond the narrow corridors of power to the American heartland, the duo set out on a tour of the Midwest to open hearts to death and dying in other lands—an Irish rock 'n' roll Paul Revere and behind him a Kennedy grandson considered by those who did not know him as little more than a Hollywood man about town.

In Bush's 2003 State of the Union address, he talked about the crisis of AIDS in Africa with the sense of urgency that Bono and Bobby had hoped to generate. "To meet a severe and urgent crisis abroad, tonight I propose the Emergency Plan for AIDS Relief—a work of mercy beyond all current international efforts to help the people of Africa," the president said. "I ask the Congress to commit fifteen billion dollars over the next five years, including nearly ten billion dollars in new money, to turn the tide against AIDS in the most afflicted nations of Africa and the Caribbean."

In September 2003 Bobby flew to Washington to meet with Bono and the DATA staff before the singer's meeting with President Bush. The activists debated how they should best push Bush to fulfill his pledge. For fiscal year 2004, Bush had committed $2 billion dollars; the AIDS relief advocates believed that at least $3 billion was needed. Although Bush had spoken with passion and commitment, the administration rationalized that there was not the infrastructure in Africa to transport and utilize more assistance. Theirs was a good bureaucratic argument, but not to Bono and Bobby, who knew that it meant tens of thousands more would die unnecessarily.

Bono went into the Oval Office to see the president and Dr. Rice and had what he called afterward "a good old row." For Bobby as for Bono this was not political gamesmanship, but life and death. "We're three years into the administration now, and not one pill from America has reached the African people," says Bobby. "It's a test of American greatness. To fail at this would be a gigantic disaster."

For all Bobby's attempts to distance himself from his family, he was full of his Uncle Jack's attitude that a problem was something you solved now, not tomorrow. Bobby thought how the Peace Corps had started. Even before money had been appropriated, there was work to be done. For the first few months Bobby's father paid for the fledgling Peace Corps with his grandfather Joe's American Express card.

Nobody was pulling out their credits cards any longer.

B obby's effort was the most significant single political achievement by any Kennedy of his generation, and he had never even run for office. His older cousin Joe was supposed to have been the standard-bearer of his generation, but he was gone from politics.

Joe's younger brother Max had moved back to Massachusetts from southern California where Max started a program in urban ecology at Boston College, and even more important, as he saw it, his wife, Vicki, had a position teaching at Harvard. Max's love for his wife was mixed with the deepest and most genuine pride in Vicki's accomplishments. He could hardly talk about himself for five minutes before the conversation sidled off into what a popular teacher Vicki was and how accomplished she was in so many ways.

Max had a spiritual aura to him and a disengagement from the world and its demands. Thirty-six-year-old Max had been addicted to alcohol, and fifteen years previously had found a second faith in AA. "Some people say there's an alcohol gene in my family," he told the *New York Times*. "All I know is, I really liked to drink."

Max's favorite book, *Quest for Identity*, was a lyrical treatise by psychoanalyst Alan Wheelis that mocked the pretensions of the author's profession.

Max told journalist Matt Bai that in his "favorite part of the book, a boy's stern and bedridden father makes him cut the lawn, one blade of grass at a time, over the course of an entire summer. For the rest of his life, no matter how hard he tries, the boy cannot feel comfortable unless he is engaged in some kind of unrelenting labor."

There was a guileless quality to him that was both one of his greatest attributes and worst difficulties, forever trusting people he shouldn't trust, and reaching out into a world that burned him. He had a puckish subtle wit, and he could turn in a phrase from ponderous seriousness to joviality.

"What about the closeness of the brothers?" an author asked solemnly the first time he talked with Max. "You brothers, I guess, are as close as ever."

"No not really," Max sighed, his voice darkened with pain. "Most of us hardly even talk."

Max listened to the deafening silence as the author tried to make sense of this extraordinary admission. "No, just kidding," Max said finally, his laughter tearing through the silence. "I can only take this seriously for a limited amount of time. Of course, we're incredibly close but you know that. I don't go a day without talking to at least one of them. There's nothing as much fun. We make an effort to spend August at the Cape to spend one week together when we're with each other from dawn until we fall asleep at night. And our children are the best of friends too."

You could not be a Kennedy without having people ask you to run for office, even if you were the most unlikely of politicians. And among the young Kennedys, no one was a more unlikely politician than Max. In early 2001, he was approached to run for the seat of the late Rep. Joe Moakley. The congressman had been a great public servant, and the candidates who sought to slip into his oversized shoes deserved to be well measured.

Max's first major appearance since his probable candidacy was at the annual breakfast of the Robert F. Kennedy Children's Action Corps in May 2001. The room was full of media to hear a Kennedy speak. Max had turned to Bob Shrum, Ted's former speechwriter turned political consultant, for his words. When Max looked at the page, he knew that these professions of virtue had been heard too often, but he had nothing of his own to give as alternatives except a tepid avowal that he wanted to go to Washington to do good.

"I remember thinking, I'm about to go up there and make a fool of myself, and there's nothing I can do about it," he admitted later. "I should have thrown out that speech and just talked." When Patrick first ran for office, he had given numerous speeches as bad as Max's, but he had at least sounded earnest. Max zoned in and out of the mush of words, tripping over a sentence as if he were reading it for the first time, then mumbling like an aged professor reading yellowed notes from decades ago, and finally laughing to himself as if the joke was on someone else.

Afterward, Ted comforted his nephew by pointing out that President Kennedy hadn't begun as a soaring orator either, but whatever else the Kennedys had done in politics, they had never sounded cavalier. A few weeks later Ted had another, tougher talk with Max.

Max backed out of the race and returned to his private life. In the summer of 2002, he went on a *National Geographic* expedition led by Robert Ballard that discovered PT-109, his Uncle Jack's famous boat in World War II. Max met Biuku Gass, the Solomon Islander who almost sixty years before had gotten Jack to carve SOS into a coconut. When the old man learned that the coconut now rests at the John F. Kennedy Library, he collapsed in tears.

Max had been told by a friend about an obscure tragedy near the end of World War II, when during the battle of Okinawa two kamikaze planes burrowed into the USS *Bunker Hill*, the aircraft carrier that served as the flagship of the task force in the Pacific. It was not the kind of subject that a would-be author chose lightly, but Max got a book contract and began serious work. He was a diligent researcher, and had found what he hoped would prove his métier. He went to Japan and though he was initially banned from a meeting of kamikaze veterans, he persisted and interviewed several pilots. He went around America talking to survivors. And Max, who had never served in war, learned things that sometimes young men learned. He met the chief engineer, who told him how during the fire and flooding, he kept getting messages from a group of sailors trapped in the radio room. "I asked him, 'What did you do?' He said, 'I cut the wire. 'Cause there was nothin' I could do to help them it was a distraction hearing them screaming.' "

W ill Smith was also contemplating running for Congress. He was no longer "Willie" but "Will," as if his name would distance him from the trial in West Palm Beach. Since his acquittal in West Palm Beach, Will had finished his medical training and moved to Chicago where he began working in rehabilitative medicine with Dr. Henry B. Betts, who in 1961 after his grandfather's stroke had overseen his care. In the years since, Betts had turned the Rehabilitation Institute into a world-renowned center, and he proved a caring mentor to Smith. Will had gotten interested in that area of medicine as a boy when his cousin Teddy Jr.'s leg had been amputated, and he worked with many patients who had gone through what his cousin had experienced.

Will also became interested in the land mine issue and was the founder of Physicians Against Land Mines (PALM). Although PALM was not founded until 1996, the organization calls itself a "co-recipient of the 1997 Nobel Peace Prize" given to Jody Williams and the International Campaign to Ban Landmines.

"We have the feeling that Dr. Smith was looking around for an issue," says Mark Perry, the former director of political affairs for the Campaign for a Landmine Free World. "He's kind of a latecomer and obviously is very proud that he is a celebrity."

Will's concern for the issue is doubtlessly sincere, but Kennedy elbows had pushed themselves to the front of a line where others had long stood. He has made trips to Afghanistan and in many ways shown his concern. "It was the most devastated place I've ever seen," Smith said of the wartorn Asian land. "We saw a lot of people who had lost limbs to the wars but we also detected a sense of hope, a feeling that now rebuilding might be possible."

One of his admirers is Jody Williams, the legitimate co-recipient of the 1997 Nobel Peace Prize for her work fighting land mines. "I know Dr. Smith and PALM well and can attest to their solid work as a member of the International Campaign to Ban Landmines (ICBL)," she says. They "are very much part of that effort—and are valued for all their work on behalf of land-mine survivors all around the globe."

In 1998 Will was a founder of the Center for International Rehabilitation. Using his Kennedy connections to raise money and attention, he turned the nonprofit organization into a major institution concerned with helping the 500 million people across the world with disabilities.

Will was not active in the Democratic Party in Chicago or in any local organizations or activities, and the Kennedy name was the most important reason he was considered a viable candidate. When his tentative candidacy came to public attention, the media focused not on the good work Will had done in the past decade but on the sordid details of that Easter weekend in Palm Beach. In the summer of 2001, Will arranged for focus groups, to evaluate his chances, especially in terms of how dark a stain the Palm Beach trial had left.

Will was unprepared for the onslaught of questions and the excursions into his past. When he decided against running, he did not grant interviews to the Chicago newspapers, as a true son of the city would have done. Instead, Will flew to New York City to appear on *Good Morning America*. "I decided over the weekend, really, that I didn't want to proceed at this time with the race, largely because of commitments I have right now to the Center for International Rehabilitation and the programs there," Will told Diane Sawyer. He could have admitted that he had underestimated how much he had been tainted by the Palm Beach trial, but all he said was that he intended to be back. "I think being chosen to serve in that capacity is a great honor," he said. "I'd love to have that experience at some point in my life. I think right now is not the time."

Sawyer also asked Will about the three women who had come forward after his rape indictment giving depositions about his sexual conduct. Max's girlfriend had been explicit about Will's behavior when he had attacked her. So was a second woman who asserted under oath that the medical student had invited her to his Georgetown home after a pool party and had "grabbed me tightly around both wrists and threw me over the back edge of the couch." The third woman alleged that Will had raped her.

"All the stories involved some women who had come to my house after a party or an evening," Will said. "The bottom line is that when they said, 'Stop,' I stopped. That was the bottom line for all of those depositions. And the fact of the matter is I don't understand how that advances anybody's case but mine." Rape is not the only sexually inappropriate behavior, and statements such as this suggested that even a decade after the trial, Will may not have grasped the extent to which his aggressiveness frightened women.

T here was yet another Kennedy planning to run for Congress in 2002. Mark Shriver had decided to run in Maryland's eighth congressional district. With a growing Democratic majority, it appeared that the popular Republican incumbent, Connie Morella, might be defeated.

Mike Feldman, a Shriver friend and adviser and an astute judge of political potential, considered Mark "the most talented political candidate I have ever seen." Feldman felt that if Mark defeated the formidable Morella, he would arrive in Congress as not only the most watched but the most talented member of his class, and that within a few years he would become a likely candidate for president or vice president.

Mark's supporters knew that he would have token opposition in the primary, but they considered Christopher Van Hollen Jr., a four-term Maryland state senator, more a nuisance than a threat. That was true, even though Van Hollen was an effective, hard-knuckled politician with intensely loyal volunteers who saw him as a battering ram fearlessly pushing a liberal agenda. Mark had his enthusiastic volunteers too and raised over $2.5 million. An independent poll commissioned by the Shriver campaign in March showed Mark ahead 45 percent to 18 percent. "That may have helped make our media people overconfident," reflects Mark. "But for fourteen months everyone worked twelve-hour days."

As the summer went by, Mark's campaign began to falter almost imperceptibly. He had tried to establish a separate identity apart from his Kennedy blood, yet almost every story referred to him as a "Kennedy cousin." His cousin Kathleen was running for governor in Maryland. Her campaign was floundering, but even if she had been far ahead, it was not helpful to have two candidates on the ballot who were both considered Kennedys, and some voters had decided to choose one or the other.

If Mark was in some ways too much of a Kennedy, in other ways he was not enough of one. If his grandfather had been alive, Mark probably would have won handily, and no one would have ever seen or realized just what Joe had done to help his grandson. The harsh, merciless political quality that had pushed the family inexorably onward was gone. Mark was a gentleman in a profession in which gentlemen often did not succeed.

The Kennedys were used to being celebrated in the editorial pages, yet every one of the five major newspaper endorsements went to Van Hollen, including the *Washington Post*. Some who watched wondered if Mark truly wanted to win, or if he was playing out the last scene in a drama others had written for him. He talked far more enthusiastically about his years with Choice in Baltimore than about his time in the legislature. Not only did the polls show Van Hollen gaining, but his teams of volunteers were perhaps a little more dedicated. In the end Mark lost 43.5 percent to 40.6 percent to an opponent who in his tough-minded ambition was more Kennedy-like than Mark.

Mark had never lost at anything he had truly wanted, and the magnitude of this loss was enormous. He had not campaigned as a Kennedy, but nevertheless, he was the first Kennedy man to lose an election, other than a presidential primary. His advocates had argued that Mark was the only candidate who could wage a strong challenge against Morella, but Van Hollen went on to decisively defeat the incumbent. Kathleen squandered her even more formidable lead. Mark helped to anoint Van Hollen as one of the most promising young Democrats in Congress, while Kathleen's defeat made governor-elect Robert Ehrlich one of the brightest stars in the Republican Party. In the aftermath, Kathleen became the president of Operation Respect, a nonprofit organization teaching children responsibility. Early in 2003 Mark joined Save the Children, the international relief and child-development charity, as vice president and managing director for U.S. operations.

CHAPTER FIFTY-TWO

Times of Testing

For all Bobby Kennedy's contributions to the environmental movement, he had an intransigent, uncompromising manner that hurt many of the issues that he cared about. In 1999 Bobby hired William Wegner to be the Hudson Riverkeeper's chief scientist. Wegner had served three years in prison for smuggling cockatoo eggs into the United States from Australia, but Bobby felt that his friend had redeemed himself. Bobby's mentor and the president of the organization, Robert Boyle, considered it unthinkable that a man convicted of betraying the basic tenets of environmentalism should be its watchdog, and he fired Wegner. To reinstate his friend, Bobby tore apart the organization that had given him a lifetime purpose. In June 2000, Boyle, the grand old man of the Hudson, resigned, along with seven other members of the Riverkeeper board.

For years Bobby had sounded the environmental alarm in bold, uncompromising terms. "Capitol Hill is essentially just indentured servants to the polluting industries," he told the *Los Angeles Times* in June 2000. Bobby was an antipolitician who in his speeches tried to scorch his enemies. For decades he had raged against factory agriculture, especially the massive hog farms that spewed pollution into the waterways. In April 2002, Bobby gave a speech in Iowa in which he claimed that "large-scale hog producers are a greater threat to the United States and U.S. democracy than Osama bin Laden and his terrorist network." His statement was little more than a burst of rhetoric, but it emboldened his enemies and troubled his friends. It showed that Bobby risked becoming tone-deaf to the nuances of the American mind.

All Bobby had to do was to apologize for his rhetorical excess and focus on the issue at stake, the environmental cost of the factory hog farms. But when he returned to the state, he repeated the charge. "Kennedy's Uncle Ted doesn't go around likening his enemies to Osama bin Laden," wrote columnist David Yepsen in the *Des Moines Register*. "Edward Kennedy has emerged as one of the greatest U.S. senators of all time, and he achieved that greatness with hard work and compromise, not name-calling."

Bobby had a disconcerting way of turning those with whom he disagreed into enemies. He did so in his defense of his cousin Michael Skakel, who in January 2000 was indicted for the murder of his Greenwich, Connecticut, neighbor Martha Moxley a quarter-century before. Skakel was not a blood Kennedy, but without the scrutiny that came with having such famous relatives, most likely he would not have found himself in criminal court so many years after the crime.

Bobby had largely broken off his relationship with Skakel, and it would have been a noble gesture to state publicly his belief in his cousin's innocence. Instead, he not only stated definitively that Skakel was innocent but savaged Dominick Dunne and Mark Fuhrman for accusing his cousin in their writings and commentary. "Michael is as honest as daylight," he told the *New York Times* when Michael was indicted. "Michael is a victim largely of Dominick Dunne's vendetta and Mark Fuhrman's greed, and there's nobody who deserves this less, principally because he's innocent, but also because he's a genuinely good and decent soul."

Bobby was there for part of the trial in the spring of 2002 that led to Skakel's conviction and sentencing to twenty years to life. As Skakel appealed, the February 2003 *Atlantic* published Bobby's eighteen-page article proclaiming Skakel's innocence. He forged onward heedless of the impact the publicity might have on his political future.

Bobby considered compromise tantamount to surrender. In the early years of the twenty-first century the environmental movement was like an apple tree in which the fruit on the lower branches had all been picked and a different means had to be used to harvest the apples far out of reach. It was hard to motivate people over complex, less dramatic problems that, unlike an oil spill or polluted air, did not touch the average American. No environmental leader was so compelling a speaker, or so good at raising money and motivating activists, but Bobby's often strident rhetoric was less successful at convincing those who were not already on his side.

Bobby remained an immensely caring human being, but with a tortured quality that the years had not diminished. In a letter Bobby wrote in Skakel's defense, he could have been writing about himself: "I have watched for 20 years as Michael has struggled heroically to overcome his addiction, and the character defects that feed it. With notable personal strength, he overcame his genetic and cultural burdens and made himself a productive member of society. . . . His face and body are battlefields of warring emotions, which he is incapable of concealing."

J oe left Washington at the end of 1998 with no bands playing. His closest friends had worked at Citizens and they were his friends no longer. Although he would never openly admit it, he was hurt deeply by how little his years of public service seemed to have been appreciated. He appeared increasingly isolated. He returned to Citizens Energy where the *Boston Globe* said that he "sits like Napoleon in exile on Elba." Soon billboards appeared asking needy citizens to call 1-800-JOE-4-OIL to get help with their fuel oil. Joe stood for a photo op in his fancy shoes pumping oil into the heating tank of a presumably needy citizen.

In Washington Joe had hated the game playing, the fake camaraderie, the schmoozing with people he didn't like, and the palavering with intrusive reporters. He was an intemperate man, and he was damned if he was going to be anything but himself any longer. His adored twin sons matriculated at Stanford. Back in the early eighties when Joe had gone to Berkeley for a semester, he was seen as a Kennedy. His sons bore the same last name, but no one gave them special deference. He was immensely proud of his sons, and no matter what might be said of Sheila and their marriage, the two of them had produced something good and strong.

Most of what Joe had created at Citizens had been peddled away, and much of what was left was no better than a name on a letterhead. In selling off many of the companies that Joe had started, Michael and Wilbur James had not only sold away his creations but they had signed noncompete clauses that prevented him from starting new businesses in the very areas where he had expertise. He was startled at the way Citizens was being run. "When I came back there were lots of different methods that people were using to get money out of the company," Joe says, always careful never to cast aspersions on his brother. "As a result of the investigation that I conducted, it became very clear to me that there was need for reform."

Joe was a very different man than he had been when he left twelve years before to enter Congress. He was happily married to a woman who was his closest friend. He was buffed up and well dressed with the patina of a high-powered executive. The new Citizens chairman and president spent tens of thousands of dollars in legal fees investigating what had gone on under the previous regime, and putting new procedures in place. Still, a person looking at the company's public filings might conclude that Joe was continuing to pay himself in the style of his immediate predecessors. In 2001, the last year in which figures are available, Joe took home $396,375 in compensation, benefit plans, and deferred compensation while his wife Beth earned another salary at Citizens overseeing charity distributions, the two of them together earning about half a million dollars. For its part, that same year Citizens gave away $3,486,351 in program services. Half of Citizens $200,500 charitable donations went to *Doubletake*, a magazine edited by Kennedy friend and counselor Robert Coles. In inflation-adjusted dollars, Citizens was giving away about half of what it had when Joe started the organization and he and his wife were earning five times the pay.

Yet Joe suggests that this may not be the best way to evaluate his efforts. "They're always comparing me to the goddamn head of the United Way,"

Joe says. "We're a business, a nonprofit business." Citizens CEO Peter F. Smith makes the point that the underlying value of the new companies Joe created does not show up in the public statements.

Joe reinvented Citizens Energy, moving it into different areas. He founded a company that bought bulk electricity and resold it to corporations; once that was up and running, he sold it and moved on. He started Citizens International that sought to provide "a new paradigm for sustainable development" including collaborative efforts of government, the private sector, foreign corporations, and communities. Joe spent several million dollars developing a new CitizensHealth medical card that would allow the uninsured to pay the same discounted prices for services as the insured. It was an idea that if it succeeded might improve the lives and health of millions of Americans.

Joe had few friends beyond his wife and his sons. He drove a Jaguar, a car he had coveted for years. He loved living well, traveling with his wife, staying in first-class hotels, associating with the upper levels of business life. He wasn't a man who asked for favors, and when the baseball playoffs came to town in October 2003, he did not pick up the phone and hit somebody up for a couple box seats. In the old days there would have been half a dozen calls offering him tickets, but not anymore, and so he watched on television.

Joe was the most stoical of all the Kennedy sons, and the world knew little about the many difficulties he had faced and overcome. He could not talk about those who had hurt him most. His father had left him and his siblings to travel to the peaks of history, and he had never returned. His mother had not been there to embrace and succor him. His own brother Michael had betrayed the ideals for which they all stood proud, and Joe could not even hint at the pain that caused him. Joe's former wife had spread out many of her worst stories of their marriage before the world. He said nothing because it was bad for his sons, and because he knew there was no solace in the public display of their private pain. And so he was quiet about many matters, and he held things within himself that no man should have been asked to hold.

🦋

A lthough Patrick is the one young Kennedy still in office, there is little talk any longer about him having a brilliant political future. When he chaired the Democratic Congressional Campaign Committee, he sometimes behaved like a petulant, self-involved politician whose time and exigencies trumped everyone else's. When visiting the Kennedy Library in Dorchester, for instance, he parked his car on the front driveway where vehicles are prohibited.

In March 2000, hurrying back from a fund-raiser on an overnight flight, he shoved a security guard at Los Angeles International Airport who was attempting to prevent him from taking luggage too large to pass through the X-ray machine directly through the archway of the metal detector. The fifty-eight-year-old security guard, making the claim that the thirty-two-year-old congressman hurt her so badly that she was unable to work for at least a year, filed a lawsuit against Patrick. It was settled out of court two years later.

That summer of 2000 Patrick chartered a forty-two-foot sloop, the *Onde Dore*, to sail from Mystic, Connecticut, to Block Island and Martha's Vineyard. A boat of that size generally requires several crew members, but Patrick sailed alone with a woman friend, Celeste Bruno. Patrick apparently broke the jib halyard that is used to raise and lower the sails. Then he managed to drain the batteries so the engine would not start. That was a calamitous beginning, and on the second day Bruno called a friend saying that she had argued with Patrick and was frightened. The Coast Guard arrived to take Bruno away. After a second woman guest arrived, Patrick attempted to continue, but he needed a tow to return to Mystic. Then Patrick set out again. Either because the boat was a lemon, as he claimed, or he was an incompetent sailor, as the charter company asserted, he broke another halyard, and after blowing the engine abandoned the disabled, dirty boat.

This was no bloody voyage of the damned, no scandal of Kennedy-esque proportion, but it suggested the pressures under which Patrick was living. If the Democrats had won enough seats in the fall election to take over control of the House, Patrick would have been celebrated as one of the primary architects of that victory. The Republicans held on to their majority, however, and Patrick rightfully shared in the blame. He returned to his state and to the duties for which his constituents had elected him. He worked on the concerns of his constituents, but polls showed that he was the least popular member of Rhode Island's congressional delegation.

In March 2000, Patrick appeared at an event in Rhode Island with Tipper Gore, the vice president's wife, who had recently admitted publicly that she suffered from depression. Patrick was as private as a public person could be, walling his life off from the world. Thus, it was extraordinary that he made a similar disclosure. "I myself have suffered from depression," he said, in an admission that went against a hundred-year family history of relentless optimism. "I have been treated by psychiatrists. . . . I'm here to tell you, thank God I got treatment, because I wouldn't be as strong as I am today if I didn't get that treatment. . . . I am on a lot of different medications for, among other things, depression."

In June 2003, the *Washington Post* reported that at a gathering of Young Democrats at the Acropolis nightclub in downtown Washington, Patrick told the crowd: "I don't need Bush's tax cut. I have never worked a [bleeping] day in my life." An observer told the *Post*'s Lloyd Grove that Patrick had "droned on and on, frequently mentioning how much better the candidates would sound the more we drank. Finally, he had to be stopped by a DNC volunteer."

Patrick was not a great leader in part because, outside of his political career, he had "never worked a [bleeping] day" in his life. He did not have Bobby Kennedy's passion and commitment. He did not have Mark's intimate, deeply felt concern for his individual constituents and their problems. He did not have Joe's charisma and energy. But he held office, and he held on to it as tightly as a life preserver in a turbulent sea.

D espite the political disappointments of other family members, Tim had recurring thoughts of moving back to Connecticut and one day running for governor, but there were so many imponderables. He had five children and heavy expenses, and as he approached his forty-fourth birthday, he was at a stage in life when most men are through taking major risks on questionable ventures. He had so much wanted to avoid being sucked into the vortex of his family, and now every morning when he walked into his office he passed walls lined with the photographs of Grandfather Joe, Grandmother Rose, Uncle Jack, Aunt Rosemary, and all the others. He had so much wanted his own life but instead had become what many considered another inheritor filling a family seat.

For the first time, in the summer of 2003 the Special Olympics World Games were held outside the United States, in Ireland. A century and a half before, Patrick Kennedy and Bridget Murphy had sailed to Boston from Wexford in search of liberty and opportunity. They had lived in a world in which those with mental retardation were considered a subspecies who usually died soon after birth or in childhood. And it had been almost precisely forty years since President Kennedy had visited the family's ancestral home for what he called the happiest three days of his life.

Tim arrived in Ireland with other family members, including Eunice, now eighty-one, Sarge, now eighty-seven, and his brothers Bobby and Mark and his sister, Maria—everyone but Anthony. Bobby had brought a contingent of Special Olympics Christmas record artists, including Bon Jovi, Quincy Jones, and Rev. Run of Run D.M.C., along with actress Ashley Judd and Roberta Armani. In his new position at Save the Children, Mark had his own interest in many of the same issues that consumed Tim and Special Olympics. Anthony's organization, Best Buddies, was such an offshoot of Special Olympics that for a time he considered merging it into the larger organization. Their sister, Maria, was not only an eloquent public voice for those with mental retardation but had written a book about them, *What's Wrong with Timmy?*, a best-seller whose royalties she donated to the cause. Her husband, Arnold, had in his work with the athletes become a hero to many Special Olympians.

A few days before the games Tim's father had written a letter to his close friends in which he called himself "the luckiest man in the world." "Recently, the doctors told me that I have symptoms of the early stages of Alzheimer's disease!" Sarge wrote. "From my point of view, this disease means one thing, and one thing only: my memory is poor. It's a handicap, and it's a challenge. But it does not mean that I am ready to stop challenging myself, *or you.* . . . If names are slow to come to me, please forgive me. But if at any moment, I seem content with things as they are, don't leave the room. Remind me of the great times we've had and of the great work waiting to be done. I'm sure I'll be eager to rise to face new challenges, whatever they may be."

Sarge was slipping away, though he still had his times of full lucidity and understanding. He would be retiring from the chairmanship of Special Olympics this week and from public life. Thus, as Tim planned for these games, he was also preparing to honor his father with a tribute.

The week before the games the athletes were housed in towns and villages across Ireland and Northern Ireland; it was as if there were not one Special Olympics but a hundred, and in some ways not two countries but one. If you drove through any town or village, you saw the banners that had gone up and the signs that schoolchildren had so proudly made, and if you walked in the streets, you might well come upon some of the Special Olympians. If you did not volunteer yourself, you surely had a friend who was one of the thirty-five thousand volunteers. If you didn't make it out to the stands to cheer on the athletes, then you watched some of the events on television.

"For a time we were not an event, we were that nation," says Tim. "That nation was united in one voice saying, we will build a place that is respectful of and accepting of and welcoming to people with mental disability. We will show that's not just for us but that we can do this on a global basis."

A s the athletes from 163 countries marched onto the field of Croke Park for opening ceremonies, there were endless waves of clapping and cheering, an emotional catharsis. And they applauded fervidly for the entertainment when U2 performed "One," the rock group's legendary song about love between different peoples.

Bono returned to the stage to introduce Nelson Mandela. The South African leader spoke not to the eighty thousand in the stadium, not to the millions watching on television in Ireland and across Europe, but to the seven thousand athletes spread before him on the great field. "You, the athletes, are the ambassadors of the greatness of humankind," he told them. "You inspire us to know that all obstacles to human achievement and progress are surmountable. Your achievements remind us of the potential to greatness that resides in every one of us."

Mandela could not stay for the whole week, but before he left Ireland he sat with Tim. "What you did was you made it clear these people are human beings," Mandela said. "They are human beings. You gave them humanity."

A fter the first day of athletic competition, about 150 Special Olympics leaders, honored guests, and athletes gathered for dinner in the Great Hall at Dublin Castle. After remarks by Prime Minister Bertie Ahern, the group watched a seven-minute movie that Tim had produced about Sarge. It was narrated by Tom Brokaw, the NBC anchor, and it celebrated a life heading the Peace Corps in its first years, leading Johnson's War on Poverty, and running for vice president as well as for the Democratic presidential nomination. Sarge then spoke for a few minutes, and though he did not stick to his notes, no one except Tim even knew. After the film the band struck up some Irish music. Sarge got up to dance, taking the hand of Rita Lawlor, an Irish Special Olympian, leading her to the floor. Lawlor danced a traditional Irish dance, and Sarge danced the jitterbug, but they were in perfect sync, and the other dancers stopped and everyone watched the two of them on the floor. "I don't know what exits are supposed to look like," says Tim, "but if you ask me, that was a pretty good one if you're going to pull down the curtain on a professional career."

L ater in the week Tim decided that he wanted to spend one night in a dormitory with some of the athletes. It was almost eleven o'clock, and time for lights-out, when Tim and his lifelong friend Billy Noonan arrived in the public room where a group of athletes from Costa Rica, Iraq, and the Philippines were hanging out. They spoke in different languages, but they communicated with laughter, gestures, songs, and games.

Tim had traveled the globe expanding the organization. He had seen the appalling conditions of those with mental retardation in many of the poorer countries. In a land where most went hungry and there was little schooling even for the brightest, what scraps of life were there for the mentally challenged? It was a miracle of understanding, and a profound statement about the meaning of life, that such countries could even field a Special Olympics team.

The Special Olympians wanted to stay up and savor every moment of this time, but it was 11:15 and enough was enough. As the young people got up to leave, fifteen-year-old Roxanne Ng motioned that she wanted to sing. The Philippine teenager didn't speak English, but as she stood alone in the center of the room she began to sing in an exquisitely tender voice "The Power of Love." As she sang her voice became stronger, and she began to dance.

Tim knew that his brothers were right. It was his mother's organization, and always would be. For Tim, that didn't matter. This was what it was all about, not accolades on a podium, not the celebrated guests, but this voice singing out on a summer's night in Dublin.

 or the Shrivers it was an extraordinary year in lives full of many extraordinary years. In August Arnold announced his candidacy for governor of California in a historic recall election. The overwhelming media attention focused not only on Arnold but on Maria and all the Shrivers.

Despite his lack of political experience, Arnold won in a landslide and the victory celebration election night at L.A.'s Century Plaza Hotel in October 2003 was riotously joyous. It was the highest honor of the campaign to stand with the candidate on the podium. Maria wanted her husband to share that moment not with high-priced consultants and Republican politicians but with her family. Upstairs in the hotel suite she asked the rest of the Shrivers to follow the governor-elect and the new first lady to the raised platform. Eunice and Sarge fell into line, as did Tim and Linda, Mark and Jeannie, Anthony and Alina, and Bobby. And this moment belonged not just to Arnold but to the Shrivers and to the century and a half history of the Kennedys in America.

Early on in the two-month-long campaign, part of the media attempted to portray a wedge between the Democratic Kennedys and the Republican Schwarzenegger. But the reality was that with the Kennedys and Shrivers, blood triumphed over politics. Privately the Kennedys saw Arnold's race not as a diminution of their clan but if anything as a method to revitalize it. Ted talked to Maria at Hyannis Port about what position the family would take on the candidacy. Arnold was a family member first and a Republican second, and when he had gone around the country promoting his party's candidates, he had stayed out of Ted's turf, and never publicly supported candidates whom Ted saw as threats. That was the way loyal family members treated each other, and although both Ted and Patrick issued statements saying that as Democrats they backed their party's candidate, Ted made it clear to Maria that privately he supported Arnold and would do whatever he could to help. Ted joked that he knew almost no Republicans in California and suggested that Arnold hire his former aide and campaign consultant Bob Shrum. Arnold did not hire Shrum. He did call Bobby Kennedy, who gave Arnold a list of names of people to contact who helped shape his progressive environmental policy. One of them, Terry Tamminen, became the head of the California Environmental Protection Agency in the new administration.

Arnold was a throwback to the willful ambition of Joseph P. Kennedy. Much of what he had done in public life could in some measure or other be attributed to the influence of his wife and her family, from working with Special Olympians, and founding the Inner-City Games Foundation, to championing an initiative for after-school education. Arnold's own parents were gone. He not only loved Eunice and Sarge, but respected the values for which they stood, and he seemed delighted that they stood there next to him on the podium with the rest of the Shrivers.

Anthony, who appeared a natural habitué of the Hollywood ambience, looked out on the room and reflected upon what an amazing country he lived in, where his brother-in-law could arrive with nothing, and end up governor of California. As for Tim, he thought this night was not just about politics but ultimately this was about family. Bobby, as was so typical, didn't even stand with the rest of the family but chose to be part of the crowd.

"You know there are so many people that I want to thank," Arnold said when the crowd finally quieted down. "And I want to start first with my wife, Maria. I want to thank her for the love and the strength she has given me. I want to thank her for being the greatest wife and the most spectacular partner. I know how many votes I got today because of you. I also want to thank my parent-in-laws, Eunice and Sargent Shriver right over here, and also all my brother-in-laws and extended family. As a matter of fact, all the people behind me are the Shrivers, OK? So I want to thank them also for coming out here. I want to thank them all for coming out here. I really appreciate their support."

A rnold had unbridled political ambition and will unlike any of the young Kennedys. While Joe and Patrick had walked away from running for higher office, Arnold had stood for governor without any administrative experience or previous public office. Whereas Joe Jr., Jack, Bobby, and Ted had all at one time in their lives assumed that they would run for president, none of the surviving sons of Camelot had such a goal. These Kennedys are not immune from the generational realities of most families, and no one should have thought that they would match the great achievements of their parents' generation. Indeed, if less had been expected, more might have been achieved. They have fallen far short of the great vision that Joseph P. Kennedy had for his sons and grandsons. Their lives have been a bewildering juxtaposition of the most exalted aspirations and notable achievements and the most spectacular failures and shortcomings. For the most part their parents' generation had been able to hide its problems, be they substance abuse, troubled marriages, or obsessions with Joe's mandate, but their children have dealt more openly with their troubles. Their attendance at Alcoholics Anonymous may be seen as mark of failure, but it is more appropriately viewed as coming to terms with their difficulties. In the early years of the twenty first century, most of these surviving sons of Camelot have made their own hard peace with the family past and the ambitions that consumed them.

Camelot was born not with the inauguration of John F. Kennedy but with his death, and that birth found expression in the figure of his only son saluting his fallen father. Camelot was an emotional repository of a people's inchoate feelings about a youthful president, a newfound elegance, a nation of endless promise, and a world of hope. Young John bore those mythic dreams, and as much as he tried to flee them, darting down Manhattan streets on his bicycle or flying above the clouds, they were always with him.

Those romantic, spiritual aspirations that drew millions to the name and ambitions of the Kennedys were born with the death of President Kennedy and much of it died with the death of his son. For years the family tried to ignore their myriad untoward acts and tragedies, pretending they were not true or did not matter, but it was a tide beating away at the foundations on which the public life of the Kennedys was built. In his fitful way, Ted tries to hold the family together but there is no one to assume his role. The family will go on, but the boundless ambition is gone, worn away by the tides of history, fate, misfortune, and the willful acts of men.

CHAPTER FIFTY-THREE

Ripples of Hope

n February 1, 2001, the newly inaugurated President George W. Bush invited the Kennedys to the White House as his first dinner guests. The new president served up a meal of hamburgers, ribs, and lighthearted hospitality, the sort of casual dinner that the Kennedys enjoyed among themselves, and screened the film *Thirteen Days* about the 1962 Cuban Missile Crisis. Bush's celebration of the Kennedy family was not just dinner in the first days of the administration and verbal tributes either. In November 2001, Bush made the extraordinary decision for a Republican president to rename the Justice Department headquarters the Robert F. Kennedy Department of Justice Building. By that gesture alone, the Republican president was saying that Bobby's legacy in the Kennedy administration was a matter no longer of partisan dispute, but was worthy of being memorialized in the concrete and steel of one of the government's most important buildings.

Bush was an astute politician, and if in part he wanted to acknowledge the Kennedys' authentic contributions to American life, he was also seeking to ingratiate himself with the senator from Massachusetts. Ted was a gate-keeper of the Senate: little of substance passed without him placing at least a few marks on it, slowing down its passage or speeding it up. Bush had vowed to become the education president, and if he won a right to that honorific title, he did so by supporting a bill that was largely what Ted and the Demo-crats wanted. Ted agreed to some minor compromises to placate the Republicans, but it was essentially his bill that passed. "A lot of my friends in Midland, Texas, are going to be amazed when I stand up and say nice things about Ted Kennedy," Bush said. "He deserves it. He worked hard on this education bill."

Ted had not been co-opted by flattery or access, but energized by it. "We cannot remain silent when the president now fails to fund his own education bill," Kennedy told a nationwide radio audience in April 2002, in the Democrats' response to the president's weekly address. "It was a wonderful promise, but it has become a hollow promise."

Ted's unyielding criticism of the president did not temper the Bushes admiration for the politician who for decades had been to many conservatives the personification of liberal excess and moral sloth. In November 2003 former President George H. W. Bush selected Ted to receive the third annual George Bush Award for Excellence in Public Service. It was a gesture of conciliation, and recognition that patriotism and admirable public service are found in women and men of all parties and political persuasions.

At seventy-one, the Senator was old for a politician at the center of American public life. He was a hefty man but his bulk seemed that of a person of physical size to match the dimensions that he merited in the world of power. He appeared to be in the place he belonged, appropriating each moment as if fully conscious of his stature in American life.

Much of the progressive politics that Ted stood for had been drained of energy. Whatever might be said about the merits of his politics, he spoke about the same issues and the same themes that had preoccupied him for forty years—the economic inequality in massive tax cuts that would benefit primarily the wealthy, a health system that risked becoming "a commodity to be rationed by ability to pay," and a forceful defense of immigration. He was a major architect of immigration laws that continued to make America a nation of nations, and many cursed him for the polyglot realities of their nation. "It is wrong to try to build a wall around our country to strengthen our security," he said. "Terrorism is the problem—not immigration."

With the other old lion of the Senate, Robert Byrd, Ted stood in March 2003 in eloquent opposition to what he considered a precipitous war in Iraq, standing against most of his colleagues and his own son Patrick, who favored the invasion. In July 2003, the senator stood up again to criticize the administration for its poor planning for the reconstruction in Iraq, and its unwillingness to involve the UN and international partners in the effort. "America won the war in Iraq, as we knew we would, but if our present policy continues, we may lose the peace," said Ted.

Since he had entered the Senate, Ted had pointedly recorded his account of events into a tape recorder or in written notes, a testimony unique in its detail and longevity. One day those tapes, notes, and transcripts will reside at the University of Massachusetts across from the Kennedy Library, and it will become even clearer that no senator in the twentieth century has had an impact on so wide a number of issues and concerns, over so long a period. Ted Kennedy has imprinted his mark on American life in ways that many of his fellow citizens barely comprehended.

T ed had the deepest, most sentimental love for his family of any of his brothers and sisters, but he had stumbled often and at times had been only occasionally helpful as a guide and mentor to his sons and nephews. In his troubled years, he had not always been there when they needed him, but he had attended the funerals of three of his nephews, testified in the rape trial of another, shepherded his own wife and two sons into drug rehabilitation, and in recent months oversaw the care of his forty-two-year-old daughter Kara who was operated upon to remove a cancerous tumor from her right lung.

Ted talked so often about the meaning of family, but he and his siblings had been the victims of family as well as the beneficiaries. His mother and father had burdened their children with overwhelming loads of expectation. Ted and his brothers had forged ahead but there had been incalculable prices to be paid. For the first time in his life, in his marriage to Vicki, Ted had a home that was a warm haven from the demands of the world. His life would have had enviable moral closure if in his second marriage he had renounced all the excesses of the past. But Ted was still a heavy drinker, and some weekends at Hyannis Port it was not an exalted statesman on view in his living room but a far lesser person. On social occasions he could be uniquely boorish, dominating a dinner party with his endless, booming certitudes. Even at fund-raising events attended by lobbyists and others who were not necessarily his supporters, he could drink too much, talk too loudly, and hardly listen.

Ted had been in the same suite at the Russell Office Building for more than a decade, and the long cream-colored walls were full of photos and other memorabilia. Another man would have found the presence of the past overwhelming, but it grounded him. Behind his desk stood an American flag that had been carried in Jack's funeral procession, and a presidential flag that had flown in his brother's White House. His eyes often went up and along the wall looking at everything from a letter from his mother criticizing her youngest son's grammar and spelling to photos of sailboats that brought forth their own recollections of summer races long ago.

"This is President Kennedy the night before exams asking to be godfather to the baby when I was born," Ted said, pointing to the scrawled note from his older brother. ("Dear Mom and Dad, It is the night before exams and I'll write more later. Love Jack. P.S. Can I be godfather to the baby.'") Ted then turned to the letter next to it on the wall. "This is Caroline's son, Jack," he said pointing to the note and reading the words: "Dear Uncle Teddy, Will you be my godfather now, Love Jack." What Ted did not say was that Jack's first godfather was his Uncle John and Uncle John was gone. Every memory, no matter how sweet, carried within it sadness.

"My father had this old saying: 'Home holds no fear for me.' Do the best you can. Do what's right. And you always can come home. And there's a place that understands, loves, and welcomes you. It was an extraordinary part, I think, of the resiliency of the family.

"I know and relive every day some of the extraordinary losses that we have had, but I think we've also been blessed in many ways as well. And I think all my sisters and brothers feel the same way.

"I was always so proud of my family, my brothers and sisters, and parents, and we were brought up to make a difference in people's lives and we've attempted to do that. My brother Bobby made a speech about ripples of hope. These small waves that go across the pond and create sort of a wave to knock down the walls of resistance of intolerance, bigotry, prejudice, and discrimination. And I think the next generation are making their marks and some will rise into greater prominence, but hopefully all of them will make some difference in people's lives. And that's really the hope that my parents would have had for this family. You don't have to be a senator to make a difference. Teachers make a difference in people's lives. Nurses make a difference in people's lives. Parents make a difference in people's lives. And that's all you have to do is make a positive difference. But I think it's the air in us, what life is really about."

Source Abbreviations

AC: Adam Clymer, *Edward M. Kennedy: A Biography* (New York: William Morrow, 1999).

AS: Arthur Schlesinger Jr., *Robert Kennedy and His Times* (Boston: Houghton Mifflin, 1978).

ASP: Arthur Schlesinger Jr. Papers, JFKPL.

BH: Burton Hersh, *The Education of Edward Kennedy* (New York: William Morrow, 1972).

BOH: Burton Hersh, *The Shadow President: Ted Kennedy in Opposition* (South Royalton, Vt.: Steerforth Press, 1997).

BP: Joan and Clay Blair Jr. Papers, University of Wyoming.

CH: Peter Collier and David Horowitz, *The Kennedys: An American Drama* (New York: Summit Books, 1984).

CK: John Cronin and Robert Kennedy, *The Riverkeepers* (New York: Scribner's, 1997).

CP: Robert Coughlan Papers.

DHP: David Heymann Papers, State University of New York at Stony Brook.

DW: Darrell M. West, *Patrick Kennedy: The Rise to Power* (Upper Saddle River, N.J.: Prentice-Hall, 2001).

FBIFOI: Federal Bureau of Investigation Freedom of Information request.

GK: Gerald Sullivan and Michael Kenney, *The Race for the Eighth: The Making of a Congressional Campaign: Joe Kennedy's Successful Pursuit of a Political Legacy* (New York: HarperCollins, 1987).

HR: Harrison Rainie and John Quinn, *Growing Up Kennedy* (New York: Putnam, 1983).

HUA: Harvard University Archives.

IHOW: Edwin Guthman and Jeffrey Shulman, eds., *Robert Kennedy: In His Own Words* (New York: Bantam, 1988).

JFKPL: John F. Kennedy Presidential Library.

JMBP: James MacGregor Burns Papers.

KLOH: JFK oral history at JFKPL.

KM: Laurence Leamer, *The Kennedy Men* (New York: William Morrow, 1999).

KW: Laurence Leamer, *The Kennedy Women* (New York: Villard, 1994).

LBJPL: Lyndon B. Johnson Presidential Library.

NHP: Nigel Hamilton Papers.

OHO: William Vanden Heuvel and Milton Gwirtzman, *On His Own: Robert F. Kennedy 1964–1968* (Garden City, N.Y.: Doubleday, 1970).

OTR: Off the record.

RB: Richard Blow, *American Son: A Portrait of John F. Kennedy Jr.* (New York: Henry Holt & Co., 2002).

RFKOH: RFK oral history at JFKPL.

RFKP: RFK papers at JFKPL.

SB: Stephen Birmingham, *Jacqueline Bouvier Kennedy Onassis* (New York: Grossett & Dunlap, 1978).

SK: Sheila Rauch Kennedy, *Shattered Faith: A Woman's Struggle to Stop the Catholic Church from Annulling Her Marriage* (New York: Henry Holt & Co., 1998).

WM: William Manchester, *The Death of a President* (New York: Harper & Row, 1967).

Notes

1. A SOLDIER'S SALUTE

1 Robert Francis Kennedy and his wife: WM, p. 617.

2 "Jack's hurt": Ibid., p. 259.

2 He had driven back: Interviews, Claude Hooton Jr. and Milton Gwirtzman.

2 broke into sobs: In his largely authoritative book, Manchester writes that EMK told JPK the following morning. In Rose Fitzgerald Kennedy's tape-recorded interviews with Robert Coughlan, for her autobiography, and in the recollections of two others present (Rita Dallas in *The Kennedy Case* and Frank Saunders in *Torn Lace Curtain*), the version told here is given.

3 "You are the oldest . . .": RFK to Joseph Kennedy II, November 23, 1963. ASP.

3 He raged blasphemously: Interview, Kerry McCarthy.

4 "natural cynicism": JPK to Lord Max Beaverbrook, October 23, 1944, NHP.

4 or meeting with mobsters: The author has in his possession a letter from a private collection addressed to JPK at the mob-controlled Cal-Neva Lodge and dated June 14, 1960. Also J. Edgar Hoover, personal memo to the attorney general, August 16, 1962, FBIFOI.

5 Americans talk about: Charles W. Collier, *Wealth in Families* (2002), p. 43.

6 He was a financier: KM, pp. 38–42.

6 When he went to Hollywood: Terry Ramsaye, "Intimate Visits to the Homes of Famous Film Magnates," *Photoplay* (1927, only date), JFKPL.

6 seduced a virginal: Interview, Janet Des Rosiers Fontaine.

6 "just tore . . .": Interview, Charles Spalding.

2. SHEEP WITHOUT A SHEPHERD

10 Every day: *Newsweek*, July 20, 1964, and *Washington Star*, May 10, 1964.

10 He remembered how he scurried: Interview, Dan Samson.

12 "Ma'am, are you . . .": Interview, Mike Howard.

12 For Easter: Sarah Bradford, *America's Queen: The Life of Jacqueline Kennedy Onassis* (2000), p. 194.

12 "We were far from home . . .": Oral history, Charles Spalding, RFKP.

13 At a dinner party: Interview, Coates Redmon.

14 "Men are not made . . .": AS, p. 618.

15 Bobby's press aide, Edwin Guthman: Edwin Guthman, *We Band of Brothers: A Memoir of Robert F. Kennedy* (1971), p. 269.

16 "I have come to understand . . .": Ibid., p. 274.

16 The American ambassador: Ibid., p. 275.

3. GAMES OF POWER

18 "bitter, mean conversation . . .": IHOW, p. 406.

18 "President Kennedy isn't . . .": Jeff Shesol, *Mutual Contempt: Lyndon Johnson, Robert Kennedy, and the Feud That Defined a Decade* (1997), p. 186.

18 "mean, bitter, vicious . . .": Ibid., p. 417.

18 "I'm not going . . .": Michael R. Beschloss, *Taking Charge: The Johnson White House Tapes, 1963–1964* (1997), p. 470.

19 "terrific": Ibid., p. 212.

19 "I think . . ." Ibid., p. 213.

19 Bobby believed that: IHOW, p. 413.

19 "Sarge really wanted . . .": Interview, Harris Wofford.

21 "You can take my job . . .": Telephone conversation, LBJ and EMK, March 30, 1964, tape WH6408.18, LBJPL.

21 "Ted, who has . . .": Interview, John F. Kennedy, JMBP.

22 On the evening before: Interview, OTR.

22 Joan watched the wedding film: Interview, Joan Bennett Kennedy, and interview, Joan Bennett Kennedy, RCP.

22 Ted and his young bride: Interview, Joan Bennett Kennedy.

23 "He [Bayh] made the decision . . .": *Roll Call Daily*, February 28, 2002.

24 broken in twenty-six places: Ibid.

24 For about a half hour: Interview, OTR.

4. THE SENATORS KENNEDY

26 "I thought that . . ." Fred Dutton, RFKOH.

26 "If I could work it out . . .": Telephone conversation, RFK and LBJ, August 12, 1964, tape WH6408.18, LBJPL.

29 "Do what you are afraid to do": Quoted in Ronald Steel, *In Love with Night: The American Romance with Robert Kennedy* (2000), p. 39.

30 "taking the child . . .": *New York Daily News*, November 8, 1964.

Notes 577

30 ring of chocolate: *New York Times*, November 8, 1964.

30 "Bob, by the way . . .": Interview, Gabe Bayz.

32 when in 1965 Ted was trying: Meg Greenfield, "The Senior Senator Kennedy," *The Reporter* (December 15, 1966).

33 Within the next two: AC, p. 443.

33 "the central principle . . .": Ibid.

33 "One day we'd crack . . .": Senator Walter Mondale, RFKOH.

33 "Is this the way . . .": Quoted in AS, p. 680.

33 "Is tomorrow's hearing an . . .": Memo, EMK to RFK, RFKP, undated 1964–66, folder: Edward M. Kennedy.

34 "would suggest that . . .": AC, p. 105.

34 "Bob was always impatient": Interview, Fred Dutton, DHP.

37 "If you're the brother . . .": Interview, David Burke.

37 she began to doubt herself: Interview, Joan Bennett Kennedy.

38 "Ted is actually not . . .": Interview, Peter Edelman.

38 "He is trying to put himself . . .": Memo, Harry McPherson to LBJ, June 24, 1965, RFK Collection, office files of McPherson, box 21, LBJPL.

5. PEAKS AND VALLEYS

40 a group of Israeli and American: Joyce Baldwin, "Genetic Link Found for the Personality Trait of Novelty-Seeking," *Psychiatric Times* (April 1996).

41 Professor Frank Farley, past president: Interview, Frank Farley, whyfiles.org.

42 "to get there, get up . . .": Interview, Malcolm Taylor, DHP.

42 "He knew that people . . .": Oral history, James W. Whittaker, RFKP.

42 "Bobby was a physical person": Interview, Martin Arnold.

42 On the Salmon River: Oral history, Whittaker, RFKP.

43 "The people my father . . .": Interview, Robert Kennedy Jr.

43 "I can remember . . .": Patricia Lawford, *That Shining Hour* (1969), p. 288.

44 The Kennedy children: Barrett Prettyman Jr., "A Quiet Sail with Robert," ASP.

45 "You should have . . .": Interview, Barrett Prettyman.

45 When Joe broke his leg: Peter Hamill, "The Woman Behind Bobby Kennedy," *Good Housekeeping* (April 1968)

45 ever-changing menagerie: Ibid.

45 built tall walls: Robert F. Kennedy Jr., "A Eulogy for Michael," *Life* (March 1998).

45 Meanwhile, Courtney: Gail Cameron, "What It Takes to Be a Kennedy," *Ladies' Home Journal* (February 1967).

46 "revolutionary hero": AS, p. 801.

46 "If you object to American . . .": OHO, pp. 166–67.

47 checked with his press secretary: Andrew J. Glass, "The Compulsive Candidate," *Saturday Evening Post* (April 23, 1966).

47 traveling with a ten-person entourage: Ibid.

47 In Argentina: Robert Marr to Steve Smith, December 17, 1965; Betty Rex-Petersen to Angela
   Novello, February 7, 1966, RFKP 1964–66, JFKPL.
47 "that the press has . . .": March 2, 1966, RFKPP 1964–66, JFKPL.
48 "as a favor to President Kennedy": Glass.
48 "Perhaps it was not . . .": Williams v. Shannon, "Said Robert Kennedy, 'Maybe We're All
   Doomed Anyway,'" *New York Times Magazine*, June 16, 1968.

6. A BROTHER'S CHALLENGE

49 "Many tens of millions . . .": Frederick G. Dutton to RFK, December 8, 1966, Senate
   1964–66, folder: Frederick Dutton, RFKP.
49 "at least one major, exciting . . .": Frederick G. Dutton to RFK, April 6, 1966, ibid.
50 "I think Bobby . . .": Charles Spalding, RFKOH.
50 "The same young . . .": Jimmy Breslin, "Kennedy's Arithmetic," *New York World Tribune*,
   December 4, 1966, FBIFOI.
50 "we thought he . . .": Murray Kempton, "Bob Kennedy's Voyages," February 1966, no publication cited, ASP.
51 "Our answer is the world's . . .": RFK, Day of Affirmation address, University of Cape
   Town, Cape Town, South Africa, June 6, 1966, RFKP.
51 "You are a generation . . .": Edwin O. Guthman and C. Richard Allen, eds., *RFK: Collected
   Speeches* (1993), p. 100.
53 "I am delighted . . .": Herbert G. Stokinger to RFK, October 4, 1966, RFNP.
53 "The party in power . . .": RFK to Joseph Kennedy II, November 14, 1966, ibid.
53 "tribune of the underclass": AS, p. 778.
53 the last American: Michael Knox Beran, *The Last Patrician: Bobby Kennedy and the End of
   American Aristocracy* (1998).
54 "play so small a role . . .": OHO, pp. 112–13.
54 "the institutions which affect . . .": Ibid., p. 90.
55 "Well, keep working on it": p. 100.
55 "special federal aid . . .": Edwin O. Guthman and C. Richard Allen, eds., *RFK: Collected
   Speeches* (1993), p. 173.
56 "My God, what . . .": Peter Edelman, RKFOH.
56 "It is Bedford-Stuyvesant . . .": "Early Years of Bedford-Stuyvesant Restoration Corporation," Bedford-Stuyvesant Restoration Corporation, Restorationplaza.org.
57 only one factory: Kirsten Moy and Alan Okagaki, "Changing Capital Markets and Their
   Implications for Community Development Finance," Center on Urban and Metropolitan Policy, The Brookings Institution, 2001.
58 "They were talking . . .": Peter Edelman, RFKOH.
59 "I think if you . . .": Ibid.

7. WAR IN A DISTANT CLIME

62 "The Vietnamese war is an event . . .": Edwin O. Guthman and C. Richard Allen, eds., *RFK: Collected Speeches* (1993), pp. 293–94.

62 "Politically and diplomatically . . .": *New York Times*, March 5, 1967.

62 "Your premise was . . .": James Reston to RFK, March 21, 1967, RFKP.

63 "He was a fervent . . ." : Interview, William vanden Heuvel.

63 "How can you say . . .": AS, p. 776.

64 "Both of us certainly . . .": Interview, Peter Edelman.

64 "They weren't respected . . .": Interview, Frank Mankiewicz.

64 "*My* feeling was that . . .": BH, pp. 277–78.

65 "Ted really did . . .": Interview, Frank Mankiewicz.

65 "We weren't that . . ." : AC, p. 105.

65 "You can't be sure . . .": Jimmy Breslin, "The Inheritance," *Sunday Herald Traveler*, April 21, 1968.

65 "People can forgive mistakes . . .": Goodwin to RFK, n.d. [autumn 1967], ASP.

66 "No, no": Jeff Greenfield, RFKOH.

66 On October 18: AS, p. 826.

66 "Is that true?": George McGovern, RFKOH.

67 "He would rather have . . .": Interview, Eugene McCarthy.

67 The group consisted: AC, p. 99.

68 "Ted was very . . .": Interview, Barrett Prettyman.

68 Chop Chai in Phu Yen: Confidential cable, Robert Komer to Undersecretary Nicholas Katzenbach, January 13, 1968, LBJPL.

69 "We must have . . .": Interview, David Burke.

69 a bomb went off: AC, p. 100.

69 Claude expected: Interview, Claude Hooton Jr.

70 "Ted looked at it . . .": Interview, David Burke.

70 "intends [to] concentrate . . .": Embassy Saigon Nodis-Back Channel for Bunker, January 24, 1968, LBJPL.

70 "that he knew the president . . .": Conversation with Larry O'Brien, January 24, 1968, LBJPL.

71 "to wonder whether . . .": Memorandum of conversation, LBJ and EMK, January 24, 1968, LBJPL.

71 "When he was asked . . .": top secret memo, H.C. Lodge to the president, January 30, 1968, LBJPL.

8. STANDING IN THE RUBICON

72 called for 206,000 more troops: AS, p. 842.

73 "most everyone . . .": OHO, p. 260.

73 "panic signal": Interview, David Burke.

73 "even though politically . . .": Milton Gwirtzman, RFKOH.

73 "Bobby's therapy is going . . .": Victor S. Navasky, "The Haunting of Robert Kennedy," *New York Times Magazine*, June 2, 1968.

74 Bobby's defenders made: OHO, pp. 383–84.

74 "we have read . . .": Ronald Steel. *In Love with Night: The American Romance of Robert Kennedy* (2000), p. 198.

74 "a little nutty": Jack Newfield, *Robert Kennedy: A Memoir* (1969). p. 225.

74 "Bobby Kennedy was getting . . .": Memo, Joe Califano to LBJ, December 8, 1967, LBJPL.

74 "The press tried . . .": Fred Dutton, RFKOH.

75 Ted suggested later: AC, p. 107.

75 Bobby was hoping: OHO, p. 303.

76 "I wasn't that eager . . .": Interview, Eugene McCarthy.

76 "Cut it as close as . . .": AS, p. 857.

76 For much of that: Interview, David Burke, and AC, p. 109.

9. A RACE AGAINST HIMSELF

77 He peppered his speeches: Wayne Thompson, "The Last Days of Robert F. Kennedy," *Old Oregon*, n.d. [1968], ASP.

77 "In retrospect, I would have . . .": Peter Edelman, RFKOH.

78 "We have to write off . . .": Jack Newfield, *RFK: A Memoir* (1969), p. 100.

78 "We're trying to . . .": *Wall Street Journal*, March 27, 1968.

78 "He could have got hurt . . .": Edwin O. Guthman and C. Richard Allen, eds., *RFK: The Collected Speeches* (1993), p. 330.

79 "I think he [Bobby] thought . . .": Charles Spalding, RFKOH.

79 elements of demagoguery: *Washington Post*, March 28, 1968.

79 "Kennedy Crowds on . . .": *New York Times*, March 25, 1968.

79 "Rowdy Teen Fans . . .": *Chicago Sun-Times*, March 25, 1968.

79 "Crowd Madness . . .": *Washington Post*, March 28, 1968.

79 Mob Scenes and Stirring . . .": *Wall Street Journal*, March 27, 1968.

79 compared the Senator: Stewart Alsop, "Good Bobby and Bad Bobby," *Saturday Evening Post* (June 15 1968).

80 Bobby told Dutton: Milton Gwirtzman RFKOH.

80 "I think they've decided . . .": *Wall Street Journal*, March 27, 1968.

81 "the impression . . .": Memo, Tom Johnston to RFK, April 20, 1968, ASP.

81 "White people [are] living better . . .": David Halberstam, *The Unfinished Odyssey of Robert Kennedy* (1969), p. 92.

81 "These colorless shopping . . .": *New York Post*, April 2, 1968.

81 "a pleasant, homogeneous, self-contained . . .": AS, pp. 903–4.

82 17th and Broadway: *Boston Globe*, May 15, 1994.

82 Walinsky, who had arrived: Adam Walinsky, RFKOH.

82 "For those of you who . . .": RFK, statement on the assassination of Martin Luther King Jr., Indianapolis, Indiana, April 4, 1968, JFKPL.

83 "You know that fellow . . .": Jeff Greenfield, RFKOH.

83 The *New York Times* wrote: Warren Weaver Jr., "Kennedy: Meet the Conservative," *New York Times*, April 28, 1968.

84 Every evening Dutton: Peter Edelman, RFKOH.

84 "I think we probably . . .": Interview, Peter Edelman.

84 "He wouldn't give you . . .": AS, p. 885.

84 "Well, they'd go . . .": Interview, Gerald Doherty.

84 "I think he had . . .": Gerald Doherty, RFKOH.

85 "The lesson of Lake County . . .": OHO, p. 100.

85 "very morally . . .": Adam Walinsky, RFKOH.

85 "Oregon is just . . .": Interview, Frank Mankiewicz.

85 The employees listened: *Washington Post*, April 20, 1968.

86 "the strangest workers . . .": Ralph Friedman, "The Disenchanted Suburbia," *The Nation* (May 8, 1972).

86 "We're gonna lead . . .": Author's personal notes.

86 "I don't know . . .": Interview, Pierre Salinger.

86 "I would never like . . .": Interview, Joseph Dolan.

86 Bobby surreptitiously: Robert Kennedy asked the author, then a graduate student at the University of Oregon, to distribute the pamphlet.

86 "It was a style . . .": Interview, Peter Edelman.

87 "But what surprised me . . .": Senator Walter Mondale, RFKOH.

10. JOURNEY'S END

88 "I wanted to make sure . . .": Adam Walinsky, RFKOH.

88 "Bobby employed what . . ." Interview, Eugene McCarthy.

89 "What Bobby did . . .": Interview, Peter Edelman.

89 "zombie": Adam Walinsky, RFKOH.

90 "Please forgive us . . .": Mike Mansfield and Everett M. Dirksen to RFK, August 24, 1966, RFKP.

90 Galland had not even: Interview, Robert Galland.

91 On Sunday, three days: Interview, Diane Broughton, and Diane Broughton, "Trouble in Bungalow 3," *Los Angeles West Magazine* (June 4, 1972).

91 Seventy-one-year-old Jack: *Washington Post*, May 3, 1968.

91 "traipsing all over . . .": *D.C. Examiner*, May 2–4, 1968.

91 A governess: AP, May 2, 1968, and *New York Post*, May 3, 1968.

92 On a trip to Montana: *Butte Standard*, October 26, 1966, as quoted in Jeanette Prodgers, *The Butte-Anaconda Almanac* (1997).

92 "When we got there . . .": Interview, Joseph Dolan.

93 "Don't ever do that": Jules Witcover, *85 Days: The Last Campaign of Robert Kennedy* (1969), p. 147.

94 "I've got to get . . .": Richard N. Goodwin, "In June," *McCall's* (June 1972).

94 "I just think he . . .": Greenfield, RFKOH.

94 That sentiment: Interview, Peter Edelman.

95 He had won almost: OHO, pp. 379–80.

97 "We have to get . . .": Interview, David Burke. Also BH, p. 330.

98 "I can't let go": BH, p. 330.

98 Bobby Jr. went out: *Washington Star*, June 6, 1968.

99 "hellish environment": CH, p. 354.

99 "I'm going to . . .": Witcover, p. 308. Also BH, p. 331.

100 Bobby's aides pushed: Interview, Dr. Herbert Kramer.

100 Bobby's eldest two sons: AP, undated clipping, RFK presidential campaign, national head-
     quarters files, RFKP.

100 Bobby Jr. helped: *Washington Star*, June 7, 1968.

100 "I was an altar boy . . .": Interview, Bob Galland.

100 In Hyannis Port: Frank Saunders, with James Southwood, *Torn Lace Curtain: Life with the
     Kennedys, Recalled by Their Personal Chauffeur* (1982), pp. 324–25.

100 "When Bobby was killed . . .": Senator Walter Mondale, RFKOH.

101 Kathleen recalls: Interview, Kathleen Kennedy Townsend.

101 David rode with: CH, p. 355.

101 He put his head: Interview, Robert Galland.

101 "Oh, yes": HR, p. 143

102 "As I took a turn . . .": Adam Yarmolinsky, "Camelot Revisited," *Virginia Quarterly Review*
     72 (Autumn 1996).

11. PORTS OF CALL

103 Ted sailed up the coast: BH, p. 333.

103 He accompanied Joe: David Lester, *Ted Kennedy: Triumphs and Tragedies* (1971), p. 178.

103 He visited Onassis: AC, p. 130.

103 who went to mass each morning: Interview, Robert Galland.

104 The convivial drinkers: BH, p. 334.

105 "My father surrounded himself . . .": Interview, Robert Kennedy Jr.

105 "James Dickey said . . .": Interview, Max Kennedy.

106 He and his brother: CK, pp. 85 and 111.

107 "You guys have got . . .": CH, p. 358.

107 with deep, painful truth: When I was starting the research on this book, Jeanne Conway
     came up to me at a party and said that if I truly wanted to understand the Kennedys, I would
     have to understand alcoholism and Alcoholics Anonymous. I accepted her offer to arrange
     for me to go to a weekly closed AA meeting. I did so for over a year, and I learned an im-
     mense amount that informs much of this book.

108 "the best of them all . . .": Interview, Lem Billings, March 31, 1972, CP.

108 "The stories he told . . .": Billings Collection, John F. Kennedy Library Foundation, 1991,
     p. 269.

109 "I remember the day . . .": CH, p. 360.

109 A seventeen-year-old woman attempted: Ibid., p. 361.

109 Galland got two of the younger children: Interview, Robert Galland.
110 Seventeen-year-old Kathleen: AP, July 15, 1968, and AP, August 18, 1968.
110 "We're Kennedys": Interview with tennis pro, OTR.
111 "You can't die" HR, p. 183.
112 In 1966 Jackie: C. David Heymann, *RFK* (1998), p. 457.
112 "He's in heaven": Interview, Patti McGinty.
113 He wore a velvet: Stephen Birmingham, *Jacqueline Bouvier Kennedy Onassis* (1978), pp. 161–62.
115 His son Bobby liked to come: Interview, Robert Shriver III.

12. TED'S WAY

118 over a million: "Biafra War," Federation of American Scientists, Military Analysis Network, http://www.fas.org/man/dod-101/ops/war/biafra.htm.
119 We hold ourselves . . .": AC, p. 130.
119 Ted defeated: Laurence Leamer, *Playing for Keeps in Washington* (1977), pp. 97–98.
119 "Charlie, I appreciate . . .": Interview, Charles Bartlett.
120 In December: BH; p. 354.
120 And he was with Ethel: Jerry Oppenheimer, *The Other Mrs. Kennedy* (1994), p. 369.
120 The boy was haunted: HR, p. 149.
120 Ted, who had been: Frank Saunders, with James Southwood, *Torn Lace Curtain* (1982), pp. 336–37.
121 "First time I've used it": BH, p. 379.
122 "So even if the time is right . . .": *New York Times*, June 8, 1969.
122 "an incredible asshole": CH, p. 372.

13. THE ROAD NOT TAKEN

123 Joan was not there: Interview, Harry Fowler.
125 "Of course, I was surprised . . .": Interview, Rose Kennedy, CP.
126 "A lot of people . . .": Ibid.
126 "Joey, what are we going to do?": Interview, Joseph Gargan.
127 "I grew up. . . .": Ibid.
127 "Oh, my God . . .": Leo Damore, *Senatorial Privilege: The Chappaquiddick Cover-up* (1988), p. 79. The most original material in Damore's book is based on hours of tape-recorded interviews with Joseph Gargan. Although at the time of publication Gargan refused any comments, critics pointed out that the most sensational revelations were neither on tape nor enclosed in quotes. Gargan talked to the author about Chappaquiddick, albeit in a more limited way than he did with Damore. About *Senatorial Privilege*, he said: "I'm not going to quibble with Leo Damore. He's dead now. And he wrote a book the way he wanted to write it and the way he saw many things. But the fact is, I think he added to the story things that he wanted to believe that were not necessarily so." Damore's conclusions and speculation are his alone, but what remains in Gargan's tape-recorded remarks is the

undeniable portrait of Senator Edward Kennedy seeking a way out of assuming responsibility for the accident.

127 "If he'd have . . .": Interview, Joseph Gargan.

128 "Markham and I understood . . .": Ibid.

129 "asked for someone . . .": Interview, Dominick Arena.

129 "In looking back . . .": Ibid.

130 "I'm in some trouble . . .": Rita Dallas, with Jeanira Ratcliffe, *The Kennedy Case* (1973), pp. 338–39.

130 "I do not understand . . .": "Personal Notes of Mrs. Joseph P. Kennedy Regarding Ted's Tragic Accident, July 1969," CP.

130 "Goddamn it": Interview, OTR.

131 "Look, Eunice, the . . .": Sarge Shriver, comments on manuscript, September 1973, CP.

132 "You'd think that . . .": Interview, Dr. Herbert Kramer.

132 taunted with the word: HR, p. 154.

132 "the media made . . .": DW, p. 4.

132 "He told me there . . .": *Boston Globe*, June 6, 1999.

14. BOYS' LIVES

134 They ran into the street: CH, pp. 374–75.

134 They threatened: Interview, Larry Newman.

134 The neighbors gossiped: Ibid.

135 "I am a widow here . . .": Interview, Harry Fowler.

136 "You're dying, just . . .": CH, p. 362.

136 "I'm sure I had it . . .": Interview, Joe Kennedy.

137 at Hickory Hill: *Washington Star*, June 8, 1969.

137 "Joe was terrific . . .": Interview, Christopher Kennedy.

138 Bobby set out: Interviews, Christopher Kennedy and Robert Kennedy Jr.

139 "It was a rich . . .": Interview, Max Kennedy,

139 "Much of my understanding . . .": Interview, Christopher Kennedy.

139 "He was engrossed . . .": Interview, Rose Kennedy, CP.

140 Rose and her children: Rita Dallas, with Jeanira Ratcliffe, *The Kennedy Case* (1973), p. 352.

140 Eight-year-old John: *Washington Post*, November 21, 1969.

15. SAILING BEYOND THE SUNSET

142 "I always felt . . .": Interview, Robert Shriver III.

143 He had gone: Interview, Neal Nordlinger.

143 "There was no room . . .": Interview, Robert Shriver III.

144 The assailant struck: CH, p. 382.

144 One of Bobby's: HR, p. 197.

144 "We feel rejected . . .": CH, p. 382.
145 "He was whining . . .": Ibid.
145 "You've dragged . . .": Ibid., p. 383.
146 "My father said, 'Look . . .' ": Interview, Robert Shriver III.
146 She swam out: Interview, Steve Smith Jr., CP.
147 When Rose came down: Interviews, Courtney Kennedy and Caroline Kennedy, CP.
147 Her grandsons approached: Ibid.
147 "being back in . . .": Interview, Joseph Kennedy II, CP.
147 "For God's sake . . .": Interview, Kerry McCarthy.
148 "said some people . . .": HR, p. 154.
148 hired a German governess: KW, p. 697.
148 "You're just giving in . . .": Interview, Joan Bennett Kennedy.
149 "What's it like . . .": HR, p. 155.
149 "a full-blown love-hate relationship": Interview, Joan Bennett Kennedy.
149 "I've seen their anger . . .": Interview, Kerry McCarthy.

16. Running Free

151 He was transferred: *Washington Post*, August 15, 1968.
152 "When required to remain . . .": Thom Hartmann, *Attention Deficit Disorder: A Different Perspective* (1993), pp. 7–8.
153 "mostly a big bore": Wendy Leigh, *Prince Charming: The John F. Kennedy Jr. Story* (1993), p. 157.
153 John did have: Peter Beard, "John F. Kennedy Jr.—Images of Summer," *Talk* (September 1999).
154 "I was told she . . .": John F. Kennedy Jr., foreword to Marta Sgubin and Nancy Nicholas, *Cooking for Madam: Recipes and Reminiscences from the Home of Jacqueline Kennedy Onassis* (1998).
154 On one occasion he and Caroline: Beard, "John F. Kennedy Jr."
155 "There you are with . . .": Jacqueline Onassis Kennedy to Richard M. Nixon, quoted in Christopher Matthews, *Kennedy and Nixon: The Rivalry That Shaped America* (1996), pp. 294–95.
155 "I really loved the dogs . . .": Quoted in ibid., p. 295.
156 One day in May 1974: Leigh, p. 162.
157 "must be allowed . . .": J. F. Walsh to Paul S. Rungle, U.S. Secret Service, May 16, 1974, www.Apbnews.com, and Sarah Bradford, *America's Queen: The Life of Jacqueline Kennedy Onassis* (2000) p. 400.

17. A Clearing in the Future

158 "waterhole in the forest": Laurence Leamer, *Playing for Keeps in Washington* (1977), p. 86.
158 If one had chanced: Ibid., p. 120.

159 "He [Haynsworth] turned out . . .": Interview, Edward M. Kennedy, by Adam Clymer. The author has a copy of portions of the interview.

159 "We don't need this," AC, p. 162.

160 "I think he was very . . .": Interview, Jean Kennedy Smith, November 12, 1973, CP.

161 Wechsler's twenty-six-year-old: James A. Wechsler, with Nancy F. Wechsler and Holly W. Karpf, *In a Darkness* (1972), p. 160.

161 "a very troubled, even tormented guy": Memo, James Wechsler to James Wechsler, Subject: Conversation with Ted Kennedy, April 13, 1971, Joseph Rauh Papers, Library of Congress.

162 "It is Asian, now . . .": Christopher Matthews, *Kennedy and Nixon: The Rivalry That Shaped America* (1996), p. 289.

162 he would not campaign: AC, p. 189.

163 sought to veto that choice: Ibid., p. 190, and interviews, OTR.

164 They had no money: The author covered the war in Bangladesh for *Harper's*, testified about the war and its aftermath before Congress, and had intimate knowledge of the lobbying campaign.

164 "It is difficult . . .": "Crisis in South Asia," report to the U.S. Senate, quoted in "Persecution of Hindus in Bangladesh, Senator. Edward Kennedy on the Hindu Genocide in East Bengal '71," Hindu Human Rights, Hinduhumanrights.org.

165 He negotiated: AC, p. 199.

166 In 1970 the president's: Ibid., p. 215.

166 "I'd like to get . . .": Ibid., p. 178.

166 "discretionary money . . .": Ibid.

166 "He will never live . . . ": Stanley I. Kutler, ed., *Abuse of Power: The New Nixon Tapes* (1997), p. 30.

166 "I want it to be damn . . ." : Ibid., p. 133.

166 "Is the Teddy story being . . .": Ibid., p. 138.

167 "He [Nixon] thinks that Ted . . .": H. R. Haldeman, *The Haldeman Diaries: Inside the Nixon White House* (1994), p. 533.

168 "The Saturday Night Massacre . . .": quoted in Matthews, p. 333.

168 "It may be self-evident . . .": Interview, Joan Kennedy.

170 "the brain can only . . .": Quoted in HR, p. 231.

171 Teddy reached: Ibid., p. 232.

171 the next day: AC, p. 207.

171 "My son was . . .": Interview, Joan Bennett Kennedy.

172 he privately told Stuart Auerbach: AC, p. 208.

172 His stomach began: HR, p. 234.

172 "Here comes the mad scientist" : Ibid.

173 "Having a child with cancer . . .": Cal Fussman, "Ted Kennedy: What I've Learned," *Esquire* (January 2003).

173 "The great lesson . . .": Ted Kennedy Jr., "A Gritty Young Kennedy Writes His Own Profile in Courage," *People* (April 27, 1981).

18. Outcasts

174 headed west to Los Angeles: HR, p. 200.

174 "I was riding . . .": CH, p. 383.

174 Bobby had been thrown out: Ibid., p. 385.

175 "spat a bit of the ice cream . . .": *Boston Globe*, September 10, 1971, HUA.

175 "It was great being . . .": CH, p. 384.

176 In the fall, David: Ibid., p. 387.

176 "work on": Ibid., p. 392.

176 "Bobby has gone . . .": Interview, Rose Kennedy, January 11, 1972, CP.

177 "Joe came home . . .": Ibid.

177 "Joe called me . . .": Interview, Diane Clemens.

182 "I want you to look . . .": Interview, Joe Kennedy.

19. The Shriver Table

183 "I had this idea . . .": Interview, Timothy Shriver.

183 "We were living on . . ." Interview, Danny Melrod.

185 "It's a blizzard . . .": Ibid.

185 "One thing people don't know . . .": Interview, Timothy Shriver.

186 "When I was a boy . . .": Interview, Mark Shriver.

186 "I got the message . . .": Interview, Timothy Shriver.

186 "It was a great joy at the table": Interview, Neal Nordlinger.

188 Eunice asked her brother: Interview, Harris Wofford.

188 When you're a kid. . . .": Interview, Robert Shriver III.

189 "It's Mrs. Shriver": Interview, OTR.

189 "the arrogantly insensitive . . .": *Yale Daily News*, April 17, 1975.

189 "In college Bobby had a polarizing . . .": Interview, Lloyd Grove.

190 "I sat down on the curb . . .": Interview, Robert Shriver III.

190 "Teddy didn't do shit . . .": Quoted in CH, p. 415.

190 the *Evening Capitol: Washington Post*, March 23, 1977.

191 "Hey, kid": Interview, Robert Shriver III.

192 Bobby asked Tim: Interviews, ibid. and Timothy Shriver.

20. John's Song

193 150-passenger replica: www.hy-linecruises.com.

193 If the tourists: Interview, William Noonan.

195 the next best thing: Ibid.

195 "When he got too full . . .": Ibid.

197 "brutal and tough": Interview, Timothy Shriver.

198 He called his mother: Interview, William Noonan.

199 "He had this beautiful, sweet, kind . . .": Interview, Sasha Chermayeff.

200 "John's friends were . . .": Interview, Ed Hill.

200 "The trouble with me . . .": SB, p. 163.

200 In June 1977, he: Wendy Leigh, *Prince Charming; The John F. Kennedy Jr. Story* (1993), p. 187.

201 "There are some people . . .": Interview, John Perry Barlow.

202 "We were having a glass": Interview, William Noonan.

203 Joe and his brothers and sisters: HR, p. 19.

204 "My father died . . .": Ibid., p. 21.

21. KEEPING THE FAITH

206 close to a hundred: BOH, p. 27.

206 In a classic example: AC, p. 227.

207 The 297-page bill: Ibid., p. 257.

208 the police in Palm Beach: Interview, Jim Connors, former Palm Beach police officer.

208 "found themselves bird-dogging . . .": BOH, p. 68.

209 "Kennedy's womanizing is . . .": Suzannah Lessard, "Kennedy's Woman Problem, Women's Kennedy Problem," *Washington Monthly* (December 1979).

209 "He'd take home briefcases . . .": Interview, Richard Burke.

210 "Kennedy right now can . . .": *U.S. News & World Report* (August 27, 1979).

210 "I think it has been . . .": *Washington Post*, June 4, 1980.

210 "almost like having . . .": Interview, Joan Bennett Kennedy.

210 When Ted was about: Richard Burke, *The Senator: My Ten Years with Ted Kennedy*. (1992) pp. 190–91, and interview, Joan Bennett Kennedy.

211 Ted liked and respected: Interview, Melody Miller.

212 Larry Horowitz talked: AC, p. 284.

212 There had been crazies: Interviews, OTR.

213 Ted had to remember: Interview, Melody Miller.

213 "And there's my daughter . . .": Quoted in HR, pp. 24–25.

213 "I'll whip his ass": *Washington Post*, June 13, 1979.

214 "I took a great deal . . .": Interview, Christopher Kennedy.

214 "ran one of the most violent . . .": *U.S. News & World Report* (December 17, 1979).

215 Ted had practiced: AC, p. 318.

215 Joan thought it was time: Interview, Joan Bennett Kennedy.

215 Ted told his wife: Ibid.

22. BOBBY'S GAMES

216 He once stood: CH, p. 388.

216 When he was caught: *Boston Globe*, March 21, 1973, HUA.

216 Bobby rarely traveled: Interview, Frank Fox.

217 "while Joe would make . . .": *Parade*, August 14, 1977, HUA.

217 "I was shocked at . . .": Interview, David Humphreville.
217 "Those guys tell stories about women": Ibid.
217 "He had such a great sense . . . : Interview, Chris Bartle.
218 "Then Lem said . . .": Interview, R. Kent Correll.
218 In the summer of 1974: CH, p. 402.
218 a journey in 1980 to Venezuela: *The Billings Collection* (1991), p. 244.
219 "How could you . . .": Interview, Chris Bartle.
219 "This isn't a rapid . . .": *The Billings Collection*, p. 250.
219 they banged heads: Ibid.
219 "the nastiest and . . .": Ibid., p. 252.
220 "The stories he told and . . .": Ibid., p. 269.
220 "During the several . . .": Ibid., p. 252.
220 Lem had grown somewhat disillusioned: CH, p. 445.
220 "Everybody in the world felt sorry . . .": Interview, Chris Bartle.
221 "I'm conscious of the burdens . . .": Interview, David Horowitz.
222 "Oh, by the way . . .": David Horowitz, *Radical Son: A Generational Odyssey* (1997), p. 332.
222 "You know, the guy was . . .": Interview, David Horowitz.
222 Chris's twenty-first birthday: James Spada, *Peter Lawford: The Man Who Kept the Secrets* (1991), p. 116.
222 Chris Lawford was busted: CH, p. 438.
223 signed his letters: Christopher Kennedy Lawford résumé, and Christopher Kennedy Lawford to Robert E. Gaauque, October 30, 1981.
223 it hurt his father immeasurably: Interview, Patricia Seaton Lawford.
223 "I used to think . . .": CH, p. 450.
223 "Bobby would shoot up . . .": Interview, OTR.
223 joined Bobby at Harvard: HR, p. 188.
224 "Look at him": Ibid., p. 189.
224 Early in 1976: CH, p. 421.
224 by prescribing Percodan: AP, January 18, 1980.
225 "The others say . . .": CH, p. 422.
225 It was there: AP, September 6, 1979.
225 "therapy by humiliation": CH, p. 439.

23. A LIFE TO BE STEPPED AROUND

226 "I want to tell your . . .": Interview, David Horowitz.
226 "I feel they should have . . .": CH, p. 440.
227 "I'm going to tell you . . .": Ibid.
228 "ability to put drugs . . .": CK, p. 90.
228 "Bobby always thought . . .": Interview, David Humphreville.
228 In May 1983, two of Bobby's: *Time* (May 30, 1983).
228 "Bobby and I were . . .": Interview, David Humphreville.
228 In April 1985, Prude shot: *Washington Post*, February 6, 1986.

228 moot in September 1983 when Bobby: *Time* . . . (September 26, 1983).

229 "My life experience . . .": CK, p. 90.

230 Two of the other patients: KW, p. 703.

230 "sexual misconduct . . .": *Adweek*, April 30, 1984.

231 "So, Prof, that's what you . . .": Interview, Diane Clemens.

232 Although the nurses: Interviews, OTR.

233 The Palm Beach County state attorney: *Palm Beach Daily News*, October 19, 1984.

233 "When David died . . .": Interview, David Humphreville.

233 He was buried: UPI, April 27, 1984.

24. THE GAMES OF MEN

235 hired several key: AC, p. 340.

235 leased office space: Interview, Melody Miller.

235 "Ronald Reagan must love . . .": AC, p. 339.

236 "Why don't we take a vote?": CH, p. 448.

236 Rader raised his hand: Dotson Rader to Laurence Leamer, March 22, 1994.

236 "The family seemed . . .": Ibid.

237 "I felt terrible . . .": *Boston Globe*, June 6, 1999.

237 he began seeing: DW, p. 11.

238 "It's not always easy . . .": Ibid.,

238 "There's a fine line . . .": Ibid., p. 12.

239 "Our relationship had been . . .": SK, p. 21.

240 "ever could happen . . .": Ibid., p. 22.

240 "We were sitting . . .": Interview, Joe Kennedy.

240 Joe was excited: *Fortune* (July 23, 1984).

240 who had just lost: GK, p. 29.

241 "I wrote with Steve's . . .": Interview, Joe Kennedy.

241 Many of the retailers: Interviews, OTR. The author talked to a number of retail dealers, all
     of whom were worried about negative ramifications if they spoke openly.

241 In October 1981, . . . Joe stood: *New York Times*, October 7, 1981.

242 He talked to the oil minister: Ibid.

242 "The net number of people . . ." : Interview, OTR.

242 "Joe likes having . . .": Interview, Fred Slifka.

243 percent of that: *Boston Globe*, February 18, 1986.

243 "The [Kennedy] magic . . .": *Boston Business* (June 1989).

244 He also received $50,216: *Boston Globe* (April 17, 1986).

244 "most of it . . .": *Forbes* (October 21, 1991).

245 "I think Wilbur James . . .": Interview, OTR.

246 The savings were: *Christian Science Monitor*, February 12, 1982.

246 "Horrors! A profit!": *Forbes*, October 12, 1981.

246 "Wilbur was very focused . . .": Interview, Joe Kennedy.

247 "I thought it was . . .": Ibid.

248 felt that this was no: GK, p. 29.
248 "It just is not . . .": *Boston Globe*, January 24, 1985.
249 At one point an aid worker: Interview, Eileen McNamara.
249 "It's there and you . . .": DW, p. 13.
249 he moved into: AC, p. 371.
250 "It boiled down . . .": *Boston Globe*, July 28, 1985.
250 He told his father: Ibid.
250 "There is no grief . . .": Ibid.

25. LEFT OUT IN THE COLD

251 "No one should . . .": *Boston Globe*, February 4, 1986.
251 In the fall: *Boston Globe*, September 15, 1985.
251 supplying only 2,300 to 5,000: *Boston Globe*, January 24, 1985 and *Fortune* (July 23, 1984).
251 "The truth is . . .": Interview, Joseph Kennedy II.
251 with $1.7 million: *Chicago Tribune*, December 16, 1986.
251 "A distracting covey . . .": *Boston Globe*, March 23, 1986.
253 Joe won 52.5 percent: GK, p. 274.
253 "a crock of baloney": *Newsweek* (September 8, 1986).
253 "We were never . . .": SK, p. 4.
254 "You like clean air . . .": Interview, Joe Kennedy.
254 "I could have raised . . .": *Boston Globe*, December 20, 1987.
254 "I mean, I know . . .": Ibid.
255 "That's the celebrity . . .": *Washington Post*, March 4, 1987.
255 "The homelessness crisis . . .": *The New Republic* (March 30, 1987).
255 "My name is Dougie . . .": Interview, Carol Fennelly.
256 That was the beginning: Interview, Douglas Kennedy.
256 He had been born: Laura Bergquist, unpublished article on Ethel Kennedy, Bergquist papers.
256 I almost gave away . . .": Interview, Douglas Kennedy.
256 "We were working . . .": Interview, Carol Fennelly.
257 there wasn't any . . .": DW, p. 15.
257 "it was time for the torch . . .": Ibid.
258 "timid like a rabbit": Ibid., p. 18.
258 At one party: Interviews, Christin and Peter Lynch.
258 "I had a very pretty . . .": Interview, Kimberly Stender.
259 It took five hours to remove: *Chicago Tribune*, April 22, 1968.
259 "every day counts": *Providence Journal-Bulletin*, October 14, 1990.
259 "I don't think I'm . . .": *Boston Globe*, June 6, 1999.
259 Skeffington was an undertaker: DW, p. 22.
259 "Hello, I'm Patrick Kenn-e-dy?" : Ibid., p. 29.
260 "Jack, I'm going to tell . . .": Ibid., p. 37.
260 Patrick won overwhelmingly: AP, September 15, 1988.
260 "None of the victories . . .": *Time* (October 3, 1988).

261 he authored and pushed: *Boston Globe*, July 27, 1989, and *The American Banker* (July 13, 1989).

261 "A number of very . . .": Interview, OTR.

261 Joe received a lesson: *Los Angeles Times*, December 7, 1988, and *Boston Globe*, December 6, 1988.

262 In December 1988: *Boston Globe*, December 21, 1988.

262 "You cannot do . . .": *Chicago Tribune*, April 6, 1988.

263 "'Don't ever, ever . . .'" *Boston Globe*, December 21, 1988.

263 "It wasn't that Teddy . . .": Interview, Joseph Kennedy II.

26. JOE JONES IN NEW HAVEN

265 "Tim's extremely . . .": Interview, Brad Blank.

265 "some do-gooding Charlie . . .": Interview, Timothy Shriver.

266 "That was the most intensive . . .": Ibid.

267 Tim went to discuss: Interview, Dr. James P. Comer.

267 "This is Tim Shriver": Ibid.

267 The salty, homespun: *Hartford Courant*, December 12, 1999.

267 A few years back: Interview, Salvatore Verderame

267 He faced a challenge: *New York Times,* October 9, 1988.

268 "the worst school . . .": Interview, John Dow Jr.

268 "He was a real Yale . . .": Interview, Salvatore Verderame.

268 On one of his first days: Interview, Timothy Shriver.

268 "Trying to teach . . .": Ibid.

269 "Oh, is that right?": Interview, Salvatore Verderame.

269 "some nice humanitarian . . .": Interview, Timothy Shriver.

270 By the time: Ibid., and interview, Salvatore Verderame.

270 "Here were these kids . . .": Interview, Timothy Shriver.

270 "Tim came into . . ." Interview, John Dow Jr.

271 "I think of myself . . .": Interview, Robert Shriver III.

272 As he looked out: Ibid.

273 Iovine wanted to memorialize: Interview, Vicki Iovine.

274 You don't know . . .": Interview, Salvatore Verderame.

274 "I was very happy . . .": Interview, Timothy Shriver.

275 "I loved that place": Ibid.

276 Second-graders learned: The author spent several days in the New Haven schools observing classes.

276 "He was never taking . . .": Interview, Karen DeFalco.

277 "After that I got . . .": Interview, Salvatore Verderame.

277 Dow had fought: *Hartford Courant*, May 7, 1992.

277 Ted could not bring: Interviews, Dr. Herbert Kramer, and OTR.

278 the decision to turn: UPI, February 7, 1992.

278 came up with a $20 million: *New York Times*, May 22, 1994.

278 "There was this gaping . . .": Interview, Timothy Shriver.

278 After dinner he handed: Interview, Karen DeFalco.

279 "One minute I'm . . .": *New York Times*, May 22, 1994.

279 when the 7,200: *Hartford Courant*, June 30, 1995.

279 The games attracted: Ibid., October 13, 1995.

27. JOHN AT BROWN

280 you could sit: Phillip Weiss, "Good Times vs. Great Books," *The New Republic* (June 9, 1986).

281 "All the ink has made . . .": Ibid.

281 "I was having to deal . . .": Interview, Robert Littell.

282 Jackie hauled into John's: Interview, Robert Reichley.

283 "weary and disgusted . . .": Interview, Matt Gillis.

283 "I asked John to drop . . .": Interview, Robert Reichley.

283 "I used to kid . . .": Interview, James Alden.

284 "That's him, that's him": Rick Moody, "A Novelist and Former Classmate of John F. Kennedy Jr. Describes Being a Sidekick to the Most Famous Teenager on the Planet," *Brown Alumni Magazine* (July–August 1999).

284 some of his fellow students: Interview, Gary Ginsberg.

284 Beiser had been willing: Interview, Professor Edward Beiser.

284 "What am I going to tell . . .": Ibid.

285 "He had a kind of . . .": Interview, Professor Charles Neu.

285 "Oh, by the way . . .": Ibid.

286 "Where's the phone?": Interview, Richard Wiese.

287 "a wonderful actor . . .": Interview, Professor John Emigh.

287 "sitting next to his . . .": Wendy Leigh, *Prince Charming: The John F. Kennedy Jr. Story* (1993), p. 218.

288 "like a Southie punk . . .": *New Paper*, April 20, 1983.

288 "What have you done . . .": Interview, OTR.

289 "I think he probably . . .": Interview, Robert Littell.

289 Truitt, whose head: Interviews, Robert Littell and Mark Rafael Truitt.

28. AN ACTOR'S LIFE

290 "You know this is fucked up": Interview, Laurence Maslon.

290 "Oh, this is a great . . .": Interview, Christian Oberbeck.

291 "Are you the real . . .": Interview, Dan Samson.

292 "We'll just blast . . .": Interview, Sasha Chermayeff.

292 In December 1981: Interview, William Noonan.

294 Robby was presented: Interview, Robert Littell.

294 "It's been the three . . .": *Washington Post*, July 21, 1986.

295 "This is a real . . .": Interview, Robert Littell.

296 "Looking at the usual . . .": Burton J. Hendrick, *Life of Andrew Carnegie* (1932), vol. 1, ch. 17.
296 told him that John: Interview, Professor Edward Beiser.
296 "He saw shit backwards": Interview, Robert Littell.
296 "I remind you . . .": *Star*, January 11, 1994. Professor Beiser confirms the authenticity of the letter.
297 She called Beiser: Interview, Professor Edward Beiser.
297 "Almost nobody knew . . .": Interview, Robert Reichley.
297 "Bringing Suzman . . .": Interview, Christiane Amanpour.
298 John could easily: Interview, Robert Reichley.
298 "I have no brilliant explanation . . .": Interview, Christiane Amanpour.
299 "We had the most . . .": Ibid.
300 "Oh, that's okay, Kissy": Ibid.
300 "Obes, come check this out": Interview, Christian Oberbeck.
301 "You know, everyone misses . . .": Interview, Gary Ginsberg.
302 "communing with his father's . . .": Interview, Christian Oberbeck.

29. SAVING GRACE

303 "retrace my steps . . .": CK, p. 90.
304 that in 1983 had hired Cronin: Ibid., p. 19.
304 "*Sports Illustrated* has . . ." Interview, Robert Boyle.
304 "the most beautiful . . .": Robert H. Boyle, *The Hudson River: A Natural and Unnatural History* (1969), p. 15.
304 "clean and wholesome . . .": Ibid., pp. 171–72.
304 "out on the river . . .": Ibid., p. 276.
304 "Bobby struck me . . .": Interview, Robert Boyle.
305 Joe Augustine . . . was an odd choice: CK, p. 95 and interviews, Linda Fehrs and Robert Kennedy Jr.
305 "Newburgh was a revelation about. . . .": Interview, Robert Kennedy Jr.
307 "I don't give interviews": Interview, Linda Fehrs.
308 Fehrs concluded: Ibid.
308 "I got no pubicity . . .": Interview, Robert Kennedy Jr.
308 "If you can't invite . . .": Interview, Linda Fehrs.
309 They had to work together: Ibid.
309 "a dope fiend": Ibid.
309 "legalized extortion": Linda Fehrs, "Kirkpatrick Blasts Cronin," *Newburgh Sentinel* [n.d.].
309 Kirkpatrick had struck: CK, p. 37,
309 Much of the money: Ibid., p. 76; *Newburgh Sentinel*, August 7, 1986.
310 In the end every one: CK, p. 110.
310 In May 1989: *St. Louis Post-Dispatch*, May 31, 1989.
310 "born alcoholic": *CBS News* transcripts, *60 Minutes*, October 19, 1997.
310 "She is his rock . . .": UPI, July 16, 1985.

310 Bobby went to night: CK, p. 119.
311 He was a brash: Interview, Ronald Gatto.
311 he said later that his bosses: *Journal News*, April 30, 2000.
311 hearing chaired by: Greenwire, October 11, 1991.
312 "We didn't choose . . .": Interview, Robert Kennedy Jr.
312 "When I've run . . .": Greenwire, October 11, 1991.

30. PETER PAN ON ROLLER BLADES

313 later in his life became closer: Interview, Sasha Chermayeff.
314 "People have all these . . .": Interview, William Noonan.
314 During his months: Interview, Christiane Amanpour.
315 "I've been reading a lot . . .": Interview, John Perry Barlow.
315 "He rolled up . . .": Interview, Bill Milliken.
315 "he and seven . . .": Alexandria to Director Boston, May 14, 1985, FBIFOI.
316 the FBI believed: File 7A-NY-254901, July 18, 1995, FBIFOI.
316 "The image of the mother . . .": Marie-Louise Von Franz, *The Problem of the Puer Aeternus* (2000).
317 "obnoxious condition": *People* (April 27, 1987).
317 "You're such a Dick": Interview, Dan Samson.
318 A few minutes later: Interview, OTR.
319 She did not have: Jessie Kornbluth, "Daryl's Winning Sheen," *Vanity Fair* (January 1990).
319 wintered in Palm Beach: Interview, Nancy Gardner.
319 "She received serious . . .": *People* (October 19, 1992).
320 "The reason it wasn't . . .": Interview, John Perry Barlow.
320 "John was as competitive . . .": Interview, Dan Samson.
322 "Do not mistake . . ." Ibid.
322 "There was a sort of . . .": Interview, Robert Littell.
323 "lessor of two evils" *People* (September 12, 1998). The author has seen the invitation.
324 "Oh, shut up!": Interview, William Noonan.
324 "Get your eyes . . .": *People* (September 12, 1988).
324 "Minned one," Interview, Robert Littell.
325 "He never bit off . . .": Ibid.
326 "His mother hung . . .": Interview, William Bradford Reynolds.
326 "Look, you know . . .": Ibid.
326 "Well, I see that . . .": *Bergen Record*, July 5, 1990.
327 took and passed: UPI, November 2, 1990.
327 "My God!": Interview, Dan Samson.
327 "The perps kept confessing . . .": Interview, Robert Littell.
327 "John asked me . . .": Interview, Christiane Amanpour.
328 "John really began . . .": Interview, Jeffrey Sachs.
329 "What we wanted to do . . .": Ibid.

329 "Whether you were rich or . . .": Interview, William Ebenstein.
330 "He knew all of us . . .": Interview, Seth Krakauer.

31. JUNGLE WASTE

331 In July 1990: Judith Kimerling et al., *Amazon Crude*, edited by Susan S. Henriksen, preface by Robert F. Kennedy Jr. (New York: Natural Resources Defense Council, 1991), pp. x–xxvi.
331 "one of the heroes . . .": NRDC videotape of press conference.
331 As an assistant attorney general: Joe Kane, "With Spears from All Sides" *The New Yorker* (September 27, 1993), p. 60.
332 She spoke no: Interview, Judith Kimerling.
332 When the Texaco executives: *New York Times*, February 26, 1991.
332 Kimerling estimated:, Kane, p. 60.
332 Part of the time: Ibid.
332 "I was basically told . . .": Interview, Judith Kimerling.
333 In the village of Shuara: *Amazon Crude*, p. xxi.
333 Scherr stood back: Interview S. Jacob Scherr.
333 Bobby moved to shake: Ibid. Kane, p. 62.
334 "with great hopes that . . .": Interview, Judith Kimerling.
334 "We saw rivers . . .": Harvard Business School, "Conoco's 'Green' Oil Strategy" (1992), p. 11.
334 49,000-acre tract: Jan Reid, "Crude Awakening" *Texas Monthly* (November 1995).
335 200 million barrels: Harvard Business School, p. 3.
335 In 1987 a Catholic bishop: Kane, p. 55.
335 "They said the right . . .": Interview, Judith Kimerling.
335 was prepared to spend: Harvard Business School, p. 8.
335 "Look": James Ridgeway, "Jungle Fever," *Village Voice*, July 9, 1991.
336 "I checked with . . .": Interview, Judith Kimerling.
336 "when half of that . . .": Robert F. Kennedy Jr., "Amazon Sabotage," *Washington Post*, August 24, 1992.
337 "I said to Bobby . . .": Interview, Judith Kimerling.
337 "unprofessional performance": Robert F. Kennedy Jr. and S. Jacob Scherr undated letter to *The New Yorker*.
337 in return for two months': Interview, Judith Kimerling.
337 "NRDC believed . . .":": Ridgeway, "Jungle Fever."
337 "black and white": Ibid.
338 "If you didn't know . . .": Ibid.
338 "I have no idea . . .": Kane, p. 79.
338 "Mistake two . . .": *Washington Post*, May 15, 1991.
339 "NRDC is committed . . .": "NRDC Confirms Opposition to Conoco Oil Drilling," NRDC press release, June 1991.

339 investing millions.: Joe Kane, in the *San Francisco Examiner*, October 29, 1995, gives the figure as $90 million, an amount that may be exaggerated.

339 almost a billion dollars: Ibid.

339 "trinkets and beads": Chris Jochnick and Paulina Garzon, "A Seat at the Table: Petroleum Industry Environmental Responsibility," NACLA Report on the Americas, January 1, 2001.

340 In 1995, YPF: Reid, "Crude Awakening."

32. A Man Apart

341 "After the '80 . . .": BOH, p. 58.

342 Ted found him an intelligent: Interview, Melody Miller.

343 "turned white as a sheet . . .": *New York Times*, October 4, 1983.

343 "seems to imply . . .": *Washington Post*, October 4, 1983.

344 "Robert Bork's America is . . .": AC, p. 418.

345 balanced a waitress: Ibid., p. 385.

345 during a luncheon: *Los Angeles Times*, April 29, 1990, *Washington Times*, October 26, 1991, and *Sunday London Times*, October 27, 1991.

345 In January 1989: AC, p. 445.

346 "I just couldn't . . .": Interview, Larry Newman.

33. Good Friday

347 "dragging the children . . .": *Toronto Star*, April 14, 1991.

347 Joe was spending: Dennis Spear, interview with Detective Rigolo, case 91-000950, April 9, 1991.

348 When a student reporter: Interview, Kimberly Stender.

349 "drove the woman . . .": Deposition of Patrick J. Kennedy in re: Investigation of William K. Smith, May 1, 1991.

349 She talked to Willie: Patricia Bowman statement to Detective Rigolo, April 8, 1991.

350 Deposition of Senator Edward Kennedy, May 11, 1991, in re: Investigation of William Kennedy Smith.

350 Cassone and Patrick: Deposition of Patrick J. Kennedy in re: Investigation of William K. Smith, May 1, 1991.

351 "This girl is really . . .": Ibid.

351 "I've been raped . . .": Christine Rigolo, Palm Beach police report, case no. 1-91-000950, March 30, 1991.

351 "there's an allegation . . .": Deposition of Patrick J. Kennedy in re: Investigation of William K. Smith, May 1, 1991.

352 "it was a violent attack": Deposition of [name withheld], October 5, 1991.

353 "We do not need . . .": *Boston Globe*, October 16, 1991.

353 "I know a bridge . . .": *New York Times*, October 16, 1991.

354 Ted had become the least: *Orlando Sentinel Tribune*, October 26, 1991.

354 "I am painfully aware . . .": *Washington Times*, October 26, 1991.

355 he came down: *Washington Post*, November 19, 1991.

355 The woman who walked: The author was in the courtroom.

356 "I was yelling . . .": *Palm Beach Post*, December 5, 1991.

357 "lost twenty five . . .": Ibid., December 7, 1991.

357 Ted wore a fine: The author was present in the courtroom.

357 "This was really . . .": Ibid.

357 "was a guest . . .": *Providence Journal-Bulletin*, April 2, 1991.

357 "Patrick hung Willie . . .": Quoted in DW, p. 68.

357 Patrick paid $2,300: Ibid, p. 69.

358 "She unbuttoned my . . .": *Palm Beach Post*, December 11, 1991.

359 "What about the three . . .": The author was present.

34. LOVE, LOYALTY, AND MONEY

360 "unreal and unsubstantiated": *Chicago Tribune*, December 11, 1991.

360 Bobby got a hurried: *USA Today*, April 24, 1994.

361 He flew to British: *Vancouver Sun*, January 21, 1995.

361 He traveled: *Ottawa Citizen*, November 26, 1994, and *Gazette* [Montreal]. September 14, 1996.

361 The Riverkeeper concept: Waterkeeper Alliance, http://www.waterkeeper.org/.

361 "I was on the river . . .": Interview, Robert Kennedy Jr.

362 $9 billion: *The Bond Buyer*, November 3, 1995.

362 "scared the organic material . . .": CK, p. 201.

362 Bobby took their side: Ibid., p. 225.

362 Bobby was the only: Ibid., p. 227.

363 "We're an Irish . . .": Interview, Douglas Kennedy.

363 Professor Peter Kreft, remembers: Interview, Professor Peter Kreft.

363 "Everything was culturalism": Interview, Douglas Kennedy.

364 "The only way . . .": Ibid.

364 The first article: Douglas Kennedy, "AIDS: A Test of Acceptance," *Nantucket Beacon*, September 4, 1991.

364 "She has noticed a spot . . .": Ibid, October 16, 1991.

365 "You're just a kid . . .": Interview, Dan Cronin.

366 "I didn't really expect . . .": Interview, Chris Bartle.

366 "You guys went . . .": Interview, Douglas Kennedy.

367 The organizational meeting: *Boston Globe*, November 28, 1993.

367 "I took that guy back . . .": Interview, Douglas Kennedy.

368 His mother sat: *Washington Post*, May 25, 1995.

368 Doug joined Fox News: Ibid., August 13, 1996.

369 In 1998 he married: *Boston Herald*, August 23, 1998.

369 he not only would pen a cover: Interview, Douglas Kennedy.

370 "Probably the best . . .": Interview, Maxwell Kennedy.

370 had a business card: Interview, Melody Miller.

371 "Chris being the executive . . .": Interview, Brad Blank.
371 who agreed to stay: Interview, Christopher Kennedy.
371 "It was the greatest": Ibid.
371 "the Mart is probably . . .": *Forbes* (October 21, 1991).

35. TEAM PLAY

373 "They all four are . . .": Interview, Maria Shriver.
374 "They do what they think . . .": Interview, Mark Shriver.
374 Mark would sometimes be was so upset: Interview, Robert Shriver III.
374 "Mark was a super team . . .": Interview, James G. Fegan.
375 "Well, you know, . . . I'm the . . .": Ibid.
375 "Clearly there was . . .": Interview, Mark Shriver.
376 "You're going . . .": Ibid.
376 $44,000 grant: *Washington Post*, August 30, 2002.
376 "The kids had choices . . .": Interview, Mark Shriver.
377 "That was difficult": Ibid.
377 "The kids know . . .": *Newsweek* (April 18, 1994).
379 over $138,000: *Washington Post*, October 23, 1994.
379 "I'm not running . . .": Ibid. August 30, 2002.
379 Mark kept as his most: Interview, Mark Shriver.

36. BEST BUDDIES

381 "Rosemary was around . . .": Interview, Anthony Shriver.
381 "On occasion we would blow . . .": Interview, Jimmy Shay.
382 For a time he worked: Interview, Carl Anthony.
382 "If you had the choice . . .": Interview, Anthony Shriver.
382 "a cool, hip . . .": *People* (February 27, 1995).
383 "There's nothing for free in life": Interview, Anthony Shriver.
383 So early in 1992: *USA Today*, June 8, 1993.
383 "The people I know . . .": Interview, Anthony Shriver.
383 Shortly after he arrived: *USA Today*, June 8, 1993.
384 "I'm not sort of going . . .": Interview, Anthony Shriver.
385 perpetually adolescent: *Chicago Tribune*, February 2, 1993.
385 "I was having a good time": Interview, Anthony Shriver.
386 "I tried to get somebody . . .": Interview, Myer Feldman.
387 "No matter what happens . . ." Interview, Anthony Shriver.
387 "I came because . . .": Interview, Timothy Shriver.
387 Tim sent out peppy: Interviews, OTR.
388 "We couldn't respond . . .": Interview, Timothy Shriver.
388 And though Pasternack had served: Interview, Bruce A. Pasternack.

37. ADRENALINE ADDICTS ANONYMOUS

391 "pulled the eventual pieces together": *Washington Post*, December 8, 1988.

392 so many at headquarters: Interviews, OTR.

393 "All I know": Interview, Wanda McDaniel.

393 "Bobby took me to yoga . . .": Interview, Russell Simmons.

394 "I was a little wound up . . .": NBC News transcripts, *The Today Show*, November 29, 1999.

395 "His adrenaline *is* his . . .": Interview, Mary Ann Dolan.

395 "Why are you doing this?": Ibid.

396 "He would have always. . . .": Interview, Robert Shriver III.

38. MICHAEL'S WAY

397 "He was the best . . .": Interview, Lorenzo di Bonaventura.

398 "We completely hit . . .": Interview, John Rosenthal.

399 "I was enamored by the . . .": Ibid.

399 an organization he founded: Ibid.

400 "Mr. Nice Guy": Michael Blanding, "Mr. Nice Guy," *Boston* (April 2003).

400 "John, look what . . .": Interview, John Rosenthal.

400 "access is the key . . .": *Boston Business* (June 1989).

400 The nonprofit company: *Boston Globe*, October 24, 1988.

401 The southern African country: *CIA World Fact Book*, "Angola."

401 "Wilbur seemed smarter . . .": Interview, Thomas Gagen.

402 "Citizens deserves . . .": *Boston Globe*, December 26, 1988.

402 "a kind of modern . . .": *Washington Post*, June 20, 1989.

402 "I made a very hard . . .": Interview, Joseph Kennedy II.

402 $662,000: *Boston Globe*, June 15, 1997.

403 it upset him: Interiew, OTR.

403 The first of these deals: *Journal of Commerce* (January 6, 1987).

403 The head of the Citizens: *Electric Perspectives* (May–June 1995), and *Boston Globe*, December 4, 1994.

403 "you can't allow this . . .": *Los Angeles Times*, October 11, 1988.

404 "Sometimes Wilbur claimed . . .": Interview, Larry Kellerman.

404 "Wilbur's got a major . . ." : Interview, James Hilliard.

404 In 1988 and 1991, unknown: *Wall Street Journal*, March 25, 1998, and interview, Larry Kellerman.

404 The *Wall Street Journal* estimates: *Wall Street Journal*, March 25, 1998.

405 something just over $3 million: Citizens Energy Corporation, press release, January 12, 1998.

405 The group was having: Interview, OTR.

406 "I watched Michael . . .": Interview, OTR.

407 Natalie told others: *Boston Herald*, July 7, 1997, and interviews, OTR.

408 "Daddy, I want . . .": Interview, OTR.

408 "mid-twentieth century . . .": Vladimir Nabokov, *Lolita* [1955] (paperback ed. 1997), p. 285.

39. A CHILD OF THE UNIVERSE

409 "So naturally the next . . .":Interview, R. Couri Hay.
409 "Most of us had had . . .": *New York Times*, July 26, 1992.
409 "We accused him of being . . .": Interview, James Alden.
410 On this trip: *New York Times*, July 26, 1992.
410 after he purchased: *New York Post*, April 1, 2000.
410 "He really didn't practice . . .": Interview, Ralph Diaz.
411 "He's about six feet . . .": Interview, Robert Littell.
411 "What the fuck, Dick": Interview, Dan Samson.
412 "He tested everyone": Interview, OTR.
412 on one of his skiing: Interview, OTR.
412 "He loved role-playing": Interview, James Alden.
414 "This is great!": Interview, Dan Samson.
414 "any lawyer's three-year-old could win": *London Sunday Times*, September 12, 1993.
414 "I remember thinking . . .": Interview, Sasha Chermayeff.
415 "You guys are the best . . .": Ibid.
415 One day about two weeks: Interview, John Henry Barlow.
416 "He was immediately . . .": Interview, Christiane Amanpour.
416 "She was surrounded . . .": CBS News transcripts, *CBS Evening News*, May 20, 1994.
416 "He was blown . . .": Interview, William Noonan.
417 "I mean, he loved . . .": Interview, Robert Littell.
417 Billy sensed that: Interview, William Noonan.
418 He was seen: *Newsday* [New York], November 24, 1993.
418 John had a friend explain: Interview, OTR.
418 "She, of all his girlfriends . . .": Interview, Robert Littell.
419 "She just blew you away . . .": Interview, Paul Wilmot.
419 It was not until: *San Jose Mercury News*, August 28, 1995.
420 "Carolyn, when she . . .": Interview, OTR.
420 John realized his wallet and keys: Interview, OTR.

40. GAMES KENNEDYS PLAY

422 "Joe, nobody's asking . . .": Interview, OTR.
422 "an inability to behave . . .": *Times of London*, September 23, 1997.
422 "I had simply become . . .": SK, p. 27.
423 English breakfast tea: *Irish Times*, September 25, 1997.
423 "My attitude up . . .": *Boston Globe*, April 24, 1997.
423 In March 1989: SK, pp. 3, 25.
423 began living with Beth Kelly: Ibid., p. 8.
424 In April 1993: Ibid. pp. 2, 7.
425 "How can you be opposed?": Ibid., p. 10.
425 in October of that same year: *USA Today*, October 25, 1993.

426 He was a wealthy man: *New Orleans Times-Picayune*, October 15, 1995.

426 in 1992 he was convicted: AP, September 26, 1992.

426 the judge ordered: *National Mortgage News*, October 4, 1993.

427 "I saw this big fish . . .": Elsa Walsh, "Kennedy's Hidden Campaign" *The New Yorker* (March 31, 1997).

428 Vicki . . . convinced her husband: AC, p. 550.

428 "Will you ride in the car with me?": Interview, David Burke.

429 Other former Washington: *Washington Post*, October 29, 1994.

429 Romney had been: *Boston Globe*, November 10, 1994.

429 "To stop all that . . .": Interview, David Burke.

430 "You could stay . . .": *Boston Globe*, November 10, 1994.

430 "I remember when . . .:

430 "Senator, before we get . . .": Interview, David Burke.

431 "Patrick had his speech . . .": Interview, Governor Bruce G. Sundlun.

432 "Young Kennedy . . .": *Providence Journal-Bulletin*, October 21, 1994.

432 In September vandals: *Washington Times*, September 19, 1994.

432 in the last week: *Chicago Tribune*, November 10, 1994.

432 "into Attila the Hun . . .": DW, p. 87.

433 Almost a hundred: *Providence Journal-Bulletin*, February 19, 1995.

433 "No father in this . . .": Ibid. January 5, 1995.

433 "led with a joke at . . .": Ibid.

434 "the most critically . . .": DW, p. 99.

434 "when she was twenty-five. . . .": *Providence Journal-Bulletin*, February 19, 1995.

434 "all that rough and tumble . . .": Ibid.

434 And there he played: Ibid.

435 "There are two ways . . .": DW, p. 99.

436 "*Vanity Fair* calls up . . .": Interview, Senator Jack Reed.

436 "a lot like high school . . .": Interview, Representative Neil Abercrombie.

436 "We'd do the *Wayne's World* . . .": Ibid.

41. HUMBERT HUMBERT

438 found Michael: *Boston Globe*, April 25, 1997, and book proposal by Michael Skakel, with Richard Hoffman, "Dead Man Walking: A Kennedy Cousin Comes Clean," and interviews, OTR.

438 Vicki later denied: *Quincy* [Mass.] *Patriot Ledger*, July 9, 1997.

438 On the very day: *People*, (February 27, 1995,) and *Philadelphia Inquirer*, February 14, 1995. *Note:* Media accounts gave the date as either January 21, 1995, or January 22, 1995. The author's sources say that Michael Kennedy left on January 21, 1995.

438 "body, mind, and soul": http://www.fathermartinsashley.com/.

438 "freaks and rehabbers . . .": *Boston Herald*, February 14, 1995.

439 "religion around recovery.": Interview, John Rosenthal.

439 Joan counted up: Interview, Joan Bennett Kennedy.

439 He was eclectic: Interview, OTR.

440 Kellerman recalls James: Interview, Larry Kellerman.

440 While Michael was: *Electrical World*, (September 1994) and *Wall Street Journal*, June 15, 1997.

440 organization sold off: *Quincy* [Mass.] *Patriot Ledger*, March 22, 1995.

441 "dirty politics": *Boston Herald*, August 1, 1996.

441 censured in 1983: *U.S. News & World Report* (August 1, 1983).

441 "taken soundings": *Boston Herald*, November 4, 1995.

441 "With Wilbur's, ego . . .": Interview, James Hilliard

442 James negotiated the purchase: *Wall Street Journal*, March 25, 1998. Most of this financial material is found in Massachusetts Office of the Attorney General, Division of Public Char ities, Form PC, Annual Report for Citizens Energy, which the author has also gone over to verify this information.

442 "Wilbur was actively working. . . .": Interview, Larry Kellerman.

442 In January 1996: Ibid.

442 "held by 25 employees . . .": *Boston Globe*, March 11, 1997.

442 During its first year: *Wall Street Journal*, March 25, 1998.

442 James's take would have: Ibid.

442 Kellerman was so distraught: Interview, Larry Kellerman.

443 "will allow us in . . .": *Boston Globe*, March 11, 1997.

443 Little Michael was standing: Interview, OTR.

444 "I was not . . .": *Boston Globe*, July 9, 1997.

445 The event would signal: Ibid. November 9, 1996.

445 "I watched the way . . .": Interview, Lorraine Paolella, and *Boston.*

446 the day after: *Boston Globe*, July 9, 1997; *Boston Herald*, May 7, 1997.

446 she climbed up: *Vanity Fair* (August 1997).

446 "under his wing" : *Boston Herald*, May 24, 2002.

446 "founded the Great Lakes . . .": Skakel and Hoffman, "Dead Man Walking."

447 "Michael and I fought. . . .": Ibid.

447 Paul married Kelley: *Boston Herald*, November 9, 1999.

447 "a long illness": *Boston Globe*, August 15, 2001.

447 "no one ever . . .": Ibid. May 10, 1997, and *Boston Herald*, May 14, 1997.

12 A TATTERED BANNER

449 sisters-in-law: *New York Post*, January 22, 1998, and interview, OTR.

449 Although the paper: *Boston Globe*, April 25, 1997.

449 "strongly hinted": *Washington Post*, May 12, 1997.

450 A reporter for the: *Quincy* [Mass.] *Patriot Ledger*, April 30, 1997.

450 "I gotta use your car, John": Interview, John Rosenthal.

450 Gifford confided to the: *New York Daily News*, June 2, 1997.

451 When Michael went to visit: Interview, John Rosenthal.

452 she was fifteen: *New York Times*, May 24, 2002.

452 "in the middle of a . . .": *Boston Herald*, May 24, 2002.

452 In the summer of 1997: *Boston Globe*, January 2, 1998.

452 he and the other two top paid employees: Massachusetts Office of the Attorney General, Division of Public Charities, Form PC, Annual Report, Fiscal Year Ending December 31, 1997, Citizens Energy.

452 $333 an hour: The board members were paid $8,000 for twenty-four hours attending meetings, ibid.

452 "than the heads of Harvard . . .": *Boston Globe*, June 15, 1997.

453 "they're less likely . . .": *Boston Herald*, April 30, 1997.

453 "cafeteria Catholic" : *Boston Globe*, April 6, 1997.

453 "Sheila Rauch Kennedy writes . . .": Ibid. March 29, 1997.

454 "I said things that I wish . . .": *New York Times*, June 9, 1997.

455 "Matt's injury wasn't serious . . .": *Boston Herald*, July 22, 1997.

455 It was serious enough: Interview, OTR.

43. "LADIES AND GENTLEMAN, MEET *GEORGE*"

457 "I always thought . . .": Interview, David Kuhn.

457 "I grew up . . .": RB, p. 220.

458 "I didn't understand . . .": Ibid.

462 "You can't tell anybody, Gary": Interview, Gary Ginsberg.

463 John secretly called in Bruce Kluger: Interviews, Bruce Kluger and Gary Ginsberg.

463 The scene in Montgomery: *The Oprah Winfrey Show*, transcript, September 3, 1996.

464 "Our first choice was Alan . . .": *Boston Herald*, September 7, 1995.

464 "Mr. Kennedy, you failed . . .": Interview, Gary Ginsberg.

44. A FATHER AND SON

468 "For me it's opened . . .": *Providence Journal-Bulletin*, October 16, 1996.

469 "Patrick, what are . . .": DW, p. 134.

469 My God . . ."; CBS Evening News, March 22, 1996.

469 "Primal scream . . ."; *Washington Post*, March 26, 1996.

470 "He doesn't really . . .": *Providence Journal-Bulletin*, October 21, 1998.

471 "Part of me worried . . .": *Boston Globe*, June 6, 1999.

471 In November 1998: *Providence Journal-Bulletin*, November 23, 1998.

472 a staff of 126: DW, p. 144.

472 "the Senate is . . .": *Irish Times*, September 21, 1995.

472 "would cause a . . .": *The American Prospect*, special supplement, "Reforming Welfare Reform; The Politics; A New Beginning; Welfare to Work" (2000).

473 "give every child access . . .": AC, pp. 591–92.

473 "It's really 3 yards . . .": Charles P. Pierce, "Kennedy Unbound," *Boston Globe Magazine*, January 5, 2002.

474 "Nothing gets through . . .": Ibid.

474 "A lot of people . . .": Elsa Walsh, "Kennedy's Hidden Campaign," *The New Yorker* (March 31, 1997), p. 81.

474 "I'm in awe at . . .": Interview, Melody Miller.

45. JOHN'S BEST SHOT

476 You've got my ring": RB, p. 96.

476 "This man who supposedly . . .": *Los Angeles Times*, August 11, 1996.

477 "God, I can't believe . . .": Interview, OTR.

477 "You went to a convention . . .": Interview, Gary Ginsberg.

478 sold 200,000 newsstand: RB, p. 106.

478 Christiane had to be: Interview, Christiane Amanpour.

478 "JFK Jr. Dumps Galpal Carolyn": *Los Angeles Times*, September 24, 1996.

479 "That's where my mother . . .": Interview, Sasha Chermayeff.

479 "He couldn't find . . .": Interview, William Noonan.

480 "They told me they . . .": The description of the wedding is based on the accounts of several of those present, including Sasha Chermayeff, Christiane Amanpour, Robert Littell, and William Noonan.

481 "This is a big . . .": *New York Daily News*, October 7, 1996.

482 "I can't see": Ibid. October 11, 1996.

483 "When summer comes . . .": Interviews, Sasha Chermayeff and William Noonan.

483 "John was patient . . .": Interview, Sasha Chermayeff.

484 "Come on, don't be ridiculous,": Interview, Dan Samson.

485 "When was the last time . . .": Interview, Sasha Chermayeff.

485 "Some have made so much . . .": Interview, Dan Samson.

485 "I felt sorry that . . .": Interview, Sasha Chermayeff.

46. POSTER BOYS FOR BAD BEHAVIOR

487 "I've learned a lot . . .": John Kennedy, "Editor's Letter," *George* (September 1997).

488 John F. Kennedy, Jr., took . . .": *Washington Post*, August 11, 1997.

488 "Ask not what . . .": *Boston Globe*, August 12, 1997.

489 "It's a very different family . . .": Interview, Kathleen Kennedy Townsend.

489 "Michael is figuring . . .": Interview, Lorenzo di Bonavantura.

491 "death to the loser": *Time* (January 12, 1998).

491 "Okay, new rule": Interview, R. Couri Hay.

491 "If you come near . . .": *Boston Globe*, January 3, 1998.

492 Fleetwood threw the ball: *Chicago Tribune*, January 13, 1998.

492 Michael had already expired: *People* (January 19, 1998).

492 John walked over from: Interview, OTR.

493 "People think that's a new concept": *Providence Journal-Bulletin*, January 4, 1998.

494 In Angola the rebel: AP, December 29, 1997; Agence France-Presse, December 29, 1997.

494 "He gave his life . . .": *Life* (March 1998).

47. CLINTON AND THE KENNEDYS

496 "That's my father, Kissy": Interview, Christiane Amanpour.

496 John could not even: RB, p. 181.

496 "John, you've got to . . .": Interview, Gary Ginsberg.

497 In the following months: Ibid.

497 "pushy": Gary Ginsberg, "The Scoop," *George* (July 1998).

497 "got into some hot water": Ibid., (March 1998).

498 "It would be great . . .": Interviews, Christiane Amanpour and James Rubin.

498 You are totally missing . . .": Interview, Gary Ginsberg.

499 "We were afraid . . .": DW, p. 144.

500 "I'm at the point . . .": *Providence Journal-Bulletin*, October 2, 1998.

500 "a series of lies . . .": Ibid.

500 "masquerading as a . . .": Ibid., January 21, 1999.

500 "His advice is always simple": AC, p. 603.

501 The previous year: *Washington Post*, April 28, 1997.

501 "something of a whore": *Boston Globe*, June 4, 1999.

501 "What does John . . .": *National Review* (June 14, 1999).

502 Moldea discovered closeted: Interview, Dan Moldea, and interview, OTR.

502 "a loathsome campaign . . .": *Washington Post*, May 5, 1999.

48. A LIFE OF CHOICES

503 forty-six hours: National Transportation Safety Board, final report on accident, NYC99MA178, probable cause approval date July 6, 2000.

503 The only magazine: Interview, J. "Mac" McClellan.

503 The owner of the company: Interview, Lloyd Howard.

504 "So I get up in it . . .": Interview, Robert Littell.

504 "I'm not coming back . . .": Interview, J. "Mac." McClellan.

505 on weekends in December 1997: National Transportation Safety Board, final report on accident, NYC99MA178,

505 "We laughed many times . . .": Interview J. "Mac" McClellan.

505 a twenty-year-old: *Palm Beach Post*, September 17, 1998.

505 "We had traveled . . .": Kyle Horst, "An Ocean of Possibility," unpublished manuscript, and interview, Kyle Horst.

506 "It was a morass": Interview, Elizabeth Kay.

506 He cried when he told: Interview, OTR.

506 Early in January 1999: Interview, Keith Kelly.

506 "searching for a . . .": *New York Post*, January 18, 1999.

507　"He [John] had been negotiating . . .": RB, p. 203.
507　"he had mixed feelings . . . :" Interview, OTR.
507　source of the leak: Interviews, OTR.
507　Most editors-in-chief: RB, pp. 135–36.
508　"The best training . . .": "Full, Unedited Q&A: John F. Kennedy Jr. on *George* and *Celebrity*," *Brill's Content Online* (March 1999).
509　"It was a fucking . . .": Interview, Gary Ginsberg.
510　the secretary of state had been impressed: Interview, James Rubin.
511　An early poll: Bulletin Broadfaxing Network, Inc., *The Bulletin's Frontrunner*, February 23, 1999.
511　A November poll: Ibid., November 12, 1998.
511　"was for this time around": *New York Times*, November 24, 1998.
511　"I don't really want . . .": Interview, Robert Kennedy Jr.
512　"the gravest threat . . .": Quoted in *Village Voice*, May 25, 1999.
512　"an agent of . . .": *New York Times*, November 10, 1999.
513　"You know we've always . . .": Interview, Dan Samson.
514　"the possibility of creating": Interview, Douglas Brinkley.
515　"Well, I'm enormously . . .": NBC News transcript, *The Today Show* July 19, 1999.
515　eighty firefighters and: *Washington Times*, June 10, 1999.
516　"Mummy, how are you?": Interview, Mark Shriver.
516　Bobby pitched the most: Interview, Robert Shriver III.
516　"I didn't know how . . .": *Montgomery Gazette*, November 10, 1999.
516　"I think she went . . .": Ibid.
516　The 7,000: AP, July 4, 1999.

49. NIGHT FLIGHT

518　"It was such a love . . .": Interview, Robert Littell.
520　The reason I broke . . .": Interview, Sasha Chermayeff.
521　He told one friend: Interview, OTR.
524　"What always stunned me . . .": Interview, Christiane Amanpour.
524　"Kissy, are you . . .": Ibid.
525　and half-carried him: Interviews, Christiane Amanpour and James Rubin.
525　"the reverse of high performance . . .": Interview, J. "Mac" McClellan.
526　fallen 30.3 percent in ad: *Advertising Age* (July 26, 1999).
526　flying to Ontario: *Toronto Star*, July 18, 1999.
526　"I think this is . . .": Interview, Gary Ginsberg.
527　"There's a chance . . .": Interview, Dan Samson.
527　"Well, Goddamn it . . .": RB, p. 254.
527　Julie Baker visited John: *New York Daily News*, December 1, 1999, and interviews, OTR.
527　She was a temptation: Interview, OTR.
528　He chanced into McClellan: Interview, J. "Mac" McClellan.
528　"You better get in another . . .": Interview, Gary Ginsberg.

530 John had lunch: RB, pp. 258–59.

530 I'm running a little . . .": Interview, Dan Samson.

532 John had wanted to do.: Ibid.

532 Of the approximately 310 hours: National Transportation Safety Board, final report on accident, NYC99MA178, probable cause approval date July 6, 2000.

532 "wanted to do it alone": Ibid.

532 "he would not have felt . . .": Ibid.

533 He had gotten halfway: Ibid.

533 he had flown so close: Ibid.

50. BEGUILED AND BROKEN HEARTS

535 "the pilot's failure . . .": *Newsday* (July 7, 2000).

536 If they buried him: Interview, OTR.

538 and only part of John's body: Interview, OTR.

538 The two brothers: Interview, Douglas Kennedy.

540 "From the first day . . .": *Buffalo News*, July 24, 1999.

51. LIFE LESSONS

541 At the end of August: Interviews, Dan Samson and William Noonan.

542 Bono came out: *Washington Post*, September 25, 2000.

543 "I am sorry that Bobby . . .": Federal News Service, November 6, 2000.

543 "I carry this commitment . . .": Ibid., March 14, 2002.

544 There were 24.5 million: "AIDS in Number," washingtonpost.com, 2000.

544 the Bush administration's response: David Corn, "So This Is the Thanks Bono Gets? Why Bush's 'Important New' Anti-AIDS Program Is an Insult to the Civilized World," Tom-Paine.com, June 21, 2002.

544 "To meet a severe . . .": Federal News Service, January 28, 2003.

545 "a good old row": *Washington Post*, September 17, 2003.

545 Bobby thought how: Interview, Robert Shriver III.

545 "Some people say . . .": Matt Bai, "Running from Office," *New York Times Magazine*, July 15, 2001.

546 "favorite part of the book . . .": Ibid.

546 Shrum, Ted's former speechwriter: Ibid.

547 laughing to himself: *Boston Globe*, May 18, 2001.

547 A few weeks later: Bai, *Boston Globe*, June 1, 2001, and interview, OTR.

547 he went on a *National*: *New York Times*, December 1, 2002.

547 When the old man: "The Islanders Who Rescued JFK," Nationalgeographic.com.

548 "It was the most devastated": Center for International Rehabilitation, 2002 Annual Report.

548 "I know Dr. Smith": Jody Williams to Laurence Leamer, November 3, 2003.

548 arranged for focus groups: *Boston Globe*, August 2, 2001.

549 "I decided over the weekend . . .": *Good Morning America*, transcript, August 1, 2001.

549 "I think being chosen . . .": Ibid.

550 "the most talented . . .": Interview, Myer Feldman.

550 raised over $2.5 million: The figure at the end of August was $2.5 million (*Baltimore Sun*, August 30, 2002), but the final figure, according to Shriver's résumé, was above $2.7 million.

550 An independent poll: *Montgomery County Gazette*, March 20, 2002.

550 "That may have helped . . .": Interview, Mark Shriver.

551 In the end Mark: Maryland Board of Elections, official results, http://www.elections. state.md.us/past_elections/2002/results/p_representative_in_congress.html.

52. TIMES OF TESTING

552 smuggling cockatoo eggs: *Time* (July 3, 2000).

552 "Capitol Hill is essentially . . .": *Los Angeles Times*, June 11, 2000.

552 "large-scale hog producers . . .": *Des Moines Register*, April 10, 2002.

553 "Kennedy's Uncle Ted doesn't go . . .": Ibid., April 14, 2002.

553 Michael is as honest as daylight": *New York Times*, January 22, 2000.

553 was like an apple tree: This metaphor comes from an interview with Terry Tamminen, executive director of Environment Now.

554 "I have watched . . .": Robert F. Kennedy Jr. to the Honorable John F. Kavanewsky Jr., August 8, 2002.

554 "sits like Napoleon . . .": *Boston Globe*, December 14, 1998.

554 1-800-JOE-4-OIL: Ibid. January 20, 1999.

555 "When I came back . . .": Interview, Joseph Kennedy II.

555 Joe took home $396,375: Massachusetts Office of the Attorney General, Citizens Energy Corp., Form PC, 2001.

555 "They're always comparing me . . .": Interview, Joseph Kennedy II.

555 Citizens CEO Peter F. Smith: Interview, Peter F. Smith.

557 When visiting the Kennedy Library: The author was present.

557 In March 2000: *Providence Journal Bulletin*, April 8, 2000, May 13, 2000, and April 12, 2002.

557 That summer of 2000: Ibid. November 3, 2000, December 10, 2000, and December 26, 2000.

558 polls showed that he: Ibid. October 28, 2001, and February 25, 2003.

558 "I myself have suffered . . .": *Boston Herald*, March 1, 2000, *Providence Journal Bulletin*, March 2, 2000.

558 "I don't need Bush's tax cut": *Washington Post*, June 27, 2003.

560 "the luckiest man in the world": Sargent Shriver to friends, June 12, 2003.

560 "For a time we . . .": Interview, Timothy Shriver.

562 "I don't know what exits . . .": Ibid.

563 Maria wanted her husband: Interview, OTR.

563 Ted talked to Maria: Ibid.

564 Anthony, who appeared: Interview, Anthony Shriver.

564 As for Tim: Interview, Timothy Shriver.

564 "I want to start first": ABC Special Report, transcript, October, 2003.

53. RIPPLES OF HOPE

566 In November 2001 : *Providence Journal-Bulletin,* November 21, 2001.

567 "A lot of my friends . . .": FDCH Political Transcripts, January 5, 2002.

568 "America won the war . . .": *Boston Globe,* July 16, 2003.

568 Since he had entered: Ed Henry, "The Senator's Tale of the Tapes," *Roll Call* (February 28, 2002).

569 "This is President . . .": Interview, Senator Edward M. Kennedy.

Bibliography

Bass, Jack. *Taming the Storm: The Life and Times of Judge Frank M. Johnson Jr. and the South's Fight over Civil Rights.* New York: Doubleday, 1993.

Beran, Michael Knox. *The Last Patrician: Bobby Kennedy and the End of American Aristocracy.* New York: St. Martin's Press, 1998.

Beschloss, Michael R. *Taking Charge: The Johnson White House Tapes, 1963–1964.* New York: Simon & Schuster, 1997.

Billings Collection, The Boston: John F. Kennedy Library Foundation, 1991.

Blow, Richard. *American Son: A Portrait of John F. Kennedy Jr.* New York: Henry Holt & Co., 2002.

Boyle, Robert H. *The Hudson River: A Natural and Unnatural History.* New York: W. W. Norton, 1969

Bradford, Sarah. *America's Queen: The Life of Jacqueline Kennedy Onassis.* New York: Viking, 2000.

Brown, Thomas. *JFK: History of an Image.* Bloomington: Indiana University Press, 1988.

Burke, Richard. *The Senator: My Ten Years with Ted Kennedy.* New York: St. Martin's Press, 1993.

Burner, David, and Thomas R. West. *The Torch Is Passed: The Kennedy Brothers and American Liberalism.* New York: Atheneum, 1984.

Canada, Geoffrey. *Reaching Up for Manhood: Transforming the Lives of Boys in America.* Boston: Beacon Press, 1998.

Clymer, Adam. *Edward M. Kennedy: A Biography.* New York: William Morrow, 1999.

Coles, Robert. *Privileged Ones,* vol. 5, *Children of Crisis.* Boston: Little, Brown, 1997.

———. *Lives of Moral Leadership.* New York: Random House, 2000.

Collier, Charles W. *Wealth in Families.* Cambridge, Mass.: Harvard University Press, 2001.

Collier, Peter, and David Horowitz. The *Kennedys: An American Drama.* New York: Summit Books, 1984.

Cronin, John, and Robert Kennedy. *The Riverkeepers.* New York: Scribner's, 1997.

Dallas, Rita, with Jeanira Ratcliffe. *The Kennedy Case.* New York: Putnam's 1973.

Damore, Leo. *Senatorial Privilege: The Chappaquiddick Cover-up.* Washington, D.C.: Regnery Gateway, 1988.

David, Lester. *Ted Kennedy: Triumphs and Tragedies.* New York: Grosset & Dunlap, 1972.

Gallagher, Mary Barelli. *My Life with Jacqueline Kennedy*. New York: David McKay Co., 1969.

Goodwin, Richard. *Remembering America*. Boston: Little, Brown, 1988.

Guthman, Edwin. *We Band of Brothers: A Memoir of Robert F. Kennedy*. New York: Harper & Row, 1971.

Guthman, Edwin, and Jeffrey Shulman, eds., *Robert Kennedy: In His Own Words*. New York: Bantam, 1988.

Halberstam, David. *The Unfinished Odyssey of Robert Kennedy*. New York: Random House, 1969.

Haldeman, H. R. *The Haldeman Diaries: Inside the Nixon White House*. New York: Putnam's, 1994.

Hallowell, Edward M., and John J. Ratey, M.D. *Driven to Distraction: Recognizing and Coping with Attention Deficit Disorder*. New York: Touchstone, 1995.

Hartmann, Thom. *Attention Deficit Disorder: A Different Perception*. Penn Valley, Calif.: Underwood-Miller, 1993.

Hellman, John. *The Kennedy Obsession: The American Myth of JFK*. New York: Columbia University Press, 1997.

Hendrick, Burton J. *Life of Andrew Carnegie*. Garden City, N.Y.: Doubleday, Doran & Co., 1932.

Hersh, Burton. *The Education of Edward Kennedy*. New York: William Morrow, 1972.

———. *The Shadow President: Ted Kennedy in Opposition*. South Royalton, Vt.: Steerforth Press, 1997.

Heymann, C. David. *A Woman Called Jackie*. New York: Lyle Stuart, 1989.

———. *RFK: A Candid Biography of Robert F. Kennedy*. New York: Dutton, 1998.

Kennedy, Edward M., ed. *The Fruitful Bough: A Tribute to Joseph P. Kennedy*. Privately printed, 1965.

Kennedy, Maxwell Taylor. *Make Gentle the Life of the World: The Vision of Robert F. Kennedy*. New York: Harcourt Brace, 1998.

Kennedy, Rose. *Times to Remember*. Garden City, N.Y.: Doubleday, 1974.

Kennedy, Sheila Rauch. *Shattered Faith: A Woman's Struggle to Stop the Catholic Church from Annulling Her Marriage*. New York: Henry Holt & Co., 1998.

Kimerling, Judith, et al. *Amazon Crude*, edited by Susan S. Henriksen, preface by Robert F. Kennedy Jr. New York: Natural Resources Defense Council, 1991.

Kutler, Stanley, ed. *Abuse of Power: The New Nixon Tapes*. New York: Simon & Schuster, 1997.

Landsberg, Ivan. *Succeeding Generations: Realizing the Dream of Families in Business*. Boston: Harvard Business School, 1999.

Lawford, Patricia, ed. *That Shining Hour*. Halliday Lithograph, 1969.

Leamer, Laurence. *The Kennedy Women*. New York: Villard, 1994.

———. *The Kennedy Men*. New York: William Morrow, 1999.

———. *Playing for Keeps in Washington*. New York: Dial, 1977.

Leigh, Wendy. *Prince Charming: The John F. Kennedy Jr. Story*. New York: Dutton, 1993.

Lester, David. *Ted Kennedy: Triumphs and Tragedies*. New York: Grosset & Dunlop, 1972.

Lippman, Theo, Jr. *Senator Ted Kennedy: The Career Behind the Image*. New York: W. W. Norton, 1976.

Liston, Robert A. *Sargent Shriver: A Candid Portrait*. New York: Farrar, Straus, 1964.

Mahoney, Richard D. *Sons and Brothers: The Days of Jack and Bobby Kennedy*. New York: Arcade, 1999.

Manchester, William. *The Death of a President*. New York: Harper & Row, 1967.

Matthews, Christopher. *Kennedy and Nixon: The Rivalry That Shaped Postwar America*. New York: Simon & Schuster, 1996.

McCullough, David. *John Adams*. New York: Simon & Schuster, 2001.

Miller, Alice. *The Drama of the Gifted Child*. New York: Basic Books, 1981.

Newfield, Jack. *Robert Kennedy: A Memoir*. New York: Dutton, 1969.

Oppenheimer, Jerry. *The Other Mrs. Kennedy*. New York: St. Martin's Press, 1994.

Rachlin, Harvey. *The Kennedys: A Chronological History, 1823 to the Present*. New York: World Almanac, 1986.

Rainie, Harrison, and John Quinn. *Growing Up Kennedy*. New York: Putnam's 1983.

Rogers, William. *When I Think of Bobby: A Personal Memoir of the Kennedy Years*. New York: HarperCollins, 1993.

Salinger, Pierre. *P.S.: A Memoir*. New York: St. Martin's Press, 1995.

Saunders, Frank, with James Southwood. *Torn Lace Curtain*. New York: Holt, Rinehart & Winston, 1982.

Schlesinger, Arthur, Jr. *Robert Kennedy and His Times*. Boston: Houghton Mifflin, 1978.

Shesol, Jeff. *Mutual Contempt: Lyndon Johnson, Robert Kennedy, and the Feud That Defined a Decade*. New York: W.W. Norton, 1997.

Steel, Ronald. *In Love with Night: The American Romance with Robert Kennedy*. New York: Simon & Schuster, 2000.

Sullivan, Gerald, and Michael Kenney. *The Race for the Eighth: The Making of a Congressional Campaign: Joe Kennedy's Successful Pursuit of a Political Legacy*. New York: HarperCollins, 1987.

Thomas, Evan. *Robert Kennedy*. New York: Simon & Schuster, 2000.

Vanden Heuvel, William, and Milton Gwirtzman. *On His Own: Robert F. Kennedy 1964–1968*. Garden City, N.Y.: Doubleday, 1970.

Von Franz, Marie-Louise. *The Problem of the Puer Aeternus*. Toronto: Inner City Books, 2000.

Wechsler, James A., with Nancy F. Wechsler and Holly W. Karpf. *In a Darkness*. New York: W. W. Norton, 1972.

West, Darrell M. *Patrick Kennedy: The Rise to Power*. Upper Saddle River, N.J.: Prentice-Hall, 2001.

Witcover, Jules. *85 Days: The Last Campaign of Robert Kennedy*. New York: Putnam, 1969.

Zelizer, Barbie. *Covering the Body: The Kennedy Assassination, the Media, and the Shaping of Collective Memory*. Chicago: University of Chicago Press, 1992.

Acknowledgments

A few years back, Janet Malcolm famously wrote that every journalist "is a kind of confidence man, preying on people's vanity, ignorance, or loneliness, gaining their trust and betraying them without remorse." I have known a few authors who employ that approach, and I also have met a few subjects who were confidence men themselves, doing most of the seduction. But in my experience the relationship is far more complicated than Malcolm suggests. Ideally, the biographer and his living subject are on a journey together toward a truthful presentation of a life. It is sometimes difficult, unpleasant, argumentative, but if the two continue together, in the end the portrait has an unquestioned resemblance to the person's life.

That approach has become increasingly difficult. In the three decades that I have been writing, people have become far more suspicious of authors as well as others they do not know. I don't blame them. I have become more suspicious myself. The Kennedys are an extreme example of this wariness, and it was not easy for them to display the remarkable candor that most of them did in their interviews for this book.

There are those who suggest that if I got the Kennedys to talk to me, I must have done so by compromising my integrity and if I had dared to write a truthful book, they would not have talked to me. I don't know quite what to say about that level of cynicism except that it tears at the very roots of journalism in a free society. A reporter is supposed to try diligently to talk to his subjects and those who know them best. A failure to attempt to do so is simply unacceptable.

The Kennedy men written about in *The Sons of Camelot* are either alive or recently enough deceased that their contemporaries are still alive. Thus, although I have found other books, articles, and transcripts helpful, for the most part this work is based on my reporting.

This is not an authorized biography, and it took many months to get some of the Kennedys to talk, but in the end I interviewed ten of the twelve living Kennedys who are major figures in *The Sons of Camelot*. Most of them I interviewed several times. I had discussions with two other Kennedy sons and interviewed two of the young Kennedy women. Surprisingly, for all that has been written about the Kennedys, the majority had never talked in detail about their lives before to a writer. I also did several hundred other interviews with friends, associates, classmates, childhood playmates, aides, and others whose lives intersected with theirs.

John F. Kennedy Jr. died in the midst of my research, and I did not interview him. I did not even begin reporting about him until a year after his death, when I thought a respectful enough time had passed to approach those who knew him. Then I contacted his closest friends. One after another they agreed to cooperate, hoping that I would write the definitive portrait. I would like to thank Christiane Amanpour, John Perry Barlow, Sasha Chermayeff, Gary Ginsberg, Ed Hill, Robert Littell, William Noonan, Christian L. Oberbeck, and Dan Samson, as well as two others who have chosen to remain anonymous. Most of these friends gave many hours to this book, talking intimately about John, suggesting other sources, and reading the manuscript. I have interviewed many other of his friends, acquaintances, business associates, and classmates from Phillips Andover Academy and Brown University. I have written only about matters for which I have responsible sources with direct knowledge of John's life. Gary Ginsberg was the first of John's friends to whom I spoke. He was always there for me during the years of this project, but more important, he was always there for John. Christiane Amanpour pushed me to do the best work I could do, not for me but for her friend. In the end, if there are weaknesses in my portrait, I am the one holding the brush.

Senator Edward Kennedy reached deep within himself and his past in our interview. I talked extensively to all four of the Shriver brothers, Robert, Timothy, Mark, and Anthony, as well as to their sister, Maria. I interviewed all five of the surviving sons of Robert and Ethel Kennedy: Joseph P. Kennedy II, Robert Kennedy Jr., Christopher Kennedy, Kennedy Maxwell Kennedy, and Douglas Kennedy and their sister, Kathleen Kennedy Townsend. John Rosenthal was Michael Kennedy's closest friend and became a friend of mine during the course of my research, and I have the greatest admiration for him.

When I was beginning work on this book, I happened to meet Jeanne Conway at a party in Palm Beach, Florida. Conway had worked with alcoholics in a corporate setting. She told me that I could not understand the Kennedys fully if I did not understand alcoholism, and if I wanted to understand alcoholism I would have to understand Alcoholics Anonymous. She arranged for me to attend a closed AA meeting, which I did for a year. My writing about the family difficulties with substance abuse is infused by what I learned.

I had help in researching parts of this book from three able colleagues: Frederic J.

Frommer, an AP reporter and coauthor of *Growing Up Baseball*; Nadine Witkin, a senior producer at CNN; and Shannon Tan, a reporter at the *St. Petersburg Times*. I would also like to thank the staffs at the John F. Kennedy Presidential Library, the Lyndon Baines Johnson Presidential Library, the Martin Luther King Jr. Library in Washington, D.C., the Library of Congress, and the Palm Beach County Library. At the Kennedy Library, I'm especially grateful to James Hill for his help in obtaining photos. I would also like to thank my lecture agent, Trinity Ray, at the American Program Bureau. And how can I forget Tom Davis's Zoot software, an indispensable tool for a serious researcher, and Don Spencer, who transcribed my interviews.

In the past couple of years there have been numerous criticisms about the overuse of anonymous sources. Anonymous sources are like garlic: use a pinch and the dish tastes better, use too much and everything stinks. I have in a few places employed anonymous sources when that seemed the only way to use the material. There are times when one of the people I have interviewed on the record asked that some information not have their name affixed to it. At other times the person was so sensitively placed and would have suffered such public damage if their name came out that I either had to use the material anonymously or not at all.

Any book is more of a collegial effort than it appears, and I have had help from many quarters. As always, Vesna Obradovic Leamer, my wife, was my researcher and constant support. In nearly two decades of marriage, she has rarely been wrong in her advice, a track record that I am loath to admit. That is a failing that makes our relationship continually diverting. She read the manuscript, as did friends of mine, including Myer "Mike" Feldman, Burton Hersh, Dr. Sheldon Stern, Herb Gray, Helen Leamer, Kerry McCarthy, Ama Neel, Kristina Rebelo Anderson, and Joy Harris. Joy Harris, my longtime agent, is another constant in my life: she is always there for me, and I am greatly in her debt. I also must give special thanks to my dear friend Mike Feldman for his many efforts on my behalf.

It is one of my major blessings to have Meaghan Dowling as my editor. She did her usual fine job reading and commenting on the manuscript before taking a leave to have her first child. As she ran out the door for a delivery far more important than my book, Meaghan handed the manuscript to Mark Bryant, who did excellent work line editing the manuscript. Rome Quesada managed all the editorial details with the aplomb of a veteran short-order cook. There are not that many first-rate copy editors; one of them is Cindy Buck, and my good fortune is that she copyedited both *The Kennedy Men* and *Sons of Camelot*. The melancholy fact is that they have left me no choice but to say that any weaknesses in *Sons of Camelot* are mine alone.

Among those who granted me interviews or helped in other crucial ways are Rep. Neil Abercrombie, Robert Adler, James Alden, Julie Amato, Hill Anderson, Carl Anthony, Patrick Apossy, Dominick Arena, Martin Arnold, Bill Arthur, Bobby Baker,

Gertrude Ball, Liz Barratt-Brown, Chris Bartle, Charles Bartlett, Gabe Bayz, Professor Edward Beiser, Kai Bird, Brad Blank, Robert Boyle, Ben Bradlee, Diane Broughton, Ham Brown, Lauri Buckingham, Charles and Claire Burke, David Burke, Richard Burke, Professor. James MacGregor Burns, Fox Butterfield, F. Caparallaro, Dennis Carleton, Loretta Claiborne, Diane Clemens, Adam Clymer, Dr. James P. Comer, Jim Connors, Jeanne Conway, R. Kent Correll, Gay Courter, Phil Courter, Dan Cronin, Elaine and Alexandre de Bothuri, Karol DeFalco, William Delahunt, Catha Deloach, Alex Dessayer, Ralph Diaz, Lorenzo di Bonaventura, Jim DiCastro, Frank Dillow, Gerard Doherty, Joseph Dolan, Mary Ann Dolan, John Dow Jr., and Dominick Dunne.

William Ebenstein, Peter Edelman, Steve Eidelman, Al Eislee, Cinda Elser, Professor John Emigh, Luis Estevez, Mark Falinga, James G. Fegan, Linda Fehrs, Myer Feldman, Carol Fennelly, Janet Des Rosiers Fontaine, Michael Foster, Harry Fowler, Frank Fox, Thomas Gagen, John Gaker, John Kenneth Galbraith, Robert Galland, Barbara Gamarekian, Nancy Gardiner, Joseph Gargan, Winston Gathings, Faye Gatto, Ronald Gatto, Matt Gillis, Bruce Ginsberg, Gretchen Long Glickman, Chuck Glynn, Molly Gorsuch, Harry Gossett, Marty Gottlieb, Arthur Grace, John Greenya, Lloyd Grove, Ed Guttman, Milton Gwirtzman, General Al Haig, Nigel Hamilton, Mark Hardman, R. Couri Hay, Bob Healy, Deidre Henderson, Burton Hersh, James Hilliard, Republican Sheila Hixson, Claude Hooton Jr., Dick Horin, David Horowitz, Kyle Horst, Lloyd Howard, Mike Howard, Ron Howard, David Humphreville, Bill Hundley, Vicki Iovine, David Israel, and Sam Juneau.

Peter Kaplan, Elizabeth Kaye, Larry Kellerman, Keith Kelly, Elmorea Kennedy, Joan Kennedy, Malcolm Kilduff, Judith Kimerling, Bruce Kluger, Lucretia Kosella, Seth Krakauer, Dr. Herbert Kramer, Professor Peter Kreft, David Kuhn, Danuta Kurzyna, Patricia Seaton Lawford, Daniela Leamer, Edward Leamer, Helen Leamer, Mary Francis Leamer, Robert Leamer, Ted Leonsis, David Linde, Christin and Peter Lynch, Lorrie Lynch, Jim Mahoney, Frank Mankiewicz, Antonio Mantilla, Mark Maremont, John Martello, Laurence Maslon, Marilyn Matlick, Senator Eugene McCarthy, Kerry McCarthy, Mary Lou McCarthy, J. "Mac" McClellan, Dick McCormack, Lisa McCormack, Wanda McDaniel, Patti McGinty, Eileen McNamara, Elizabeth Mehren, Danny Melrod, Melody Miller, Bill Milliken, J. Moakley, Dan Moldea, Emily Moye, Mark Mullin, Evelyn Murphy, Mauray Murphy, Ama Neel, Professor Charles Neu, Larry Newman, and Neal Nordlinger.

Brian O'Connor, Phillip O'Connor, James O'Neill, Ivanka Ostojic, Lorraine Paolella, Bruce A. Pasternack, Mark Perry, Mary Beth Postman, Alex Pozzy, Barrett Prettyman, Jamie Price, Joan Quick, Michael Radutzky, Harrison Rainie, Coates Redmon, Senator Jack Reed, Jewel Reed, Robert Reichley, Judge Stephen Reinhardt, Joel Reynolds, William Bradford Reynolds, Marilyn Riseman, Bill Robinson, Raleigh Robinson, Maurice Rosenblatt, James Rubin, Gus Russo, Jeffrey Sachs, Pierre Salinger, S. Jacob Scherr, Marian Schlesinger, John Seigenthaler, Ira Shapiro, Jimmy Shay, Ed

Shorter, Russell Simmons, Fred Slifka Bernie Smith, Edward Smith, Peter F. Smith, former governor Crocker Snow Jr., Charles Spalding, Gene Sperling, Jeff Stein, Kimberly Stender, Jennifer Sulentic, Bruce G. Sundlun, Terry Tamminen, F. Tavares, former Maryland House Speaker Casper R. Taylor Jr., Lee Tomic, Michael Tomic, Mark Rafael Truitt, William vanden Heuvel, Salvatore Verderame, Sue Vogelsinger, Paul Wachter, Adam Walinsky, Rhoda H. Weinman, Professor Roger Weisberg, Nancy Weiss, Richard Wiese, Tony Williams, Mark Willie, Don Wilmeth, Paul Wilmot, Harris Wofford, Jennifer Woolford, and Professor Abraham Zaleznik.

Index